Lecture Notes in Networks and Systems

Volume 266

The series "Lecture Notes in Networks and Systems" publishes the latest developments in Networks and Systems—quickly, informally and with high quality. Original research reported in proceedings and post-proceedings represents the core of LNNS.

Volumes published in LNNS embrace all aspects and subfields of, as well as new challenges in, Networks and Systems.

The series contains proceedings and edited volumes in systems and networks, spanning the areas of Cyber-Physical Systems, Autonomous Systems, Sensor Networks, Control Systems, Energy Systems, Automotive Systems, Biological Systems, Vehicular Networking and Connected Vehicles, Aerospace Systems, Automation, Manufacturing, Smart Grids, Nonlinear Systems, Power Systems, Robotics, Social Systems, Economic Systems and other. Of particular value to both the contributors and the readership are the short publication timeframe and the world-wide distribution and exposure which enable both a wide and rapid dissemination of research output.

The series covers the theory, applications, and perspectives on the state of the art and future developments relevant to systems and networks, decision making, control, complex processes and related areas, as embedded in the fields of interdisciplinary and applied sciences, engineering, computer science, physics, economics, social, and life sciences, as well as the paradigms and methodologies behind them.

Indexed by SCOPUS, INSPEC, WTI Frankfurt eG, zbMATH, SCImago.

All books published in the series are submitted for consideration in Web of Science.

More information about this series at http://www.springer.com/series/15179

Christine Leitner · Walter Ganz ·
Debra Satterfield · Clara Bassano
Editors

Advances in the Human Side of Service Engineering

Proceedings of the AHFE 2021 Virtual Conference on The Human Side of Service Engineering, July 25–29, 2021, USA

Springer

Editors
Christine Leitner
Centre for Economics and Public Administration
London, UK

Debra Satterfield
California State University
Orange, CA, USA

Walter Ganz
Fraunhofer IAO
Stuttgart, Baden-Württemberg, Germany

Clara Bassano
University of Naples Parthenope
Avellino, Italy

ISSN 2367-3370 ISSN 2367-3389 (electronic)
Lecture Notes in Networks and Systems
ISBN 978-3-030-80839-6 ISBN 978-3-030-80840-2 (eBook)
https://doi.org/10.1007/978-3-030-80840-2

This Springer imprint is published by the registered company Springer Nature Switzerland AG
The registered company address is: Gewerbestrasse 11, 6330 Cham, Switzerland

Advances in Human Factors and Ergonomics 2021

AHFE 2021 Series Editors

Tareq Z. Ahram, Florida, USA
Waldemar Karwowski, Florida, USA

12th International Conference on Applied Human Factors and Ergonomics and the Affiliated Conferences (AHFE 2021)

Proceedings of the AHFE 2021 Virtual Conference on The Human Side of Service Engineering, July 25–29, 2021, USA.

Advances in Neuroergonomics and Cognitive Engineering	Hasan Ayaz, Umer Asgher and Lucas Paletta
Advances in Industrial Design	Cliff Sungsoo Shin, Giuseppe Di Bucchianico, Shuichi Fukuda, Yong-Gyun Ghim, Gianni Montagna and Cristina Carvalho
Advances in Ergonomics in Design	Francisco Rebelo
Advances in Safety Management and Human Performance	Pedro M. Arezes and Ronald L. Boring
Advances in Human Factors and Ergonomics in Healthcare and Medical Devices	Jay Kalra, Nancy J. Lightner and Redha Taiar
Advances in Simulation and Digital Human Modeling	Julia L. Wright, Daniel Barber, Sofia Scataglin and Sudhakar L. Rajulu
Advances in Human Factors and System Interactions	Isabel L. Nunes
Advances in the Human Side of Service Engineering	Christine Leitner, Walter Ganz, Debra Satterfield and Clara Bassano
Advances in Human Factors, Business Management and Leadership	Jussi Ilari Kantola, Salman Nazir and Vesa Salminen
Advances in Human Factors in Robots, Unmanned Systems and Cybersecurity	Matteo Zallio, Carlos Raymundo Ibañez and Jesus Hechavarria Hernandez
Advances in Human Factors in Training, Education, and Learning Sciences	Salman Nazir, Tareq Z. Ahram and Waldemar Karwowski

(continued)

(continued)

Advances in Human Aspects of Transportation	Neville Stanton
Advances in Artificial Intelligence, Software and Systems Engineering	Tareq Z. Ahram, Waldemar Karwowski and Jay Kalra
Advances in Human Factors in Architecture, Sustainable Urban Planning and Infrastructure	Jerzy Charytonowicz, Alicja Maciejko and Christianne S. Falcão
Advances in Physical, Social & Occupational Ergonomics	Ravindra S. Goonetilleke, Shuping Xiong, Henrijs Kalkis, Zenija Roja, Waldemar Karwowski and Atsuo Murata
Advances in Manufacturing, Production Management and Process Control	Stefan Trzcielinski, Beata Mrugalska, Waldemar Karwowski, Emilio Rossi and Massimo Di Nicolantonio
Advances in Usability, User Experience, Wearable and Assistive Technology	Tareq Z. Ahram and Christianne S. Falcão
Advances in Creativity, Innovation, Entrepreneurship and Communication of Design	Evangelos Markopoulos, Ravindra S. Goonetilleke, Amic G. Ho and Yan Luximon
Advances in Human Dynamics for the Development of Contemporary Societies	Daniel Raposo, Nuno Martins and Daniel Brandão

Preface

If there is any one element to the engineering of service systems that is unique, it is the extent to which the suitability of the system for human use, human service, and for providing an excellent human experience has been and must always be considered. The International Conference on the Human Side of Service Engineering (HSSE) was organized within the framework of the International Conference on Applied Human Factors and Ergonomics (AHFE) as an Affiliated Conference. This conference included over 60 presentations during the three-day event in sessions which focused on the study of people as the key variable in service systems, with service innovation as the primary goal to invent, adapt, and engineer better skills, technologies, organizations, information resources, and value propositions to address the needs, wants, and aspirations of diverse people in service systems. Newly included topics for HSSE relate to determinants of re-balancing benefits in the digital age in the wake of the COVID pandemic and trust: standards, experience, public policy, participatory governance, privacy, and ethics. All of these are types of shared information resources that influence trust between actors and will be impacted by advances in artificial intelligence (AI).

The book is divided into eleven major sections as follows:

1. Re-Balancing Benefits in the Digital Age
2. Addressing Complexity in the Emerging Platform Society
3. Enhancing Interdisciplinary in Service Innovation
4. Resilient Design for Service-Oriented Value Creation
5. Emerging Research Innovations in AI, User Experience, and Design
6. Human-Centered Service Design
7. Augmenting Service Capabilities in the Post-Pandemic Era
8. Education, Training, and Technology Facilitating a Future Workforce
9. Artificial Intelligence-Based Co-Creation
10. Creating Value in Teams and Organizations
11. Creating Value in Communities and Society

A major area of expansion in this conference is the intersection of AI, big data, and human, social, and knowledge science in the human side of service engineering, and also various interdisciplinary contributions to the field of service innovation and co-creation of value, calling for rethinking existing service systems and the role of technology in human–machine co-creation processes in the digital age.

We are deeply indebted to all session chairs for their contributions to HSSE 2021. We hope that you find this volume useful and interesting and invite all to join us for next year's meeting. Our sincere thanks and appreciation goes to the board members listed below for their contribution to the high scientific standard maintained in developing this book. In particular, we would like to thank Jim Spohrer, Louis Freund, and Wojciech Cellary for their dedication and support to the HSSE community.

M. Warg, Germany
C. Zagel, Germany
A. Zimmermann, Germany

July 2021

Christine Leitner
Walter Ganz
Debra Satterfield
Clara Bassano

Contents

Re-Balancing Benefits in the Digital Age

The Human-Side of Service Engineering: Advancing Technology's Impact on Service Innovation

Wojciech Cellary[1(✉)], Louis E. Freund[2], Stephen K. Kwan[2], Christine Leitner[3], and Jim Spohrer[4]

[1] Poznan University of Economics and Business, Niepodleglosci 10, 61-875 Poznan, Poland
cellary@kti.ue.poznan.pl

[2] San José State University, 1 Washington Square, San José, CA 95192, USA
{louis.freund,stephen.kwan}@sjsu.edu

[3] Centre for Economics and Public Administration Ltd., 111 Buckingham Palace Road, London SW1W 0SR, UK
christine.leitner@cepanet.eu

[4] IBM, USA, IBM Research – Almaden, 650 Harry Road, San Jose, CA 95120, USA
spohrer@us.ibm.com

Abstract. As advanced Information Technology (IT) drives changes in digital service and service systems across industries, what are the implications for the Human-Side of Service Engineering (HSSE)? This work updates previous HSSE frameworks and extends the boundaries of Service Science, Management, Engineering, and Design (SSMED). Newly included are HSSE areas related to determinants of trust: standards, skills, public sector, privacy, and ethics. All of these are types of shared information resources that influence trust between actors. These updates are needed, as the impact of IT on service systems continues to expand, and AI technologies successfully emulate more human capabilities. The purpose of this paper is to consider possible future directions of HSSE using observations and examples. The paper also presents a discussion of challenges to society and its fundamental values coming from recent advances in AI.

Keywords: Service science · Service standardization · T-shape experience · Participatory governance · Privacy and ethics

1 Introduction

This paper provides an updated introduction to the area of study known as the Human-Side of Service Engineering (HSSE). HSSE is the area within the broader field of service science that focuses on the performance and accommodation of people as key actors within service systems and emerging service innovations. Service innovations include the results of inventing and adapting engineering technology, organizations, and information systems to better support people's needs, wants, and aspirations. In the original HSSE paper [11], the service research foundations upon which our efforts are

C. Leitner et al. (Eds.): AHFE 2021, LNNS 266, pp. 3–17, 2021.
https://doi.org/10.1007/978-3-030-80840-2_1

built were discussed, including Service Science, Management, Engineering, and Design (SSMED) [27], Service-Dominant Logic [34–36], and a proposed framework for Service Innovation [15].

HSSE is concerned with the human-centric issues encountered in designing, developing, and operating better service systems for our work and daily lives. It includes topics ranging from ergonomics and human factors engineering to the determinants of trust.

In Sect. 2, the original framework is summarized. Section 3 presents an extended framework. Sections 4, 5, 6 and 7 discuss the proposed extensions related to standards, skills, public sector and privacy, respectively. Section 8 provides summary and future research directions.

2 Summary of Original Framework

2.1 Background

In [11], Freund and Spohrer proposed a framework for the factors contributing to the Human-Side of Service Engineering consisting of seven interrelated concepts: Service, Service Systems, Value Propositions, Modern Service Systems, Service System Entities, Engineering, and Service Engineering. Figure 1 (developed for this paper) presents a depiction of this Framework showing the interrelationships among these concepts.

As depicted in the lower right quadrant of Fig. 1, The Human-Side of Modern Service Systems is an area of study within Modern Service Systems focusing on the people who are providing services as well as the people being served by a service system, e.g., citizens, customers, entrepreneurs, employees, etc. The goal of this area of study is to apply the principles of service science, human factors and the engineering methods of ergonomics to make and scale diverse types of service systems that are efficient, reliable, effective, and safe, at all times for all participants, as well as all others who

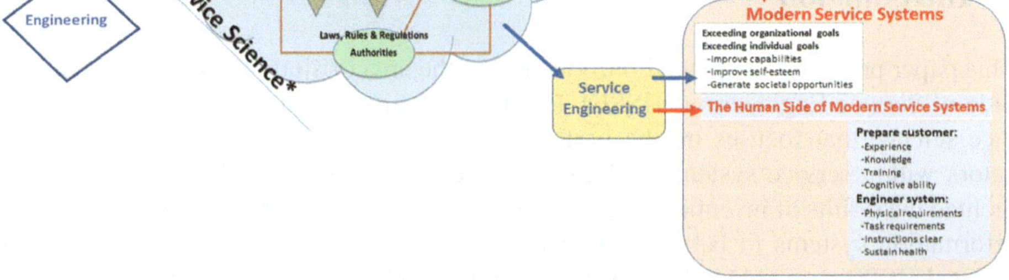

Fig. 1. Framework as described in [11]. Service Science portion of the diagram from [28]

may be impacted by the system. It focuses on preparing the customer in every way to efficiently and safely receive services, as well as engineering service systems to be adaptive and to safely deliver the services as intended and expected.

2.2 Designing Service Systems for the General Population

Attention to human capabilities and limitations is at the core of the Human-Side of Service Engineering, and rigorous systems requirements analysis. To address this, service engineering and service design take advantage of the science of ergonomics, which focuses on the physical capacities, anthropometric features, and biomechanical aspects of humans. For example, aircraft, buses, trams, and other transportation service systems require seating that fits most people and properly designed ingress and egress passages and procedures, for both routine and emergency operations.

Similarly, the domain of human factors ranges from the cognitive (mental workload, memory, information processing, multitasking, etc.), perceptual, and communication (controls, displays, haptics, augmented reality, and other technological interfaces) interaction capabilities to the performance impacts of environmental factors (such as lighting, temperature, and noise) as well as social factors.

2.3 Designing to Serve Special Populations

Populations with special needs such as the elderly, children, people with disabilities, pregnant women, people with allergies, gender identity, etc., give rise to human factors design challenges for service engineers. For example, especially challenging are theme parks, because their employees may wear costumes that limit visibility and hearing, while their visitors are very diverse.

Accommodation means respecting the dignity of individuals while providing a real and non-demeaning approach for access and engagement. In some major urban areas, for example, some segments of the population may not be accommodated well by service systems that are not designed to serve homeless people suffering from mental illnesses.

The Human-Side of Service Engineering is continuously challenged by the contextual and dynamic nature of the capabilities, constraints, rights, and responsibilities of persons with special needs.

2.4 Ensuring Safety and Protecting People

Ensuring safety and protecting people requires avoiding accidents and errors. For example, ATMs available day or night from a street need cameras and proper lighting that are necessary to protect customers and provide them with safety. Service systems such as airports require security screening on entry to exclude weapons and explosives, followed by specialized equipment and trained personnel to operate it. Shootings in schools, places of worship, cinemas, commercial centers, etc. have increased awareness of the threat of violence in service systems. New technologies such as smartphones, wearables, car cameras, and internal house cameras are documenting incidents of violence as they unfold, leading to rethinking approaches to safety in service systems.

2.5 Exceeding Individual and Organizational Goals

Humans play different roles daily within diverse organizations. Therefore, service engineering must include the social and behavioral aspects of people as goal-driven individuals. For example, human resource professionals measure and try to enhance the levels of employee engagement at work. Employees and students look for success, so they improve, exceed, and advance according to measures of success. Gamification is a technique aiming at the recognition of the hierarchy of needs, wants, and aspirations of individuals in many social contexts and to help them to improve performance. The Human-Side of Service Engineering includes maximizing enhanced healthy choices, and minimizing unhealthy options, from vending machines at school to arrangement of items on shopping websites. Even the household and family can be viewed through the lens of the Human-Side of Service Engineering in encouraging healthy, smart, and wise choices that achieve diverse goals.

2.6 Assuring Information Security and Privacy

Assuring security requires both protecting the systems and protecting the privacy of customers and citizens. For example, mobile apps provide customers and service providers with opportunities to interact more frequently but also have also opened the door for nefarious practices that result in intrusions and lost data. Bad actors use the acquired personal information of others for their own gain, from manipulating online behavior to stealing assets and identity theft. The Human-Side of Service Engineering benefits from advanced capabilities such as smartphones, apps, social media, but new challenges arise as well. The General Data Protection Regulation (2016) in the European Union is an essential step towards protecting and securing personal information and reducing unintended usage of personal data.

2.7 Capability and Opportunity Enhancement

Modern service systems enhance the capabilities and opportunities of customers, employees, and other stakeholders for value co-creation. For example, loyalty programs allow customers to earn higher levels of service, as they consume more of an offering. New AI technologies can augment human capabilities and enhance human performance in certain areas, e.g., health care, where AI is applied in almost every field analyzing massive data sets to discover new drug therapies, new treatment approaches, and new strategies for operating delivery systems efficiently and effectively [1].

3 Extending the Framework of the Human Side of Service Systems

As described above, the original HSSE framework addressed issues related to the diversity of populations and individuals. The extensions to that framework proposed in this section address issues related to standards, talent and skills, participatory governance in the public sector, and privacy and ethics in service systems. Figure 2 adds these elements to the HSSE Framework presented in Fig. 1.

The extensions to the original framework are represented in Fig. 2 by the white boxed elements. Each is discussed in detail in the following sections.

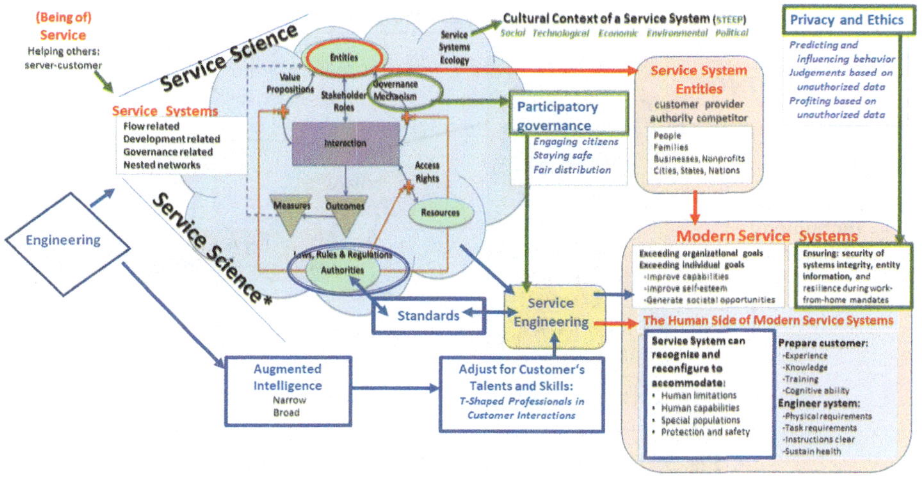

Fig. 2. Extended Framework as described in present paper. Service Science diagram from [28]

4 Extensions Related to Standards

Manufacturing industries – especially those operating on a global scale – use standards extensively to facilitate cooperation between entities by composing value chains, and benefit from the resultant economy of scale. Service providers have similar needs to use standards. Unfortunately, the development and publishing of service standards lags far behind the growth of the service economy. The service sector in the U.S. is estimated to be 80% of the GDP in 2017 [32].

Weissinger and Kwan [37] found that most of service standards published by the International Standards Organization (ISO) have been related to the back stage of service provider's operations (see Fig. 3) or processes related to production as enablers of services [30]. The back stage received more attention since, historically, standards were developed mostly for technical specification and interoperability of products.

Weissinger and Kwan [37] also noted that the challenges of standards development for services at the front stage come from the frequent involvement of humans in service delivery, as well as customer interaction as part of their service experience. Despite those challenges, currently ISO, EU and national standardization organizations (European Committee for Standardization (CEN), American National Standards Institute (ANSI), German Institute for Standardization (DIN)) are actively promoting standards for the service sector. For example, ISO/IEC had updated its Guide for Service Standards Development ISO/IEC Guide 76 [16]. The Committee on Consumer Policy (COPOLCO) of ISO has also been active in promoting consumer-related standards projects. Many service standards concerning quality, governance and management have been developed to accompany product specification or performance standards. Many of those, often referred to as horizontal standards, are closely related to service operations.

In their research, the authors (op. cit.) looked for correlation and synergy between service science research and service standards development. They noted that the number of service science researchers increased as the service sector started to dominate

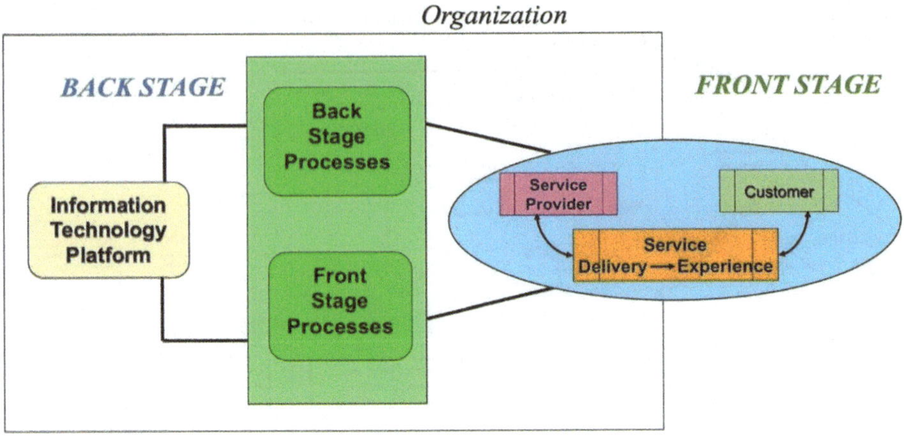

Fig. 3. Front and back stage of a service organization

in national economies. This led to a growing number of scientific publications that impacted industry standards and public policy. The authors found, however, that one of the fundamental concepts of service science: "service as value co-creation" had not yet been adopted by developers of service standards. There were marginal references to the concept of a customer's value proposition. The authors found that for standard developers a service is mostly understood as the unidirectional delivery of an intangible good to a customer usually as part of a product. In the lexicon of Service-Dominant Logic, such understanding is characteristic of goods-dominant logic as opposed to that of Service-Dominant Logic (op. cit.).

In Fig. 4, a schema of cooperation and exchanges between service science researchers (academic) and service standardization professionals (industry) is presented. Such cooperation would help establish a more solid foundation for service standards. For standardization professionals, a closer cooperation with service researchers could be of benefit with the availability of a relatively consistent common theoretical framework that could provide a bridge and common language between different sectors and areas of service standardization. The adoption of service science concepts would permit the needed move from goods-dominant logic to Service-Dominant Logic in service standards development. Service-Dominant Logic would provide service standards with a Human-Side framework for more customer-driven orientation and, as a result, alignment with customer value expectations and needs fulfillment.

For service science researchers, involvement in service standardization processes would provide a "real world test" of the applicability of theoretical concepts in the service industry. Standards as a tool of knowledge dissemination and innovation, could contribute to increases in the spread and influence of ideas, concepts and methods that have originated in service science.

The observations and recommendations discussed above are becoming even more relevant as AI applications are proliferating in both market-based and non-market-based (such as public) services (see Sect. 6 below). As the technology matures, industry is driving the development of pertinent standards. The ISO/IEC JTC1 SC 42 Artificial

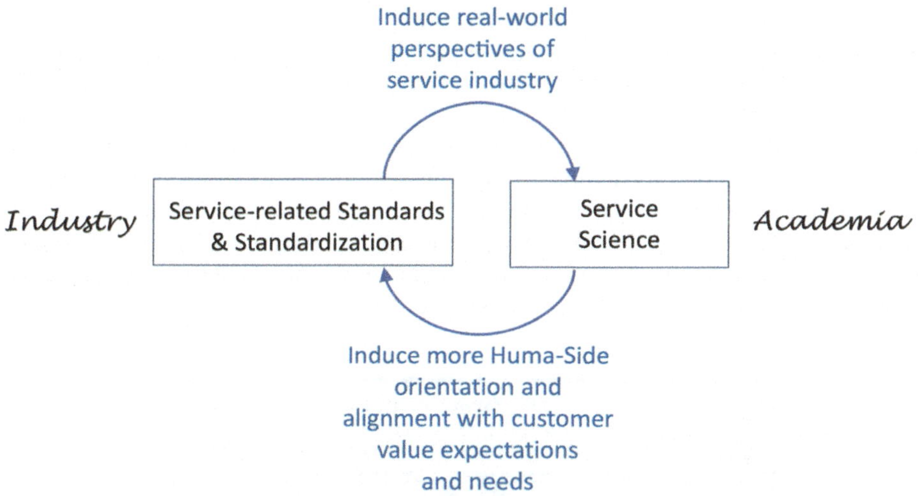

Fig. 4. Exchange and closer cooperation

Intelligence [17] was formed in 2017 and there are now twenty-nine member nations working on sixteen standards related to concepts and terminology, bias, trustworthiness, framework, use cases, big data, governance and other matters.

5 Extensions Related to Customer's Talents and Skills

There is a widely recognized need for new skills in the digital economy [5, 20], but it remains unclear as to whether increasing the user's skills, alone, will satisfy the immense and growing potentials of AI in service system designs. Today, we easily can envision a future where aspects of each human's work, educational, and personal experiences (including all skills) are continuously entwined with multiple service systems simultaneously, providing information to these systems and tailored guidance to the user for each activity that he/she undertakes. Point-of-sale transaction terminals connected to cloud-based processing ensure that myriads of retail and professional services can be widely deployed with confidence in the security and validity of each transaction. Web-based communication services, such as Skype, WhatsApp, Webex, Zoom, and others have enabled new constructs for collaborations within and between organizations globally in real-time. Services aimed at transporting, enriching, or caring for individuals such as car-on-demand, financial planning, medical diagnosis and treatment, education, and travel planning and many others are now broadly tied to, and dependent on computing and data storage technologies.

5.1 AI Maps with the T-Shape Representation of Experience

Service Systems designed and operated with the capabilities and limitations of the humans they will be serving will be better able to support us, as individuals, if we allow each system to learn more about us as we go about our daily professional lives, and additionally share aspects of our personal lives that may, in turn, result in better service. We may be able to establish a framework for this communication pathway by using the T-shaped paradigm to represent an individual's personal and professional achievement and experience [6, 13, 29]. According to this model, the stem (or vertical part) of the T reflects the depth of expertise a person has gained in a discipline or a system, while the top (or crossbar) of the T reflects a person's breadth of experience in disciplines, responsibilities, cultures, organizations and other professional and personal broadening experiences (Fig. 5). Further explorations of the horizontal bar/crossbar of the T-shaped adaptive innovator continue to highlight the importance of a strong focus on the human aspects of service system knowledge [7, 8].

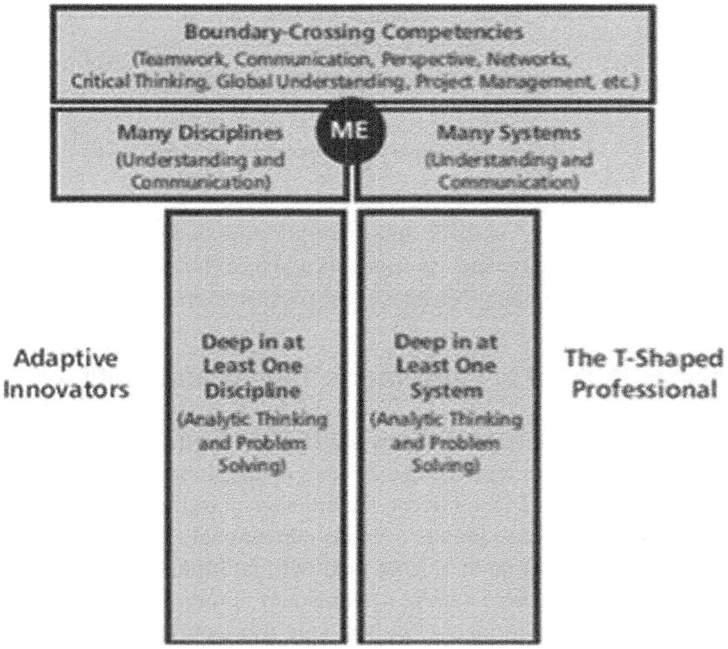

Fig. 5. The T-Shaped representation of experience [21]

The T paradigm has been proposed to represent the current state of any person at any point in their career path [12]. This paradigm defines the T-stem in terms of a person's: (1) Memberships, authorships, and recognitions. (2) Education, degrees and certifications. (3) Operations responsibilities and expertise. (4) Software/Device Proficiencies. (5) Methods/Skills and Proficiencies. Conceivably, based on a person's documented T stem history to date, a service system supported by AI could configure and supplement the expertise of that person for purposes of assuring that the system and the person are optimally integrated.

The paradigm's T-top represents an individual's "breadth" of experience by the following categories: (1) Project management. (2) Organizational design. (3) Communications. (4) Critical thinking. (5) Teamwork. (6) Networking. (7) Empathy. (8) Perspective. (9) Global understanding.

5.2 Enhancing the Human-Side the T-Shape Representation of Experience

People develop experience through both specialized and broad life experiences. Today some of these experiences and outcomes are documented in platforms such as LinkedIn. It seems likely that AI and the Human-Side of Service Systems could meet and exchange information using the T-shape framework as a platform. As individuals advance their T-shape, by completing degrees, taking on new organizational roles, traveling to a different culture, or learning new software, the information could be passed to an AI module supporting a service system that they are interacting with. The information would be used by the AI to configure the interface for that person appropriately based on the record of accomplishments and experiences registered in his or her T-shape. In other words, an AI supported service system could conceivably modify the user's system interface in real-time for each different person based on his or her T-shape metric record.

Additionally, by "knowing" the T-shape metric records of a person, AI could be a much more robust support for systems interaction. Instead of knowing only the person's past interactions with the system to propose guidance and support (as Amazon now guides us based only on our past interactions with Amazon), AI access to something akin to the T-shape metric would be able to assist and suggest to us in much more broader contexts, based on what we do, where we have been, where we are planning to go, and myriads of other ways. As we build out our T-shape profiles, the contributing elements of expertise and experience would enable interfaced AI to configure our interactions with service systems accordingly.

As these concepts evolve, the Human-Side of service systems works together with the AI supporting the systems to produce an enhanced service experience. It is as though the human, as reflected by his/her T-shape constructs is enhanced by the T-shape enhanced service system.

6 Extensions Related to Participatory Governance in the Public Sector

In the current century, digital platforms have changed the way service systems in the public sector operate. They serve not only individual clients: citizens, enterprises and administrative units, but also provide the general public with 'public goods' in the public interest. Public administrations at all levels gather, process and deliver information of different kinds and forms – detailed and statistical; numerical, textual and graphical; this information is subsequently stored in databases and document repositories. Governments and citizens are interrelated by a 'social contract' that includes the delivery of public goods and is reflected in the rule of law, the protection of citizens' rights and civil liberties, legal certainty, political and economic stability, security, public order and the provision of basic infrastructures. While the social contract is based on continuity, IT is characterized by streams of breakthrough innovations followed often by disruptions and discontinuity. This aspect of IT must be identified and managed, always putting the objectives of the social contract and the citizen at the center.

In recent years, the spread of advanced IT and its applications on mass scale, such as increased network throughput and availability, web services, cloud computing, mobile apps, social media, and artificial intelligence stimulated deep changes within the public sector. Evidenced-based policy-making and digital services are currently widely applied in the public sector. New governance models have emerged, with a focus on citizen participation (e.g. in participatory budgeting), 'co-creation', and 'co-production'. The Internet of Things (IoT) is driving Smart City development, while AI, machine learning, automated systems, and robots are replacing routine processes (e.g. the granting of building permits), taking over hazardous activities (e.g. the use of robots to clean up hazardous waste) or forecasting (e.g. disaster management and pandemics prevention). Therefore, public servants can focus on tasks that provide more value for citizens that are focused on citizen interactions, while generating more high-quality jobs and increasing job satisfaction among public sector employees. To this end, continuously upgrading the T-shape experiences of the public workforce in both depth and breadth, as discussed in the previous section, will be crucial.

Three main questions arise when considering the deployment of new technologies in the public sector: (1) Engaging Citizens: How can governments best engage citizens while addressing the challenges of the digital transformation of government? (2) Staying Safe: How to ensure security, and mitigate cybercrime, disinformation, and civil unrest? (3) Fair Distribution: How to ensure more equal distribution of benefits and costs? Each question will be discussed below:

Engaging Citizens: Now, about two decades after the beginning of 'e-government' [18], public administrations are going increasingly 'digital by default' or "digital first". However, currently citizens require governments to be even more focused on the 'human side' [22, 31] due to increasing awareness of the potential negative impacts of digital technologies on individuals and society (for example, mass manipulation during elections). In democratic societies, citizens ask their government to fulfill the social contract also throughout their offerings in digital services, security, the protection of citizens' rights and the provision of public service offerings based on citizens' needs (and wants).

Staying Safe: Cyberattacks by criminals against people and enterprises, and by local or foreign state-sponsored actors in pursuit of a political agenda are a serious concern. Cybersecurity risks to the critical infrastructure, which is already a major concern for traditional networks, are likely to be amplified by the deployment of the IoT. Networks serving private and government systems are expected to be populated with millions of sensors and actuators, most of them quite small, with limited processing power, so the systems will have reduced possibilities of 100% defense unless there are interfaces permitting human monitoring and threat detection. The mass cyberattacks on critical infrastructure and public institutions in Estonia in 2007 are now considered to be the beginning of an era of the use of digital technology as a powerful and destructive weapon.

In recognition of the potentially catastrophic consequences of a cyberattack on critical infrastructure, such as power grids, communication networks, financial systems, traffic systems, or – more recently during the COCID-19 pandemic – hospitals/health care providers, governments of the world have started to work on improving the resilience of these essential networks. However, open and interconnected societies are exposed to a large variety of other types of attacks such as espionage, data theft, fraud, and the targeted dissemination of disinformation via the electronic media.

Social cohesion is menaced when IT is used to increase tensions between different segments of a society. The phenomenon of 'filter bubbles' in social media contributes to this phenomenon. A 'Filter bubble' is a combination of user profiling and search algorithms applied by digital media operators that may lock users into self-contained, self-referential and self-reinforcing networks of like-minded participants where entrenched views and positions are rehashed and reasserted [23]. Online discourse within 'filter bubbles' becomes fragmented and constrained, so public online discourse becomes no longer effective as an integrative, inclusive medium for public deliberation and consensus-building.

Fair Distribution: When certain social groups bear a disproportionate share of the costs and negative side effects of adopting a new breakthrough technology, the social contract comes under additional strain. Forecasts of large reductions in the number of job positions due to deployment of AI and robotics tend to accrue disproportionately to a small segment of the population, but unsettle large parts of the population. The above remark is valid in both a national and international context. All these undermine acceptance of the existing social contract and renders citizens susceptible to populist agitation [19].

7 Extensions in Privacy and Ethics

In the current stage of development and deployment of IT, we observe the integration of the physical world with the digital world giving birth to the cyber-physical world [3]. Due to the combined effect of advances in hardware (data storage, processors), software (virtualization, data analytics) and infrastructure (broadband capacity), data in general, and personal data in particular, has become a storable and tradeable commodity enabling companies to devise new business models and governments to re-design public service offerings for businesses and citizens. However, collecting data through smart devices connected via IoT platforms and human-operated platforms like e-commerce and social media raises new questions related to privacy and ethics [38]. Personal networked devices, whether they be baby monitors, fitness trackers, smartwatches, or medical alert sensors for seniors, transfer sensitive personal information via their network connection. Smart devices that monitor public spaces most likely are collecting information about individuals without their knowledge or consent. More and more of these data and the tools and infrastructure to process it in large quantities is concentrated in the hands of ever fewer data repository and data analytics players.

In the cyber-physical world, after extracting new knowledge from collected big data, AI is used to influence people either directly through the internet of people or indirectly through the internet of things that surround people [33]. Today, stored data reflect not only a person's possessions, actions, and relationships with other people, but even their wishes, intentions, and emotions. Data collected in real-time about a person may be combined with those cumulated over a long time. Data collected for a particular reason (with consent) may be re-used for a very different reason.

Finally, data describing a particular person may be processed in the context of data describing different, but somehow similar persons, circumstances, and events [2]. Predicting human behavior in complex service systems is an important aspect of the Human-Side of Service Engineering [10]. The availability of big data and AI tools opens great opportunities to not only predict human behavior, but also to then attempt to influence the future behavior of people [2]. Consequently, an important privacy and ethics issue relevant to the Human Side of Service Engineering concerns the intentions of persons and organizations that make such predictions. The potential good uses include providing people with services that are personalized and well adapted to their needs, as well as more effectively and quickly protect against hazards. The potential nefarious uses include forming personalized restrictions of individual rights and the erosion of the foundations of trust inherent in a democratic society [9].

In the future cyber-physical world inundated with data, the problem of people's privacy will become crucial [4, 24, 26]. It is generally agreed that private data should be kept secured for at least four different reasons. (1) To reduce the possible distress caused by the change in social relations: a person who has lost some aspect of his or her privacy can consequently be subject to judgment by other people, hardly ever favorable. (2) To reduce a person's vulnerability to business-related attacks such as aggressive marketing, refusing to enter certain contracts, or aggravating contractual provisions. (3) To minimize the probability of criminal attacks. Private data is aggressively sought by cybercriminals to target potential victims and to minimize their risk when planning a crime. (4) To minimize vulnerability to identity theft. Identity theft has severe consequences for a victim. It is tough to prove that decisions, such as bank transfers, were made by an identity thief, instead of an actual bank client, when the credentials used in the transaction appeared in all respects to be authentic.

Business, unsurprisingly, is interested in maximizing the profitable use of the personal data of its current and potential clients to reduce its risk of presenting unwanted offers of goods and services. Such offers could both disturb customers and generate costs that do not provide the business with profits. A business may use its customer's data to detect and forecast his/her possible needs and potential vulnerability to arguments and suggestions to purchase its goods and services to meet those needs. This ability for forming predictions is a kind of win-win situation in the business – customer relationship.

In the European Union, the General Data Protection Regulation (GDPR) [14] restricts collection and use of personal data to protect people from abuses. Personal data about a person may be collected only for a clearly communicated purpose and only if that person explicitly agrees. Any person may ask any business, governmental or non-governmental unit operating in the EU about personal data collected that concern him/her, as well as may require updating or the removal of that data. Personal data may only be retained for as short a time as possible.

GDPR is in line with many scholars and policymakers who call for ensuring that algorithms governing our lives are transparent, fair, and accountable [25].

8 Summary and Future Work

New information technologies strongly stimulate service innovation. As technologies advance, so do rules, such as standards, policies, and laws. We discussed the growth of service and AI-related standards that is driving the need for more mutually beneficial collaboration between industry and service science research in academia. In addition, we discussed how the advance of augmented systems will enable human system interfaces that are uniquely synergized with each person's skills, knowledge, and abilities. Formulation of systems interfaces for people based on models such as the T paradigm will improve safety, efficiency, and satisfaction.

The Covid-19 pandemic has accelerated the shift to digital-first in government across the globe to ensure resilient public services. As highlighted in Sect. 6, this transformation calls for a strong focus on the Human Side in digital government solutions. Decision makers are confronted with three key challenges. (1) Ensuring safety and security of all citizens in the digital sphere. (2) Balancing costs and benefits for all. (3) Making sure all citizens can exercise their democratic rights effectively and safely.

As technology continues to permeate all aspects of business and society, the need to establish governance structures and regulatory frameworks increases accordingly. This process should not be perceived as an impediment to innovation but as a precondition to assure the legitimacy and acceptance of technology by society at large.

Advancing technological capabilities such as AI will become a two-edged sword augmenting our ability to scale service to everyone, while also challenging personal privacy as businesses and governments create better models of customers and citizens to serve them better. Unfortunately, better model may also serve for better manipulation of customers and citizens. Therefore, it is necessary to widely use the privacy-by design approach. Responsible technology innovation will require people who can collaborate across disciplines consistent with the T paradigm as well.

As our data and experience become enmeshed with our IT helpers, future research directions of the HSSE community must include work on: new methodologies and tools, service robots, and platforms and service ecosystems.

The study of human memory and creativity in the age of smart machines should be made a priority. Increasing capabilities of individuals and families in the era of AI could give rise to self-study methodologies and tools.

The deployment of service robots in businesses as well as in households is an important HSSE topic. Designs for some service environments may try to make some robots invisible, while other environments may require robots that are increasingly human-like. The ability to transform a service robot to provide personalized service is an area for future research. How might government service robots lead to new forms of computational democracy? How will service robots in transportation and logistics, in healthcare and education, impact the quality of life?

The growing economic impact of platforms in all walks of life, most accessible from smartphones that almost always in the position of an individual, create diverse opportunities for HSSE studies. Digital marketplaces, circular economy, multi-sided markets, government as a platform and more and more interactions that are made by our AI agents on our behalf will lead to unexpected consequences that are fertile areas of study for HSSE researchers and practitioners alike.

References

1. Ahmed, M.N., Toor, A.S., O'Neil, K., Friedland, D.: Cognitive computing and the future of health care cognitive computing and the future of healthcare: the cognitive power of IBM Watson has the potential to transform global personalized medicine. IEEE Pulse **8**(3), 4–9 (2017)
2. Alpaydin, E.: Introduction to Machine Learning, 3rd edn. The MIT Press, Cambridge (2017)
3. Alur, R.: Principles of Cyber-Physical Systems. The MIT Press, Cambridge (2015)
4. Angwin, J.: Dragnet Nation: A Quest for Privacy, Security, and Freedom in a World of Relentless Surveillance. Henry Holt and Company, LLC., New York (2014)
5. Bakhshi, H., Downing, J., Osborne, M., Schneider, P.: The Future of Skills: Employment in 2030. Pearson and Nesta, London (2017). https://www.nesta.org.uk/report/the-future-of-skills-employment-in-2030/
6. Barile, S., Saviano, M.: Dynamic capabilities and T-shaped knowledge: a viable systems approach. In: Barile, S. (ed.) Contributions to Theoretical and Practical Advances in Management. A Viable System Approach (VSA), pp. 39–59. ARACNE Editrice S.r.l., Rome (2013)
7. Barile, S., Franco, G., Nota, G., Saviano, M.: Structure and dynamics of a "T-Shaped" knowledge: from individuals to cooperating communities of practice. Serv. Sci. **4**(2), 161–180 (2012)
8. Barile, S., Saviano, M., Simone, C.: Service economy, knowledge, and the need for T-shaped innovators. World Wide Web **18**(4), 1177–1197 (2014). https://doi.org/10.1007/s11280-014-0305-1
9. Draft Ethics Guidelines for Trustworthy AI. The European Commission's High-Level Expert Group on Artificial Intelligence (AIHLEG) (2018). https://ec.europa.eu/digital-single-market/en/news/draft-ethics-guidelines-trustworthy-ai
10. Freund, L.E., Cellary, W. (eds.): AHFE 2017. AISC, vol. 601. Springer, Cham (2018). https://doi.org/10.1007/978-3-319-60486-2
11. Freund, L.E., Spohrer, J.C.: The human side of service engineering. Hum. Factors Ergon. Manufact. Serv. Ind. **23**(1), 2–10 (2012)
12. Freund, L.E.: MyT-me – your personal T-shape scoring system. In: Moghaddam, Y., Demirkan, H., Spohrer, J.C. (eds.) T-Shaped Professionals: Adaptive Innovators, pp. 107–115. Business Expert Press, New York (2018)
13. Gardner, P., Estry, D.: A Primer on the T-professional. Collegiate Employment Research Institute, Michigan State University, MI, USA (2017). https://www.ofic.org/s/Monograph-Primer-on-the-T-professional.pdf
14. General Data Protection Regulation (GDPR), Regulation (EU) 2016/679 of the European Parliament and of the Council of 27 April 2016 (2016). https://gdpr-info.eu/
15. IfM, IBM: Succeeding Through Service Innovation: A Service Perspective for Education, Research, Business, and Government. University of Cambridge Institute for Manufacturing, Cambridge UK (2008). https://www.ifm.eng.cam.ac.uk/insights/servitization/succeeding-through-service-innovation/
16. ISO/IEC Guide 76: Development of service standards – Recommendations for addressing consumer issues (2020). https://www.iso.org/standard/73717.html
17. ISO/IEC Joint Technical Committee 1 Sub Committee 42 Artificial Intelligence (2017). https://www.iso.org/committee/6794475.html
18. Leitner, C.: eGovernment in Europe: the state of affairs. European Institute of Public Administration EIPAScope, no. 1, pp. 36–48, Maastricht, The Netherlands (2004)
19. Leitner, C., Stiefmueller C.: Disruptive technologies and the public sector: the changing dynamics of governance. In: Baimenov, A., Liverakos, P. (eds.) Public Service Excellence in the 21st Century. Palgrave, Singapore (2019)

20. Markow, W., Hughes, D., Bundy, A.: The New Foundational Skills of the Digital Economy: Developing the Professionals of the Future. Burning Glass Technologies and Business Higher Education Forum Report, Boston, MA, USA (2018). http://www.bhef.com/sites/default/files/BHEF_2018_New_Foundational_Skills.pdf
21. Moghaddam, Y., Bess, C., Demirkan, H., Spohrer, J.C.: T-Shaped: The New Breed of IT Professional, Agile Product Management and Software Engineering Excellence Executive Update, vol. 17, no. 8 (2016). https://www.cutter.com/sites/default/files/APM/apmu1608.pdf
22. Going Digital: Shaping Policies, Improving Lives – Summary, OECD, Paris (2019). www.oecd.org/going-digital/going-digital-synthesis-summary.pdf
23. Pariser, E.: The Filter Bubble: What the Internet Is Hiding from You. Penguin Press, New York (2011)
24. Pasquale, F.: The Black Box Society, Harvard University Press, Cambridge, MA, USA (2015). https://www.hup.harvard.edu/catalog.php?isbn=9780674970847
25. Rahwan, I.: Society-in-the-loop: programming the algorithmic social contract. Ethics Inf. Technol. **20**(1), 5–14 (2017). https://doi.org/10.1007/s10676-017-9430-8
26. Schneier, B.: Data and Goliath: The Hidden Battles to Collect Your Data and Control Your World. W.W. Norton & Company, New York (2015)
27. Spohrer, J.C., Kwan, S.K.: Service science, management, engineering, and design (SSMED): an emerging discipline – outline and references. Int. J. Inf. Syst. Serv. Sector **3**(1), 1–31 (2009)
28. Spohrer, J.C., Kwan, S.K.: Fundamental concepts and premises of service science. Paper presented at the 1st International Conference on Serviceology (ICServ 2013), Tokyo, Japan (2013)
29. Spohrer, J.C., Maglio, P.P., Bailey, J., Gruhl, D.: Steps towards a science of service systems. IEEE Comput. **40**(1), 61–77 (2007)
30. Teboul, J.: Service is Front Stage – Positioning Services for Value Advantage. INSEAD Business Press Palgrave McMillian, New York (2006)
31. The Digital Compass (2021). https://digital-strategy.ec.europa.eu/en/policies/digital-compass
32. The World Fact Book (2021). https://www.cia.gov/the-world-factbook/countries/united-states/#economy
33. Unemyr, M.: The Internet of Things – The Next Industrial Revolution Has Begun: How IoT, big data, predictive analytics, machine learning and AI will change our lives forever. Kindle Edition (2017)
34. Vargo, S.L., Lusch, R.F.: Evolving to a new dominant logic for marketing. J. Mark. **68**(1), 1–7 (2004)
35. Vargo, S.L., Lusch, R.F.: It's all B2B…and beyond: toward a systems perspective of the market. Ind. Mark. Manag. **40**(2), 181–187 (2011)
36. Vargo, S.L., Lusch, R.F.: Institutions and axioms: an extension and update of service-dominant logic. J. Acad. Mark. Sci. **44**(1), 5–23 (2015). https://doi.org/10.1007/s11747-015-0456-3
37. Weissinger, R., Kwan, S.K.: Service science research and service standards development. In: Satzger, G., Patrício, L., Zaki, M., Kühl, N., Hottum, P. (eds.) IESS 2018. LNBIP, vol. 331, pp. 330–343. Springer, Cham (2018). https://doi.org/10.1007/978-3-030-00713-3_25
38. Wojtowicz, A., Cellary, W.: New challenges for user privacy in cyberspace. In: Moallem, A. (ed.) Human Computer Interaction and Cybersecurity Handbook, pp. 77–98. CRC Press – Taylor and Francis, Boca Raton (2018)

New Ecosystem Based on Big Data for More Digital Impact

Loris Schmid(✉) and Christoph Glauser

IFAAR, Institute for Applied Argumentation Research, Marktgasse 55, 3011 Bern, Switzerland
{Schmid,Glauser}@ifaar.ch

Abstract. This article introduces a practical approach of managing a public- and a business- market network of two B2B organisations and creates an independent empirical overview in a newly drawn ecosystem for "digital impact" measures. The results enabled the management to take solid data science based decisions. Starting with an unstructured database, containing 14 years of customer relations and 5 years of digital online interactions, various technological and statistical methods like in a "popular census" are applied to draw an innovative ecosystem of B2B customers which are actively interested in improving their own "digital impact". The results of the study allowed the stakeholders and the management to fully restructure their B2B approach and to learn new priorities on how to rethink their own contact activities. It has revealed that on the commercial business level all kinds of Agencies, followed by food and nonfood, insurances and banks are the most impact oriented industries. On the institutional level there are Federal Governments, Universities and all sorts of Media among the most actively interested bodies, followed by various Consultants. Based on these results and having visualized the data of the empirical research, the organizations call their ecosystem now the "digital impact ecosystem".

Keywords: Ecosystem · Business network · CRM · B2B · Digital impact

1 Introduction

This article describes an empirical approach and the final results of a study which was conducted on behalf of an online market research company (B2B) who wants to know more about their customers, their business network and their contacts. A recent overview over the four main types of ecosystems – business, innovation, entrepreneurial, and knowledge ecosystems has been previously described very broadly in literature [1].

The current empirical ecosystem combines aspects of these ecosystem types in an empirically novel way. The company intends to 1) learn more about their own "digital impact B2B market". 2) The results should enable more focused decisions in marketing their solutions in the future. 3) Finally the knowledge about the industries should be extended by visualizing big data sources as for example extracts of the CRM data with customer contacts and industry references as well as with first contact customers using a tool, which provides real-time online market insights.

C. Leitner et al. (Eds.): AHFE 2021, LNNS 266, pp. 18–26, 2021.
https://doi.org/10.1007/978-3-030-80840-2_2

2 Definitions

There are four main directions of ecosystem development in literature. Scaringella and Radziwon [1] provide a table of various definitions. The current work is an attempt to mirror and visualize the theoretical bases with empirical data of a particular online B2B market environment.

For instance, the present work is referring to the business ecosystem, for which Iansiti and Levien [2] provide the following definition:

> "Loose networks – of suppliers, distributors, outsourcing firms, makers of related products or services, technology providers, and a host of other organizations – affect, and are affected by, the creation and delivery of a company's own offerings. Like an individual species in a biological ecosystem, each member of a business ecosystem ultimately shares the fate of the network as a whole, regardless of that member's apparent strength."

The other main concepts are the innovation ecosystem, defined by Adner [3], the entrepreneurial ecosystem, defined by Isenberg [4] and the knowledge ecosystem, defined by Clarysse et al. [5].

As digitalization progresses, the measurement of digital impact grows more important. This measuring of impact on the internet can only be achieved by cooperation of the company, the agency providing the communication strategy and a provider of a measurement tool. Therefore, the current work considers the loose network of these impact oriented agents and displays the network in a private and an institutional ecosystem representation. The method presented here, can be applied to other companies in order to create their own visual display of a network.

3 Methodology

The process starts with systematic CRM data of the former 14 years. Every B2B contact originating online is linked to a person representing its respective company or organization. The explicit goal of this study is to provide a valid and reliable graphical representation of the user-, customer-, provider-, and even reseller-driven business network, which should be ideally displayed in the form of an ecosystem. Following the logic of the business network interconnections on the level of the ecosystem, it furthermore should allow a ranking, showing the relevance and importance of the different industry players to the company, with the explicit goal of supporting the management in their strategic B2B marketing decisions. The study is grounded on the general assumption that users contacting the company by digital applications or forms usually have various interests around *digital impact* and improving *digital online performance*. The ecosystem is designed to combine in one single visual presentation three different contextual levels of the ecosystem at the same time – environment (digital impact), network (industry) and actors (users) [6].

The CRM database is offering a highly differentiated picture of B2B customer contacts, which has grown during the past years by marketing and networking means. The active involvement of end user accounts for almost half the basic data are used as user-driven data in the empirical setup. Literature points to the fact, that especially for B2B applications and cases, ecosystems have unfortunately not often been applied so far [7]. This may be due to the fact that B2B relations strongly relay on personal-, or benefited business. Such individuals rarely share their unique market knowledge and views.

3.1 Research Design

In a first step all the CRM data with its industry references, attributes and characteristics is analyzed and further regrouped into entire industry or branch features. The data is already sorted according to the systematically collected features assigned to every entry in the database. The most prominent distinction is driven by the data which divides between contacts in the institutional market and contacts in the commercial market. This ultimately leads to the two different ecosystems. Other features include the type, scope, branch, and origin of the contacts as well as the relationships this B2B contact has with the company.

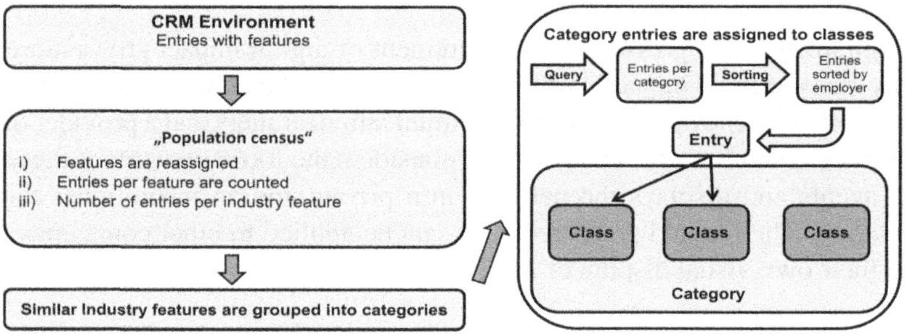

Fig. 1. The Process illustration of the "population census" shows how to extract from raw CRM data entries to categories containing classes. (color)

The process described in the following is illustrated in Fig. 1. In order to sort all the entries and elements, the features out of the CRM are reevaluated and reassigned respectively following this logic:

i) Once the correct assignment of features is finalized ii) the numbers of entries for each feature are counted by use of queries on the database. Like in a "population census" the different entries linked to a personnel contact are iii) categorized by the number of entries per existing industry feature. The "population census" data is kept in German and only the final ecosystem results are translated. Based on the industry feature entries, larger categories are created by grouping industry features. To form the category "Education" for example, the features "school" and "educational institution" are bundled. By looking at the "census" results, every entry in a category is assigned to one or multiple classes. For example, a B2B customer can be an Agency which is in the advertising agency class *and* in the PR-agency class. One of the largest categories in the institutional market is the category "Public Administration" which includes classes like "Federal Governments (on Ministerial level)" and "Municipalities" or "cities". The assignment of entries is done in the following three steps. First, a query is made that counts every entry of one or several features in the database. Thus all the entries of a category are calculated. Second, the entries are sorted by name of the employer. This gives the opportunity to assign similar entries to the same classes. Third, every entry is assigned to one or several classes, which create for the respective class a count of one per entry. Thus a company with 3 contact entries counts as 3 entries in the classes that are assigned to the company. In the end, the number of entries per class is counted and saved in order to weight the classes later.

After a first review of the findings by the scientists and the company, the categories are finalized and the allocation of each B2B customer is clearly defined. Every category is displayed in a bubble containing the respective classes represented by circles. Figure 2 shows an example of such a bubble. In this case it is the bubble of the "Agencies" category. The bubble representation serves to group all the classes together in a category. The size of the circles is given by their respective weight among their category. In order to weight the classes, five intervals are set for each category. The class with the most entries provides a guiding value for the upper end of the interval for the biggest circle, while the class with the least entries provides a guiding value for the lower end of the interval for the smallest circle. The intervals are constant for the respective categories.

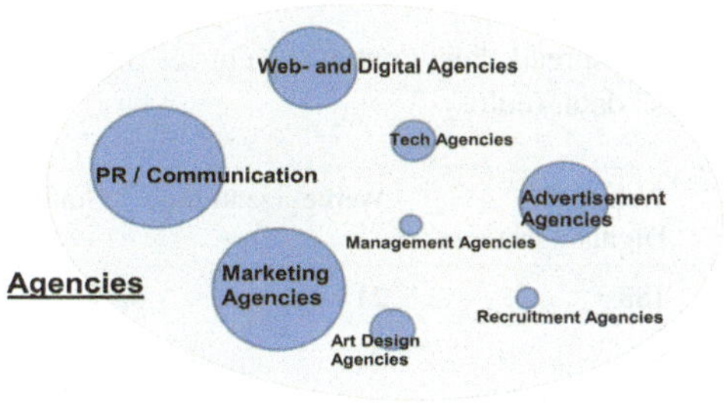

Fig. 2. Bubble of the category "Agencies" containing circles of classes. The circles are sized according to the number of B2B customer entries assigned to the corresponding classes. (color)

In the graphical representation of the ecosystem the circles are ordered in size according to the interval in which the number of CRM-entries of the class belongs to the considered category. This gives an overview of which classes are most prominent in which category. While marketing agencies are the most prominent class of agencies with 278 entries, recruitment agencies are only present with 18 entries.

The overall comparison of the classes is given by a ranking. Here the classes are sorted into intervals over the whole ecosystem. The intervals are created in a similar manner as before. This ranking then gives an overview of which classes are most prominent over the whole ecosystem. As Nambisan et al. [8] are claiming: digital platforms and ecosystems (DPE) are "affording new ways of internationalization, as facilitating new ways of building knowledge and relationships, and as enabling new ways of creating and delivering value to global B2B customers".

4 Data Collection and Categorization

The population of B2B customers is 6'037 entries in the CRM system used by the company. The software is used by the company to collect contact and activity data of contact persons since 2006. Contact persons include B2B customers, individual customer, suppliers, consultants and resellers. The data collected includes the name and address of the company, the position, phone number, e-mail address of the contact person as well as any activities with the contact person itself. Activities can be offerings, orders, assignments or e-mail communication, phone calls, meetings, workshops, data deliveries etc. In 2015, the company has developed and launched a new online access tool where customers can discover digital markets on their own behalf. Users have to register on the tool and thereby, with their consent, also provide their name and e-mail address and their position in their B2B firm to the CRM. From the 6'037 B2B entries, 2'042 contacts were generated directly via online tool. Of the entries generated this way, 446 are linked with the institutional market, while 1'596 are linked with the commercial market. In a process similar to a population census, there were 1'042 B2B entries counted, which definitely had to be assigned to the institutional market. The entries assigned to the commercial market count 3'494 B2B entries. The total sum of B2B entries assigned to classes is therefore 4'536.

Table 1. Sample excerpt of the spread sheet containing entries of the "Agencies" category taken out of the "population census" data. (color)

Klasse	Tech-Agenturen	Web- und Digital-agenturen	Werbe-agenturen	Grafik-agenturen	PR Komm
Anzahl	**75**	**188**	**213**	**86**	**265**
Kreisgrösse (1–5)	**2**	**4**	**4**	**2**	**5**
		Agency A			Agency A
		Agency B			Agency B
		Agency B			Agency B
		Agency C		Agency C	
			Agency D		Agency D

Data reorganization of the B2B customers led to two ecosystems: One shows the commercial, the other shows the institutional B2B market players. Table 1 displays a sample of a table of the "population census" data which is taken out of the spread sheet concerning the category "Agencies". This table illustrates how the contact person entries are assigned to classes and how the intervals are created. In respect of the agencies' privacy the names of the agencies are anonymized by calling them Agency A, Agency B, etc. The first row contains the names of classes such that every class has its column. In the second row, the number of entries assigned to the respective class is counted and the third row indicates the number linked to the interval in which the class is situated. During the process of the "population census" every entry of a category is assigned to one or several classes, as described in Sect. 3.1. Here this process is illustrated in the way that the entry "Agency A" is assigned to the classes "Web- and Digital Agencies" as well as ""PR/Communication (agencies)". This gives both classes a count of 1. Every row in the table refers to a single entry in the CRM database. If there are to contact persons that are affiliated with the same agency, each contact gives the assigned class a count of 1. Every class is transformed into a circle for the visual presentation in the ecosystem. The size of the circle is then given by the number representing the interval with 5 being the largest circle.

5 Results/Discussion

The following figures display the commercial- and institutional ecosystem with circles as well as industry category bubbles.

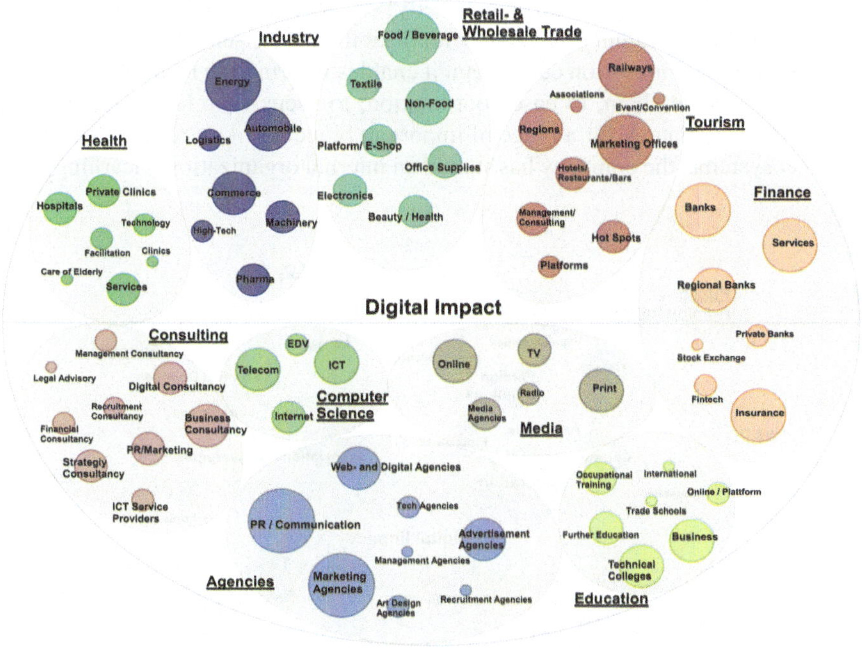

Fig. 3. The Commercial Ecosystem visualizes the intra-category prevalence of classes. Categories are represented by bubbles and classes by circles. (color)

While operational Figs. 3 and 4 allow a comparison of circles inside the respective category, the rankings which are not shown here due to page limitations concentrate on a comparison of circles over the whole ecosystem.

One of the most surprising findings: There are many smaller B2B customers made visible, for example individual Consultants, which have had almost no priority before, even though they secretly have become an important group for the company. The customer segment of Media, especially TV customers, is a much bigger segment than perceived by the management. On the other end of the ranking, health care and educational systems are typically rather "loud" and seemingly very important industries, but have shown to occupy a much less powerful position than other industries. The findings have led to decisions on how the offerings of the company can be combined much more efficiently by focusing on relevant B2B industry players and customer circles which are actively interested in having corporate "digital impact". Among the discovered priority industries, the most prominent are many different types of Agencies (Marketing, PR, Advertising, and Web) followed by food & nonfood, insurances and banks. Among the institutional B2B customer contacts, the ecosystem discovers rather surprisingly that Federal Governments, Universities as well as Media of all kind and Consultants or Corporate Associations are the most interested segments here. They need to assure digital impact even on a legal bases (citizen-centricity) or "access for all" legal regulations by the convention on the rights of persons with disabilities [9]. As Adner [3] describes in the definition of the innovation ecosystem, it enables information technologies to better understand the value-chain, to ease coordination, to focus on relevant industries and develop a growth strategy for a range of important branches. As a result of the analysis of these ecosystems, the company has started an internal organizational learning process [10].

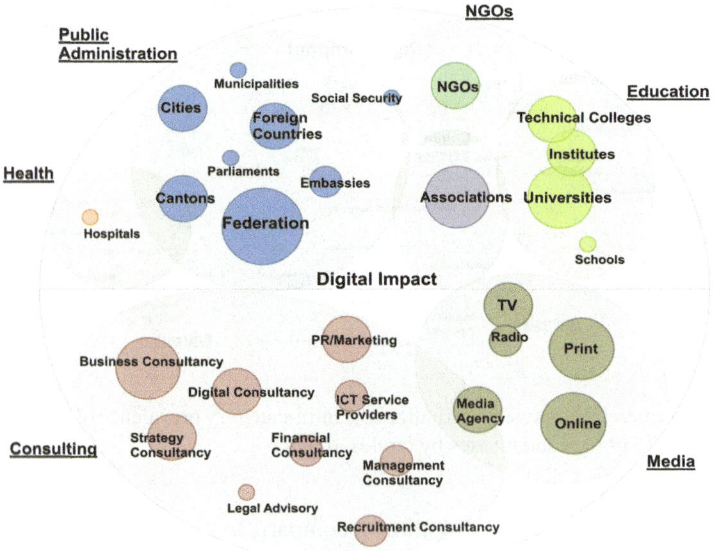

Fig. 4. The Institutional Ecosystem visualizes the intra-category prevalence of classes. Categories are represented by bubbles and classes by circles. (color)

6 Conclusions

The empirical results of the scientific work displaying a new ecosystem retrieved from largely unstructured data of a complex B2B-CRM system finally allows all the stakeholders to discover, identify, visualize, prioritize and understand their new *"digital impact ecosystem"*. These valuable results are widely taken up by the sales people, the management and even the board of the company. Management realizes based on the ecosystem of their existing business network, how the B2B customer landscape is structured and how it can be addressed much more efficiently as an existing but newly discovered market playground. The current work has developed a method on how to proceed from B2B-CRM customer contact data to a graphical display of industry importance in the form of an ecosystem.

The company has reorganized most of its B2B marketing activities considerably. The developed method can be applied repeatedly to show the upcoming trends, emerging markets, or even disappearing B2B players who are actively interested in digital impact. As a Chinese saying tells us: "What is big today, can become small tomorrow and what is small today, can become big tomorrow." There is no better way than to display this in an ecosystem. The results show an ecosystem which has delivered well known as well as surprising results to the management of the company that offers fully digitalized market research and a paid service for online market research based on scientific digital market demand volumes via APIs.

In further research, it would be interesting to include even more of the CRM data in order to display ecosystem concentrating for example on the actions taken. The weight of a contact could be drawn from the activity intensity of the company with a specific contact. Also, a reproduction of the study considering another company, perhaps in another industry, would show if the presented method can be applied to any company.

References

1. Scaringella, L., Radziwon, A.: Innovation, entrepreneurial, knowledge, and business ecosystems: old wine in new bottles? Technol. Forecast. Soc. Change **136**, 59–87 (2018)
2. Iansiti, M., Levien, R.: Strategy as ecology. Harvard Bus. Rev. **82**(3), 68–78 (2004)
3. Adner, R.: Match your innovation strategy to your innovation ecosystem. Harvard Bus. Rev. **84**(4), 98 (2006)
4. Isenberg, D.J.: How to start an entrepreneurial revolution. Harvard Bus. Rev. **88**(6), 40–50 (2010)
5. Clarysse, B., Wright, M., Bruneel, J., Mahajan, A.: Creating value in ecosystems: crossing the chasm between knowledge and business ecosystems. Res. Policy **43**(7), 1164–1176 (2014)
6. Möller, K., Halinen, A.: Managing business and innovation networks—from strategic nets to business fields and ecosystems. Ind. Mark. Manag. **67**, 5–22 (2017)
7. Hein, A., Weking, J., Schreieck, M., Wiesche, M., Böhm, M., Krcmar, H.: Value co-creation practices in business-to-business platform ecosystems. Electron. Mark. **29**(3), 503–518 (2019). https://doi.org/10.1007/s12525-019-00337-y
8. Nambisan, S., Zahra, S.A., Luo, Y.: Global platforms and ecosystems: Implications for international business theories. J. Int. Bus. Stud. **50**(9), 1464–1486 (2019)

9. United Nations Convention on the Rights of Persons with Disabilities (CRPD). https://www.un.org/development/desa/disabilities/convention-on-the-rights-of-persons-with-disabilities.html
10. Schön, D., Argyris, C.: Organizational Learning II: Theory, Method and Practice. Addison-Wesley Publishing Company, Boston (1996)

New Frontiers in Cyberspace – Recent European Initiatives to Regulate Digital Finance

Christian M. Stiefmueller(✉)

CEPA | Centre for Economics and Public Administration Limited, 111 Buckingham Palace Road, London SW1W 0SR, UK
christian.stiefmueller@cepanet.eu

Abstract. On 24 September 2020, the European Commission presented its Digital Finance Strategy for the EU, a programmatic roadmap accompanied by a package of legislative proposals. This initiative forms part of the Commission's broader strategy for 'Shaping Europe's Digital Future', as announced by President von der Leyen in her speech on 19 February 2020, and aims at creating a legal framework for the adoption of recent technologies, such as online platforms, distributed ledger technology (DLT), cloud computing, in the field of financial services. In conjunction with other relevant regulatory initiatives, most notably the Digital Services and Digital Markets Acts (DSA and DMA), it also seeks to address the impact of digital technologies on the structure and competitive dynamics of a variety of markets including, but not limited to financial services, and to provide safeguards against potential risks to financial stability and the welfare of citizens and retail customers. This contribution looks at the regulatory approach that informs the proposed legislation, in particular from an international perspective, and examines its potential impact on the development of digital financial services in the EU.

Keywords: Human side of service engineering · Financial services · Innovation · Digitalisation · Competition · Data protection · Regulatory policy

1 Introduction

On 19 February 2020, the European Commission issued a communication on 'Shaping Europe's Digital Future'[1], which set out a three-pillar strategy for the digital transformation in Europe:

– The first pillar, *'technology that works for people'*, calls for the development and deployment of technology that *'makes a real difference to people's daily lives'* and envisions *'a strong and competitive economy that masters and shapes technology in a way that respects European values.'*

[1] Communication from the Commission to the European Parliament, the Council, the European Economic and Social Committee and the Committee of the Region on Shaping Europe's Digital Future, COM(2020) 67 final, 19 February 2020.

C. Leitner et al. (Eds.): AHFE 2021, LNNS 266, pp. 27–36, 2021.
https://doi.org/10.1007/978-3-030-80840-2_3

- The second pillar of the Commission's digital strategy calls for a *'fair and competitive economy'* and *'a frictionless single market, where companies of all sizes and in any sector can compete on equal terms'*.
- The third pillar – *'an open, democratic and sustainable society'* calls for a *'trustworthy environment in which citizens are empowered in how they act and interact, and of the data they provide both online and offline.'* It postulates a *'European way to digital transformation which enhances our democratic values, respects our fundamental rights, and contributes to a sustainable, climate-neutral and resource-efficient economy'*.

On 24 September 2020, the European Commission presented its Digital Finance Strategy for the EU[2], a programmatic roadmap for the digital transition of the financial services in Europe based on the February 2020 communication. The 'digital finance package' comprises three major legislative proposals: a regulation on Markets in Cryptoassets (MiCA)[3], a regulation on Digital Operational Resilience (DORA)[4], and a regulation on a pilot regime for market infrastructures based on Distributed Ledger Technology (DLT Pilot)[5], accompanied by a blueprint for a Retail Payments Strategy[6].

- MiCA sets out a comprehensive regulatory framework for cryptoassets. It provides a broad legal definition that is intended to capture most existing and potential new types of 'cryptoassets' and defines certain sub-categories ('asset referenced tokens', 'e-money tokens', and 'utility tokens'), which are subject to specific regulatory regimes. A fourth category, 'securities tokens', is defined separately in an amendment to the 2nd Markets in Financial Instruments Directive (MiFID 2)[7], which expands its scope so that tokens that represent/replicate financial instruments are covered by the existing EU securities legislation. For market participants who provide services in connection with cryptoassets, such as placement, trading, and custody, but are not licensed credit institutions or investment firms ('cryptoasset service providers', CASPs) MiCA sets out a framework for governance and market conduct and assigns supervisory responsibilities to national authorities and the European Banking Authority (EBA).

---

[2] Communication from the Commission to the European Parliament, the Council, the European Economic and Social Committee and the Committee of the Regions on a Digital Finance Strategy for the EU, COM (2020) 591 final, 24 September 2020.

[3] Proposal for a Regulation of the European Parliament and of the Council on markets in crypto-assets, COM (2020) 593 final, 24 September 2020.

[4] Proposal for a Regulation of the European Parliament and of the Council on digital operational resilience for the financial sector, COM (2020) 595 final, 24 September 2020.

[5] Proposal for a Regulation of the European Parliament and of the Council on a pilot regime for market infrastructures based on Distributed Ledger Technology, COM (2020) 594 final, 24 September 2020.

[6] Communication from the Commission to the European Parliament, the Council, the European Economic and Social Committee and the Committee of the Regions on a Retail Payments Strategy for the EU, COM (2020) 592 final, 24 September 2020.

[7] Directive 2014/65/EU of the European Parliament and of the Council of 15 May 2014 on markets in financial instruments (MiFID 2); OJ L 173, 12 June 2014, pp. 349–496.

- DORA aims at establishing a harmonised set of rules for managing operational risk connected to ICT systems in the financial sector. It represents a more comprehensive and stricter, sector-specific framework that complements the existing Network and Information Security Directive[8]. DORA applies to a broad range of financial institutions, including banks, insurers, and investment firms, operators of financial infrastructures and cryptoasset service providers. It imposes requirements for ICT governance, risk management, incident reporting, and resilience testing, and sets out rules for managing risk that arises from outsourcing services that support critical functions to third-party ICT service providers.
- The DLT Pilot Regulation specifies a dedicated legal framework for the small-scale, experimental deployment of DLT infrastructures. This regulatory regime applies exclusively for multilateral trading facilities (MTFs) and securities settlement systems using distributed ledger technology ('DLT market infrastructures') and is restricted to the trading and settlement of equities and bonds, other than sovereign bonds, with small issue sizes (€ 200 mn for equities, € 500 mn for bonds) and limited liquidity.

On 15 December 2020, the Commission presented a second legislative package consisting of two comprehensive, cross-sectoral proposals, the Digital Services Act (DSA)[9] and the Digital Markets Act (DMA)[10]. While these regulations are not aimed specifically at the financial sector they both seek to establish a framework for the governance of online platforms acting as intermediaries in digital markets for goods and services.

- DSA updates and expands the existing legal framework in the EU for providers of digital services, such as internet service providers, cloud and webhosting providers, and, in particular, online platforms that host and intermediate online services. It clarifies the rules that govern liability for online content, and provides new protections for consumers and citizens against illegal content, goods and services. Very large online platforms (at least 45 mn average active users per month in the EU) will also be required to undertake annual assessments to identify and address risks, e.g. of their services being used for the dissemination of illegal content or manipulated intentionally to disrupt civic discourse or electoral processes, or jeopardise public security. These operators will be obliged to conduct independent audits, and submit to enhanced supervision and enforcement by the Commission.
- DMA is a cross-sectoral regulation that applies to a broad range of 'core platform services' including, in particular, online intermediation services, search engines, social networking and messaging services, video-sharing platform services, online advertising services, app stores, operating systems, and cloud computing services. Payment services are covered by DMA if they are 'ancillary services' offered in the context

[8] Directive (EU) 2016/1148 of the European Parliament and of the Council of 06 July 2016 concerning measures for a high common level of security of network and information systems across the Union (NIS Directive), OJ L 194, 19 July.2016, pp. 1–30.

[9] Proposal for a Regulation of the European Parliament and of the Council on a Single Market in Digital Services (Digital Services Act), COM (2020) 825 final, 15 December 2020.

[10] Proposal for a Regulation of the European Parliament and of the Council on contestable and fair markets in the digital sector (Digital Markets Act), COM (2020) 842 final, 15 December 2020.

of or together with core platform services. Providers of core platform services that meet a given set of quantitative and qualitative criteria (e.g. EEA revenues of € 6.5 bn or more, market capitalization of € 65 bn or more) are designated as 'gatekeepers'. DMA imposes on gatekeepers a set of specific obligations including, in particular, the compartmentalisation of data generated in different business lines into 'data silos' that cannot be shared; an obligation to ensure interoperability so that competitors are not excluded from offering services on the platform that compete with the gatekeeper's own; and a ban on 'self-preferencing' that requires gatekeepers to adjust their search algorithms so that competitors' offers receive the same level of prominence as their own. The European Commission, as the principal EU competition authority, is granted new monitoring, investigative, and enforcement powers to implement these obligations.

2 Conditions and Dynamics in Relevant Markets

Until fairly recently, payment services – mainly bank transfers, debit and credit cards – were regarded as the archetypal 'flow business', a business of large volumes and small margins, dominated largely by bank-owned and -operated networks and a number of large global card services firms. This situation was highlighted in 2014 by the *Mastercard* case[11], which triggered the subsequent review of the Payment Services Directive (PSD 2)[12] and ushered in reforms aimed at breaking up the payments supply chain and stimulating competition. PSD 2 tried to address the incumbency advantages of traditional banks granting regulated third parties access to customers' bank accounts, either to initiate payments on their behalf or to provide an overview of their account information to others service providers.

Only a few years later there is a sense among many observers and policymakers that these reforms may have overshot the target and tipped the scales too far in favour of new entrants and, especially, 'Big Tech' operators of platform services [13]. Thanks to their dominance in smartphone operating systems, with an estimated installed base of ca. 4 bn iOS and Android devices, Apple and Alphabet/Google already have a strong presence in the fastest-growing segment of the market, mobile, smartphone-enabled payments. Amazon and Apple, meanwhile, have entered the much larger, and still growing, credit card market. Facebook has taken a different approach altogether and plans for a proprietary global payment system that would allow its 2.8 bn active users to make low-cost, peer-to-peer payments using the Diem (formerly Libra) stablecoin, a 'cryptocurrency based on, and exchangeable into one or several major global currencies. The announcement of this project in June 2019 was perceived as a direct challenge to central banks' traditional monopoly on money issuance and monetary policy. A number of major jurisdictions, including the EU, have responded in the meantime by proposing legislation to regulate stablecoins and similar privately-issued 'currencies', and by fast-tracking their own projects for 'central bank digital currencies (CBDCs).

[11] European Court of Justice: Judgment of 11 September 2014, *MasterCard Inc. and Others vs. European Commission*; C-382/12 P, EU:C:2014:2201.

[12] Directive 2015/2366/EU of the European Parliament and of the Council of 25 November 2015 on payment services in the internal market; OJ L 337, 23 December 2015, pp. 35–127.

Beyond payments, online platform operators, notably Amazon, are exploring lending to small businesses. Probably the most prominent example is Ant Group in China, which has expanded successfully from payments (AliPay) to consumer lending. It offers unsecured consumer micro-loans granted based on largely automated credit assessment using its customer profile to derive a credit score. Loans originated through the platform are not kept on the platform operator's own balance sheet but either underwritten by partner banks or securitised.

As digital platform providers increasingly interpose themselves between traditional banks, investment firms, and retail customers investment and savings products and services that are today sold directly, or through dedicated channels, such as independent financial advisers and retail brokers, are likely to become intermediated by platform providers [3]. Again, a look at the Chinese market could prove instructive. Within five years, the Yu'e Bao money market fund, launched by Ant Group in 2013, grew to become the largest in China, and one of the largest worldwide. Most major mutual fund companies, insurers, banks and securities companies in China are now partnering with Ant Group to have their products distributed through its platform.

In each one of these cases, digital platform providers are leveraging their brand recognition, direct access to customers' devices, massive data stores and user bases, and the benefits of network effects [7, 12] to enter new markets. Although regulators in China have stepped in recently to tighten the regulatory framework and impose stricter controls on Ant Group's fast-growing activities, and conditions in the EU financial markets differ from those in China in many respects, lessons from this precedent have not been lost on financial institutions and regulators in Europe. While some institutions are considering partnerships with large platform providers, others are wary of losing control of their customer relationships and being reduced to competing between themselves as suppliers of commoditised products and services to the platform.

3 Competition and the 'Level Playing Field'

Competition in the digital domain has long been a concern for EU authorities and member states. In a series of high-profile anti-trust cases against global technology firms, including Microsoft[13], Alphabet/Google[14], and Apple[15], over the course of the last twenty years, the European Commission has established principles and precedents that continue to inform competition law doctrine and practice internationally. Preserving or, if necessary, restoring a 'level playing field' in the Digital Single Market is one of the priorities of the Commission's digital strategy.

[13] e.g. *Microsoft (Interoperability)* (Case COMP/37.792) Commission Decision C(2004) 900, OJ L 32, 06 February.2007, pp. 23–28; *Microsoft (Tying)* (Case COMP/39.530) Commission Decision C(2009) 10033, OJ C 36, 13 February2010, pp. 7–8.

[14] e.g. *Google Search (Shopping)* (Case AT.39740) Commission Decision C(2017) 4444, OJ C 09, 12 January 2018, pp. 11–14; *Google Android* (Case AT.40099) Commission Decision C(2018) 4761, OJ C 402, 28 November 2019, pp. 19–22; *Google Search (AdSense)*, (Case AT.40411) Commission Decision C(2019) 2173, OJ C 369, 03 November 2020, pp. 6–10.

[15] e.g. *Apple (iTunes)* (Case 39154) Commission Announcement IP/08/22, 09 January 2008; *Apple (Mobile payments)* (Case AT. 40452) (ongoing).

In the past, slow and inconsistent competition enforcement policies have proven costly, especially in the financial sector. In the run-up to the global financial crisis, throughout the 1990s, the EU and US banking sectors experienced a wave of bank mergers, creating 'systemically important' banks that still define the landscape today. Rescuing the giant institutions that emerged from this consolidation process, on both sides of the Atlantic, left the global financial system on the brink of collapse in 2008. Exponential growth and a tendency towards concentration on the supply-side have pushed the pace of development in digital markets to a degree that processes that took nearly two decades in banking may now take place in only a few years. There is little doubt that digital platform providers and 'critical third-party ICT service providers' could become a source of risk to the stability of the financial system, i.e. 'systemically important' entities, themselves.

The economics of two-sided platform markets where product and services are intermediated, and their impact on competition in markets for information and communication services has been studied by economists for some time [7, 12]. They pose a number of peculiar problems in the context of antitrust law, e.g. regarding the methodologies and tools traditionally used to define markets, identify dominant market positions, and measure the impact on consumers and other market participants. Whereas classic econometric modelling has been developed and updated apace, other, more recent insights regarding the decisive influence of 'bounded rationality' on the behaviour of market participants, and hence the competitive dynamics of the markets, are not yet incorporated adequately into the design of competition law, and the processes and tools applied by competition authorities [6, 10].

DMA proposes to grant new supervisory powers to the European Commission, which complement and, in some respects, mirror the EU's existing competition rules, in particular Art. 102 TFEU[16] and the merger regulations[17]. They would invest the Commission with new monitoring, investigative, and enforcement powers, including one-off fines and periodic penalty payments, to sanction anti-competitive behaviour.

Finally, switching costs for users should be kept as low as possible in order to keep digital markets competitive and protect users from being locked in to a particular platform. It is critical that citizens enjoy effective ownership and portability of their personal data, and that data remain interoperable between different platforms [10].

4 Extraterritoriality and 'Digital Sovereignty'

A recurring, and remarkable feature of the EU's proposed legislative initiatives is their ambition to reach out beyond its borders to protect the rights of EU citizens and businesses in the digital domain. They contain several provisions that are modelled on the

[16] Treaty on the Functioning of the European Union (TFEU), OJ C 326, 26 October 2012, pp. 47–390.

[17] Council Regulation (EC) No 139/2004 of 20 January 2004 on the control of concentrations between undertakings (EC Merger Regulation), OJ L 24, 29 January 2004, pp. 1–22; and Commission Regulation (EC) No 802/2004 of 07 April 2004, OJ L 133, 30 April 2004, pp. 1–39 (Implementing Regulation).

groundbreaking General Data Protection Regulation (GDPR)[18]. Art. 3 GDPR defines the territorial scope of its application on the basis of two main criteria: the 'establishment criterion' (Art. 3(1)), and the 'targeting criterion' Art. 3(2)). While the establishment criterion – entities established on the territory of the EU should be bound by its laws and jurisdiction – is largely undisputed among scholars of international law the 'targeting criterion' – GDPR should apply to all entities that process personal data of EU citizens for the purpose of offering goods or services or to monitor their behaviour – remains contested [1]. Its justification is grounded largely in the jurisprudence of the European Court of Justice (ECJ):

- in its *Google Spain* decision of May 2014[19], which predates GDPR, the ECJ held that the processing of personal data of EU citizens by Google in the US was deemed to be carried out in the context of the activities of its EU subsidiary, established in Spain, due to the economic link between the EU establishment and the US processing entity. Therefore, the US entity was deemed to be bound by EU law when processing EU data on US territory;
- in two landmark decisions, *Schrems I*[20] in October 2015 and *Schrems II*[21] in July 2020, the ECJ invalidated the agreements between the European Commission and the U.S. Department of Commerce ('Safe Harbour Privacy Principles' and 'EU-US Privacy Shield') that had facilitated and governed the transfer of personal data between the EU and the USA, on the grounds that European citizens' fundamental rights to privacy and the protection of personal data (Art. 7 and 8 Charter of Fundamental Rights of the European Union) could not be forfeited by the Commission when the relevant overseas jurisdiction did not offer adequate protections.

The 'targeting criterion' features prominently in the Commission's recent proposals:

- according to its Art. 1(3), DSA should apply *'to intermediary services provided to recipients of the service that have their place of establishment or residence in the Union, irrespective of the place of establishment of the providers of those services'*;
- according to its Art. 1(2), DMA should apply *'to core platform services provided or offered by gatekeepers to business users established in the Union or end users established or located in the Union, irrespective of the place of establishment or*

[18] Regulation (EU) 2016/679 of the European Parliament and of the Council of 27 April 2016 on the protection of natural persons with regard to the processing of personal data and on the free movement of such data (General Data Protection Regulation; GDPR); OJ L 119, 04 May 2016, pp. 1–88.

[19] European Court of Justice, Judgment of the Court (Grand Chamber) of 13 May 2014, *Google Spain SL and Google Inc. v Agencia Española de Protección de Datos (AEPD) and Mario Costeja González*, Case C-131/12, ECLI: EU:C:2014:317.

[20] European Court of Justice, Judgment of the Court (Grand Chamber) of 06 October 2015, *Maximillian Schrems vs. Data Protection Commissioner*, C-362/14, ECLI: EU:C:2015:650.

[21] European Court of Justice, Judgment of the Court (Grand Chamber) of 16 July 2020, *Data Protection Commissioner vs Facebook Ireland Ltd. and Maximillian Schrems*, C-311/18, ECLI: EU:C:2020:559.

residence of the gatekeepers and irrespective of the law otherwise applicable to the provision of service'; and
- Art. 43 MiCA stipulates *'that any e-money token that references the currency of an EU member state shall be deemed to be offered to the public in the Union'.*

It is not surprising that the proposed legislative initiatives by the European Commission have come under close scrutiny, especially from abroad. Most, if not all of the 'gatekeepers' the DMA has in its regulatory sights, as well as the potential issuers of global, Euro-based stablecoins targeted by MiCA, are non-European, mostly US-based corporations. From their perspective, the EU's regulatory proposals are depicted mainly as a thinly veiled exercise in digital protectionism. In the EU, by contrast, the objective to restore a 'level playing field' in digital markets is frequently linked with the notion of defending, if not reclaiming, 'digital sovereignty' [5]. The legal proposals should not be viewed as protectionist industrial policy, but as a way to draw a jurisdictional perimeter and (re)assert jurisdictional control in the digital domain. In that vein, the Commission's February 2020 communication emphasises that *'European technological sovereignty is not defined against anyone else, but by focusing on the needs of Europeans and of the European social model.'*

As the internet develops and matures the recognition has grown that it is, despite its decentralised design and global reach, ultimately bound to the same territorial and jurisdictional constraints as any other information and communications network. In recent years, national governments and authorities have proven repeatedly that they are determined, and able, to varying degrees, to control, filter and censor cross-border online traffic, or disconnect domestic users from the internet altogether. In China[22] and Russia[23], legal provisions explicitly require operators of internet exchange points to be able to monitor and, if necessary, interrupt international data traffic. In 2019 alone, governments in Bangladesh, the Democratic Republic of Congo, Egypt, India, Indonesia, Iran, Iraq, Sudan, Myanmar and Zimbabwe shut down the internet in all or some parts of their countries. This should not be read as an endorsement of such practices but as a reminder that the popular notion of a 'cyberspace' that eschews regulation and jurisdictional boundaries is largely fictional.

5 Conclusions

Overall, the EU's recent regulatory initiatives reflect the organic development of an evolving and maturing internet towards a transparent, rules-based environment designed to provide stability and legal certainty, a level competitive playing field for businesses, and legal protection for individual citizens.

[22] People's Republic of China, Provisional Regulations on the Management of International Networking of Computer Information Networks. State Council Decree No. 195 of 01 February 1996 (as amended by Decision of the State Council on 20 May 1997).

[23] Russian Federation, Federal Law No. 90-FZ of 01 May 2019 'on Amendments to the Federal Law on Communications and the Federal Law on Information, Information Technologies and Information Protection'.

With its emphasis on 'European values' and fundamental rights the EU's digital strategy has a strong 'citizen-centric' foundation. In practice, e.g. in the field of financial services, a careful balance has to be struck between these principles and the desire to create favourable conditions for European businesses to compete on level terms with global competitors that thrive on the ready availability of 'big data'. While it appears sensible to suggest, in the context of PSD 2, that data-sharing obligations between financial institutions and platform providers should become mutual, any such exchanges should always remain caveated by the principles of 'purpose limitation' and 'data minimization', and subject to the strict customer consent requirements enshrined in GDPR.

When considering the competitive position of gatekeepers and other platform providers in financial services, in particular, it will not be sufficient to equip supervisors with new competencies. Competition authorities, too, will be called upon to be vigilant, and act decisively to prevent one set of dominant incumbents merely being replaced by another. Business models, commercial strategies, and economics of digital platforms are shaped driven fundamentally by behavioural insights obtained from the continuous aggregation and analysis of customer data. It would appear sensible, therefore, to incorporate this line of analysis more prominently into the toolset of competition authorities. Acquisitions of emerging competitors by 'Big Tech' platform operators, as well as by traditional incumbents, should be examined critically. and from a forward-looking perspective to prevent nascent competition from being eliminated early on by entrenched market players.

The European Commission's proposals focus primarily on delineating a regulatory perimeter and setting rules, for the Digital Single Market. For the effective implementation and enforcement of these rules on the global internet, however, cooperation with international partners will be essential. That citizens' fundamental rights should be protected across national borders should be common ground for democratic societies and could serve as a point of departure for further international cooperation. With the passage of the GDPR, the EU set a widely recognised global standard for the protection of personal data. It remains to be seen if its latest regulatory initiatives in the digital domain will gain similar levels of global acceptance.

References

1. Azzi, A.: The challenges faced by the extraterritorial scope of the general data protection regulation. J. Intellect. Prop. Inf. Technol. E-Commerce Law **9**(2), 126–137 (2018)
2. Crémer, J., de Montjoye, Y.-A., Schweitzer, H.: Competition Policy for the Digital Era. Final Report to the European Commission (Directorate-General for Competition). European Commission, Brussels (2019)
3. European Commission: Final Report of the Expert Group on Regulatory Obstacles to Financial Innovation. 30 Recommendations on Regulation, Innovation and Finance. European Commission, Brussels (2019)
4. Financial Stability Board: Big Tech in Finance: Market Developments and Potential Financial Stability Implications, 09 December 2019
5. Floridi, L.: The fight for digital sovereignty: what it is, and why it matters, especially for the EU. Philos. Technol. **33**(3), 369–378 (2020)
6. Huffman, M.: Marrying neo-chicago with behavioral antitrust. Antitrust Law J. **78**(1), 105–145 (2012)

7. Katz, M.L., Shapiro, C.: Systems competition and network effects. J. Econ. Perspect. **8**(2), 93–115 (1994)
8. Organisation for Economic Cooperation and Development (OECD): Rethinking Antitrust Tools for Multi-Sided Platforms. OECD, Paris (2018)
9. Organisation for Economic Cooperation and Development (OECD): Start-ups, Killer Acquisitions and Merger Control. OECD, Paris (2020)
10. Organisation for Economic Cooperation and Development (OECD): Digital Disruption in Banking and its Impact on Competition. OECD, Paris (2020)
11. Morozov, E.: Think again: the internet. Foreign Policy **179**, 40–44 (2010)
12. Rochet, J.-C., Tirole, J.: Platform competition in two-sided markets. J. Eur. Econ. Assoc. **1**(4), 990–1029 (2003)
13. Stiefmueller, C.M.: Open banking and PSD 2: the promise of transforming banking by 'empowering customers.' In: Spohrer, J., Leitner, C. (eds.) AHFE 2020. AISC, vol. 1208, pp. 299–305. Springer, Cham (2020). https://doi.org/10.1007/978-3-030-51057-2_41

emyConnect: Building a Smart Public Service for Young Mobile Europeans

Andreas Martin[1](✉), Mohammad Allagha[2], and Yuri Misnikov[3]

[1] Austrian Institute of Technology GmbH, Giefinggasse 4, 1210 Vienna, Austria
Andreas.Martin@ait.ac.at

[2] Politikos – Wir müssen reden!, Thayagasse 1A, 1210 Vienna, Austria
mohammad.allagha@politikos.at

[3] Kaunas University of Technology, K. Donelaičio St. 73, 44249 Kaunas, Lithuania
yuri.misnikov@ktu.lt

Abstract. The paper presents and discusses the results of a European project 'Empowerment of mobile youth in the EU'. It studied the information and communication needs of European mobile youth (EMYs) living in Austria and Estonia to help them better participate in social and political life in their new host countries. These results have been used to develop an emyConnect chatbot as a tool to facilitate such participation and eventually help realize the European citizenship with the help of modern technology. The project outcomes suggest that the chatbot technologies suit well meeting the personalized public service in the domain of life events. From that perspective, chatbot technology demonstrates potential to effectively bridge the information deficit and communication gaps that mobile young people face on a daily basis when living away from home.

Keywords: Chatbots · Public services · Political participation · Mobile youth · Software engineering

1 Introduction

Young people are the most mobile age group within the EU living in large numbers outside their home countries, of who mobile students constitute a significant proportion. Recent evidence gathered during the implementation of the European Commission's project 'Empowerment of mobile youth in the EU' implemented in Austria and Estonia[1] found that, for example, young Europeans study at as many as 32 educational institutions in Austria alone. Yet they avoid active engagement in local social and political life of the countries of their current residence and in effect surrender their European citizenship rights granted by law. Deepening young Europeans' democratic participation was recognized as a priority of the EU youth strategy 'Renewed framework for European cooperation in the youth field (2010–2018)'.[2] Deployment of entry points and structures

[1] See more about the EMY project at https://europeanmobileyouth.eu/emy-project, Grant Agreement (GA) number 826686.

[2] See more at https://eur-lex.europa.eu/legal-content/EN/TXT/PDF/?uri=CELEX:32009G1219(01)&from=EN.

C. Leitner et al. (Eds.): AHFE 2021, LNNS 266, pp. 37–44, 2021.
https://doi.org/10.1007/978-3-030-80840-2_4

for involving young people in decision-making and review of the quality of participatory mechanisms has become a policy priority for all Member States. According to the 2013 Eurobarometer survey on 'European Youth: Participation in Democratic Life' [1], several EU Member States saw a significant decrease in the proportion of young people voting in elections. In a similar vein, the turnout in the European Parliament elections was declining continuously over the years (until 2019), with only 28% of young voters (18–24) casting their vote in the 2014 EU elections [2]. Apart from the issues of low levels of political trust, it is believed that if voters are still insufficiently informed about the EU in general and local politics, or rather that such in-formation does not always reaches young people through adequate communication channels, as the media consumption patterns change. Studies show that among 15–24-year-olds mobile internet access as is considered the most important communication tool in their daily lives [3]. Young people are also twice as likely as the average EU citizen to use instant messaging or post content on a social media on a daily basis [4]. However, millions of mobile EU citizens have been neglected as a specific target group in elections and other forms of democratic participation across all levels of government. Young mobile citizens especially do not fully exercise their rights to engage in political life at the EU and local levels in the country they live in.

In this paper we would like to share our findings in implementing the abovementioned project and describe the emyConnect communication tool to connect European mobile people to the sources of credible sources of political information. First, we analyses the reasons of the lack of local political participation on the part of young mobile. After that we explain their communication habits and propose digital solutions to meet the EMYs' information needs. The paper ends with the description of the informational chatbot prototype developed by the project for young mobile people to connect them directly with relevant content.

2 Democratic Participation of Young Mobile Citizens

The Europeans mobile youth – EMYs – are not a monolithic target group: they comprise of disparate sub-groups that are differentiated byfactors like age, country of origin, level and place of study or occupation, and duration of stay. It is essential to consider these differences because their communication habits, communication channels, the use of digital devices use, general interest in democratic participation differ. Mobility is especially vital for students who benefit from student exchange programs. Our project has found that young Europeans are not served by public administrations adequately to meet their information needs when living abroad. For example, as many as 85% of mobile students in Austria did not register to vote locally in the European Parliament elections in 2019, with 40% of them indicating registration procedures as the single biggest obstacle to voting in the host country.

A desk analysis of over 30 different communication platforms and website that provide information about participation to young Europeans showed that these are primarily informational instruments helping find other opportunities like internships, studies, job vacancies, among others. Other platforms that focus on mobile youth, assisting them in adapting to new living and social arrangements in their host countries, but do not usually address democratic participation issues. Platforms that focus on EU elections, by contrast, tend to address EU citizens in general, but not specifically mobile youth.

The analysis shows that the information on participation is generally available but is scattered across many competing sources. Even though the massive amount of information is technically accessible either via the websites or their social media pages, it is challenging to navigate through numerous sites. To focus better on the mobile youth' needs, firstly, the available content needs to be restructured to benefit from new communication technologies, such as messengers and related chatbots, and correspond the young people's habits to use them and consume different media. Secondly, new content should be generated be relevant information owners to bridge the existing address gaps and meet unmet information and communication demands. The following key findings and gaps have been identified also drawing on the official EU documents [5, 6].

2.1 Participation Gaps

- Relevance of the information provided for the target group varies; information is not sufficiently complete and specific to encourage participation.
- Mobile citizens are a subset of the voting population in a host country; information and participation needs are not addressed by local political community.
- Most of the available information is supplied on the assumption that mobile users are actively searching for it; that is, they are not approached pro-actively.
- Credibility of data and information sources (e.g. website) is sometimes difficult to be assessed by EMYs.
- EMYs are considered as a single target group, while there are disparate subsets with specific information and communication needs.

2.2 Communication Gaps

- No linkages exist across different communication channels or between platforms except for social media;
- All platforms require authentication to digitally connect with others, which complicate seamless communication;
- Sharing across different platforms and building communities is difficult.
- Technologies to connect communities with ownership and transparency are not offered; social media plugins exist but ownership belong to a third party.
- Social media discussion groups may cover similar content, whereas valuable information that would be relevant to all mobile youth is not shared groups.
- EMY community is fragmented and dispersed; information needs are granular and differ for host country.
- There are no resources that specifically cater for EMYs who just arrived in their respective host countries and consolidate relevant information in an accessible, user-friendly way.
- Most offerings use English as their main language, with little content translated into other European languages.

These findings demonstrate that the EMYs as a special diverse group of European citizens is not a recipient of dedicated digital services in a systematic and comprehensive manner.

3 Smart Digital Public Service for EMYs

At the start of 2021, the European Commission has launched a public consultation exercise to discuss the roadmap for a new edition of the European Interoperability Framework (EIF) – a cornerstone policy for setting up digital public services that should be seamlessly exchanged both between government institutions within the EU Member States and across their borders. This is part of the policies aimed at making digital solutions a powerful driving force of a European society for the years to come. The consultation will result in mandatory common rules and standards for secure and borderless public sector data flows and services. The undertaken in 2020 impact assessment of the current interoperability framework specifically emphasizes the personalized and user-friendly public service to reduce administrative burden for both citizens and public administrations. The importance of such rules and standards cannot be overestimated for young Europeans for whom mobility across the EU has become a way of live. Smart public services can make their life less complicated with the implementation of the common pan-European Catalogue of Services that contains the Core Public Service Vocabulary Application Profile (CPSV-AP) as a data model to implement digital services focused on specific needs of EMYs.

However, such services need to take a better account of the type of media they are built into. The rapid increase over the past several years in popularity of smartphone-based mobile platforms and messaging applications apps – such as Facebook Messenger, WhatsApp, Twitter, Instagram, Viber, Telegram and others – has led to radical shift to a distributed media environment that changes traditional communication channels that citizens used in the past to find information and visit virtual spaces to participate in politics [7]. The overall human networking landscape continues to transform as 'mobile messaging apps and news aggregators are becoming increasingly central to people's media use" [8]. The website-based medium lose their popularity among young people, as they spend every day some five hours by using different media mobile apps and just 13 min on other 'older' traditional media, such as Internet websites [7]. It's also been found that social networking via smartphone helps build social capital and offer additional entry points to political participation [9]. It is also argued that the role of software applications become more important and are in fact start dominating the overall digital media environment making it a place for significant sociocultural transformations on a wider scale [10].

In more techno-philosophical terms, Adrian Mackenzie defines the new mobile network culture as wirelessness which erodes the very edges of the Internet and mobile telecommunications and designates an experience trending toward entanglements with things, objects, gadgets, infrastructures, and services, and imbued with indistinct sensations and practices of network-associated change. Wirelessness affects how people arrive, depart, and inhabit places, how they relate to others, and indeed, how they embody change. In floating such an awkward term as wirelessness, I would invite readers to attend mostly to the suffix ness [11: 5].

The rising importance of mobile digital services prompts to establish a stronger link between the medium type and the content (and functionality) the medium transmits, including in the field of social and political participation. New mobile technologies make Marshal McLuhan's metaphorical slogan 'The media is the message' relevant as never since its invention in the television era of the 1960s.

The popularity of the mobile messaging services among young people suggests to take a closer look at chatbot technologies to implement more personalized public services for mobile citizens.

4 Building emyConnect Prototype as a Public Service

4.1 Objectives of emyConnect

The emyConnect communication and information tool as a software prototype has been developed to enhance the participation potential for EMYs for their civic and political life while living abroad. The studies preceding the design and development of the emyConnect prototype indicated the need to connect EMYs directly to the trustworthy source of the primary content, such as official authorities or other trusted institutional stakeholders. The tool should also provide a digital space for EMYs to interact and exchange information among themselves about opportunities for democratic participation in the local context. For this reason, the emyConnect prototype has a messenger-based bot architecture which can be incorporated in Telegram Groups and on Facebook pages.

The tool is based on two main building parts. The messenger user interface to be accessed by EMYs and a multi-user front-end web interface for verified EMY stakeholder (e.g. official authorities). This front-end can be used to publish relevant information and interaction possibilities on political participation opportunities which then are pushed via the bot to the receiver. A possibility to incorporate the bot into EMYs' personal communication channels enables the stakeholder to remove the 'information bubbles' that often surround EMYs.

4.2 Design and Development

To better understand the emyConnect target groups – EMYs and their stakeholders – the design and development process exploited the user-centred design principles to fully incorporate the users' feedback [12, 13]. The first phase of a design process examined the service needs of EMYs with the help of an online survey among 379 individuals, well as through four focus groups. Based on the identified needs, a first high level architecture and personas, as well as anti-personas were developed. The concept of personas is used to develop user requirements by identifying typical and specific behaviour, characteristics, needs of the target group and summarizing these to fictional user groups [14]. The expected benefits of personas lie in an improved communication about the target demographics. The concept of personas was developed via a multi-stage process [15]. Afterwards, themes and persona characteristics were jointly formulated for and discussed with the EMYs and stakeholders. These themes included the host-country integration aspects, willingness to seek for information, as well as the usage and choice of preferred information channels. EMYs prioritized information quality that stems from the level of credibility of information sources. The next phase synthesised these characteristics with persona narratives linked to respective life event scenarios like search for participation opportunities, information seeking, or continuous host-country engagement.

4.3 emyConnect User Groups

The very name of the tool – emyConnect – places the emphasis on the need to digitally bridge multiple disconnects that EMYs experience with the host country's social and political life on an every-day basis. The EMYs aged 16–20 years are the main target group. Additional characteristics include the length of stay (short stay for less a year; long stays for loner a year), as well as the purpose of stay (e.g. student exchange, work). These helped determine EMYs information needs and communication habits to be covered by the prototype. The second user group are the key stakeholders, such as public authorities in the host country, student unions or other representative bodies. These stakeholders are among the first contact points for EMYs when arriving in the host country and therefore most suited to introduce arriving EMYs to emyConnect.

4.4 emyConnect Architecture

The two main components of the prototype are: (a) the multi-user graphical web front-end (and back-end) to publish curated information by representatives of trustworthy stakeholders and (b) the client bot component to receive these information by the means of a messenger-based chat bot connected to the back-end. The emyConnect front-end enables EMY stakeholders to publish curated information by means of WEB User Interface. Its main function is the content publishing interface and administration of credible information provided to EMYs. The emyConnect's front-end further allows the involved stakeholders to add, delete, discard or edit information facilitating EMYs' local engagement.

The emyConnect bot's enables interaction with the platform through mobile devices (smartphone/computer) through two modes of communication. The first one is a direct message mode, the second one is in a group chat. In direct message mode the user has the possibility to browse categories and corresponding topics and FAQs. The interaction with the bot is realised by means of button like items that the user can simply tap to navigate and display events and other details. The interaction with the bot is realised by means of button like items that the user can simply tap to navigate and display events and other details (Fig. 1).

The group chat mode is used to display push notifications published by the EMY stakeholder. These notifications are solely for announcing new topics or updates to existing ones. The actual interaction with the bot (e.g., browsing topics and FAQs) is designed to take place in the direct message mode. This is done in order to hide the user-bot interaction from the other group users and limit the amount of content published by the bot in a group. The usual interaction on smartphones imposes some limitations in terms of the amount of content provided.

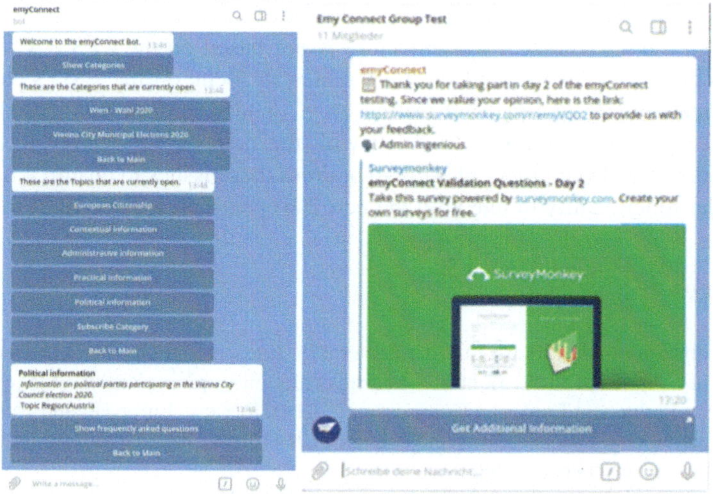

Fig. 1. emyConnect interface.

From a user experience standpoint, it is recommended to expose no more than ten categories and ten topics per category. This reduces the amount of data the bot pushes to the chat and reduces the risk of overwhelming the user with too much information. The topic and category names as well as the topic details should be short and concise. Topic details should provide links to more detailed information as well as contact details that are relevant in that scope.

4.5 Validation of the emyConnect Prototype

The emyConnect bot was tested during the municipal election of the City of Vienna in October 2020. Individual EMYs and a representative of the City of Vienna were invited to test emyConnect's content and functionality for stability, functionality and appropriateness of system and content. Feedback highlighted interest in a search and reply functionality. Commenting or content editing appeared less important for EMYs than it was originally planned. Content dealing with European citizenship, information about voting procedures, other life event information of administrative and practical character, as well as more general content devoted to local political life was prioritized. The feedback also indicated that short video clips and infographics are preferred over text format. Besides, the testing highlighted additional privacy and security requirements.

5 Conclusions

Anestis Stamatis, Alexandros Gerontas, Anastasios Dasyras and Efthimios Tambouris provide a useful review of the research and practice in the field of chatbot technologies as a basis for personalized public services focused on life events with the help of CPSV-AP [16]. According to them, life events services fit best the definition of the personalized user-centric public service through chatbots, the use of which is currently underexploited. Even though the emyConnect's architecture is not based on the CPSV-AP standards, it was designed bearing in mind the importance to focus on the EMYs' life events. The experience of developing and testing emyConnect confirm the centrality of the personalized public service implemented with the help of the chatbot technologies. The next software development phase will address these issues to ensure a full-fledged cross-border functionality in line with the new European Interoperability Framework.

References

1. European Commission: Flash Eurobarometer 375: European Youth: Participation in Democratic Life, April 2013 (2015). http://ec.europa.eu/public_opinion/flash/fl_375_en.pdf
2. European Parliament/European Commission (2014). Review: European and National Elections Figured Out. Special Edition – November 2014 (2014). http://www.europarl.europa.eu/pdf/elections_results/review.pdf
3. European Commission: Standard Eurobarometer 84 (2015)
4. European Commission: Special Eurobarometer 438 (2015)
5. European Commission: EU citizenship consultation. Common values, rights and democratic participation (2015). https://ec.europa.eu/info/sites/info/files/public-consultation-booklet_2015_en.pdf
6. European Commission: European Youth Report, COM (2015) 429 final (2015). http://ec.euro-pa.eu/youth/policy/implementation/report_en.htm
7. Newman, N., Fletcher, R., Nielsen, K., Kalogeropoulos, A.: Reuters Institute Digital News Report 2019, Reuters Institute for the Study of Journalism, Oxford University (2019)
8. Kalogeropoulos, A., Fletcher, R., Nielsen, K.: News brand attribution in distributed environments: do people know where they get their news? New Media Soc. **21**(3), 583–601 (2019)
9. Park, K.-G., Han, S., Kaid, L.L.: Does social networking service usage mediate the association between smartphone usage and social capital? New Media Soc. **15**(7), 1077–1093 (2013)
10. Light, B., Burgess, J., Duguay, S.: The walkthrough method: an approach to the study of apps. New Media Soc. **20**(3), 881–900 (2018)
11. Mackenzie, A.: Wirelessness: Radical Empiricism in Network Cultures. MIT Press, Cambridge (2010)
12. Veryzer, R.W., de Mozota, B.B.: The impact of user-oriented design on new product development: an examination of fundamental relationships. J. Prod. Innov. Manag. **22**(2), 128–143 (2005). https://doi.org/10.1111/j.0737-6782.2005.00110
13. Maguire, M.: Methods to support human-centred design. Int. J. Hum. Comput. Stud. **55**(4), 587–634 (2001). https://doi.org/10.1006/ijhc.2001.0503
14. Faily, S., Flechais, I.: Persona cases: a technique for grounding personas. In: Conference on Human Factors in Computing Systems - Proceedings, pp. 2267–2270 (2011). https://doi.org/10.1145/1978942.1979274
15. Miaskiewicz, T., Kozar, K.A.: Personas and user-centered design: how can personas benefit product design processes? Des. Stud. **32**(5), 417–430 (2011). https://doi.org/10.1016/j.destud.2011.03.003
16. Stamatis, A., Gerontas, A., Dasyras, A., Tambouris, E.: Using chatbots and life events to provide public service information. In: Proceedings of the 13th International Conference on Theory and Practice of Electronic Governance (ICEGOV 2020), Athens, Greece, 23–25 September 2020 (2020)

The Engagement Catalyst Initiative: How One Global Organization Activates and Energizes Employee Engagement

Laura Challman Anderson[1](✉), Michal Jacovi[2], Jonathan Lenchner[3], and Jade Nguyen Strattner[3]

[1] IBM Research – Almaden, 650 Harry Road, San Jose, CA 95120, USA
lca@us.ibm.com
[2] IBM Research – Haifa, Haifa University Campus, Mount Carmel, 3498825 Haifa, Israel
michal.jacovi@il.ibm.com
[3] IBM Research – Watson, 1101 Kitchawan Road, Yorktown Heights, NY 10598, USA
{lenchner,jade.n.strattner}@us.ibm.com

Abstract. Employee engagement is often highlighted as a critical factor to achieve innovation, responsiveness to customers, and numerous other desirable business outcomes. Less clear in the literature, and in today's business enterprises, are approaches to reliably and sustainably drive increases in employee engagement. We share our perspective that a successful approach to positively impact employee engagement is one that is a hybrid mix of top-down and bottom-up initiatives, with diverse stakeholders who share execution responsibility. We detail our approach, the operational principles and work cycles, the resulting projects and interventions, our study findings, and the resulting outcomes in our organization. Finally, we describe future research so that our work will continue to activate an enhanced culture of engagement in our organization.

Keywords: Employee engagement · Work engagement · Personal engagement · Organizational culture

1 Introduction

In pursuing this research on the nature of employee engagement, we are often asked the question: "If one has captivating work to do, what more is necessary?" However, we know that work cannot always be captivating, and even if possible, might not be enough to ensure employee engagement. As human beings, we are engageable on many levels. For example, if we believe in the mission of our company and feel that we have a voice in crafting that mission, we may be engaged at a higher level. Indeed, engagement is a balancing act of many factors. We do not attempt to completely unlock all sources of engagement, but rather ask the practical research question: **what interventions would create a spark to produce the positive outcome of enhanced employee engagement in an organization?** In this paper, we review a novel approach to positively impact employee engagement in a large global enterprise using a hybrid approach and highlight the outcomes.

C. Leitner et al. (Eds.): AHFE 2021, LNNS 266, pp. 45–53, 2021.
https://doi.org/10.1007/978-3-030-80840-2_5

2 Literature Review

Employee engagement is a persistent topic of interest, a focus of both popular leadership and management publications and scholarly literature. The topic crosses many disciplines, including business management, organizational culture, human resources management, and psychology. William Kahn [1] is widely viewed as the "father" of employee engagement, a concept further developed and operationalized through metrics deployed in enterprise employee surveys by the Gallup organization over the past 20 years [2]. Adjacent related engagement concepts have emerged along the way with overlapping and fuzzy definitions, including job engagement [3], organizational engagement [4], personal engagement [1] and work engagement [5, 6]. For the purposes of this paper, we focus on employee engagement and use this definition [3], p. 499: "Employee engagement is [therefore] the level of commitment and involvement an employee has towards his or her organization and its values."

Employee engagement can be seen as a kind of holy grail for an organization, although it is not one thing: engagement is experienced in an individual way and is influenced by a constellation of complex, multi-level and multi-dimensional personal, interpersonal, and workplace factors. Notable factors affecting engagement in the literature include meaningfulness [1], collaboration, personal development [3], company values, leadership [4], social environment [7], financial rewards [8], talent management [4], and work environment [8]. The literature provides some insights on our basic research question, from two angles: 1. top-down leadership or Human Resource Management (HRM)-driven initiatives and 2. bottom-up initiatives. Top-down initiatives tend to focus on programs of manager and leadership assessment and skill-building [6, 9, 10, 11] and on building personal capabilities and well-being [6, 12, 13]. Bottom-up initiatives, that is, grassroots efforts directly initiated by employees, were predominantly focused on the shaping of one's own job by employees [5, 6] Intrapreneurship [14] was also viewed as a bottom-up mechanism to activate employee engagement.

Our paper addresses two gaps in the literature. First, scarce in the literature are studies utilizing a "hybrid" co-creative approach for intervention. By "hybrid" we mean a combination of interventions that are initiated from multiple points: the top (what we refer to as "top-down") and from the bottom (what we refer to as "bottom-up") with respect to the organizational hierarchy. Both types of interventions engage additional organizational stakeholders to affect employee engagement and associated effectiveness [6]. A second gap addressed by our study is linking employee engagement with organizational sustainability and organizational culture [15], with a focus on fostering an engagement culture through a persistent framework for employee engagement initiatives, and an approach to pump engagement energy into an organization.

3 The Engagement Catalyst Initiative

3.1 Inception

Employee engagement has been a multi-year focus item for the company's management and is considered a key aspect of an overall transformation that included strategy, portfolio and operating model changes. In early 2020 a new executive role was created and the

Engagement Catalyst (EC) initiative was launched. First introduced in an executive team discussion, the EC Initiative was guided by the results of an annual employee survey as well as by popular business literature. The EC Initiative created a team of "engagement catalysts" – a team composed through a mix of executive nomination, invitations by fellow ECs and volunteers who would take on specific engagement-catalyzing roles in addition to their regular responsibilities.

Early discussions guiding the Engagement Catalyst initiative also included reviews of popular business literature. The importance of organizational culture, such as the Open Organization championed by Jim Whitehurst [16] was an influential idea. The Open Organization concept echoes larger societal changes seen in the last few decades that reflect the expectations of employees for both an environment of transparency and their own agency to influence the workplace. Also considered was the work of Patrick Lencioni [17], who focused a lens on meaningful work and how to effectively nurture operational teams; Gibbons' work on organizational change [7]; and Simon Sinek's ideas about leadership [18].

The initial EC cohort was composed of 38 site representatives and 11 people from the line organizations (e.g., HR) who identified, self-selected and drove specific projects. The cohort increased to over 100 people during 2020. These dynamic projects morphed, merged, were prioritized or de-prioritized, completed, or carried over to 2021. A critical success factor in the initiative was to establish cross-organizational teams in a way that allowed engagement catalysts to work side-by-side with the line organizations; the functional owners of processes to be transformed, such as HR, Information Technology and the Communications professionals. The engagement catalysts were chartered to be "a global coalition of advocates who serve as the voice of the community and work together to build awareness, skills and collaboration across the organization. Our goal is to amplify our organizational impact by creating a vibrant workplace where everyone feels connected and valued."

3.2 Design of the Approach

Three operational principles were established at the beginning: that the work be:

1) reflective, learning-oriented, iterative and evolutionary,
2) co-created with the community through an intentional process, and
3) executed as a shared responsibility between EC Initiative stakeholders.

The iterative approach consisted of 'Listen', 'Act', 'Communicate' and 'Assess Impact', a repeating set of actions to build understanding about the changes implemented and enable course-corrections and/or further changes. This design of how the global catalysts would work together as a team, and how they would each work with their populations, was key to maintaining enthusiasm and momentum. Also important was the capability to ensure relevance of the work by shifting priorities of specific projects while keeping the overall initiative moving forward.

Two accompanying objectives were defined based on employee feedback from the annual employee survey, roundtables, and interviews: to activate and enable the communities; to communicate, capture feedback and act. These objectives helped to ensure the relevance of the initiative through the engagement of each catalyst with their community, emphasizing two-way communication coupled with tangible outcomes. The operational principles and objectives, and the lived experience of the catalysts, ensured that everyone's investment in the work was worthwhile.

3.3 Project Execution and Work Cycles

The engagement catalysts worked in two directions: directly with the population they represented; and together as a global team, subdivided into specific projects. Table 1 shows a representation of the projects undertaken in 2020. These 18 projects, focusing on collective needs, are coded to indicate the category of the work (column 1), who initiated, designed, approved and implemented (columns 2–5), and the project outcome in 2020 (column 6). There is a wide range in project topics and variety in stakeholder participation from initiation, design, approval, and implementation. The projects also spanned a broad range of topics - from employee career to communications to operations. Continuous and dynamic prioritizing of projects was done to focus limited resources on key issues, and project execution was a combination of global and local resources. Top-down priorities were set by focusing on change in areas where the leadership team thought that the broadest positive impact would be realized on organizational effectiveness.

Table 1. Engagement catalyst projects in 2020.

Project category	Initiated by	Designed by	Approved by	Implemented by	Outcome in 2020
COVID-19	EC	EC	EC	EC & Line org.	Y
FAQs	EC	EC	L	Line organization	Y
Organizational Clarity	EC	EC	L	Line organization	Y
Organizational Clarity - career	EC & L	EC	L	EC & HR	Y
Organizational Clarity - project metrics	EC & L	EC	L	EC & Line org.	Y
Organizational Clarity - role	EC & L	EC	L	EC & HR	Y
Organizational Culture	L	EC & L	L	EC, HR, & L	Y
Organizational Culture - diversity & inclusion	L	EC & TF	L	L	Y
Organizational Culture - local site	EC	EC	EC	EC	Y
Organizational Processes	EC & L	EC & TF	L	Line organization	Y
Employee Recognition	EC & L	EC	L	Line organization	Y
Client focus	EC	EC	n/a	n/a	Ongoing
Strategic Area Communication	EC	EC	n/a	n/a	Recommendations not implemented
IT Tools	EC	EC	n/a	n/a	Ongoing - merged with another project
New Hire Experience	EC	EC	n/a	n/a	Ongoing
Organizational Clarity - Strategic Goals	EC	EC	n/a	n/a	Merged with ongoing project
Physical Work Environment	EC	EC	n/a	n/a	Work paused due to COVID-19
Wellness	EC	EC	n/a	n/a	Work paused due to COVID-19

Legend: EC = Engagement Catalyst, L = Leadership, HR = Human Resources, TF = Task Force, Y= actions implemented in 2020.

In addition to the projects in Table 1 there were many other local EC projects – such as listening and then acting on changes at the local level. Listening strategies included conducting over 100 roundtables across the global organization and implementing spot surveys to gather additional information and define follow-on actions.

4 Findings and Discussion

This section is framed in the context of how our research addresses the gaps in the literature and how it answers the question of **what interventions would create a spark to produce the positive outcome of enhanced employee engagement in an organization?** Positive results from the EC Initiative have been multi-dimensional and include:

(1) The annual employee survey showed a quantitative improvement in the engagement index, and a trend change to a positive trajectory. The engagement index is calculated from employees' response to the following three assertions indicating how strongly they agree or disagree on a 5-point Likert scale:

   - I am proud to be an employee of the company.
   - For me, the company is a great place to work.
   - I would recommend the company as a good place to work.

   The score for "I would recommend the company as a good place to work" increased by a notable margin.

(2) More than 50% of the global projects resulted in tangible outcomes. Of the 18 projects identified in 2020, eleven results in action. All of the projects developed as the catalysts worked on them, and split or merged with each other, in a natural and organic evolution. The remaining seven projects didn't result in 2020 actions for various reasons, such as being a low priority, timing, the lack of a good match with catalysts to drive the work, and projects that are still underway. The breadth of project results shows the natural selection process, and the capability of the EC framework to filter and prioritize projects while accommodating a wide range of project types.

(3) Project lifecycles during 2020 showed (i) grassroots, executive leadership, and line organization investment, (ii) cross-organizational co-creation, and (iii) collaboration between geographically distant and cross-organizational catalyst teams.

(4) The vitality and ongoing viability of the initiative can be seen in the number of catalysts in the EC Initiative: 49 in January 2020, growing to over 100 by year-end.

(5) Finally, the EC catalyst team was in place and positioned to respond quickly to the COVID-19 pandemic, an unplanned challenge. Existing projects were de-prioritized so that a set of global and local actions could be deployed to support employees this time. These actions included support for rapid sharing of practical ideas on Slack™ (https://slack.com/) channels, such as remote work strategies for effectiveness and well-being.

Additional major observations include:

- Successful projects required active involvement of both engagement catalysts and the responsible line organizations.
- Leadership was required to approve almost all the projects.
- Employees would benefit from more effective strategy area communication. We also found that improved communication related to the Engagement Catalyst Initiative

itself would strengthen the initiative, as sometimes employees did not know that the changes were being led and driven by engagement catalysts.
- Improved metrics are also needed to measure the impact of individual interventions in this multi-dimensional space.

There can be a middle gap in organizations as new strategies are defined by organizational leaders, and specific projects and milestones need to be adjusted. The engagement catalysts filled this gap by developing detailed implementation plans and activities to support higher level goals and objectives of defined strategies. We also found the EC framework to be effective in the formation of sub-organization catalyst "squads" who accumulated important issues and solution ideas affecting engagement in that community. This included the need to better articulate what was special about working at this company, which is important for employee well-being, recruiting and to highlight specific accomplishments. The EC structure and persistent working relationship with the executive team was helpful to rapidly raise other issues that arose during the year, and enabled leadership and catalysts to make progress together on other issues, resulting in empowering outcomes.

By pumping "engagement energy" into the organization by involving people from all locations and all parts of the organization, we reduced the peer distance between each employee and between employees and leadership. While not strictly measurable, we also saw an increase in the quantity and impact of the problems addressed in 2020.

4.1 The Unique Design of the Employee Engagement Intervention and Evaluating Effectiveness

Catalysts came to the EC Initiative as volunteers and have expressed feedback about experiencing empowerment through their project work. They were energized and felt valued. Quotes by catalysts expressing the impact of the EC Initiative include:

- "We invited people to get involved in what they are passionate about and empowered them to contribute and make a difference by allowing the projects to self-direct but with support from leadership."
- "It's a mechanism to enact the best of the creative ideas so generously contributed by so many who took the risk to state them."
- "I got the opportunity to make a difference – my voice was heard by the EC team, and our voice as a team was heard by leadership and our community."

The hybrid nature of the EC Initiative, with projects identified and nurtured from multiple angles, and worked on by a cross-organizational team was a key design feature which produced tangible positive results across a wide spectrum of topics related to employee engagement. The initial project selection was collaboratively determined by EC Initiative stakeholders: senior leadership and employees. This joint stakeholder model empowered the grassroots employees working on these initiatives, with executive support when needed, and provided the opportunity for catalysts to work together on projects thus creating cohesion. Projects were able to be dynamically reprioritized when the unexpected occurred (as it always does), such as the COVID-19 pandemic. We found that, throughout 2020, the first year of the EC Initiative, prioritization of projects needed to be dynamic, driven by continuous listening/discovery and changing business needs, and additional resources sometimes needed to be assigned to some projects that were important to all stakeholders.

We now turn to evaluating the impact of the EC Initiative through consideration of the aspects of the EC Initiative that are critical success factors. First, engagement was nurtured in the participants themselves through their EC work and their experience of empowerment and appreciation. This positioned them to nurture engagement, in turn, in the populations that they represented. Employees at all levels of the organization need to be empowered and enabled to lead change. This takes more than just telling them that they are empowered; the lived experience of practices that can be replicated in other work contexts is a powerful, although intangible, force. We saw many examples of catalysts as leaders of projects, really listening and acting. Second, the iterative design of the overall initiative and some of the projects (agility, user-centric principles of design-thinking with regular scrums, read-outs, testing) is important. Practicing agile methods enables teams to more effectively solve identified engagement problems, especially if the problems are complex. Executive support for the approaches and outcomes driven by the engagement catalysts was critical. Finally, it is a powerful capability to have a dedicated coalition to dynamically deploy on important challenges.

4.2 The Power in Linking Employee Engagement Activities with Organizational Sustainability

The EC Initiative provided a comprehensive, persistent framework for employee engagement interventions, and for new projects to be included as time goes by from hybrid sources. We believe that this is superior to an ad-hoc set of tactics and approaches. To create sustained employee engagement at scale, organizations need to have a systematic strategy and approach, and a framework that persists and is adaptable to change over time. The strategy of pulling-in the responsible business support organizations to participate as members of the EC project teams provided subject matter expertise for the work. This approach also served to smooth the transition of the ideas into the infrastructure, and transfer ideas into their home infrastructure functions in an evolutionary manner. The leader of the EC Initiative provided important guidance along the way, including finding the right people for projects, and providing reinforcement needed by the project team, such as sponsorship, project management, and agile expertise.

5 Conclusion

The framework and model of the Engagement Catalyst Initiative is a novel and successful approach to improve employee engagement in a global organization, to initiate transformational activities in business organizations, and to be a continuing force for a sustainable and evolutionary change. By empowering employees in the work of the EC Initiative, a subset of the population has expanded their focus from engaging with their day-to-day work to engaging with the mission and strategy of the organization, and in this way engaging themselves to a greater degree of depth and intensity.

Future plans include continued work on additional important dimensions of this employee engagement initiative, to create lasting culture change and to initiate an in-depth study of leadership and employee engagement.

Acknowledgments. We are very grateful to Will Green, Rodrigo Neumann Barros Ferreira, and Donny Rose for their many constructive comments on paper drafts, and to the global engagement catalysts, organizational leaders and managers who work together to bring engagement to life every day in the company.

References

1. Kahn, W.A.: Psychological conditions of personal engagement and disengagement at work. Acad. Manag. J. **33**, 692–724 (1990)
2. Mann, A., Harter, J.: The worldwide employee engagement crisis. Gallup Bus. J. **7**, 1–5 (2016)
3. Andrew, O.C., Sofian, S.: Individual factors and work outcomes of employee engagement. Procedia-Soc. Behav. Sci. **40**, 498–508 (2012)
4. Chandani, A., Mehta, M., Mall, A., Khokhar, V.: Employee engagement: a review paper on factors affecting employee engagement. Indian J. Sci. Technol. **9**(15), 1–7 (2016)
5. Bakker, A.B., Demerouti, E., Sanz-Vergel, A.I.: Burnout and work engagement: the JD-R approach. Annu. Rev. Organ. Psych. Organ. Behav. **1**(1), 389–411 (2014)
6. Knight, C., Patterson, M., Dawson, J.: Work engagement interventions can be effective: a systematic review. Eur. J. Work Organ. Psychol. **28**(3), 348–372 (2019). https://doi.org/10.1080/1359432X.2019.1588887
7. Gibbons, P.: The Science of Successful Organizational Change: How Leaders Set Strategy, Change Behavior, and Create an Agile Culture. FT Press, Upper Saddle River (2015)
8. Rao, V.: Innovation through employee engagement. In: Second Asia Pacific Conference on Advanced Research, pp. 1–9 (2016)
9. Decuypere, A., Schaufeli, W.: Leadership and work engagement: exploring explanatory mechanisms. German J. Hum. Resour. Manag. **34**(1), 69–95 (2020)
10. Barik, S., Kochar, A.: Antecedents and consequences of employee engagement: a literature review. Int. J. Latest Technol. Eng. Manag. Appl. Sci. **6**(4), 33–38 (2017)
11. Sahu, S., Pathardikar, A., Kumar, A.: Transformational leadership and turnover: mediating effects of employee engagement, employer branding, and psychological attachment. Leadersh. Organ. Dev. J. **39**(1), 82–99 (2018)
12. Norcross, M.A., Farran, P.: Improving engagement during times of change. In: Dhiman, S., Marques, J. (eds.) New Horizons in Positive Leadership and Change. MP, pp. 249–267. Springer, Cham (2020). https://doi.org/10.1007/978-3-030-38129-5_14
13. Pattnaik, L., Jena, L.K.: Mindfulness, remote engagement and employee morale: conceptual analysis to address the "new normal". Int. J. Organ. Anal. (2020)
14. Gawke, J.: Employee intrapreneurship and work engagement: a latent change score approach. J. Vocat. Behav. **100**, 88–100 (2017)
15. Nandan, S.: Organizational culture dimensions as drivers of employee engagement for business sustainability: towards a conceptual framework. In: Vanka, S., Rao, M.B., Singh, S., Pulaparthi, M.R. (eds.) Sustainable Human Resource Management: Transforming Organizations, Societies and Environment, pp. 109–132. Springer, Singapore (2020). https://doi.org/10.1007/978-981-15-5656-2_7
16. Whitehurst, J.: The Open Organization: Igniting Passion and Performance. Harvard Business Review Press, Brighton (2015)

17. Lencioni, P.M.: The Advantage: Why Organizational Health Trumps Everything Else in Business. Wiley, Hoboken (2012)
18. Sinek, S.: How great leaders inspire action [Video file] (2009). https://www.ted.com/talks/simon_sinek_how_great_leaders_inspire_action

Addressing Complexity in the Emerging Platform Society

Exploring New Digital Age Challenges

Clara Bassano[1(✉)], Sergio Barile[2], Paolo Piciocchi[3], Marialuisa Saviano[1], and James C. Spohrer[4]

[1] Department of Pharmacy, University of Salerno, 84132 Fisciano, SA, Italy
{cbassano,msaviano}@unisa.it
[2] Department of Management, Sapienza University of Rome, Rome, Italy
sergio.barile@uniroma1.it
[3] Department of Political Science and Communication, University of Salerno, 84132 Fisciano, SA, Italy
p.piciocchi@unisa.it
[4] IBM, Almaden Research Center, San Jose, CA, USA
spohrer@us.ibm.com

Abstract. Digital age challenges are intertwined and STEEP (Social-Technical-Economic-Environmental-Political). The hallmark of the digital age is the platform society. Platforms provide connected individuals personalized service (news, communications, transportation, lodging, health, education, gig jobs). Platforms cut costs, bypassing cumbersome institutions in exchange for personal data. Data feeds algorithms at scale to improve AI-based service, hence platforms compete for more users and collaborators (ecosystems). The four digital age challenges explored in this paper are: (1) technological over dependency, (2) marginalization of unconnected, (3) trust of platforms, and (4) competing for collaborators. We explore these four challenges from an integrated Service Science (SS) and Viable Systems Approach (*vSa*) perspective.

Keywords: Service science · Viable systems approach · Platform society · Interpretative schemes · Challenges · Transdisciplinarity

1 Introduction

Before the digital platform society, in the industrial world and even in the ancient world of Rome - people depended on supply chains for many items, especially spices and metals, and other luxury items of the day. Today, the connection is an information supply chain that digitally connects us to others. Moreover, in these pandemic times, how could we survive without digital connections which avoid face-to-face contact? And yet simultaneously, disconnecting from devices to simplify and become more self-sufficient are also on people's minds again. Lifestyles reflect the distribution of knowledge and know-how in a society, impacting business models and investment opportunities, and how we can think about value co-creation and capability co-elevation capabilities in society [1].

Until now, value growth logic has been primarily seen as an outcome of investing simply in more powerful technical competencies, as it is for Artificial Intelligence (AI).

C. Leitner et al. (Eds.): AHFE 2021, LNNS 266, pp. 57–66, 2021.
https://doi.org/10.1007/978-3-030-80840-2_6

However, human-side investments matter as well. For example, learning to read and write is big investment and hard work, and business and society have made the decision that knowledge in everyone (nearly) leads to better outcomes [1].

Nevertheless, too often the focus is on AI and smart technologies, but few studies address the human-side of AI adoption perspective. The human-side of AI adoption is Intelligence Augmentation (IA) [2–7]. Both perspectives are important to wiser service systems as people aspire to work together to solve more worthy and ambitious challenges [1].

The hallmark of the digital age is easy access to personalized service offerings via online platforms feeding data hungry AI systems, and bypassing cumbersome traditional institutions [8]. This paper explores four new digital age challenges: (1) technological over dependency, (2) marginalization of unconnected, (3) trust of platforms, and (4) competing for collaborators; all four are explored from an integrated Service Science (SS) and Viable Systems Approach (*vSa*) perspective.

By means of this integrated approach, we will highlight that to manage wise decision-making complexity it is fundamental the centrality of the human decision maker and the greater criticality of technology in the changing scenarios of sustainability, where (AI) supports complicated decisions that must be Smart (Smart cities) and (IA) supports complex decisions that must be sustainable (Wise cities) [9–11]. So, our focus will point out that new levels of complexity require new pattern or schema (Barile, 2009; Barile and Saviano, 2018) to confirm that the choice of this rather than that possible solution, or rather the approach to a resolutive hypothesis rather than another, significantly dependent on the role played by the general schemes.

2 Design/Methodology/Approach

Our methodology is interpretative, exploring four digital age challenges (dependency, marginalization, trust, competition) from an integrated service science and viable systems approach perspective.

From a Service Science (SS) perspective, entities interact normatively to achieve value co-creations outcomes [12, 13]. Technology (including platforms) and organizations/institutions help entities gather data and analyze data about each other, entities with little data may have less trust, and more difficulty identifying collaborators for value co-creation interactions [14]. In the past, winner-take-all dynamics have dominated, but by balancing that dynamic with improve-weakest-link dynamics a wiser outcome could result than at any previous time in history. The key to wiser outcomes maybe re-imagining value co-creation interactions with future versions of self and future generations [15]. However, data centers consume increasing amounts of energy, and the environmental costs of platforms is increasing.

From a Viable Systems Approach (*vSa*) perspective, to manage wise decision-making complexity, the new intelligence conceptualization (IA) favors a path of constant and progressive optimization of computational models by virtue of the centrality of the human decision maker and the greater criticality of technology in sustainable decision-making processes [7]. In short, interaction with (AI) changes the way people mature their rational and emotional intelligence: for example, compared to peers ten years ago, digital natives experience technology as an essential and indispensable component in many processes of their life. An important implication of this change is that the way in which humans interact with technology is much more relevant than the technology itself [5].

Both perspectives derive from General Systems Theory (GTS). We propose that leaders should understand that efficient systems use AI (Smart cities) and sustainable systems use IA (Wise cities) [9, 11]. Conditions of complexity require decision-making capabilities based on the use of AI for efficiency and IA for sustainability [4, 6, 7].

From the *VSA*, IA results from effective integration and interaction between human and AI entities; a "collaborative integration" of intellectual processes that generates a positive differential of "Intelligence Augmentation" [7]. In particular, at certain conditions, IA allows the system to evolve from an intelligent configuration to a wise one, in which the rational component (prerogative of AI) integrates with the emotional one (prerogative of human intelligence). So, the concept of *Intelligence Augmentation* is not synonymous with the concept of *the feeling economy* introduced by Huang and Rust [16], it is an extension and broader concept [6]. IA qualifies as an interpretative approach to the problem that, according to the (*VSA*) Information Variety model (IVM), the use of the interpretative schemes and value categories progressively make the human-machine system wiser in the decision-making processes. In fact, the action of the schemes and value categories in a collaborative environment, opens up a range of resolution options with greater potential for resonance in the context. By this means, *intelligence* is interpreted as the ability to approach a solution by changing the system's endowment of knowledge, i.e. its Information Variety that can be described as follows [2–7, 17–19]:

$$V_{inf}(K) = \left(U_{inf}(k), S_{int}(k), C_{val}(k)\right).$$

where:

V_{inf} (K) = Information variety of viable system k
U_{inf} (k) = Information units of Informative variety of viable system k
S_{int} (k) = Interpretative schemes of Informative variety of viable system k
C_{val} (k) = Value categories of Information variety of viable system k

In particular:

- The *information units* are all perceived elements that contribute to generating knowledge.
- The *interpretative schemes* are cognitive models through which incoming information and knowledge is processed by the system.
- The *value categories* are the deeply rooted strong beliefs that affect humans' behaviors and choices through the use of interpretative schemes allowing effective decision and creation of knowledge.

In Fig. 1, IA is an intelligence that can result from effective integration and interaction between human and AI entities. In particular, through this model it is possible to represent a system's knowledge endowment as an information variety. The model allows

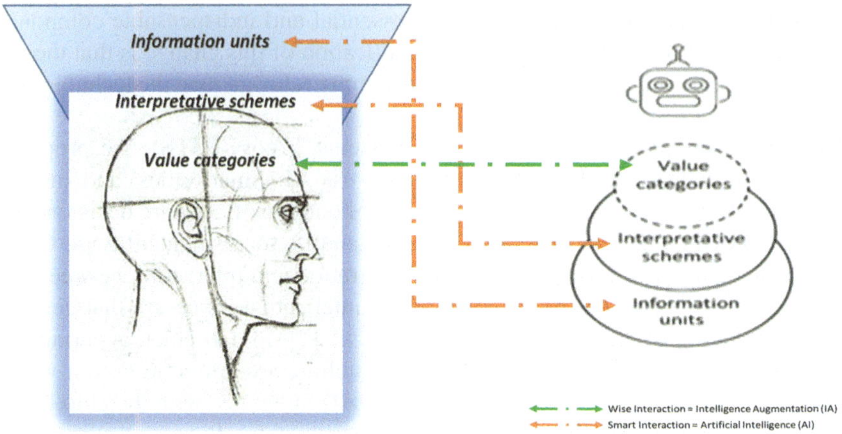

Fig. 1. Information variety between human and AI entities - Authors' elaboration from Barile 2009. www.asvsa.org.

comparing any kind of system's information varieties and analyzing the conditions for effective interaction [19, 20]. Being intelligent means acting not just on the basis of data and objective information but also and mainly on the basis of interpretative schemes and values that make knowledge an 'intelligence' able to take the best decision in specific circumstances. In these pandemic times, when we are more digitally connected, our ability to survive if platforms fail depends on *wisdom* and local intelligence combined in a choral manner [2]. From interactions, a cognitive transformation emerges [21]. This transformation is at the base of the intelligence augmentation (IA) since it modifies the processes of human thought, through the optimization of new interpretative schemes [7, 11]. In sum, AI supports resource efficiency (smartness) by removing people, while IA supports entity sustainability (wisdom) by including people.

3 Implications: Re-localization of Knowledge

AI is an enabling technology accelerating progress in many fields [22–24]. Dealing with the complexity of accelerating progress requires new patterns or schema [17, 18] to formulate and re-design the ecology, identifying appropriate approaches for businesses or government enterprises and for the whole system [25, 26]. Consequently, it may be necessary to re-define the role of AI where configurations of people, organization and technologies work together for a mutual benefit [1]. In these human-centered systems (like families, companies, cities, nations), it is fundamental to pay attention to the integration of machines with the social (i.e. human) environment in which they operate.

Better framing of the research problem could shed new light on future research indicating key elements to consider in the *re-localization of knowledge* that is happening, as the pendulum swings back - due to robotics, 3D printing, new local energy sources. Two to three hundred years ago, nearly everything was produced within a few hundred-mile radius of a city - not so today.

From this perspective, technology choices can be thought in terms of knowledge flow choices and sustainable quality of life above all if the platform fails. To this end, we propose to analyze the four digital challenges as a learning phenomenon and through the *v*SA (IVM) we try to highlight a relevant distinction between *general* schemes, *specific* schemes and *implicit* schemes that help to clarify the role of AI and IA. In particular, among the interpretative schemes, it is possible to distinguish implicit schemes, specific schemes and general schemes:

a) *implicit* schemes for experiential and not formal capacity of people;
b) *specific* schemes for interpretative competences through diffusive models;
c) *general* schemes for contexts and consonance on shared value categories.

For example, an expert carpenter uses implicit schemes to realize wood stuffs based on his long life experiences; if the carpenter would like to transfer to others (apprentices) his expertise it becomes necessary to formalize his expertise through an instruction manual; in this way, implicit schemes become specific schemes as a potential of replicability, consonant with the differential style of carpenter; moreover, if the reputation grows and becomes a benchmark for other carpenters as general schemes.

Implicit schemes help one entity, specific schemes increase the system's know how potential, while the general schemes increase the system's know why potential, necessary for wise decisions. Without such endowment, any new information can only increase confusion, hence, entropy [17, 18]. Consequently, new levels of complexity require new pattern or schema [19] as illustrated in Fig. 2. Both solutions and hypotheses significantly dependent on the role played by general schemes.

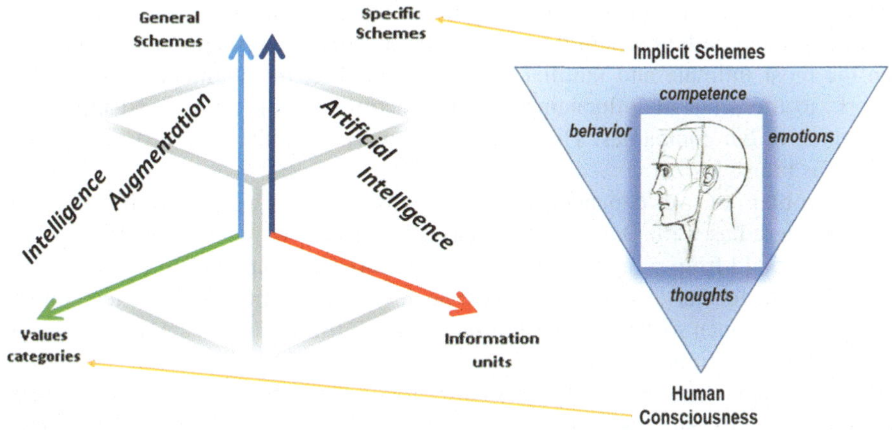

Fig. 2. Implicit vs Specific vs General Schemes - Authors' elaboration from Barile, Di Nauta and Iandolo 2016. www.asvsa.org.

In Fig. 2, AI can help people in the transaction from implicit schemes into specific schemes. Technology facilitates the formalization of implicit schemes into specific schemes. However, value categories pertain to human beings and to their endowment of consciousness/awareness. IA derives from the evolving of specific schemes into general schemes thanks to the enrichment of value categories (re-humanization process of intelligence). So, according to our perspective AI and users of AI technology interact exchanging information units through the use of interpretative schemes and increase the overall potential of the system to manage data and information transforming them into useful knowledge on the basis of the interpretative schemes. So, while AI can empower the interpretative schemes' understanding potential, IA acts more deeply empowering the system's wise decision-making capability through the use of general schemes and values categories.

Until today, the value was strictly connected to competencies. Previous studies [4, 6, 7] interpret value as "empowering value" which means assuming a value co-creation logic in the digital age that becomes capability-based in the sense that can be interpreted as the potential to realize a different type of project in terms of consensus capacities and reaching agreements between wise entities and not merely task competences [4]. This justifies the need to shift from the concept of AI connected to the context and the competence to do better what you can do to the innovative concept of (IA) [2, 3]. In short, what the VSA community hypothesized is that *when humans effectively interact with intelligent machines, like AI, a learning effect arises that empowers the humans' interpretative capability: an IA effect* [6, 7].

4 New Digital Age Challenges

4.1 Technological over Dependency

Digital not only surrounds us with its tools and screens that we constantly wade in-to, but it permeates every aspect of our life. Homo digitalis acts in symbiosis with digital technologies and participates in the transformations that are reshaping, in a short time, even the most intimate and familiar aspects of our life, in sentimental relationships, at work, to the point of influencing politics, institutions, operations and methods of functioning of our companies. The boundary between the public and private aspects of our lives is blurred.

The problem of over dependence on platforms and what happens if they fail needs to be considered as well. One systems approach to technology failure is the capability to rapidly rebuild from scratch Open source is knowledge commons that created better building blocks for startups - so they can create new configurations and new technologies more easily - so open source enables inventors and innovators. Open source also enables rapidly rebuilding from scratch.

How to be able to humanize, to personalize for example the intimate relationship that the historian builds with his sources and through his investigations, when these sources are made of bits and pixels? Perhaps we have lost, through the digital mutation, the sense of history and its representations.

History today cannot do without digital, which has transformed its DNA made up of concrete information processing practices like other humanistic subjects. Research, teaching and communication of results are today inextricably linked to digital. It is a question here of history with the digital which is thus the result of various practices that are fully digitized and computerized up to the communication of results and the production of research data. Furthermore, elaborating a research project to obtain funding is no longer conceivable without adding precise considerations on the good management of the research data in compliance with a FAIR use, from English, Findable, Accessible, Interoperable, and Reusable which underlines the need to open access to scientific research results to all.

4.2 Marginalization of Unconnected

As a major consequence of this changing world, a new existential paradigm is emerging and needs to be understood, in which we see the evolution from Homo Sapiens Sapiens to Homo digitalis. This passage outlines our future impossibility to do anything by ourselves. In fact, we will increasingly need to be connected to others through platforms. On the other hand, the digitized world will inevitably create conditions of exclusion and marginalization of large parts of populations with no access to the digitized world. In other words, in the very next future having no access to the platforms of the digitized world will mean being cut off from society or unable to participate to social, economic, and politic life, hence becoming irrelevant.

All this leads us to reflect:

a) "if" the disruptive paradigm of AI has the seed to orient itself towards sustainability and therefore commit to create sustainable service systems (wise systems) characterized by widespread welfare, or to expand stress and economic and social inequalities due to oligarchic systems (smart systems);
b) "how" in viable service systems the man-machine relationship can represent an effective "moment of physiological transformation" (wise systems), or a "phase of pathological restructuring" of prevalent efficiency (smart systems) with imaginable, even if undesirable effects of socio-economic inequality and concentration at city, regional, national and global levels.

4.3 Trust of Platform

The rise of the platform economy has led to the emergence of "multisided digital frameworks that shape the terms on which participants interact with one another" [27]. Given the paramount important role of platforms' design choices for many of today's social and economic interactions, Timm Teubner a David Dann [28] provide a practical overview and empirical insight into the ways platforms build trust between their users. While recent research has considered the effects of trust-building mechanisms on different platforms separately (e.g., how profile images and star ratings affect users' trust and booking intentions on Airbnb; [29], the overall platform landscape as a whole has received much less attention.

Drawing on Annette Baier's philosophy of trust, we argue that trust can only exist between people. We do not trust technologies like we trust people, rather we rely on them and they can succeed or fail. Technology disappoints us but does not betray us when it fails. Baier defines trust as a special case of reliance, where one party is relying specifically on the good will of the other party. Reliance, on the other hand, is defined as a continued relationship on the basis of one party's dependable habits toward the other [30]. According to our perspective, what is needed to be argued in future research is that reliability is a mediator of trust toward the makers of a technology and that relationships of reliance mediate our relationships of trust [31].

4.4 Competing for Collaborators

The Platform society determines and makes necessary a radical change not only in the substance but also and mainly in the form of what we can define the schemes useful to interpret our world. All the main concepts that characterize our environmental, social, and economic world are re-codified and reinterpreted on the basis of new 'algorithm and platform' knowledge schemes. Consequently, further questions arise: first, how science is facing these digital challenges? Second, is disciplinary knowledge adequate to address the complexity of these changes? That is why referring to competing for collaborators as digital challenge we propose a collective synergy with new areas of knowledge referring to service systems and viable systems scientists to increase society requisite variety and deal with the actual level of complexity. In other words, according to our perspective, competing for collaborators is strictly connected to our need to explore in future research people's wide variety of possible interactions in order to increase society's requisite variety to deal with the challenges of a changing world and of the emerging platform society.

Regarding to general schemes, probably, it would be necessary – maybe not hope for human – that IA will be able to develop autonomous value categories....as in the Blade Runner – Ridley Scott fantasy movie of 1982 – where Rachael, who turns out to be non-human, too is a replicant produced by the Corporation, but she herself ignores this fact first: for this reason she demonstrates that she has consciousness, feelings as such as a real human...so does she have value categories and general schemes as humans does?

5 Concluding Remarks

Platforms connect. Connected entities interact and form supply chains. Challenges have always included (1) entities surviving broken connections, (2) injustice of unequal connectedness, (3) lack of trust breaking connections, (4) and entities switching allegiance. People depend on others to survive. How can interconnected entities become better future versions of themselves? People, businesses, universities, and nations are all interconnected entities. Service systems and viable systems, from a service science and viable systems perspective respectively, are being transformed by AI progress. The promise of AI is smarter systems; somewhere between the euphoria of efficiency and anguish of joblessness, there is an opportunity for deeper understanding. Will Intelligence Augmentation advance our understanding fast enough to make wiser decisions? Nothing is yet settled, and more research is needed.

Websites
http://www.asvsa.org/
https://www.slideshare.net/spohrer/spohrer-icer-20150810-v1
https://www.postlean.com/ and https://www.postlean.com/events/
https://ssrn.com/abstract=3266472 or https://doi.org/10.2139/ssrn.3266472

References

1. Spohrer, J., Maglio, P.P.: The emergence of service science: toward systematic service innovations to accelerate co-creation of value. Prod. Oper. Manag. **17**(3), 1–9 (2008). https://www.slideshare.net/spohrer/spohrer-icer-20150810-v1, https://www.postlean.com/and https://www.postlean.com/events/
2. Barile, S., Ferretti, M., Bassano, C., Piciocchi, P., Spohrer, J., Pietronudo, M.C.: From smart to wise systems: shifting from artificial intelligence (AI) to intelligence augmentation (IA). In: Poster in International Workshop on Opentech AI in Helsinki, 13–14 March (2018)
3. Barile, S., Piciocchi, P., Bassano, C., Spohrer, J., Pietronudo, M.C.: Re-defining the role of artificial intelligence (AI) in wiser service systems. In: Ahram, T.Z. (ed.) AHFE 2018. AISC, vol. 787, pp. 159–170. Springer, Cham (2019). https://doi.org/10.1007/978-3-319-94229-2_16
4. Barile, S., Piciocchi, P., Saviano, M., Bassano, C., Pietronudo, M.C., Spohrer, J.C.: Towards a new logic of value co-creation in the digital age: doing more and agreeing less. In: Gummesson, Polese, Mele (eds.) The 10th Naples Forum on Service, 1–13, Ischia, Napoli 4–7 (2019)
5. Barile, S., Bassano, C., Lettieri, M., Piciocchi, P., Saviano, M.: Intelligence augmentation (IA) in complex decision making: a new view of the VSA concept of relevance. In: Spohrer, J., Leitner, C. (eds.) AHFE 2020. AISC, vol. 1208, pp. 251–258. Springer, Cham (2020). https://doi.org/10.1007/978-3-030-51057-2_35
6. Bassano, C., Barile, S., Saviano, M., Cosimato, S., Pietronudo, M.C.: AI technologies & value co-creation in a luxury cotext. In: Proceedings of the 53th Hawaii International Conference on Systems Sciences, pp. 1618–1627, scholarspace.manoa.hawaii.edu (2020)
7. Barile, S., Bassano, C., Piciocchi, P., Saviano, M., Spohrer, J.C.: Empowering value co-creation in the digital age. J. Bus. Ind. Mark. (2021, in press)
8. Van Dijck, J., Poell, T., De Waal, M.: The Platform Society: Public Values in a Connective world. Oxford University Press, Oxford (2018)
9. Spohrer, J., Golinelli, G.M., Piciocchi, P., Bassano, C.: An integrated SS-VSA analysis of changing job roles. Serv. Sci. **2**(1–2), 1–20 (2010)
10. Barile, S., Polese, F.: Smart service systems and viable service systems: applying systems theory to service science. Serv. Sci. **2**(1–2), 21–40 (2010)
11. Spohrer, J., Bassano, C., Piciocchi, P., Siddike, M.A.K.: What makes a system smart? Wise? In: Ahram, T., Karwowski, W. (eds.) Advances in the Human Side of Service Engineering, pp. 23–34. Springer, Cham (2017). https://doi.org/10.1007/978-3-319-41947-3_3
12. Spohrer, J.C., Maglio, P.P., Bailey, J., Gruhl, D.: Towards a science of service systems. Computer **40**(1), 71–77 (2007)
13. Spohrer, J., Kwan, S.K.: Service science, management, engineering, and design (SSMED): an emerging discipline-outline & references. Int. J. Inf. Syst. Serv. Sector (IJISSS) **1**(3), 1–31 (2009)
14. Pakkala, D., Spohrer, J.: Digital service: technological agency in service systems. In: Proceedings of the 52th Hawaii International Conference on Systems Sciences, scholarspace.manoa.hawaii.edu (2019)

15. Bassano, C., Piciocchi, P., Spohrer, J.C., Pietronudo, M.C.: Managing value co-creation in consumer service systems within smart retail settings. J. Retail. Cons. Serv. **45**, 190–197 (2018)
16. Huang, M.H., Rust, R.T.: Artificial intelligence in service. J. Serv. Res. **21**(2), 155–172 (2018)
17. Barile, S.: Management sistemico vitale. Giappichelli, Torino (2009)
18. Barile, S., Saviano, M.: Complexity and sustainability in management: insights from a systems perspective. In: Barile, S., Pellicano, M., Polese, F. (eds.) Social Dynamics in a Systems Perspective. NEW, pp. 39–63. Springer, Cham (2018). https://doi.org/10.1007/978-3-319-61967-5_3
19. Barile, S., Di Nauta, P., Iandolo, F.: La decostruzione della complessità. Studi MOA, Collana Studi di Management e Organizzazione Aziendale, Editrice Minerva Bancaria s.r.l., Roma (2016)
20. Barile, S., Saviano, M.: An introduction to a value co-creation model, viability, syntropy and resonance in dyadic interaction. Syntropy **2**(2), 69–89 (2013)
21. Carter, S., Nielsen, M.: Using artificial intelligence to augment human intelligence. Distill **2**(12), e9 (2017)
22. Minsky, M.L.: The Society of Mind. Simon and Schuster, New York (1986)
23. Stone, P., Brooks, R., Brynjolfsson, E., et al.: Artificial Intelligence and Life in 2030. One Hundred Year Study on Artificial Intelligence. Report of the 2015–2016 Study Panel. Stanford University, Stanford, USA (2016)
24. Demirkan, H., Bess, C., Spohrer, J., Rayes, A., Allen, D., Moghaddam, Y.: Innovations with smart service systems: analytics, big data, cognitive assistance, and the internet of everything. CAIS **37**, 35 (2015)
25. Piciocchi, P., Bassano, C., Paduano, E., Galvin, M.: The viable systems approach (VSA) for re-interpreting network business dynamics. In: Grabis, J., Kirikova, M. (eds.) BIR 2011. LNBIP, vol. 90, pp. 304–320. Springer, Heidelberg (2011). https://doi.org/10.1007/978-3-642-24511-4_24
26. Spohrer, J., Piciocchi, P., Bassano, C.: Three frameworks for service research: exploring multilevel governance in nested, networked systems. Serv. Sci. **4**(2), 147–160 (2012)
27. Kenney, M., Zysman, J.: The rise of the platform economy. Sci. Technol. **32**(3) (2016)
28. Teubner, T., Dann, D.: How Platforms Build Trust (2018). https://doi.org/10.2139/ssrn.3266472https://ssrn.com/abstract=3266472
29. Ert, E., Fleischer, A., Magen, N.: Trust and reputation in the sharing economy: the role of personal photos in Airbnb. Tour. Manag. **55**, 62–73 (2016)
30. Baier, A.: Trust and antitrust. Ethics **96**(2), 231–260 (1986)
31. Deley T., Dubois, E.: Assessing Trust Versus Reliance for Technology Platforms by Systematic Literature Review. Social Media + Society. April-June, 1–8 (2020)

Reducing Industry Complexity with International Standards: Current Efforts for Services, E-commerce, Artificial Intelligence

Stephen K. Kwan[1](✉) and Jim Spohrer[2]

[1] Lucas College and Graduate School of Business, San José State University, San José, CA, USA
stephen.kwan@sjsu.edu
[2] IBM Research – Almaden, San José, CA, USA
spohrer@us.ibm.com

Abstract. The global business environment has grown increasing complex due to economic, cultural, demographic, legal, trade, and political (to name a few) changes, differences, and disparity among nations and geographical regions. This complexity is exacerbated by rapid advances in technology which could be employed appropriately as accelerators or inappropriately as barriers to beneficial change to society. One of the ways industries seek to reduce complexity is through the development and use of standards that cover many areas of manufacturing, hardware, software, telecommunication, consumer products, business processes, construction, etc. In the last few decades, the advent of service innovation, e-commerce, and now digital platforms powered by AI have all given rise to international standardization efforts. In this paper we discuss ***some*** of these standardization efforts that are service-related and representative which are of interest to the Service Science community with the aim of informing and encouraging stakeholders to participate, contribute, and produce synergy among academic and industry partners.

Keywords: International standards · Services · E-commerce · Artificial Intelligence · Reducing industry complexity · Service Science

1 Introduction

Standards are all around us from the units of measures we use, to the side of the road we drive on, to the type of monetary currency we exchange, to the languages we speak, and of course, the shape of the electrical plugs we use every day. Standards reduce the complexity by increasing coordination capabilities of entities. From a service science perspective, standards are a type of *shared information resource* in service systems [1]. Responsible entities comply with standards and work together to co-create new and improved standards when needed; confidence and trust in an effective standard in an emerging technology area has the potential to dramatically co-elevate the collective capabilities of entities interacting [2]. While standards improve the performance of

The contents of this paper represent the opinions of the authors only.

C. Leitner et al. (Eds.): AHFE 2021, LNNS 266, pp. 67–76, 2021.
https://doi.org/10.1007/978-3-030-80840-2_7

entities seeking to interact in a coordinated manner and co-create value (service), they can also be a powerful force for international competition in contesting for control and advantage [32].

Since standards are such a common characteristic and beneficial part of service system performance, and yet little studied in the service research literature, the motivation for this paper is to introduce the service science community more broadly to the process of international standards creation and invite further study as well as participation in this important activity that shapes our world, lives and livelihoods. The body of this paper presents a current snapshot of the international standards efforts for Services, E-Commerce, Artificial Intelligence. In the remainder of this introduction, we further motivate why the service research and especially the Service Science community should both study and seek to engage with international standards organizations as they do their work.

First, the economic payoff for entities that successfully establish standards or gain a first mover advantage can be a significant source of competitive advantage [3]. For example, shipping containers of a standard size for trucks, railroads, and cargo ships grew the global economy, and also helped to establish Singapore as a leading link in the global supply chain [4]. Standards can force costly learning curves and rework for competitors, and give an ongoing, increasing returns, innovation advantage to a leader and their ecosystem partners. Businesses often compete to create de facto standards by rapidly growing ecosystems of adopters, faster than international standard organizations can achieve consensus. National standards bodies may take actions to solidify formal standards around the de facto standards of their own national businesses. In the game of establishing standards the payoffs for winners can be large and hence successful competition for collaborators is key [5].

Second, the phenomenon of standards in service system performance is truly complex and significant, especially with respect to measures of compliance and role of authorities. Within an ecosystem, compliance with standards can improve the performance of service systems – productivity, quality, and smooth the adoption of innovations. Therefore, the authority role can establish and enforce standards, wherever standards are most needed in a service ecosystem. Providers and customers must conform to those standards or face penalties. Authorities are allowed to use coercion in their value propositions (e.g., comply or pay a fine, or even comply or go to prison if convicted of compliance violations of some laws; in fact, laws are another example of shared information and standards).

Third, and perhaps more fundamentally, the phenomenon of standards is intended to reduce variance and improve coordination of entities interacting. It is well known that during the 1970's Japanese manufacturers used standard work practices to remove variance from manufactured products such as cars leading to greater competitiveness of their offerings – by squishing variance. Earlier, in the late 1940's MacDonalds fast food restaurants developed the "Speedee Service System" model, again using standard work practices to remove variance and increase quality and competitiveness of their service operations and offerings [6]. Even earlier, the USA developed universal education as a way to standardize reading, writing, and arithmetic capabilities and elevate the capabilities of large numbers of citizens as workers in the early industrial era, improving the competitiveness of a whole nation in that time of fundamental work transformation. Examples of the benefits of standardization are numerous and well-known.

Nevertheless, in spite of the improved performance goal of standardization and these success stories, the old joke about standards is "the wonderful thing about standards is that there are so many to choose from!" In fact, complex systems tend to operate "best" within a zone that balances order and chaos, stability and change, standards and diversity [7]. The study of standards may shed light on this fundamental and important dichotomy. From a Service Science perspective, the world of people, businesses, and nations are all service systems [1]. Service systems are defined as dynamic configurations of resources (people, organizations, technology, and shared information) interconnected internally and externally by value propositions. Shared information includes language, laws, measures, money, standards, signage, software and much more. In particular, the study of standards may be a fertile domain of study to connect Service Science and the older, yet still emerging, areas of study know as complexity economics [8].

In the remainder of this paper, we present a current snapshot of the international standards efforts for Services, E-Commerce, Artificial Intelligence, and Open Source, and then outline future research directions.

2 Service Standardization

Service standardization has been undertaken by different standards development organizations and in different industrial sectors. The definition of service is varied and covers micro (e.g., API-based services) to macro level of details (e.g., economic-sector-based services such as healthcare, education, etc.) An important overview of service standardization is *ISO/IEC's Guide 76: Development of service standards* [9] which contains recommendations for addressing consumer issues that would resonate with the interest of the Service Science community. The Guide is relevant to the full range of service types mentioned above regardless of whether the provision is free (such as public services) or is based on contract and financial considerations.

Guide 76 was first published by the ISO Committee on Consumer Policy (COPOLCO) in 2008 and updated as a second edition in 2020[1]. The new edition emphasizes the consumers as key stakeholders in the standards development process and acknowledges the increasing importance of diverse types of service in the economy. It considers that consumers have different needs and characteristics and that standards developers should "take the needs of all consumers into account, regardless of their circumstances or ability, including those who have different needs" (*ibid*, pg. v). The Guide defines service as "activity, or set of activities, provided for the benefit, or to meet the needs, of a consumer" (*ibid*, pg. 3). Of note is that the Guide no longer includes the phrase "product service" but clarifies that a service could be stand-alone or bundled with a product [10].

The Guide provides a clear context of What, Who, To whom, Where, When and How a service is provided. It also explains how to apply key (Human-Side) consumer

[1] Author 1 participated in the revision of Guide 76 as a member of the ISO/COPOLCO/WG 18 – Consumer Issues in Services.

principles of choice, access, information, safety, redress (more on this in Sect. 3), sustainability, representation, quality, privacy and data protection, across common elements of service. One of the sections in the Guide delineated service elements with consumer principles and related consumer questions such as "what added value/value proposition does this service have for me?" (*ibid*, pg. 10). In following the guiding principle of "value co-creation" in service activities, these consumer principles should be considered by providers in formulating value propositions for the customers to reflect what is of importance and value to the customers (e.g., see [11, 12] for a discussion of value components of a value proposition and [13] for a formal representation of a value proposition).

The Guide was updated partly due to an increase in the development of service standards during the past decade. A recent study [14] found that most of the standards published by the ISO have been related to the backstage [15] of service provider's operations or processes related to the production as enabler of service activities (74%) and the definitions of "service" were quite varied. It was not surprising since historically standards were developed mostly for technical specification and interoperability of products. The authors also found that developers of international standards were actively promoting service standards (e.g., ISO/COPOLCO, ANSI, DIN, CEN[2], etc.) to meet the needs of an expanding service sector worldwide. In Sect. 6 we will discuss how the Service Science community can participate and benefit from such efforts.

3 E-commerce Standardization

E-Commerce has been growing tremendously from the early days of the Internet [16]. As the underlying technologies and business transformations grew more complex and expansive [16, 17], so did the business need for international standards including those related to quality management [18] and data quality [19]. E-Commerce differs from traditional retail and E-Business in terms of high volume and smaller values in the buy/pay/ship/delivery cycle [20] giving rise to need for and the development of many international standards. During the current pandemic crisis, in some regions E-Commerce has grown many folds in reaching the anticipated growth of multiple years in a few months. Even with E-Commerce's transformation of retail business[3] and the current spike there are still barriers particularly in cross-border transactions where consumers are not familiar with vendors, policies, financial considerations, risks, product quality, trustworthiness, dependability, etc. Furthermore, such transactions have to comply with disparate national regulations that involve different agencies such as customs, border security, health and safety, agriculture, etc. Businesses that have to conduct cross-border transactions welcome international standards that set a harmonized level playing fields for participants instead of having to deal with regulations and divergent national standards of each country involved. In some severe cases, the WTO frowns on rigid and harsh regulations that might be intended to give advantages to national products over

[2] ISO Committee on Consumer Policy (COPOLCO), American National Standards Institute (ANSI), German Institute for Standardization (DIN), European Committee for Standardization (CEN).

[3] See [14] for a Service Science perspective of the transformation of retail business by E-Commerce.

those of other countries without there being legitimate "essential security interest" [21]. Consumer complaints about E-Commerce transactions have also increased in the past few years. Consumer Federation of America reported in 2019 that internet sales were one of the fastest-growing areas of complaints among the top ten in frequency [22]. In this section, we will report on an effort in reducing these barriers: ISO/TC 321 Transaction Assurance in E-Commerce[4] [23].

This Technical Committee is managed by the National Bodies of People's Republic of China and France. Its scope of work is standardization in the field of "transaction assurance in e-commerce related upstream/downstream processes" [23]. It has nineteen participating members including the US, UK, Japan, India, etc. It is working on two international standards: Foundations and Framework to be followed by Online Disputes Resolution, Guidelines for Exchange of Quality Evaluation Information, and Sharing of Product Traceability Information. The stakeholders of these standards include E-Commerce companies and platform providers, retail merchants, and consumers. Some large E-Commerce companies and platform providers have their own well-established business procedures and trusted by their customers through long and consistent practices. Their competitors that are venturing into new markets[5] have to convince third party merchants to sell on their platforms and new customers to trust their business procedures. The standards being developed by TC 321 will have more Human-Side considerations that will benefit providers by offering more palatable value propositions to their partners and customers.

One of the major barriers to customers patronizing an unfamiliar or foreign E-Commerce site is its return/refund policy and procedures for handling disputes (i.e., when a recovery is needed after a transaction failure, one of the elements of a customer-facing value proposition [11]). Big E-Commerce companies and platform providers could afford to handle such disputes with internal staff or third-party partners[6]. In some cases, they can even afford to have customers keep the merchandise in question because it is too costly to process the return/refund [24]. The work in TC 321 on Online Dispute Resolution will standardize some of these practices that are practical, transparent and explainable to provide more assurance and foster acceptance by customers.

4 Artificial Intelligence Standardization

After many years of research and development in academic and business organizations, Artificial Intelligence (AI) technologies are maturing and are moving from laboratories to deployment with some success in many industry sectors and regions. The impacts of these technologies are being recognized and their potentials are being explored and exploited in a wide variety of applications. As in many maturing technologies, the proliferation of AI technologies had led to challenges of interoperability, lack of transparency,

[4] Author 1 was an observer of this Technical Committee.

[5] For example, cross border commerce or doing business as a subsidiary of the parent company in a foreign country such as AliExpress and Rakuten in the US.

[6] For example, Gongdao.com is a partnership between Alibaba Group and the municipal government of the City of Hangzhou tasked with processing disputes arising from E-Commerce and other transactions that are within their purview.

lack of common terminology, inability to identify what is an AI application, etc. resulting in misunderstanding, mistrust and hesitancy in adoption. The international standards community recognized these problems and had begun to develop standards that could alleviate some of these problems and lead to more interoperability and community acceptance so that the technology could deliver its promises of benefits to society. We will report on such standardization efforts from the International Organization for Standardization (ISO), International Electrotechnical Commission (IEC), and IEEE Standards Association (IEEE SA).

The leading effort in international standardization of AI is undertaken by ISO/IEC JTC 1 Sub Committee (SC) 42 Artificial Intelligence[7] [25]. This committee was created in 2017 and holds plenary meetings twice a year. The scope of the standardization is 1) serves as the focus and proponent for JTC 1's standardization program on Artificial Intelligence; 2) provides guidance to JTC 1, IEC, and ISO committees developing Artificial Intelligence applications. In the following we will provide a status report on SC 42 activities as of its plenary meeting in October 2020.

Figure 1 shows the organizational structure of SC 42's parent organizations and their relationship to National Bodies members and their Technical Advisory Groups (TAG) which are, in essence, mirror committees of SC 42 in each participating country. The US TAG is hosted by INCITS and has close to forty members from enterprises, federal government agencies, academic institutions, HW, SW and consulting companies, etc. Figure 1 also shows the organizational structure of SC 42 where the standards development work is done at the working group level by experts. Note that SC 42 also has a liaison relationship with many other committees and organizations who have interest in AI and could contribute existing standards and/or potential use the standards developed by SC 42 in their own domains. The extent of the program of work undertaken by SC 42 can be seen in the committee's website [25] (6 published standards and 23 standards under development).

SC 42 develops foundational AI standards for definitions, technical concepts and technical topics such as computational approaches, system engineering, etc. (WG 1, WG 5, AG 2). The Subcommittee also develops standards related to data, lifecycle, quality, etc. that feed into AI systems (WG 2), AI System Lifecycle and Guideline for AI Applications including services (WG 4). Many of these standards will help to reduce the complexity of developing and interoperate this technology. On the other hand, SC 42 also undertakes the development of standards that are related to non-technical (Human-Side) customer/user concerns related to apprehension, acceptance and adoption of the technology such as trustworthiness, safety, risk, bias, ethical/societal issues, and governance[8] (WG 3, AHG). Just as described in Sect. 3 about E-Commerce, these standards "will benefit providers in offering more palatable value propositions to their partners and customers".

There are many AI-related standards being developed by other Standards Development Organizations (SDO). One that would be of interest to the Service Science community is the development of Responsible AI License (RAIL) by members of IEEE

[7] Both authors are participants of SC 42 by way of their organization's membership in INCITS.

[8] Some of these challenges were cited as leading barriers to AI adoption in a recent survey of the media and consulting company reports by the authors.

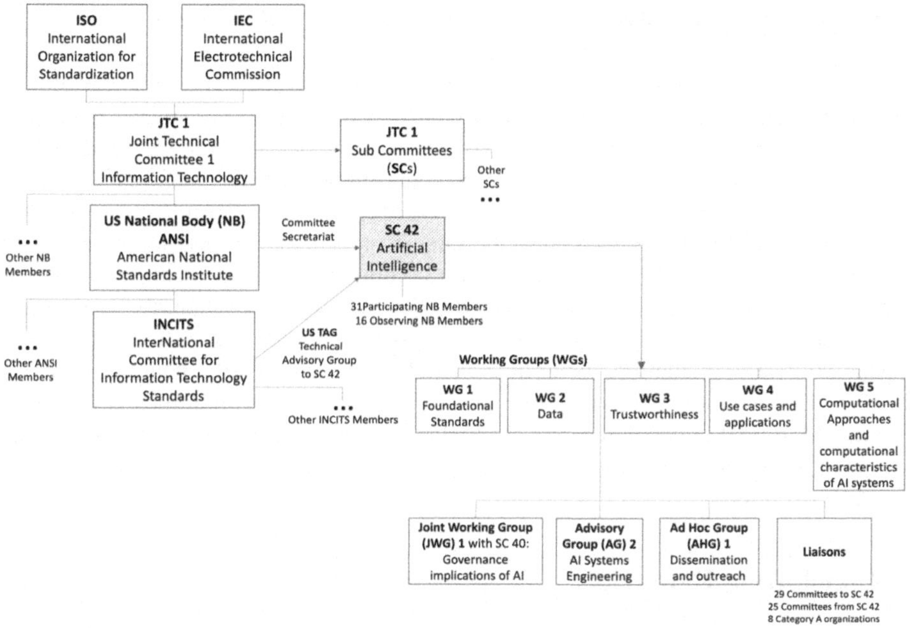

Fig. 1. SC 42, its parent organizations, membership, and committee structure

P2840 project[9] [26]. This is a work-in-progress and the group attempts to address the permissible use of AI software by licensing terms (which in essences is part of the value proposition to the customer). This project potentially could influence the structuring of End-User Licensing Agreements (EULA) which has certain capabilities, in this case, AI. One concern is that such licenses are "adhesion contracts" that claims control of the consumer's use behavior. The group is also concerned about defining the terms when the AI software is licensed to an end-user versus the software is offered as AI-as-a-service (AIaaS). Another group that is also working in the area is the LF AI & Data Foundation (part of the Linux Foundation). Of interest is one of their projects on defining the LFAI Principles of Trusted AI – Reproducibility, Robustness, Equitability, Privacy, Explainability, Accountability, Transparency and Security.

The Service Science community had recognized that AI technology is extending the automation spectrum and augmenting enterprise and personal decision making [27, 28]. AI technology is increasingly being embedded in service systems and in many cases fundamentally changed the relationship between the service provider and consumer [29]. It is thus incumbent on the Service Science community to heed the work undertaken by the standards development community described in this section and seek to provide positive influence on its progress.

[9] Both authors are members of IEEE P2840.

5 Future Research Directions: Open Source Software

This section briefly explores the role of standards developed by industry consortia in the era of platforms and ecosystems, many powered by the rise of open-source software. The rise of open-source software in business is creating de facto standards, where speed of adoption can be measured in weeks and months, rather than years to establish international standards [30]. The growth of open-source software is also apparent in Microsoft's acquisition of GitHub, the largest repository of open-source software, and IBM's acquisition of Red Hat the largest commercial success measured in annual revenues of any open-source based company. The equivalent of standards formation in international standards organizations happens in vendor-neutral foundations, such as the Apache Software Foundation, the Eclipse Foundation, the Linux Foundation and its LF AI & Data Foundation. The study of open-source software communities and projects provides additional opportunities for the Service Science community to study and participate in the creation of shared information resources that are shaping the world, our lives, and livelihoods.

6 Concluding Remarks

A schema of cooperation and exchanges between Service Science researchers (academia) and service-related standardization professionals (industry) is presented in Fig. 2. Such cooperation would help establish a more solid foundation for service-related standards. for standardization professionals, a closer cooperation with service researchers could be of benefit with the availability of a relatively consistent common theoretical framework that could provide a bridge and common language between different sectors and areas of service-related standardization. The adoption of service science concepts would provide service-related standards with a human-side framework for more customer-driven orientation and, as a result, alignment with customer value expectations and needs fulfillment.

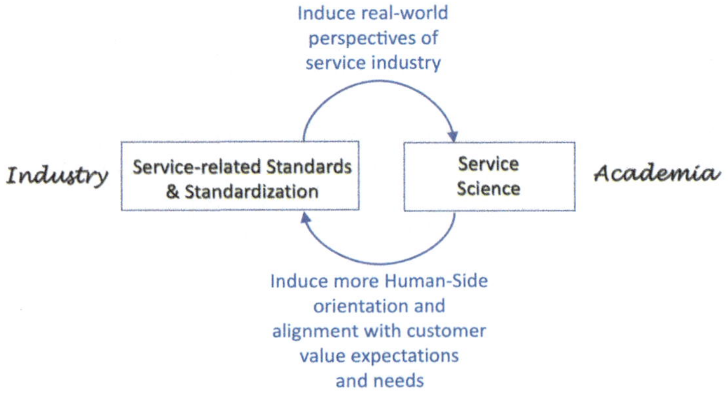

Fig. 2. Exchange and closer cooperation (from [31])

For service science researchers, involvement in service-related standardization processes would provide a "real world test" of the applicability of theoretical concepts in the service industry. Standards as a tool of knowledge dissemination and innovation, could contribute to increases in the spread and influence of ideas, concepts and methods that have originated in Service Science [31].

References

1. Spohrer, J., Kwan, S.K.: Service science, management, engineering, and design (SSMED): an emerging discipline – outline and references. Int. J. Inf. Syst. Serv. Sect. **3**(1), 1–31 (2009)
2. Schiff, D., Ayesh, A., Musikanski, L., Havens, J.C.: IEEE 7010: a new standard for assessing the well-being implications of artificial intelligence. In: 2020 IEEE International Conference on Systems, Man, and Cybernetics (SMC), pp. 2746–2753. IEEE (2020)
3. Su, H.C., Dhanorkar, S., Linderman, K.: A competitive advantage from the implementation timing of ISO management standards. J. Oper. Manag. **1**(37), 31–44 (2015)
4. Levinson, M.: The Box: How the Shipping Container Made the World Smaller and the World Economy Bigger. Princeton University Press (2006)
5. Spohrer, J., Kwan, S.K., Fisk, R.P.: Marketing: a service science and arts perspective. In: Handbook of Service Marketing Research. Edward Elgar Publishing (2014)
6. Heineke, J., Davis, M.M.: The emergence of service operations management as an academic discipline. J. Oper. Manag. **25**(2), 364–374 (2007)
7. Kauffman, S.: At home in the Universe: The Search for the Laws of Self-organization and Complexity. Oxford University Press, Oxford (1996)
8. Arthur, W.B.: Foundations of Complexity Economics. Nat. Rev. Phys. (2021)
9. ISO/IEC Guide 76:2020 Development of service standards – Recommendations for addressing consumer issues. https://www.iso.org/standard/73717.html
10. https://isotc.iso.org/livelink/livelink/Open/8389141 (2020)
11. ISO 9000:2015(en) Quality management systems—Fundamentals and vocabulary; 3.9.1 modified (2015)
12. Kwan, S.K., Hottum, P.: Maintaining consistent customer experience in service system networks. Serv. Sci. **6**(2), 136–147 (2014)
13. Kwan, S.K., Müller-Gorchs, M.: Constructing effective value propositions for stakeholders in service system networks. In: Proceedings of SIGSVC Workshop (2011)
14. Winkler, M., Dosoudil, V.: On formalization of the concept of value proposition. Serv. Sci. **3**(3), 194–205 (2011). https://doi.org/10.1287/serv.3.3.194
15. Weissinger, R., Kwan, S.K.: Service science research and service standards development. In: Satzger, G., Patrício, L., Zaki, M., Kühl, N., Hottum, P. (eds.) IESS 2018. LNBIP, vol. 331, pp. 330–343. Springer, Cham (2018). https://doi.org/10.1007/978-3-030-00713-3_25
16. Teboul, J.: Service is Front Stage – Positioning services for value advantage. INSEAD Business Press, Palgrave McMillian (2006)
17. Hänninen, M., Kwan, S.K., Mitronen, L.: From the Store to Omnichannel Retail: Looking Back over Three Decades of Research. The International Review of Retail, Distribution and Consumer Research, November 2020
18. Hänninen, M., Mitronen, L., Kwan, S.K.: Multi-sided marketplaces and the transformation of retail: a service systems perspective. J. Retail. Consum. Serv. **49**, 380–388 (2019)
19. Tan, B., Lin, C., Hung, H.: An ISO 9001:2000 quality information system in e-commerce environment. Ind. Manag. Data Syst. **103**(9), 666–676 (2003). https://doi.org/10.1108/02635570310506089
20. ISO 8000 Data Quality (many parts) (2020). https://eccma.org/iso-8000/

21. Probert, S.: What standards for eCommerce? Hangzhou Forum, United Nations Centre for Trade Facilitation and Electronic Business (UN/CEFACT), 16 October 2018
22. World Trade Organization (WTO). Technical Barrier to Trade (TBT) Agreement
23. https://www.wto.org/english/tratop_e/tbt_e/tbt_e.htm (1995)
24. Consumer Federation of America: 2019 Consumer Complaint Survey Report, 27 July 2020
25. ISO/TC 321 Transaction assurance in E-commerce (2020). https://www.iso.org/committee/7145156.html
26. Nunez, A.: Amazon, Target, and Walmart forgo return processing and let consumers keep unwanted products in cost-saving push. Business Insiders, 13 January 2021
27. ISO/IEC JTC 1/SC 42 Artificial Intelligence (2020). https://www.iso.org/committee/6794475.html
28. IEEE P2840 Responsible AI Licensing Working Group. IEEE Computer Society Standards Activity Board Standards Committee (C/SAB SC) (2020). https://standards.ieee.org/project/2840.html
29. Kwan, S.K., Pietronudo, M.C.: Artificial intelligence – extending the automation spectrum. To be presented at 2020 INFORMS Conference on Service Science, 19–20 December (2020)
30. Rouse, W.B., Spohrer, J.C.: Automating versus augmenting intelligence. J. Enterpr. Transf. 1–21 (2018)
31. Huang, M., Rust, R.T.: Artificial intelligence in service. J. Serv. Res. **21**(2), 155–172 (2018)
32. Ing, D.: Open Innovation Learning: Theory Building on Open Sourcing While Private Sourcing. Coevolving Press (2018)
33. Cellary, W., Freund, L.E., Kwan, S.K., Leitner, C., Spohrer, J.: The human-side of service engineering: advancing technologies impact on service innovation. Presented at the Naples Forum on Service, Ischia, Italy, 4–7 June (2019)
34. Pop, V., Hua, S., Michaels, D.: From lightbulbs to 5G, China battles west for control of vital technology standards. Wall Street J. 7 February 2021

Ecosystems Transformation for Social Change: How to Challenge Emergency Through Emergence

Francesco Polese, Mara Grimaldi, and Orlando Troisi(✉)

Department of Management and Innovation Systems, University of Salerno, Via Giovanni Paolo II, 132, 84084 Fisciano, Salerno, Italy
{fpolese,margrimaldi,otroisi}@unisa.it

Abstract. The study explores the remodelling of actors, interactions and relationships due to the different use of technologies that can enable contemporary organizations, conceptualized as service ecosystems, to comply with the posed by Covid-19. The aim is to investigate how the adoption of technology can lead to the readaptation of interactions between users to tackle the health "emergency" through the "emergence" of ecosystems transformation by creating innovation and social changes. After the proposition of a theoretical framework, through the interpretative lens of service ecosystems view, Higher education is reread an ecosystem to identify the different technological touchpoints, relational modalities and resources integrating practices that can be implemented in teaching and learning processes to foster the emergence of new knowledge, value and social practices.

Keywords: Service ecosystems · Human-technology interactions · Emergency · Emergence · Social change

1 Introduction

Challenging COVID-19 pandemic requires companies to redesign their business models [1], to refocus their strategies, to redefine their interactions with stakeholders. Even if the effects of the global crisis cannot be evaluated definitively, it is acknowledged that the use of smart technology can be considered as one of the key enablers to manage emergency situations [2, 3]. Lockdowns all over the world lead companies, public institutions and non-profit associations to adapt to the digitalization of business practices, which is dramatically changing people's daily experiences and business contexts.

Through an integrated set of digital technologies available across multiple touchpoints with users, organizations can maintain and renew the relationships with and between actors and can exploit new ways to experience interactions, by establishing a continuous tension for transformation, change, innovation.

However, the "simple" use of technology could be necessary but not a sufficient condition for the fast readaptation of business processes and for the development of

C. Leitner et al. (Eds.): AHFE 2021, LNNS 266, pp. 77–88, 2021.
https://doi.org/10.1007/978-3-030-80840-2_8

innovation [4]. During a state of emergency, the economic, technological and knowledge gaps between companies that own the right skills and attitude to use technology meaningfully to comply with users' needs and companies that do not (or cannot) adopt technologies strategically can be emphasized. The adoption of technology could be necessary but not a sufficient condition for the effective readaptation and redefinition of the organizational models [5] dictated by the need to overcome the health emergency.

Recent studies show that technology use may also have negative implications on users' wellbeing, by determining a real stress toward the application of ICTs (*information and communication technologies*) to daily lives [6, 7]. To fully integrate technologies with habits and routine, users should learn to manage technological tools and should refocus their cognitive strategies to accomplish with the cultural and social requirements related to their use. Moreover, technology per se does not foster the development of innovation but it is human activation of technology [8], mediated by the political and institutional context [9], that can foster the co-creation of values, knowledge and novelties.

Therefore, the economic, relational and social transformations determined by the different strategies implemented to attain an active resolution of Covid-19 can change, probably definitively, the nature of interactions and collaborations between users and providers [10], by emphasizing that the application of human intervention (e.g., the attitude, the adhesion, the propensity, the smart orientation and the willingness to use technologies of individuals-citizens) is the only way to use technology effectively to manage unexpected phenomena [11, 12].

Contemporary networked organizations can overcome emerging challenges by investing in relationships and value creation strategies that can give birth to the creation of new knowledge, rules and institutions [13], which determines the evolution of networks into ecosystems [14]. For this reason, the adoption of service ecosystems view [15] can offer the right multi-level perspective to analyze organizations as embedded systems of actors who exchange resources more easily through technology and based on the constant redefinition of institutions and rules that coordinate exchanges [16, 17].

However, in service ecosystems research, there is still the need to explore the insights offered by a multilayered perspective to the analysis of pandemic [18]. Moreover, despite the investigation of the "social side" of service ecosystems [19, 20], the mechanisms that foster value co-creation and the creation of social value and practices have not been yet identified [21, 22].

Therefore, the work seeks to address the following research questions:

RQ1: which are the key enablers of ecosystem's remodelling for a smart use of technologies?

RQ2: how can the readaptation of interactions between users lead to the "emergence" of societal transformation and social changes to challenge public health "emergency"?

The study introduces a framework that can help the exploration of ecosystems as networks of actors who co-develop innovation to meet relevant social challenges (from skill's enhancement to educational innovation to sustainability and well-being). After the presentation of the framework, a potential application to Higher Education ecosystem is performed to detect: 1) the transformations introduced in actor's relationships, human-technology interactions, resource integration and value co-creation practices to adapt to the new rules imposed by the health emergency; 2) how these transformations can redefine the "old" rules, practices and institutions by activating innovation and changing social modalities.

The case of Higher Education shows how actor's relationships through technology-enabled interactions can not only contribute to challenge the negative effects of the Pandemic but can give birth to new value, interaction modalities by determining innovation and social change.

The development of the framework and its practical application contribute to enhance organizations' understanding of how to readapt their traditional practices and processes to comply with society's needs and evolution and how to turn the social, economic and political crisis into an opportunity for innovation.

2 Theoretical Background

A conceptual overview highlights the usefulness of Service-dominant logic [23, 24] and Service ecosystems view for a deep understanding of how companies can reframe the interactional modalities between actors and their resources integration processes through a wise use of technology, which can produce new value, knowledge and institutions to adapt to the transformation dictated by the pandemic and produce innovation and social change.

Therefore, to lay the theoretical foundations for the further resolution of the research questions, the current section analyses conceptually: 1) the key ecosystem's enablers of innovation; 2) the potential drivers for ecosystem's readaptation and re-institutionalization for social change.

2.1 Innovation in Service Ecosystems

Service ecosystems view is a systems perspective introduced in Service-dominant logic (*S-D Logic*, [24]), which understands service as the glue of resource integration among engaged actors that, through a complex set of technology and ICTs-enabled interactions, can co-create value [25]. The advancement of service ecosystems view stems from the need to explore the multi-levelled and networked dynamics of the service exchange and to take into account the embeddedness of contexts [15] in which multiple actors and resources are engaged.

The main dimensions of ecosystems, which can be intended as enablers of value co-creation are: 1) institutions: coordination mechanisms, norms, social rules and shared practices that encourage actors to integrate resources and to pursue shared purposes [26]; 2) value propositions: the selection and the subsequent sharing of common values among actors; 3) technologies: technological tools that improve interactions and actor's engagement [27].

Service ecosystems, as a set of technologies, actors and skills connected with multiple relationships, are based on the exchange of material and immaterial resources across three contextual levels [28, 29]: 1) *micro-level*: set of individuals with given intentions, attitudes, cognitive processes, value perception, skills and resources; 2) *meso-level*: networks of relational and social connections between actors; 3) *macro-level*: ecosystem's general community (public administration, institutions, legal system, etc.). Value co-creation is intended as a process emerging from the combination of multi-levelled transformations at micro (individual), meso (relational) and macro (institutional) levels of exchange [28] that can enable the co-development of new value, new practices, innovation [30].

S-D logic and social approach to ecosystems observe that technology per se cannot allow the automatic development of innovation [31]. Thus, the right application of human skills and knowledge and the social and cultural variables are considered as drivers of innovation, together with the (still) relevant technological adoption.

Thus, extant literature on service innovation broadens the focus from the study of technological innovation to the adoption of a systems orientation [28] that identifies the need to integrate technological dimension with human component [32, 33] and the broader social system in which actors are embedded [34].

There is the shift from a *technology-driven approach*, that considers technology as the main enabler of value co-creation and innovation [35] and defines technological tools (platforms, IT and ICTs based systems) from a structural point of view [36, 37], to a *knowledge-based approach*, in which the sharing and integration of knowledge between actors is viewed as the essential driver to use technology efficiently and to improve services or create new ones [38]. Lastly, a *social-oriented approach* conceptualizes the relevance of the social sphere in the shaping of value co-creation and innovation [28, 31] and of the role of social connections, rules and power relations that govern technology-mediated interactions in the enhancement of well-being and innovation.

2.2 From Innovation to the Co-development of Social Change

The dynamic combination of service ecosystem's enablers (institutions, value propositions, technology) allows the development of value co-creation, that can be intended as a multi-levelled cultural and social phenomenon that can lead to the emergence of new value and, thus, potential innovation. Through the key notion of institutionalization, ecosystem's transformation is defined as the complex result [28] of: 1) *maintenance* (readaptation of ecosystem's structural elements); 2) *disruption* (recombination of extant skills in use to foster the emergence of new value; 3) *change* (creation of new value and discovery of solutions for new or existing problems).

The transition from maintenance to disruption and change is based on the proactive exchange of resources (human component, actor's attitude, digital competencies, willingness to use technology etc.) that can act as key drivers for the transformation of crisis into opportunities for innovation [39] and social change [40].

The investigation of the main features of ecosystems restructuring and adaptation can contribute to detect the transition from the resolution of emergency to the emergence of innovation for the (co-) development of social change. If the emergence allows the development of new, unexpected elements, understanding how to manage emergence can be a source for innovation but also for social change. Therefore, challenging the emergence by learning to manage the appearance of new elements can make it possible to transform the unknown, the crisis into opportunity. The new co-created values, knowledge, beliefs and meanings are re-institutionalized through the emergence of new elements and properties that do not represent the simple "sum" of the input resources but have an added value.

Service ecosystem views can contribute to shed light on the mechanisms of this process by exploring the role of technologies and human factors and the enablers of an optimal management of knowledge exchange processes through the key elements of ecosystems (technology, actors, integration resources, value propositions). Hence, the exploration of the different ways to reframe human's interactions mediated through technology to challenge global epidemic can help the identification how social, economic and health emergencies can enable the attainment of continuous societal transformation. The investigation of the main features of ecosystems restructuring and adaptation can contribute to detect the transition from the emergence of innovation to the development of innovation and social change (see next paragraph).

3 A Conceptual Framework for Ecosystems Innovation and Social Change

To address RQ1 (*which are the key enablers of ecosystem's remodelling for a smart use of technologies?*) the main ecosystem's enablers for a smart use of technologies, which can foster in turn innovation, can be identified across micro, meso and macro levels. Transitioning from micro to macro context, the movements from subjective, intersubjective [41] and collective value creation can be observed and the transformation of value across multi-layered contexts to develop innovation incrementally can be investigated.

In the micro-level, actors can develop their ability and propensity in the use of technology (platforms, IT systems, etc.), therefore their attitude towards technology adoption and individual skills (subjective dimension). In the meso-level, the actors become integrators and transformers of knowledge (intersubjective dimension) and foster the emergence of new knowledge, therefore unexpected value, to co-create new practices, new interactive modalities, redefine in progress actor's relational and interactional modalities. At the macro level, the new methods for interaction and co-creation practices flow into the social-collective sphere, by generating new roles for the actors, meanings and symbols, therefore a new culture resulting from the accumulation of knowledge.

It is essential to spread the tension for change throughout the three ecosystem's contexts, in which different strategies can be identified. The new co-created value can give rise to the creation of new knowledge [13], new rules and institutions, or ways to interact, to experience social relationships, whose combination and constant renewal determine the perpetual evolution of systems.

To describe the different types of ecosystem change that can be implemented to challenge global emergency and answer RQ2 (*how can the readaptation of interactions between users lead to the "emergence" of societal transformation and social change to challenge public health "emergency"?*), the maintenance-disruption-change model introduced by the Vargo et al. [28] can be associated with the three contexts of service ecosystems.The different "changes" of status in the ecosystem and can be connected with micro, meso and macro contexts to detect the different ecosystem "level" on which organizations should act to make the change.

Through *maintenance*, at micro-level organizations can readjust their pre-existing value propositions and institutions to adapt the skills (in general and related to the use of new technologies), the business idea, the activities to user's needs and attitude. In the *disruption* phase, at meso-level, the existing knowledge, rules and practices are recombined and re-designed to give shape to patterns of connections between actors based on new interactive-communicative methods in a smart perspective. Transformation at systems macro-level can *change* the organizational structure, culture, the relationships between actors and their roles by co-creating new shared meanings and innovation.

The novelties co-created give birth to meanings and symbols that create a unique and cohesive culture by renewing organizational structure, processes, and culture, at micro, meso and macro-levels. As Fig. 1 shows, it follows that innovation can be reframed as a process emerging from the combination of multi-levelled transformations across different ecosystem's context.

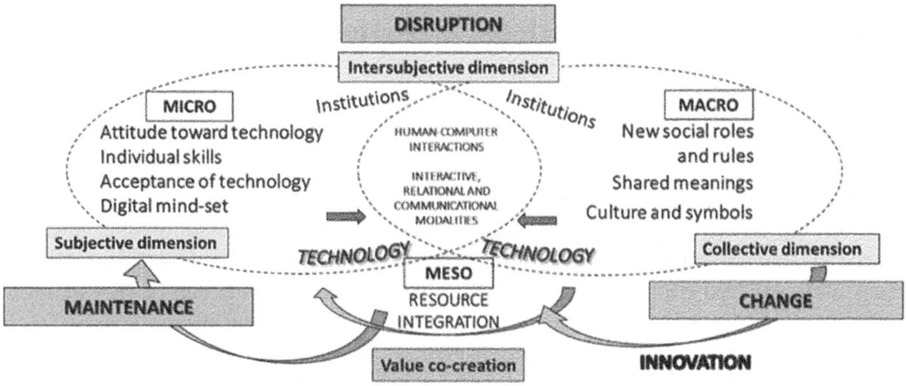

Fig. 1. A framework for ecosystems transformation

For this reason, social change can be defined as a complex process through which innovation is spread across ecosystem's contexts, by enveloping individual intentions [42] and orientation (subjective dimension), social networks of interactions (sphere dimension) and ecosystem culture (collective dimension). Therefore, innovation in ecosystems can act as an engine for social change and transformation. The propagation of innovation in ecosystems reformulates the way in which actors face social challenges, in which they co-create value and transform integrated resources and practices into change and social development. Human's adoption of technology is a key driver for the co-creation of value and knowledge, which can be renewed over time and can transform actors into performers of social roles, by modifying social structure and culture.

4 Discussion: An Application to Higher Education

Therefore, the association of micro, meso and macro-levels with maintenance, disruption and change can allow the investigation of: 1) the different enabling elements that can be used in each level to foster resource integration to maintain, disrupt and change the ecosystem (RQ1); 2) the emergence of innovation and social change as a result of continuous ecosystem's transformation to overcome the challenges posed by the pandemic (RQ2).

Higher education system can be reread as an ecosystem, a network of industrial, institutional and research actors who collaborate and develop innovation projects to meet relevant social challenges (from knowledge exchange to the creation and renewal of culture to the improvement of digital skills, quality of life, well-being and educational innovation).

Starting from the identification of the basic elements of the three contexts of HE ecosystem, the different kinds of institutions, value propositions and technologies employed for teaching and learning at the micro, meso and macro-levels (RQ1) and the different practices that can lead to innovation and social change (RQ2) can be identified. Therefore, the framework introduced in Fig. 1 can be applied to the context of education to identify the circular process that depicts how learning/teaching processes can be enhanced through technology (see Fig. 2). It can be noticed that the exchange of knowledge is not performed only during the delivery of service, but even before and during the delivery in line with the aim of continuous improvement. Technology-enhanced teaching, learning and evaluation at each level can improve the service progressively, by "adding" incremental value and enriching the value co-created in each step.

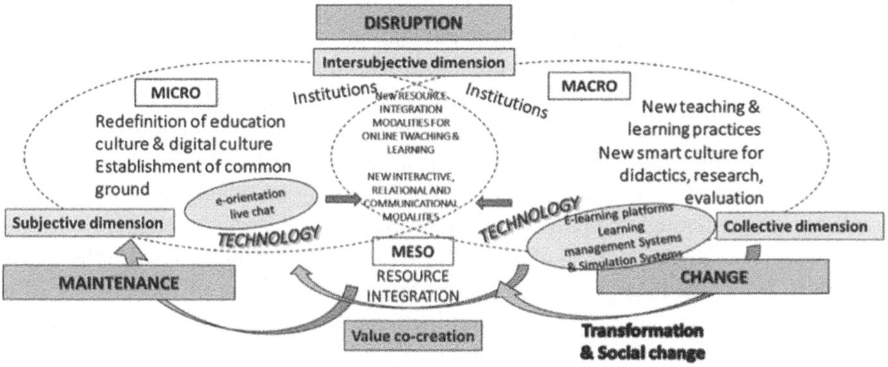

Fig. 2. An application to education ecosystem

At micro- level, University as an organization, with the support of teachers, establishes a common ground with students by co-creating with team guidelines, rules and the program of education offering through technological engaging tools, aimed at providing teachers and students with training activities to use e-tools and psychological support, through live chat for assistance and for student orientation. Two key activities can be realized: 1) alignment of teachers' skills and attitude with students' background and willingness to engage; 2) change in the entrepreneurial mind-set of university. In the first activity, management tries to assess the ability-propensity of students and teachers in the use of technologies based on their background and level of digitization and their possible resistance in the use of telematic tools in work, study and of daily life. The provision of online teaching implies a change not only in the methods of digital service provision but also in the identification of objectives and the division of tasks (through the enhancement of autonomy and power distribution). University managers should rethink the service/business digitally, as a new way of doing education and culture.

At meso-level, the learning technology tools for course interaction (such as Teams, Zoom) can support teacher and students in creating an interactive atmosphere in class. The teachers can use various functions to enable the teacher-student interaction function and to let all students "raise their hands" or answer questions through different devices. After having uploaded student's answers, the teacher can select the works to observe, share and discuss in the same room class. In this way, new communication and relational methods for technology-enhanced learning are created, based on timely information and on the possibility of being in contact 24 h a day, through tools such as live chat, one-to-one and personalized assistance to students, etc. "Raising hand" activities permit to collect simultaneously ideas from every student, to store and to re-elaborate them for the proposal of new service, extension of the programme. By providing their insights and by visualizing other students' suggestions, a process of co-innovation can start in which each member can improve the service.

After the lesson, e-learning platform can help teachers and students handle homework assignments and share course materials. These tools permit to provide students with assignments that can be completed in real time. Students can share their feedback during the lessons through learning management systems such as "Moodle" and can use advanced assessment settings for different question types, such as shuffling the items and their options, using sequential or free navigation.

The integrated use of technology-enhanced learning tools and evaluation tools can facilitate the sharing of experience, tacit and codified knowledge to obtain new knowledge and co-create learning (*co-learning*). Cooperative learning, a common concept in education, can be translated into co-learning (co-created learning) according to the principles of S-D logic. In this way, new ways of exchanging knowledge and creating new knowledge are generated. Students and digital natives can provide teachers with their experience, especially in the field of technology, due to their familiarity with the Internet and with ICTs. Thus, these tools can permit the full realization of the essence of co-learning: education does not imply the unidirectional sharing of knowledge (from teachers to students) but empowers the enrichment of both students and teachers experience, know-how, tacit knowledge, culture and beliefs.

The introduction of new teaching, learning and evaluation practices for scholars and students can develop constantly opportunities to change and pursue continuous improvement. The novelties emerged at macro-level are: 1) new methods for educational service provision, which can become stable practices over time and can be maintained even after the restarting of activities in the presence; 2) a new smart culture for training, learning, didactics and research that redefines languages and shared meanings between students, teachers and staff.

The tools employed to assess students' and teachers' performances and students' opinion on the education service are online surveys on the satisfaction of courses and exams and on University reputation, monitoring of students access in user areas, tracking of users behaviour on internet, visualization techniques and dashboard to collect data on students and on teachers' performance.

Starting from the collection of students' opinions and behaviours, teachers can make decisions based on statistical data and adjust teaching according to the information extracted. Data on teachers and student's performance are analyzed through data mining techniques to discover and obtain knowledge from databases to support the analysis of student learning processes and the evaluation of the effectiveness and usability of online courses.

The circularity in the framework is stressed through the bidirectional arrows that connect the three ecosystems contexts, which can represent the so-called downward effect (Peters 2016). At the beginning, the emergence of online delivery method for lessons and exams is tactical, forced by contingencies (at meso-level); then, it is institutionalized at the macro-level and becomes an established and accepted practice again that can "come back" to the micro-level, in which it can become an integral part (a rule, a section of the programme, a new condition to make online exams and thesis dissertation, or to evaluate scientific publications) of the educational offering.

The key findings of the conceptual analysis can be translated into two propositions, which address the three research questions of the study:

P1: the key enablers of ecosystem's remodelling for a smart use of technologies are the adoption of a digital mind-set and a positive attitude toward technologies, which increase the possibility to experience, co-learn, co-create and improve services through the wise use of a set of touchpoints across micro, meso and macro-levels.

P2: ecosystems transformation (maintenance, disruption and change) can lead to the re-institutionalization of the new rules, interaction modalities and coordination mechanisms for resource integration, that can lead to the emergence of social change to tackle the technological challenges posed by the emergency across three interconnected processes:

1) *The redefinition of actors' attitude, orientation willingness to use technology at s subjective level (micro);*
2) *The co-learning and co-creation of new knowledge at the intersubjective level of interactions and resource integration, in which new modalities of teaching and learning emerge (meso);*
3) *The re-institutionalization of the new emerging modalities, practices and rules and their acceptance as established ecosystem's elements at a collective level (macro).*

5 Concluding Remarks and Implications

The study explores the key enablers for the redesign of actor's interactions, service provision modalities and technology use in contemporary organizations which aim at complying with the challenges posed by Covid-19.

Through the interpretative lens of service ecosystems view, Higher education is reread an ecosystem to detect: 1) the main drivers for ecosystems re-adaptation to challenge the technological requirements imposed by the pandemic; 2) how these drivers can be activated to enable societal changes and innovation as a result of the complex transformation of the crisis into innovation opportunities.

The work analyses the case of Higher Education by producing theoretical advancements on the classification of the different technological tools that can support the provision of educational services in the different moments of service provision and across the different resource integration and knowledge exchange involved in value co-creation and co-learning processes. The identification of the enablers of societal changes in smart education and the revealing of the new interaction modalities and main strategies to challenge the pandemic can help scholars and practitioners to shed light on the key drivers to overcome social and economic crisis. The elaboration of a framework that analyzes how technology can redefine human's interactions and can foster social change in smart cities can address a gap in literature related to the absence of studies exploring the role of technologies in reframing traditional cities management and social innovation [43].

The adoption of ecosystem perspective allows to investigate: 1) the "which" dimension: what are the different technologies, interactive methods, knowledge, skills, orientation to technology use capable of transforming the crisis into opportunities; 2) the "how" dimension: the way in which the enablers can be combined to develop a proactive and resilient attitude, which mediates between propensity for change and consolidation of practices, between flexibility and elasticity (over time), in order to allow the transition from the challenge to global emergency to the emergence of innovation.

Thus, ecosystems view permits not only to grasp the different types of technologies and institutions that can enable innovation but also to grasp the most appropriate strategies for the exchange of resources and for the continuous readaptation of organizations, enabling the transition from innovation to social change. Thus, the framework proposed can help identify how social, economic and health emergencies can enable the achievement of continuous social transformations and changes within ecosystems by developing multiple innovation processes to be regenerated over time. Education managers can understand: 1) how the use of different kind of technologies can help redefine the interaction modalities between and among students, teachers and community to challenge the global epidemic; 2) the key ecosystem's enablers for social changes and the development of different innovation opportunities through crisis resolution.

Further studies can employ the framework proposed through grounded theory according to a constructivist approach [44, 45], a technique usually employed in HE, which provides researchers with the possibility to explore how the enrichment of knowledge can be enhanced gradually through multiple research steps and different phases of observation, in which the new knowledge acquired and the changes in the relationships between actors can be assessed in each step according to an incremental methodological procedure.

References

1. Seetharaman, P.: Business models shifts: Impact of Covid-19. Int. J. Inf. Manag. **54**, 102173 (2020)
2. Palmieri, F., Ficco, M., Pardi, S., Castiglione, A.: A cloud-based architecture for emergency management and first responders localization in smart city environments. Comput. Electr. Eng **56**, 810–830 (2016)

3. Ivanov, D., Dolgui, A.: Viability of intertwined supply networks: extending the supply chain resilience angles towards survivability. A position paper motivated by COVID-19 outbreak. Int. J. Prod. Res. **58**(10), 2904–2915 (2020)
4. Gupta, M., George, J.F.: Toward the development of a big data analytics capability. Inf. Manage. Res. **53**(8), 1049–1064 (2016)
5. Danquah, M., Amankwah-Amoah, J.: Assessing the relationships between human capital, innovation and technology adoption: evidence from sub-Saharan Africa. Technol. Forecast. Soc. Chang. **122**, 24–33 (2017)
6. Hauk, N., Göritz, A.S., Krumm, S.: The mediating role of coping behavior on the age-technostress relationship: a longitudinal multilevel mediation model. PloS one **14**(3), e0213349 (2019)
7. Nimrod, G.: Technostress: measuring a new threat to well-being in later life. Aging. Ment. Health **22**(8), 1086–1093 (2018)
8. Spohrer J., Bassano C., Piciocchi P., Siddike M.A.K.: What makes a system smart? wise?. In: Ahram T., Karwowski W. (eds.) Advances in The Human Side of Service Engineering. Advances in Intelligent Systems and Computing, vol. 494. Springer, Cham (2017). https://doi.org/10.1007/978-3-319-41947-3_3
9. Kummitha, R.K.R.: Smart technologies for fighting pandemics: the techno-and human-driven approaches in controlling the virus transmission. Gov. Inf. Q. 101481 (2020)
10. Velotti, L., Murphy, P.: Service and value co-production and co-creation in emergency services and emergency management. Int. J. Emerg. Serv. **9**(1), 1–7 (2020)
11. Azoulay, P., Jones, B.: Beat COVID-19 through innovation". Science **368**(6491), 553 (2020)
12. Kunzmann, K.R.: Smart cities after covid-19: ten narratives. disP Plan. Rev. **56**(2), 20–31 (2020)
13. Polese, F., Payne, A., Frow, P., Sarno, D., Nenonen, S.: Emergence and phase transitions in service ecosystems. J. Bus. Res. **127**, 25–34 (2021)
14. Gervilla, M.J.Q., Díaz-Mendez, M., Gummesson, E.: Balanced centricity and triads: strategies to reach ecosystem equilibrium in the arts sector. J. Bus. Ind. Mark. **35**(3), 447–456 (2019)
15. Chandler, J.D., Vargo, S.L.: Contextualization and value-in-context: how context frames exchange. Mark. Theory **11**(1), 35–49 (2011)
16. Vargo, S.L., Lusch, R.F.: It's all B2B… and beyond: Toward a systems perspective of the market. Ind. Mark. Manage. **40**(2), 181–187 (2011)
17. Spohrer, J., Piciocchi, P., Bassano, C.: Three frameworks for service research: exploring multilevel governance in nested, networked systems. Serv. Sci. **4**(2), 147–160 (2012)
18. Brodie, R.J., Ranjan, K.R., Verreynne, M.L., Jiang, Y., Previte, J.: Coronavirus crisis and health care: learning from a service ecosystem perspective. J. Serv. Theory Pract. (2021)
19. Edvardsson, B., Tronvoll, B., Gruber, T.: Expanding understanding of service exchange and value co-creation: a social construction approach. J. Acad. Mark. Sci. **39**(2), 327–339 (2011)
20. Letaifa, S.B., Edvardsson, B., Tronvoll, B.: The role of social platforms in transforming service ecosystems. J. Bus. Res. **69**(5), 1933–1938 (2016)
21. Bartoli, G., Fantacci, R., Gei, F., Marabissi, D., Micciullo, L.: A novel emergency management platform for smart public safety. Int. J. Commun. Syst. **28**(5), 928–943 (2015)
22. Rihova, I., Buhalis, D., Moital, M., Gouthro, M.B.: Conceptualising customer-to-customer value co-creation in tourism. Int. J. Tour. Res. **17**(4), 356–363 (2015)
23. Vargo, S.L., Lusch, R.F.: Evolving to a new dominant logic for marketing. J. Mark. **68**(1), 1–17 (2004)
24. Vargo, S.L., Lusch, R.F.: Service-dominant logic: continuing the evolution. J. Acad. Mark. Sci. **36**(1), 1–10 (2008)
25. Vargo, S.L., Lusch, R.F.: From repeat patronage to value co-creation in service ecosystems: a transcending conceptualization of relationship. J. Bus. Mark. Manag. **4**(4), 169–179 (2010)

26. Akaka, M.A., Vargo, S.L., Lusch, R.F.: The complexity of context: a service ecosystems approach for international marketing. J. Int. Market. **21**(4), 1–20 (2013)
27. Storbacka, K., Brodie, R.J., Böhmann, T., Maglio, P.P., Nenonen, S.: Actor engagement as a microfoundation for value co-creation. J. Bus. Res. **69**(8), 3008–3017 (2016)
28. Vargo, S.L., Wieland, H., Akaka, M.A.: Innovation through institutionalization: a service ecosystems perspective. Ind. Mark. Manage. **44**, 63–72 (2015)
29. Frow, P., Nenonen, S., Payne, A., Storbacka, K.: Managing co-creation design: a strategic approach to innovation. Br. J. Manag. **26**(3), 463–483 (2015)
30. Grieco, C., Cerruti, C.: Managing co-creation in innovative business models: the case of sharing economy. Sinergie Italian J. Manag. **36** (2018)
31. Lusch, R.F., Nambisan, S.: Service innovation. MIS Q. **39**(1), 155–176 (2015)
32. Yu, E., Sangiorgi, D.: Service design as an approach to implement the value cocreation perspective in new service development. J. Serv. Res. **21**(1), 40–58 (2018)
33. Piciocchi, P., Bassano, C., Pietronudo, M.C., Spohrer, J.C.: Digital workers in service systems: challenges and opportunities. In: Maglio, P.P., Kieliszewski, C.A., Spohrer, J.C., Lyons, K., Patrício, L., Sawatani, Y. (eds.) Handbook of Service Science, Volume II. SSRISE, pp. 409–432. Springer, Cham (2019). https://doi.org/10.1007/978-3-319-98512-1_18
34. Siltaloppi, J., Koskela-Huotari, K., Vargo, S.L.: Institutional complexity as a driver for innovation in service ecosystems. Serv. Sci. **8**(3), 333–343 (2016)
35. Löbler, H., Lusch, R.F.: Signs and practices as resources in IT-related service innovation. Serv. Sci. **6**(3), 190–205 (2014)
36. Ramaswamy, V., Gouillart, F.: Building the co-creative enterprise. Harvard Bus. Rev. **88**(10), 100–109 (2010)
37. Breidbach, C.F., Brodie, R.J.: Engagement platforms in the sharing economy: conceptual foundations and research directions. J. Serv. Theory Pract. **27**(4), 761–777 (2017)
38. Maglio, P.P., Srinivasan, S., Kreulen, J.T., Spohrer, J.: Service systems, service scientists, SSME, and innovation. Commun. ACM **49**(7), 81–85 (2006)
39. Polese, F., Sarno, D., Vargo, S.L.: The role of emergence in service systems. In: Proceedings of the 53rd Hawaii International Conference on System Sciences (2020)
40. Visvizi, A., Lytras, M.D., Damiani, E., Mathkour, H.: Policy making for smart cities: innovation and social inclusive economic growth for sustainability. J. Sci. Technol. Policy Manag. **9**(2), 126–133 (2018)
41. Heinonen, K., Strandvik, T., Voima, P.: Customer dominant value formation in service. Eur. Bus. Rev. **25**(2), 124–133 (2013)
42. Peters, L.D., Nenonen, S., Polese, F., Frow, P., Payne, A.: Viability mechanisms in market systems: prerequisites for market shaping. J. Bus. Ind. Mark. **35**(9), 1403–1412 (2020)
43. Lytras, M.D., Visvizi, A.: Who uses smart city services and what to make of it: toward interdisciplinary smart cities research. Sustainability **10**(6), 1998 (2018)
44. Charmaz, K.: Qualitative interviewing and grounded theory analysis. In: Gubrium, J.F., Holstein, J.A. (eds.) Handbook of Interview Research: Context and Method, pp. 675–694. Sage, Thousand Oaks (2002)
45. Gummesson, E.: Case Theory in Business and Management: Reinventing Case Study Research. Sage, London (2017)

The Role of Technological Platforms in Co-creating Symbiotic Relationships Between Firms and Society

Pierpaolo Testa[1(✉)], Luigi Cantone[1], Giuseppe Fabio Cantone[1], and Jay Kandampully[2]

[1] Department of Economics, Management, Institutions, Federico II University of Naples, Naples, Italy
p.testa@unina.it

[2] College of Education and Human Ecology, Ohio State University, Columbus, OH, USA

Abstract. The role of society in a firm's competitive advantage has often been underestimated. The relationship between firms and society has traditionally been considered as responding more to a firm's moral duty or social responsibility. Some of the literature referenced in this article demonstrates that, in practice, society has often been seen, albeit implicitly, as a source of competitive and comparative advantage for successful firms and/or newly established ones. Technological platforms therefore become an elective tool for engaging with society, both for the economic actors who manage them and for platform-dependent entrepreneurs who use them to increase their ability to stay in tune with the public.

Keywords: Symbiotic relationships between firms and society · Technological platforms · Business platforms

1 Introduction

The theme of the relationship between business and society is a long-debated one in the spheres of Corporate Social Responsibility (CSR), business ethics, corporate citizenship, stakeholder management, and sustainability [1]. However, it lacks a unique and integrated framework [2, 3]. In terms of the aims of this proposed paper, some encouraging evidence has emerged from a pre-study analysis investigating how a selected sample of Italian firms (Olivetti, Barilla, Egea, Unicredit, Kiton, Pasta Garofalo, and Marinella) and a group of SMEs and emerging entrepreneurs based in the Campania Region perceive their relationship with society, its impact on their business model, and competitive advantage. Other evidence has emerged directly from the literature and from well-known case histories.

In what follows, some important research streams or topics will be cited to illustrate the role that society plays in the competitive advantage of both well-established firms and newly emerging ones. The proposal then goes on to explain how technological platforms, especially business ones, can help foster the ability of individual firms or the business

C. Leitner et al. (Eds.): AHFE 2021, LNNS 266, pp. 89–96, 2021.
https://doi.org/10.1007/978-3-030-80840-2_9

platform itself to engage with society, contributing to the creation of new symbiotic value. There follows a description of the methodological aspects and the abductive reasoning applied to the research aim. Lastly, some emerging key points on the role of society in creating competitive advantage for firms will be discussed before concluding with a reflection on the limitations of the research and some managerial implications.

2 Theoretical Background: Society as an Implicit Variable Impacting on the Economic Advantage of the Firm

Large Italian firms often start life valorizing and industrializing the artisan skills found in the society and territory of reference [4, 5]. For instance, the Emilia Romagna region in northern Italy has a strong tradition in hand-made pasta and typical pasta recipes. From its inception, Barilla has industrialized the process and become the leader on the home and international markets. Gragnano—a small hill town located between a mountain ridge and the Amalfi coast, within the Naples Metropolitan Area—has a strong tradition as an ideal district for the production of high-quality pasta. Here, the Pasta Garofalo brand has recently been industrialized and has acquired 7% of the Italian market. At the same time, the tailoring skills that have spread throughout the Campania and Tuscany regions thanks to their centuries-old tradition have spawned some important players (Kiton, Marinella; Ferragamo, Gucci) in the worldwide luxury market.

If we look at some renowned wine-producing firms in Bourgogne in France, or Tuscany, Friuli Venezia Giulia, and Campania in Italy, it is clear that they have not only interpreted the agricultural vocation of their respective territories, but they have also coevolved [6, 7] with the societies and places in which they operate. In the literature, the term "terroir" has been coined to denote this vocation, which takes on a difficult to define meaning that encompasses a combination of several factors, such as "climate, soil, cultivar and human practices interacting each other" [8]. This social concertation [9, 10] and cognitive logic [11] is perhaps well represented in the Italian sectoral literature by industries like mechanics, shoes, textiles, and agri-food, which are expressions of long-standing traditions and relationships among several local entrepreneurs. Other societal factors, such as institutional coordination, consuming cultures, and events and synergies with other local actors lead the rationale of geographical cluster development [12, 13]. Similarly, the international scenario indicates that start-ups are often founded upon the premise of facing and solving society's problems [14, 15]. The factual thinking of entrepreneurs generally activates reasoning around the key problems in society at a given time (unemployment, caring, garbage collection and disposal, connecting people, food delivery) with the aim of solving them by proposing key value propositions and emerging business models, sometimes benefitting from digital technologies.

Successful and well-established companies (like Olivetti in Italy) frequently try to solve the problem of disadvantages for groups in society and mitigate integration problems [5, 16, 17]. Thinking of Olivetti, the role of the factory was not subservient to profit but nurtured a community of individuals who, finding purpose in their lives, produced different social interests that blended through the culture and progress of the people themselves.

In accordance with resource-based theories (resource-based theory, competence-based theory, knowledge-based theory, dynamic capabilities), the advent of the new economy has prompted companies to search for elements of comparative advantage among the macro-environmental PESTEL factors [18, 19]. Opportunities for, and threats to, newly established firms have often arisen from variables evolving at the macro-environmental level; an example is the digital born companies (Google, Facebook, Amazon, Yahoo) that have benefitted from the evolution of information and communication technologies at planetary level.

Some digital oriented companies (Coca Cola, Apple, Dove) seek clear identity value positioning opportunities [20]. In fact, ICT has increased the opportunities for traditional firms or digital pure players to leverage communication initiatives to distinguish themselves in the digital environment from the corporate values standpoint. Studies on brand iconicity [21] have shown the high relevance and profitable longevity of brands that have managed to intercept and solve social tensions (Apple, Starbucks, Barilla, Mulino Bianco, Alfa Romeo, to name a few). Brands can become icons when they interpret a given mythology and adopt a specific ideology and cultural codes that allow them to solve a specific tension in society at a given moment, producing what has been described as "social disruption".

Academic research on social media is highlighting the impact of the social dimension of competitive advantage, especially in the digital context [22, 23]. Social media are able to leverage social networks in order to create a competitive advantage for them and for the other partner firms operating on the same media. Some firms, furthermore, see their activity predominantly as developing services and solutions for the territory or for the whole of the society [14, 24] in which they operate (such as multi-utility companies, sport teams, cinemas, multimedia, certification bodies, and third-party organizations). If we consider the advantage for entire nations and places nurtured by some strategic service providers (multi-utilities, railways, airports), we can appreciate that sometimes the competitive advantage of some strategic service providers reflects a reciprocal social advantage for the territory in which they are located. In developing countries, base-of-pyramid business models [25, 26] are widespread; these are always low-cost solutions developed by firms operating in those countries in order to permit the citizens living there to have access to more accessible prices for the same benefits as those in developed countries (low-cost cars and mobile-phones, budget hotels, etc.).

This evidence shows that society is seeing growing relevance as a source of creative inspiration for the fundamental levers of the business model and competitive advantage. This calls for an evolution in firms' resource-based theory [27, 28] and dynamic capabilities [29] for competitive advantage. Our suggestion is that we have been looking at the problem from the wrong angle in the past. Studies on corporate social responsibility tend to fall short of this aim. What is stated here differs from the usual claims made in CSR literature. Porter and Kramer (2006) realized that CSR initiatives are often unproductive because "they put business against society when the two are interdependent" (p.78) and "to say broadly that business and society need each other might seem like a cliché, but it is also the basic truth that will pull companies out of the muddle that their current corporate-responsibility thinking has created" (p. 82) [30]. They recognize the existence of a social dimension of competitive advantage in terms of "strategic philanthropy that leverages capabilities to improve salient areas of competitive context" (p. 89). Here, on the other hand, the contention is that society can be a driver of the competitive advantage of a successful company or an inspiration for the entrepreneurial conduct of an infant company, not only for philanthropic reasons but also for economic ones.

3 The Role of Technological Platforms: A Catalyst for Business and Social Relations

Recently, the management literature is increasingly reflecting awareness of a shift toward a platform economy [31], which is also contributing to a transformation of capitalism [32], and society alike. In fact, it is hypothesized that in the future platforms will have five areas of potential application: 1) the innovation platform, using the creative dialogue platforms of large companies in the traditional economy to manage strategic outsourcing relationships for new product development [33], e.g., Enovia for Boeing 787 Dreamliner development; 2) the social platform to manage relationships and content in digital contexts, e.g., Google, Facebook; and 3) the transaction platform, a pure business model (e.g., Amazon, Uber), especially in the US economy [32, 33]. The platform business model is characterized by firms that massively leverage strategic outsourcing and the network of actors in the business ecosystem (Amazon, Uber, Facebook, Google) and, thanks to the use of algorithms and datamining, it is able to reduce prices for customers or increase the level of service; 4) the E-business platform: to connect (one-to-one or many-to-many) the business actors and consumers operating within the same supply chain [34] or business network (E-Supply Chain, E-Demand Chain, E-corporation, E-marketplace) like Pirelli in Italy for instance; 5) the Governance platform has begun to emerge recently for the sustainable development and governance of complex relationships between citizens and the authorities in a given locality [35], including institutions, resident actors (businesses, consumers, resident citizens), and non-resident actors (tourists, investors, multinational companies, etc.).

Gawer and Cusumano (2014) distinguish between "internal (company or product) platforms as a set of assets organized in a common structure from which a company can efficiently develop and produce a stream of derivative products" and "external (industry) platforms as products, services, or technologies that are similar in some ways to the former but provide the foundation upon which outside firms (organized as a "business ecosystem") can develop their own complementary products, technologies, or services" [36] (p. 418). The same authors clarify that "Internal platforms allow their owners to achieve economic gains by reusing or redeploying assets across families of products developed by either the firm or its close suppliers. By contrast, industry platforms allow firms to manage a division of innovative labor that originates beyond the confines of the firm or its supply chain" [36] (p. 428).

These platforms will only make companies' search for advantage deriving from social variables more virtuous in terms of the economic and non-economic actors of the business ecosystem. "[I]t could be argued that firms aspiring to grow and to become significant actors in the market must sooner or later attain a business platform" [37] (p. 3). Thus, ".... if a business platform is not attained, the firm will sooner or later go under and will disappear from the market—at least as independent actor" [37] (p. 6).

Van Dijck et al., (2018) identify some risks arising from the spread of business platforms in society. Public values may be under threat from powerful digital platforms traditionally hostile towards government regulation and whose primary goal is sales. The leverage mechanisms of business platforms—datafication, commodification, and selection—permit these business actors to seize a near-monopolistic position from where they can lucratively exploit the markets, collecting big data without sharing their power, capabilities, and profits with the rest of society [38].

According to Monahan, et al., (2010; p. 1) "Digital platforms become dominant social structures in their own right, subordinating other institutions, conjuring or sedimenting social divisions and inequalities, and setting the terms upon which individuals, organizations, and governments interact" [39]. They are a new form of capital accumulation able to perform pervasive surveillance over other social actors, but they also have the power to disclose multiple possibilities, some of which are more hopeful than others [39].

Another relevant aspect is the impact of the platform on the network of innovation to produce complements that make a platform more valuable [36]. Cutolo and Kenney (2020) affirm [40] that platform-dependent entrepreneurs -constellations of producers, sellers, and specialized service providers that have emerged to earn through these platforms -"experience a great power imbalance in relation to the platform owners, who can unilaterally enforce changes in the competitive conditions on the platform [41, 42] (p. 4)" and run the risk of becoming increasingly dependent upon precisely these technological actors. Thus they demonstrate that Schumpeter's theory and managerial practice of new entrepreneurship are not valid in this context, so a new theoretical reconceptualization is urgently called for. These platforms, only if properly leveraged and subject to clear worldwide agreement on effective rules, can increase the symbiotic value created among the social and business actors at system level, thus increasing opportunities to achieve and sustain a competitive and comparative advantage for firms and, at the same, increased social value for the community.

4 Research Design and Methodology

This article is a part of an ongoing research project aiming to develop the topic as part of an ideal literature-practice conversation, following abductive research reasoning [43]. Evidence from the pre-study has highlighted the important role of society, place, local tradition, and aptitude in nurturing the competitive advantage of the informant firms and/or in underpinning new business adventures. Naturally, the fundamental role of ICTs and emerging digital platforms in determining great opportunities and threats also emerged during the interviews. Thus, we looked back in the literature, examining several key research topics from the past that implicitly or explicitly acknowledged the role of society in creating value for businesses and entrepreneurs. This allowed us to explain and rework the findings regarding the role of society in business from a new perspective, also in light of new technological developments and the spread of digital platforms in society. Lastly, leveraging the counter-intuitive evidence, we propose a reconceptualization of the theme of the relationships between firms and society, aided by new technological platforms.

5 Findings: Businesses and Society Co-evolving and Creating Symbiotic Value

It can therefore be stated that, by interpreting and blending the latent variables behind the determinants of competitive advantage of a social nature, society can affect the competitive advantage of firms' through these social mechanisms: the territory, the aptitudes of its inhabitants, its network of skills, its ideology and collective culture, problems of a collective nature or those rooted in specific communities; the ability of firms to ride the evolutionary social trends and their key dimensions; the ability of certain firms to globally aggregate business actors and consumers. Therefore, two transversal themes emerge from our empirical-theoretical conversation: a) the concept of symbiotic value: a firm and a societal context can co-evolve together through digital mediated relationships. "When working in collaboration with different stakeholders and bringing in all the relevant knowledge and skills, the firm and its stakeholders can achieve collective impact in society and create what we call "symbiotic value" that would not be possible alone" [44] (p. 1046). This symbiotic value seems to feed both the entrepreneurial attitude of the firm as well as the competitive advantage. b) the role of technological platforms has allowed greater societal permeability and, at the same time, has increased the ability of firms to deal with and tune in to society and several of its key actors. Malthouse, et al. (2019) argues [45] "that in the evolving network economy, data and value diffusion to all stakeholders in the network (platform) are critical for the long-term growth and competitive advantage of service firms. Specifically, we suggest that, for data to become a source of competitive advantage, there must be a symbiotic relationship among all the stakeholders of the data ecosystem" (p. 508).

6 Research Limitations and Future Research Trajectories

The main investigation in this research is related to multi-industry contexts and follows a twofold practitioner-theoretical approach. The methodology adopted is qualitative and interpretive. In the future, a multi case-study approach will be carried out at the business platform level to explore their social impact in qualitative and quantitative terms, focusing on the symbiotic co-created value for the various actors involved (business and non-business ones).

References

1. Schwartz, M.S., Carroll, A.B.: Integrating and unifying competing and complementary frameworks: the search for a common core in the business and society field. Bus. Soc. **47**, 148–186 (2008)
2. Jones, T.M.: An integrating framework for research in business and society: a step toward the elusive paradigm? Acad. Manag. Rev. **8**, 559–564 (1983)
3. Kuhn, T.S.: The Structure of Scientific Revolutions. University of Chicago Press, Chicago (1970)
4. Fabricant, M.: The industrialization of social work practice. Soc. Work. **30**, 389–395 (1985)
5. Dyer Jr., W.G., Whetten, D.A.: Family firms and social responsibility: preliminary evidence from the S&P 500. Entrep. Theory Pract. **30**, 785–802 (2006)

6. Varaldo, R., Ferrucci, L.: The evolutionary nature of the firm within industrial districts. Eur. Plan. Stud. **4**, 27–34 (1996)
7. Giuliani, E., Bell, M.: The micro-determinants of meso-level learning and innovation: evidence from a chilean wine cluster. Res. Policy. **34**, 47–68 (2005)
8. Van Leeuwen, C., Seguin, G.: The concept of terroir in viticulture. J. Wine Res. **17**, 1–10 (2006)
9. Ottati, G.D.: Social concertation and local development: the case of industrial districts. Eur. Plan. Stud. **10**, 449–466 (2002)
10. Becattini, G.: Italian industrial districts: problems and perspectives. Int. Stud. Manag. Organ. **21**, 83–90 (1991)
11. Rullani, E.: The Industrial District (ID) as a Cognitive System. In the Technological Evolution of Industrial Districts. Springer, Boston (2003)
12. Porter, M.E., Ketels, C.: Clusters and Industrial Districts: Common Roots, Different Perspectives. A Handbook of Industrial Districts, pp. 172–183 (2009)
13. Rocha, H.O., Sternberg, R.: Entrepreneurship: the role of clusters theoretical perspectives and empirical evidence from Germany. Small Bus. Econ. **24**, 267–292 (2005)
14. Eggers, W.D., Macmillan, P.: The Solution Revolution: How Business, Government, and Social Enterprises are Teaming up to Solve Society's Toughest Problems. Harvard Business Review Press, Boston (2013)
15. Dees, J.G.: A tale of two cultures: charity, problem solving, and the future of social entrepreneurship. J. Bus. Ethics. **111**, 321–334 (2012)
16. Zahra, S.A., Rawhouser, H.N., Bhawe, N., Neubaum, D.O., Hayton, J.C.: Globalization of social entrepreneurship opportunities. Strateg. Entrepr. J. **2**, 117–131 (2008)
17. Zahra, S.A., Gedajlovic, E., Neubaum, D.O., Shulman, J.M.: A typology of social entrepreneurs: motives, search processes and ethical challenges. J. Bus. Ventur. **24**, 519–532 (2009)
18. Ginter, P.M., Duncan, W.J.: Macroenvironmental analysis for strategic management. Long Range Plann. **23**, 91–100 (1990)
19. Yüksel, I.: Developing a multi-criteria decision-making model for PESTEL analysis. Int. J. Bus. Manag. **7**, 52–66 (2012)
20. Husted, B.W., Allen, D.B., Kock, N.: Value creation through social strategy. Bus. Soc. **54**, 147–186 (2015)
21. Holt, D.B.: Jack Daniel's America: iconic brands as ideological parasites and proselytizers. J. Consum. Cult. **6**, 355–377 (2006)
22. Grier, S.A., Deshpandé, R.: Social dimensions of consumer distinctiveness: the influence of social status on group identity and advertising persuasion. J. Mark. Res. **38**, 216–224 (2001)
23. Dessart, L., Veloutsou, C., Morgan-Thomas, A.: Consumer engagement in online brand communities: a social media perspective. J. Prod. Brand. Manag. **24**, 28–42 (2015)
24. Kotler, P.T., Lee, N.R.: Up and Out of Poverty: The Social Marketing Solution. Pearson Prentice Hall, Hoboken (2009)
25. Gebauer, H., Haldimann, M., Saul, C.J.: Business model innovations for overcoming barriers in the base-of-the-pyramid market. Ind. Innov. **24**, 543–568 (2017)
26. Goyal, S., Sergi, B.S., Kapoor, A.: Understanding the key characteristics of an embedded business model for the base of the pyramid markets. Econ. Sociol. **7**, 26–40 (2014)
27. Grant, R.M.: The resource-based theory of competitive advantage: implications for strategy formulation. Calif. Manage Rev. **33**, 114–135 (1991)
28. Penrose, E.T.: The Theory of the Growth of the Firm (with a new Foreword by the Author) , 3rd edn. Oxford University Press, Oxford (1995)
29. Teece, D.J., Pisano, G., Shuen, A.: Dynamic capabilities and strategic management. Strateg. Manag. J. **18**, 509–533 (1997)

30. Porter, M.E., Kramer, M.R.: Strategy & society. The link between competitive advantage and corporate social responsibility. Harv. Bus. Rev. **84**, 78–92 (2006)
31. Kenney, M., Zysman, J.: The rise of the platform economy. Issues Sci Technol. **32**, 61–69 (2016)
32. Rahman, K.S., Thelen, K.: The rise of the platform business model and the transformation of twenty-first-century capitalism. Polit Soc. **47**, 177–204 (2019)
33. Teece, D.J.: Dynamic capabilities and (digital) platform lifecycles. In: Furman, J., Gawer, A., Silverman, B., Stern, S. (eds.) Entrepreneurship, Innovation, and Platforms. Advances in Strategic Management, vol. 37. Emerald Publishing Ltd. (2017). https://doi.org/10.1108/S0742-332220170000037008
34. Frohlich, M.T., Westbrook, R.: Arcs of integration: an international study of supply chain strategies. J. Oper. Manag. **19**, 185–200 (2001)
35. Janowski, T., Estevez, E., Baguma, R.: Platform governance for sustainable development: reshaping citizen-administration relationships in the digital age. Gov. Inf. Q. **35**, S1–S16 (2018)
36. Gawer, A., Cusumano, M.A.: Industry platforms and ecosystem innovation. J. Prod. Innov. Manage. **31**, 417–433 (2014)
37. Davidsson, P., Klofsten, M.: The business platform: developing an instrument to gauge and to assist the development of young firms. J. Small Bus. Manag. **41**, 1–26 (2003)
38. Van Dijck, J., Poell, T., De Waal, M.: The Platform Society: Public Values in a Connective World. Oxford University Press, Oxford (2018)
39. Monahan, T., Phillips, D.J., Wood, D.M.: Surveillance and empowerment. . Surveill. Soc. **8**, 106–112 (2010)
40. Cutolo, D., Kenney, M.: Platform-Dependent Entrepreneurs: Power Asymmetries, Risks, and Strategies in the Platform Economy. Acad. Manag. Perspect. (2020). https://doi.org/10.2139/ssrn.3372560
41. Kapoor, R., Agarwal, S.: Sustaining superior performance in business ecosystems: evidence from application software developers in the IOS and android smartphone ecosystems. Organ. Sci. **28**, 531–551 (2017)
42. Wen, W., Zhu, F.: Threat of platform-owner entry and complementor responses: evidence from the mobile app market. Strateg. Manag. J. **40**, 1336–1367 (2019)
43. Kovács, G., Spens, K.: Abductive reasoning in logistics research. Int. J. Phys. Distrib. Logist. Manag. **35**, 132–144 (2005)
44. Aksoy, L., Alkire, L., Kandampully, J., Kemppainen, L., Kong, L., McClelland, L.E.: The role of service firms in societal health: the case for symbiotic value. J. Serv. Manag. **31**, 1041–1058 (2020)
45. Malthouse, E.C., Buoye, A., Line, N., El-Manstrly, D., Dogru, T., Kandampully, J.: Beyond reciprocal: the role of platforms in diffusing data value across multiple stakeholders. J. Serv. Manag. **30**, 507–518 (2019)

Digital Transformation in the Era of Covid-19

Marco Tregua(✉), Cristina Mele, Tiziana Russo-Spena, Maria Luisa Marzullo, and Adriana Carotenuto

Department of Economics, Management, Institutions, University of Naples Federico II, 80126 Naples, Italy
{marco.tregua,cristina.mele,tiziana.russospena, marialuisa.marzullo,adriana.carotenuto}@unina.it

Abstract. Digital transformation attracted scholars' attention in various scientific domains, as business models, strategies, human resources management, and, more in general, social sciences. Additionally, digital transformation drove changes in both businesses and society as a whole. Due to its relevance in recent studies and daily activities, we observed the innovation brought by digital transformation at the time of pandemic to understand the contribution it offered. Some of the solutions proposed worldwide were observed, leading to understand the role of artificial intelligence, machine learning, and blockchain – among others – in the fight against Covid-19. The analysis offers a perspective on the ways smart technologies enabled suitable outcomes in the activities performed by specialists, firms, and people either to counteract the further spreading of the pandemic or to take care of the patients affected by the virus.

Keywords: Digital transformation · Practice-based studies · Texture of practice · Covid-19 · Innovation

1 Introduction

Digital transformation started to be a main concern for businesses in last years and a key element of their actions; indeed, about 80% of managers in a McKinsey Global Survey stated that their organizations started paying attention to and acting towards digital transformation since 2014 [1]. Anyway, the effects of digital transformation are yet to come in these businesses and in most cases, performance have not yet been affected and the essence of transformation is still very limited. The complexity of the transformation, the lack of proper backgrounds – including people leading such a transformation process –, the need to further empower people inside firms and along supply chains, and the necessity to upgrade the new integrated tools are listed by McKinsey as the most problematic issues, leading to a limited operationalization of the advantages that firms can get from digital transformation.

Additionally, there is a lack of clarifications on what digital transformation is; Tabrizi et al. [2] investigated the perception of executives about the implementation of digital transformation in their businesses. It was considered as a concern more than an opportunity to get results; this consideration was enforced by the about 900 billion $ thrown

C. Leitner et al. (Eds.): AHFE 2021, LNNS 266, pp. 97–105, 2021.
https://doi.org/10.1007/978-3-030-80840-2_10

away out of the 1,3 trillion $ spent on digital transformation in 2018. The lack of a mind-set, flaws in businesses processes, and the vagueness of business strategy were figured out by the executives as the crucial reasons for such a big failure.

Conversely, digital transformation is meant to be an avenue for relevant changes, thus it attracted scholars' attention in various scientific domains in business studies, such as business models [3], strategies [4], human resources management [5], and the effects on society as a whole [6]. Furthermore, it is interesting to observe the effects on innovation debate too, as Hinings et al. [7] notice digital innovations lead to new agents, actors' constellations, practices, values, and beliefs. By adopting a wider and institutional perspective [8], the changes brought by digital transformation may be observed in societies. We chose to analyze the solutions enabled by smart technologies during the pandemic, aiming at observing their contribution to people, businesses, and society as a whole [9]. As a consequence, next section reviews literature on digital transformation, including a focus on digital transformation during pandemic era, then methodology is introduced. Findings are based on about 60 cases of solutions presented during 2020 to counteract the critical issues depending on the pandemic. Discussion and implications for both scholars and practitioners conclude this contribution, together with limitations and suggestions for further research.

2 Theoretical Background

2.1 Digital Transformation and Technologies: A Brief Overview

Nowadays digital transformation permeates several businesses and industries, and it is meant as a process affecting firms at all levels and not just a mere technological transition [10]. Indeed, scholars describe it as a disruptive process of change that affects firms in strategizing, in redefining their business models, in upgrading the customer experience, and in operations management. To sum up this vision, Hess et al. [4] consider digital transformation as a priority challenge for managers, to integrate and exploit the potential of smart technologies in business. Similarly, Reddy and Reinartz [11] highlight the 'sea of change' ahead depending on the revolutionary impact digital transformation may have as it can lower costs, change information-based processes, lead to new products and services, and bring flexibility to firms, also at societal level, due to the efficiency it can offer to public administration and in the provision of public services.

Due to the pervasiveness of digital transformation in business, scholars pay attention to the creation of a mindset and of the proper conditions to achieve this transformation fruitfully. For instance, Guinan et al. [12] highlight the role of digital project team in creating and developing the right conditions to make this transformation happen. Indeed, they stressed the need to set a team that makes a firm able to "adapt, respond, and position itself" (p. 717) in a rapid process of change. When dealing with the skills and features a team should have, the authors stress the abilities in defining new value propositions, testing new solutions, experimenting effects on other stakeholders, and providing feedback about the change brought by smart technologies.

Smart (i.e., digital and cognitive) technologies are at the core of digital transformation and they greatly affect business strategy; this impact has been observed in several domains and scholars (e.g., [13]) pay increasingly attention to this phenomenon and to the opportunity these technologies may bring to business and to their development. Mele, Russo Spena, Kaartemo, and Marzullo [14] conceptualize smart nudging as uses of cognitive technologies to affect people's behaviour predictably, without limiting their options or altering their economic incentives. Several choice architectures and nudges affect value co-creation, by (1) widening resource accessibility, (2) extending engagement, or (3) augmenting human actors' agency.

Smart technologies enable a process to achieve a continuous optimization of a firm activities, meant as a reaction to market changes and to innovate what a firm does [15]. In any case, it is important to stress that these technologies may lead to a transformation that cannot happen just by chance, but it should be part of a digital strategy a company implemented. Indeed, lots of scholars stated that smart technologies are more than just tools in the process of implementation of new technologies in a firm (e.g., [2, 16]), but digital transformation as an approach and the related technologies as tools are part of a process run in order to transform the way of thinking embedded in a company and how operations are affected by the implementation of new technologies.

2.2 Digital Transformation and Covid-19: First Studies

As a process affecting both business and society, scholars consider digital transformation in counteracting pandemic and the deriving limitations. Indeed, scholars observe the acceleration in the implementation of digital transformation in relation to contextual conditions. These studies consider different topics, as small and medium enterprises (SMEs) and their business model [17], service provision [18], and business in general [19] among others.

The transformation-to-be is highlighted in the context of SMEs by Priyono et al. [17] to categorize firms' readiness in the achievement of such a transformation; the authors show the main features of digitally mature firms, the choice of sales transformation in firms with a limited technology readiness or low funding, the impossibility to activate a transformation due to high levels of technology limitations. Bartsch et al. [18] investigate the effects of leadership on service workers while working in a virtual environment. Surprisingly, the authors report no negative effects on work performance nor reduction in team cohesiveness. An overview of digital transformation in business is offered by Soto-Acosta [19], as he clarifies the difference between digital projects and transformational projects and the support novel technologies may offer to counteract the economic collapse and the other negative effects of Covid-19. The author observes the puzzling consequences of some digital projects, partially leading to concrete improvements at all organizational levels, besides Covid-19 accelerated the transformation itself.

Furthermore, Abdel-Basset et al. [20] investigate the disruptive effect of smart technologies in the fight to the pandemic, focusing on safety and monitoring as the most evident results; anyway, they stress the need for further research due to the novelty of the topic and the yet-to-be-expressed potential of these technologies.

3 Research Aim and Process

Due to the ties between digital innovations and society and the challenges society is experiencing at the time of pandemic, we observe the texture of practice [21] emerged from the implementation of digital innovations as a response to Covid-19, in line with previous studies on innovation in healthcare (e.g., [22]); moreover, Abdel-Basset et al. [20] call for further research on the use of smart technologies in the fight to pandemic, therefore, we aim at expanding the assessment of the use of these technologies in the current scenario, with people, health organizations, governments, and societies trying to counteract the pandemic and its effect. Due to this, we investigate about 60 cases of firms recently proposing novel solutions to stop the further spreading of the virus or to offer a way to contrast its negative effects. Multiple sources were used, as scientific journals, media, firms' official website and other online channels, reports, and governments statements, leading to have a wide perspective on the solutions adopted in various areas of the world and referred to several smart technologies, as artificial intelligence (AI), bluetooth, drones, chatbot, robots, and deep learning, among others.

The research approach described as texture of practice is in line with the research aim due to its being related to innovation, the multiple connections it can leverage on, and the social-contextual nature of changes that can describe.

4 Findings

Findings are articulated in four challenges that affected the world during 2020 and that still represent a continuous struggle in 2021 to get out of the pandemic. The cases analyzed are described based on the solution they may offer, namely the support they offer to prevention, diagnosis, treatment, and research during the pandemic. Some examples are presented in next sub-sections to show the main effects already available and yet-to-be-achieved via novel opportunities brought by smart technologies.

4.1 Preventing

Preventing means to take care of the conditions and people that may be affected by contagion; therefore, performing predictive analyses and support the decision-making process are two key tasks for health institutions acting to prevent the further spreading of the pandemic. Smart technologies that can be considered for prevention are AI, Bluetooth, chatbot, computer-aided design (CAD), drones, robots, and wearables. In next lines some examples are proposed to show how concretely technology support people and society, as well as to describe the actions performed.

IBM Watson assistant is a conversation platform operating via AI to process data. A cloud service allows the analysis of data from multiple sources and create a set of answers to common questions about the virus. The service is offered via website and mobile chat, therefore it is provided to companies with no time constraints and aims at reducing the inquiries for information on the pandemic.

StoppCorona is an initiative launched in Austria by the local Red Cross; through an app people may communicate via Bluetooth to digitally handshake one another in an era where contacts are very limited and at the same time information may be collected to track encounters among friends, colleagues, and families and detect the risk of having been infected. If one gets infected, the app allows to send a notification to the people he/she has been in touch with to inform them about the risky condition.

Terradrone is a device featured by a drone and a GPS tracking. The drone flies towards areas where foods and drugs are needed, as well as other supplies. Thus, delivery time is reduced, there are no human interactions in line with the restrictions to personal contacts and the system allows to continuously monitor where the drone is and the surrounding area.

4.2 Diagnosing

The diagnostic evaluation of people conditions has been very challenging in the first months of the spread of the virus since there was limit knowledge on how the virus showed its effects on patients. Indeed, symptoms were various and accurate information missed. Innovative technologies helped in identifying the symptoms, doing that quickly and sometimes independently from health institutions, as well as favoring the isolation of people affected by the virus.

The first example we classified as 'diagnosing' is Memora Health a chatbot based on AI that can run a risk assessment of patients' conditions based on symptoms and exposure profile. People can ask questions through the chatbot, therefore they avoid staying in crowded areas of hospitals, reducing the risk of contagion.

AI also helped in setting more accurate instruments to run medical tests; Qure.ai is an example of such an implementation, as a software for medical imaging was combined with a learning technology for medical analysis and AI assisted the process of analyzing X-rays and tomographies. The interpretation of results was quicker than with traditional instruments and useful in determining the effect virus had on lungs.

Quickness was one of the most scouted features in new solutions and Seegene offers an example in this sense; indeed, an intelligent system was set to interpret genetic material, leading to quick diagnosis of symptoms. The novelties are not limited to the fast interpretation of symptoms, as a molecular test greatly decreased the time needed to have a way to test patients. A new algorithm was set in just three weeks, allowing the tests to be performed in a quick way since the first weeks of the pandemics.

4.3 Treating

The treatment of patients affected by Covid-19 required relevant efforts due to the limited knowledge about the virus and the difficulties in identifying the symptoms as showed in the previous challenge. Additional knowledge was needed to identify new therapies, limit the risk of mistakes in implementing a therapy, and look for solutions in drugs used to contrast similar diseases. Technologies were implemented in traditional medical processes to speed up the healthcare and increase its efficiency.

Micromedex is a software embedding AI to support health specialists in their decision-making about a drug therapy. The software allows to generate proper information about how to use drugs, the risks related to the specific health conditions of a patient, and all the precautions to be considered. Micromedex increased safety in administering a therapy and validate also experimental therapeutic pathways.

The use of novel technologies also affected physical instruments used by health specialists, as it was with the Charlotte Valve, designed by Isinnova. Through CAD and

3D-printing technology a full-face snorkeling mask was transformed into an emergency respiratory mask with valves assisting patients in ventilation. This new instrument was used in sub-intensive therapy and the quick process enabled by 3D printing technology dramatically reduced both times and costs for construction and made the use of these masks easy, due to the similarity with other masks.

Taking care of people did not take place just in hospital but also at home, therefore instruments to assist people during their at-home therapy were needed. XR Health is a device embedded with virtual reality and creating an engaging experience through which patients can participate to the treatment process via lenses and joypads to follow the instructions of specialists. Additionally, specialists may monitor people in quarantine and avoid stay in contact with them, as well as greatly reducing the time to be spent to support dozens of patients.

4.4 Researching

The use of novel technologies also supported the research process to get knowledge about the virus and about the vaccine. Indeed, new technologies were infused in the research process to identify therapies, drugs, and vaccines. This task is particularly challenging, due to the lack of information depending on a totally new typology of virus, with effects on people hard to be detected since the beginning and consequences not easy to be identified.

Blockchain was implemented – together with AI – by Innoplexus to set a data analysis system with certified data. This process supported the study of pharmaceutical solutions with both structured and unstructured data and ensured a knowledge basis to be constantly updated while conducting research on new solutions.

Cloud computing was used by Benevolent-AI as a tool to favor the creation of data archive on medical information. Cloud computing was combined with AI and machine learning to analyze scientific literature as well as new data and look for potential solutions, with a focus on how to block the cellular development of the virus.

Finally, also deep learning was implemented in research process for new medical solutions, as Insilico Medicine set a model to evaluate drugs and the testing of these drugs was made quicker by deep learning, through simulations that made this process quicker, more effective, and cheaper compared to extant solutions.

5 Discussion, Implications, and Further Research

The evidence above describes the actions shaping each challenge and offer an in-depth observation on how digital transformation may affect the fight to pandemic.

First of all, innovative technologies may favor prevention via two concrete actions, namely monitor and protect. As it regards monitor, smart technologies can observe and verify how the phenomenon evolves through data analysis and also check the vital parameters of people to stop the ones that can further spread the virus. Additionally, protect is the action that can limit the effect of the viral agents and reduce the risk of contagion.

Secondly, diagnosing helps people in self-evaluating their health conditions and generally accelerated the process of data analysis to evaluate symptoms. Therefore, the two actions describing diagnosing are self-evaluation and acceleration of the process of analysis of symptoms. Self-evaluation favors the assessment for citizens and led to collect data on a wide range, making the intervention of local health actors less risky than with personal visits and more effective due to time reduction. The acceleration of symptoms evaluation facilitates people working with the disease to analyze the advance of the virus in people and identify changes in vital parameters, also enabling the identifying of therapeutic steps, as it will be further explained in next challenge.

As it regards treating, it occurs as a combination of two actions, namely improve and handle. With improve we refer to the enrichment of medical knowledge about a virus with several obscure features since the first weeks of contagion. Similarly, knowledge about how to deliver care services and how to identify best practices was acquired thanks to new technologies. The second action – namely handle – support both health specialists and patients in carrying on the therapies while avoiding the risks of direct contact with other people and favored the availability of medical instruments as protective equipment, valves, and safer contexts for treatments, without endangering the effectiveness of a therapy.

Finally, researching took place through study and develop. Study is the action run to analyze the features of the virus and the disease it generates, as well as to find useful information needed to look for solutions. Develop is the action useful in defining the features of molecules, compounds, drugs, and other solutions to be implemented in the fight against the virus; additionally, it describes the activities run to create new drugs and quickly test them to make them available worldwide.

In line with our research approach, the four challenges and the eight actions describe how digital transformation impact on healthcare services in the era of Covid-19; indeed, the changes in role of actors, contexts of service provision, and the activities carried on, new practices emerge [21, 22] and they can be described as a texture of practice due to three reasons, as (a) they are related to innovation and interrelated between them with mutual effects; (b) generate multiple connections among people, including patients, healthcare specialists, healthcare institutions, research centers, and government, with all of them being infusing novel ways of acting in their traditional process as a way to contrast the ongoing scenario; and (c) the observed changes are social-contextual in nature as they occur in society and this represents also a change to the usual context in which healthcare services are provided, since digital transformation changed the service context in several cases and brought in new actors previously slightly related – or even unrelated – to the healthcare services. Additionally, the results achieved in this research complement the recent understanding of Abdel-Basset et al. [20] on the disruptive effect performed by smart technologies against the pandemic. Our evidence aligns with the chance to increase safety – as in preventing – and provide opportunities for monitoring – as in diagnosing – while the findings related to treating and researching expand the range of opportunities brought by digital transformation in healthcare services.

Therefore, the further implementation of digital transformation in healthcare services mirrors the 'sea of change' proposed by Reddy and Reinartz [11] and describes changes affecting society at a whole, since it is not a mere implementation of technologies, but the widening of roles, ways of doing, and activities.

Finally, this research has been performed as a desk analysis of multiple cases via multiple sources, therefore additional knowledge may emerge from interviews with the entities operating in service domains, as well as the observation of further implementation of digital transformation and its tools in empirical settings.

References

1. McKinsey: Unlocking success in digital transformations (2018). https://www.mckinsey.com/business-functions/organization/our-insights/unlocking-success-in-digital-transformations. Accessed 29 Jan 2021
2. Tabrizi, B., Lam, E., Girard, K., Irvin, V.: Digital transformation is not about technology. Harvard Bus. Rev. **13**, 1–6 (2019)
3. Berman, S.J.: Digital transformation: opportunities to create new business models. Strat. Lead. **40**(2), 16–24 (2012)
4. Hess, T., Matt, C., Benlian, A., Wiesböck, F.: Options for formulating a digital transformation strategy. MIS Q. Exec. **15**(2), 123–139 (2016)
5. Eden, R., Jones, A.B., Casey, V., Draheim, M.: Digital transformation requires workforce transformation. MIS Q. Exec. **18**(1), 4 (2019)
6. Rowe, F.: Being critical is good, but better with philosophy! From digital transformation and values to the future of IS research. Eur. J. Inform. Syst. **27**(3), 380–393 (2018)
7. Hinings, B., Gegenhuber, T., Greenwood, R.: Digital innovation and transformation: an institutional perspective. Inf. Organ **28**(1), 52–61 (2018)
8. Orlikowski, W.J., Barley, S.R.: Technology and institutions: what can research on information technology and research on organizations learn from each other? MIS Q. 145–165 (2001)
9. Mele, C., Russo-Spena, T., Kaartemo, V.: The impact of coronavirus on business: developing service research agenda for a post-coronavirus world. J. Serv. Theor. Pract. (2020, forthcoming)
10. Westerman, G., Bonnet, D., McAfee, A.: The nine elements of digital transformation. MIT Sloan Manage Rev. **55**(3), 1–6 (2014)
11. Reddy, S.K., Reinartz, W.: Digital transformation and value creation: Sea change ahead. Market. Intell. Rev. **9**(1), 10–17 (2017)
12. Guinan, P.J., Parise, S., Langowitz, N.: Creating an innovative digital project team: levers to enable digital transformation. Bus. Horizons **62**(6), 717–727 (2019)
13. Mugge, P., Abbu, H., Michaelis, T.L., Kwiatkowski, A., Gudergan, G.: Patterns of digitization: a practical guide to digital transformation. Res. Technol. Manage. **63**(2), 27–35 (2020)
14. Mele, C., Russo-Spena, T., Kaartemo, V., Marzullo, M.L.: Smart nudging: how cognitive technologies enable choice architectures for value co-creation. J. Bus. Res. (2020, forthcoming)
15. Gobble, M.M.: Digitalization, digitization, and innovation. Res. Technol. Manage. **61**(4), 56–59 (2018)
16. Bowersox, D.J., Closs, D.J., Drayer, R.W.: The digital transformation: technology and beyond. Supply Chain Manag. Rev. **9**(1), 22–29 (2005)
17. Priyono, A., Moin, A., Putri, V.N.A.O.: Identifying digital transformation paths in the business model of SMEs during the COVID-19 pandemic. J. Open Innov. Technol. Mark. Compl. **6**(4), 104 (2020)
18. Bartsch, S., Weber, E., Büttgen, M., Huber, A.: Leadership matters in crisis-induced digital transformation: how to lead service employees effectively during the COVID-19 pandemic. J. Serv. Manage. **32**(1), 71–85 (2020)

19. Soto-Acosta, P.: COVID-19 pandemic: Shifting digital transformation to a high-speed gear. Inform. Syst. Manage. **37**(4), 260–266 (2020)
20. Abdel-Basset, M., Chang, V., Nabeeh, N.A.: An intelligent framework using disruptive technologies for COVID-19 analysis. Technol. Forecast. Soc. **163**, 120431 (2021)
21. Gherardi, S.: Situated knowledge and situated action: What do practice-based studies promise. The SAGE handbook of new approaches in management and organization, pp. 516–525 (2008)
22. Russo-Spena, T., Mele, C.: Innovating in practices. In: Toivonen, M. (ed.) Service innovation. TSS, vol. 6, pp. 129–148. Springer, Tokyo (2016). https://doi.org/10.1007/978-4-431-54922-2_6

Management of Smart City in Lens of Viable System Approach

Leonard Walletzký[1(✉)], Luca Carrubbo[2], and Františka Romanovská[1]

[1] Faculty of Informatics, Masaryk University, Botanická 68a, Brno, Czech Republic
{qwalletz,xromanov}@fi.muni.cz

[2] Department of Medicine, Surgery and Dentistry, University of Salerno, Fisciano, SA, Italy
lcarrubbo@unisa.it

Abstract. In Smart Cities, there are various contexts and dynamic changes that create complexity and uncertainty for designing the city services. The lack of contextual analysis can cause this to understand how a city should be prepared to accommodate various contexts. Therefore, in this paper, we take the lens of viable systems approach and analyze how the systems and contexts are interconnected and affecting each other. The understanding of the way of connectivity among systems and context will be presented as a critical aspect of city resilience; however, the paper will reflect the position of resilience in the environment of other goals and services of the city. This paper further revisits the smart city layers and services to highlight the connections among different services. To address the contextual changes, to cover the problems with the management of such a complicated environment, IT Service Management applications will be presented.

Keyword: Smart city · Multicontextual view · Smart services · Service-dominant logic · ITIL · Smart Services Management

1 Introduction

To deepen the new reflections on the concept of S.M.A.R.T. (Acronym proposed by IBM researchers to mean something "specific, measurable, agreed - agreed-, realistic and timely", [1]), many scholars have tried in recent years to study all possible service applications, defined "on stage", referring to the practical evidence of something truly iterative, interactive, interconnected, intelligent [2], and which was truly representative of a more intelligent planet [3], as in the sample of Smart Cities [4]. Indeed, in smart cities, it is interesting to understand which aspects are related to the multitude of actors (workers, citizens, producers, suppliers, authorities, consumers, users, etc.) and facilitators (retail sales, large retailers, networks information, financial services, public health administration), which are fundamental for improving the ability to manage and implement collaboration strategies [5]. In this sense, the service systems (smart, complex, ecothat they are) do not derive from simple intuitions but they result from systematic methods of continuous learning, data collection, rational innovation, social responsibility, and network governance [6].

C. Leitner et al. (Eds.): AHFE 2021, LNNS 266, pp. 106–114, 2021.
https://doi.org/10.1007/978-3-030-80840-2_11

By applying an intelligent service to intelligent practice, inserted in a contest of intelligent cities, constituted by intelligent organizations, through intelligent operations, to obtain intelligent results, there can be some important changes in every aspect of our daily life [7].

Many stakeholders and researchers relate the Smart City development to the non-trivial implementation of information and communication technology (ICT). Sometimes, Smart City itself is confused with ICT – the city leaders believe that the achievement of "smartness" depends on the number of IT devices or solutions they are using. It is based on ignoring or misunderstanding the facts mentioned above. This phenomenon often leads to wasting public funds when the cities are paying for the solution that does not fit into their environment or is not needed by the citizens or stakeholders in general. Indeed, ICT plays a unique, significant role in the Smart City implementation. The critical part is understanding the final clients (customers) of the city's services – in their understanding of the value they can get. This perspective is transforming ICT to the role of a tool necessary for the achievement of the city services value.

In this paper, we first analyze the Smart City as viable multi contextual environment of the services and suggest how this complicated system can be managed. In the next chapter, we present the inspiration in ICT services management methodology, called ITIL. We offer the main ideas of its possible adaptation and analyze the issues in this process. In the discussion, we present the possible further development and practical application of this promising research.

2 Viability and Multi Contextual Environment

The Viable Systems Approach (VSA) meta-model allows us to identify critical factors, influencing contingences, consonant aspects, relevant elements and other systems determinants useful for a better managerial practice in competitive contexts [8]. Thanks to VSA it is then possible to re-draw smart cities in a more organic, holistic and cybernetic manner [9], give that the only lens to approach service systems is trans-disciplinar [10]; particularly, VSA FCs [11] lead scholars to outline insightful suggestions for interpreting surrounding conditions in complex contexts [10]. Here, two points of view have been used, in order to insightfully bridge between theory and practice.

From the theoretical point of view:

Environment/Context dichotomy (FC n.2) in VSA shows us the path in which we live and operate as something subjectively perceived by everyone and localized through the mutable set of relationships activated by the studied actor while it is a system [6]. Changing or updating strategies, partnerships, contracts, ways of doing, etc. any actor shapes its own context consequently [12, 13]. Similarly, being part of a context (as a whole) led any actors to be influenced by several external issues, as they time-to-time recognize in a sort of unique interconnected pattern [14]. The shared space cannot be seen as a "third" with respect to the actors of the exchange and their practices, but inevitably it is "immersed" and inescapable with respect to it [15]; any event, when relevant, can temporarily influence or more/less determine such a behavior of every invested and interested entity, as stated by Contingency Theory [16, 17]; it definitely means that relationships firstly matter, all the rest just a bit less.

Consonance/Resonance dichotomy (FC n.7) in VSA helps us in deeply interpreting dynamics between actors and their context by exploiting the potential compatibility (as the systems' consonance is) of actor's features and its mode in actions with those of other stakeholder working in the same space in which the service is experienced [18]. Under this systems perspective, this allows the emergence of structural pre-conditions for the success of business strategies, launch of new solutions, innovation processes, and partnerships [19]. The effective harmony, if reached, shows the resonance of operations and a sort of sustainable action performed dynamically by actors while they interact [20]. In this sense, the same phenomena we want to focus on can show different contexts through several actors' own perspectives, as they are involved in a defined process, directly or not [21]; this led to a multi-contextual lens to have a complete overview of any investigation.

Furthermore, the evolving conditions affecting the balance of any contextualized ground contribute to complicating the analysis of investments, efforts, and results [22]. This can be useful to better understand what is really going on in practice.

From the practical point of view:

- there is who makes the asset available, who requests it and provides a fee to find it and to have it, who favors the research, profiles the user, classifies/qualifies the actors and their behaviors [23]. In smart cities, new practices take place due to the existence of a common way of dealing with the issues, that is, because the philosophy of "sharing" is combined, with respect to mode and times, making this a self-feeding process [24].
- Contingencies favor new reflections for decision-making processes in organizations, as in performance measurement [25], supply chain management [26], talents' place to do well in their current role [27], capacity planning [28], management of organization's complexity [29] and many others.
- Experiences, know-how, best-practices, and knowledge management help organizations in outperforming successful operations over time, by leveraging on adaptive, reactive, and proactive approaches to properly face with changing smart cities contexts [30]. This is true in the continuity and viability of any action fostering organization in plan and conduct several solving options like adjustments, transformations and reconfigurations (fitting with different depth levels of interventions) needed during the time to survive in the long run [8].
- Resource integration, equifinality, cognitive alignment, resiliency are some of supportive features to be used to accelerate the convergence of intents [31]; the "way" you share soon is more important and it is exchanged than the "what" is exchanged [32]; there are different spaces in which to find the required solutions, but it is not always possible to actively participate; simple requests and offers are similar to "fundamental knowledge" to which today "semantic knowledge" is preferred because the same requests and offers would be interpreted differently when changing one or more conditions [33].

3 Smart City and Its Services

Every city is undoubtedly connected with many services, and it is especially prominent in smart cities. The difference is that smart cities use ICT to realize or support their services and gather valuable information. This information can further improve the services themselves and create new ones to enhance the quality of life of its stakeholders.

Two types of services are present in smart cities – IT services representing a smart city's technical part and smart services. Smart services, unlike the IT ones, can bring value to all the stakeholders a city has. However, it does not mean IT services bear no value for stakeholders at all. While their function is mainly the foundation of smart city (e. g. networks and sensors), they may be proposed as valuable to many specialized stakeholders, primarily to specialized companies or academic purposes. On the other hand, citizens probably will not be able to utilize IT services to their full potential.

It is important to note that while smart city uses ICT and provides services, not every service has to use ICT directly. The usage of ICT might be hidden if the current service uses the result of another one, which uses the ICT to provide its value. Despite this fact, every city service is considered to be a smart service.

Traditionally, smart city services were structured and classified in the same way as departments of a city's government [34]. Best known is the classification into six categories – smart economy, smart people, smart governance, smart mobility, smart environment, and smart living [35].

However, this representation does not take into consideration the dependencies of services [4]. The figure represents the fact that services are interconnected but do not describe their meaning, role, or dependencies. Furthermore, it portrays smart people (citizens) on the same level as the other services, even though smart people are the clients of services and not a service itself. Despite this, smart people are essential for a smart city because they are the end-users of the provided services, and it is their quality of life that the city is trying to improve.

4 IT Service Management and Its Viable Adaptation

The missing links among services and the inaccurate representation led to creating a layered model of smart city services [4]. There, the services are represented in layers according to their functionality in the smart city environment. While some services bring value directly to the citizens, others serve as support or can be valuable to other stakeholders. On the bottom layers, there are IT Services – hardware and software – mentioned above.

This structurization of services allows cities to manage the services more holistically across all departments. Departmentalism is generally a problem in many cities, smart or not. Nevertheless, the harm is crucial to the idea of a smart city. A smart city should be interconnected not only on the level of technologies but also on the city's management and the services level. Without proper communication between the departments, finances will be wasted, and new solutions to problems will be limited. If the department of environment wants to install CO2 sensors on streetlamps, it must inevitably consult the department taking care of the street lighting. However, excluding other departments from this communication can prevent or make more problematic further public lightning usage.

Typically, smart solutions are not isolated solutions. They are affecting each other in many ways and, according to the facts from Chap. 2, we can apply the viable approach.

Viable systems are entities adaptable to their continually evolving environment and therefore capable of surviving. The smart city services' behavior is the same – and we can identify them as viable systems. The viable system is an abstract description of cybernetics, applicable to all autonomous organizations; from this derives the study of the structure of regulations, of reactions to changes, of feedback, or of everything that allows the system to adapt to the new conditions of the surrounding area [36].

Then, each viable system must be able to manage its external relationships, in compliance with shared rules, roles, and responsibilities, in an attempt to satisfy the common interests of the systemic entities involved in the same ecosystems [37].

That means it is necessary to adapt an existing management system to a smart city environment. Such as inspiration we can take the IT Service Management and its globally known and used framework – ITIL [38]. It is a very complex library of the best practices and processes of developing, maintaining, and improving IT services. The main advantage is the holistic view of services – it takes them as the system, which can react to changes, which correspond with viability.

Currently, no resources are trying to adapt the ITIL framework to the smart city environment as it possesses many complications. Furthermore, adopting such an immense framework to an entirely new environment calls for long-term research. However, a basic introduction to the new perspective of ITIL was proposed in [39]. Thanks to the ITIL framework's flexibility, it is deemed suitable for adaptation to the smart city environment, mostly thanks to its ability not to be used as a whole. Thus, it is possible to adapt and implement only part of the framework and add what would be considered by the smart city as the most useful. The adaptation can be presented in the five main strategies:

Strategy	Meaning in ITILv3	Limitations for Smart Cities
Service Strategy	Service Strategy organizes information about services, serves as a vision of service. Allows a quick reaction to a change	Several processes of this stage have a foundation that is not suitable for the smart city environment. An example is a financial management because of the difference a city and a company have regarding their goals. While a company wants to earn money, a city wants to provide services
Service Design	Service Design ensures the creation and development of services that are in alignment with a company's strategy	The main limitations for cities in this stage may be Service Level Agreements, a contract between a provider and a client, because it may be unattainable to have SLA with all of the stakeholders. Similarly to financial management, capacity management in cities should not ensure the maximisation of profit

(*continued*)

(*continued*)

Strategy	Meaning in ITILv3	Limitations for Smart Cities
Service Transition	Service Transition is applied anytime a service changes its stage to Operation or is retired. It manages the resources necessary for release or to successfully discharged a service	A restrain can be located in the process of change evaluation where a smart city should not use primarily financial metrics but rather consider the value of the change for stakeholders. Furthermore, another limitation can be in the process of knowledge management because of the need to educate all the stakeholders using a service. This mass education can be problematic, especially in regard to large number of users (e.g. citizens)
Service Operation	Service Operation contains services in the production environment that bring value to customers. It includes their management and ensures compliance to the set SLAs. Furthermore, this stage includes four functions taking care of the services and customers, for example, a Service Desk	Service Operation is present in every smart city because every city provides services. While systematic management of this stage could bring immense value to the smart city, there are several limitations. These are especially the interconnection of different services and departmentalism. Both of these problems limit the potential of a smart city if not managed properly. Also, the function Service Desk would need to undergo significant change in order to be usable in smart cities
Continual Service Improvement	The ITIL framework defines a process to help improve services and their effectiveness and usefulness. The continual service improvement process occurs in all of the stages and has seven steps	There are no identified limitations regarding Continual Service Improvement. However, it is important to note that cities may not be able to employ the same metrics as a company

Even though the above limitations of ITIL adaptation were identified, using the basic ideas adapted to the environment of the city seems to be valuable.

5 Conclusion

The methodology of Smart Services management is very complicated. The services are not isolated and always behave as a system – where one change can have many unexpected consequences. Therefore, the traditional ways of management cannot work.

According to the studies on Systems, the (objective) environment, in which the (subjective) context of each specific system develops and is consolidated, can affect the daily strategies undertaken and pursued, enhancing or developing the attitudes and the relative survival skills (for example of plants, but also of businesses and organizations in general) in the long run of all systems (more or less rooted, more or less contextualized). To survive, organizations cannot fail to relate to their subjective context of reference and expect the constraints and rules of the objective over-determined environment. For a given system, the context is therefore that set of exchange objects whose characteristics influence and are influenced by the behavior of a system.

The existing IT Service management system (ITIL) cannot be taken "as is" because of many limitations. Smart Services' system is more complicated than IT Services – it contains more layers, relationships, and actors. There is no doubt about the "viability" of such a system of services.

ITIL is also too profit-oriented. In the city environment, we can identify different values (benefits) than profit, and this fact must also be reflected. The whole service ecosystem must be analyzed and developed together with close cooperation among all stakeholders (municipalities, academia, government, business entities etc.).

But using of complex methodology of complex service management (as ITIL) can bring many advantages to the municipalities. The actual request of multidisciplinary solutions and projects with overlaps to other domains, the holistic vies to Smart Services is the key to success. Isolated solutions cannot succeed in this complex environment.

References

1. I. F. M. I.B.M.: Succeeding through Service Innovation: A Service Perspective for Education, Research, Business and Government. University of Cambridge Institute for Manufacturing, Cambridge (2008)
2. Maglio, P.P., Srinivasan, S., Kreulen, J.T., Spohrer, J.: Service systems, service scientists, SSME, and innovation. Commun. ACM **49**(7), 81–85 (2006)
3. Maglio, P.P., Weske, M., Yang, J., Fantinato, M. (eds.): ICSOC 2010. LNCS, vol. 6470. Springer, Heidelberg (2010). https://doi.org/10.1007/978-3-642-17358-5
4. Walletzký, L., Bühnová, B., Carrubbo, L.: Value-driven conceptualization of services in the smart city: a layered approach. In: Social Dynamics in a Systems Perspective, pp. 85 98. Springer, Cham (2018). https://doi.org/10.1007/978-3-319-61967-5_5
5. Walletzký, L., Carrubbo, L., Ge, M.: Modelling service design and complexity for multi-contextual applications in smart cities. In: Proceedings 2019 23rd International Conference on System Theory, Control and Computing (ICSTCC) (2019)
6. Badinelli, R., Barile, S., Ng, I., Polese, F., Saviano, M., Di Nauta, P.: Viable service systems and decision-making in-service management. J. Serv. Manage. **23**, 498–526 (2012)
7. Spohrer, J., Anderson, L., Pass, N., Ager, T.: Service science e service dominant logic. Otago Forum **2**, 4–18 (2008)
8. Barile, S., Polese, F., Carrubbo, L.: Il Cambiamento quale Fattore Strategico per la Sopravvivenza delle Organizzazioni Imprenditoriali. In: Barile, S., Polese, F., Saviano, M. (eds.) Immaginare l'innovazione, pp. 2–32. Giappichelli, Torino (2012)
9. Golinelli, G.M., Barile, S., Saviano, M., Polese, F.: Perspective shifts in marketing: toward a paradigm change? Serv. Sci. **4**, 121–134 (2012)
10. Polese, F., Barile, S., Loia, V., Carrubbo, L.: The demolition of Service Scientists' cultural-boundaries. In: Maglio, P.P., Kieliszewski, C.A., Spohrer, J.C., Lyons, K., Patrício, L., Sawatani, Y. (eds.) Handbook of Service Science, Volume II. SSRISE, pp. 773–784. Springer, Cham (2019). https://doi.org/10.1007/978-3-319-98512-1_34

11. Barile, S., Polese, F.: Smart service systems and viable service systems: applying systems theory to service science. Serv. Sci. **2**, 21–40 (2010)
12. Barile, S., Carrubbo, L., Iandolo, F., Caputo, F.: From 'EGO' to 'ECO' in B2B relationships. J. Bus. Market Manage. **6**(4), 228–253 (2013)
13. Peters, L.D., Nenonen, S., Polese, F., Frow, P., Payne, A.: Viability mechanisms in market systems: prerequisites for market shaping. J. Bus. Ind. Market. 1–10 (2020)
14. Barile, S., Lusch, R., Reynoso, J., Saviano, M., Spohrer, J.: Systems, networks, and ecosystems in service research. J. Serv. Manage. **27**(4), 652–674 (2016)
15. Mele, C., Pels, J., Polese, F.: A brief review of systems theories and their managerial applications. Serv. Sci. **1**, 126–135 (2010)
16. Woodward, J.: Management and technology: Problems and progress in technology, vol. 3. Her Majesty's Stationery Office, London (1958)
17. Fiedler, F.E.: A contingency model of leadership effectiveness. Adv. Exp. Soc. Psychol. **1**, 149–190 (1964)
18. Polese, F., Carrubbo, L., Bruni, R., Maione, G.: The viable system perspective of actors in eco-systems. TQM J. **29**, 783–799 (2017)
19. Polese, F., Payne, A., Frow, P., Sarno, D., Nenonen, S.: Emergence and phase transitions in service ecosystems. J. Bus. Res. **127**, 25–34 (2021)
20. Iandolo, F., Barile, S., Armenia, S., Carrubbo, L.: A system dynamics perspective on a viable systems approach definition for sustainable value. Sustain. Sci. **13**(5), 1245–1263 (2018). https://doi.org/10.1007/s11625-018-0565-2
21. Polese, F., Tronvoll, B., Pels, J., Carrubbo, L., Bruni, R.: A4A relationships. J. Serv. Theory Pract. **27**, 1040–1056 (2017)
22. Walletzký, L., Carrubbo, L., Toli, A., Ge, M., Romanovská, F.: Multi-contextual View to Smart City Architecture. In: Spohrer, J., Leitner, C. (eds.) AHFE 2020. AISC, vol. 1208, pp. 306–312. Springer, Cham (2020). https://doi.org/10.1007/978-3-030-51057-2_42
23. Barile, S., Piciocchi, P., Bassano, C., Spohrer, J., Pietronudo, M.: Re-defining the role of artificial intelligence (AI) in wiser service systems. In: Ahram, T.Z. (ed.) AHFE 2018. AISC, vol. 787, pp. 159–170. Springer, Cham (2019). https://doi.org/10.1007/978-3-319-94229-2_16
24. Polese, F., Tommasetti, A., Vesci, M., Carrubbo, L., Troisi, O.: Decision-making in smart service systems: a viable systems approach contribution to service science advances. In: Borangiu, T., Drăgoicea, M., Nóvoa, H. (eds.) IESS 2016. LNBIP, vol. 247, pp. 3–14. Springer, Cham (2016). https://doi.org/10.1007/978-3-319-32689-4_1
25. Rejc, A.: Toward contingency theory of performance measurement. J. East Eur. Manage. Stud. **9**, 243–264 (2004)
26. Stonebraker, P.W., Afifi, R.: Toward a contingency theory of supply chains. Manag. Decis. **42**, 1131–1144 (2004)
27. Northouse, P.G.: Contingency Theory. Leadership: Theory and Practice, pp. 113–126. Sage Publications, Inc., Thousand Oaks (2007)
28. Tenhiälä, A.: Contingency theory of capacity planning: the link between process types and planning methods. J. Oper. Manage. **29**, 65–77 (2011)
29. Franklin, M.L.: Chaos, complexity, and contingency theories: a comparative analysis and application to the 21st century organization. J. Bus. Adm. Res. **9**, 44–51 (2020)
30. Barile, S., Polese, F., Carrubbo, L., Caputo, F., Wallerzky, L.: Determinants for value co-creation and collaborative paths in complex service systems: a focus on (smart) cities. Serv. Sci. **10**(4), 379–477 (2018)
31. Drăgoicea, M., et al.: Service design for resilience: a multi-contextual modeling perspective. IEEE Access **8** 185526–185543 (2020)
32. Polese, F., Di Nauta, P.: A viable systems approach to relationship management in S-D logic and service science. Bus. Adm. Rev. Schäffer-Poeschel **73**, 113–129 (2013)

33. Tronvoll, B., Barile, S., Caputo, F.: A systems approach to understanding the philosophical foundation of marketing studies. In: Barile, S., Pellicano, M., Polese, F. (eds.) Social Dynamics in a Systems Perspective. NEW, pp. 1–18. Springer, Cham (2018). https://doi.org/10.1007/978-3-319-61967-5_1
34. Lee, J., Lee, H.: Developing and validating a citizen-centric typology for smart city services. Gov. Inf. Quart. **31**, 93–105 (2014)
35. Giffinger, R., Gudrun, H.: Smart cities ranking: an effective instrument for the positioning of the cities. ACE: Archit. City Environ. **4**, 7–26 (2010)
36. Beer, S.: The Heart of Enterprise. Wiley, Chichester (1979)
37. Wieland, H., Polese, F., Vargo, S.L., Lusch, R.F.: Toward a service (eco)systems perspective on value creation. Int. J. Serv. Sci. Manage. Eng. Technol. **3**, 2–25 (2012)
38. Cannon, D., Wheeldon, D.: I. T. I. L. Service Strategy, Hanna, A. (ed.) TSO (The Stationer Office) (2011)
39. Romanovská, F.: Adaptation of ITIL for Smart City implementation. Brno (2021)

Enhancing Interdisciplinarity in Service Innovation

Qui legit, ingenium veterum mirabile laudet: How to Introduce New Knowledge into Curriculum Using the Example of Alcuin's *De dialectica*

Hauke Hasenknopf(✉)

University of Applied Sciences Coburg, Friedrich-Streib-Straße 2, 96450 Coburg, Germany
Hauke.Hasenknopf@hs-coburg.de

Abstract. This paper argues that the way Alcuin implemented new material into curriculum is essentially the same as we do today. The only major difference lies in the relationship of the new material to the new material.

Keywords: Didactics · Carolingian renaissance · Alcuin

1 Introduction

How does one make new knowledge available for students? This question has been important for as long as there was education.

In the Early Middle Ages, education was a monopoly of the clergy. Without exception, after the utter destruction of Rome, her society and social order in Justinian's Gothic War (535–554 AD[1]) and the following conquest of Italy by the Langobards (568) learned men were clergymen. A lot of antique books were destroyed in the wars of the Migration Period [1], as well as book-burnings [1–3]. Although members of the nobility, and certainly members of catholic Europe's royal families, received some education, they do not appear as learned men. This assessment stands, even though ruler may sometimes appear as referees in theological debates [4]. Though this implies a certain degree of theological knowledge and training in the art of dialectic and logical thinking, this is far more a portrayal of those rulers as theologically knowledgeable by biographers and contemporary clergymen than it purports to actual fact. The decision of Visigothic king Reccared I in 587 to abandon Arianism and convert to Catholicism is just as much a decision of *machtpolitik* as Charlemagne's judgment to disavow adoptianism in 792/794 [5]. Sure enough, both these rulers quoted Scripture and Church Fathers (*patres*) to support their respective verdicts and were admired by their contemporaries for their wisdom and knowledge [4, 5]. Indeed, Alcuin (* c. 735–†804), says that the Frankish Empire needs the holy authority of the King to be kept in the right path [5]. Thus, it was his

Qui legit, ingenium veterum mirabile laudet:—"He, who reads this, may praise the old, wonderful genius."

[1] If not specified otherwise, every date is AD.

C. Leitner et al. (Eds.): AHFE 2021, LNNS 266, pp. 117–125, 2021.
https://doi.org/10.1007/978-3-030-80840-2_12

responsibility as a learned man to equip the King with the proper knowledge and wisdom to adequately defend Christian orthodoxy [6].

Still, while members of nobility had a certain a degree of education – even necessarily so [5] – and occasionally appear as referees in theological debates, we can still safely discount them as members of an *intelligentsia.* The arguments they used were presented to them, they did not think of them themselves. Whether or not there existed educated laymen, we can not determine, since they neither appear as authors nor are they mentioned in other sources. To argue that they didn't exist would be an *argumentum e silentio,* but we can certainly say that even if they did, they did not play any major role in Early Medieval societies. This is paralleled by the contemporary understanding that a sinless way of life that was oriented towards God was a necessary condition to advanced education [7].

Hence, whenever I use the term "education" in this paper, I mean clerical, ecclesiastical, monastic education.

The common view on the Early Middle Ages is that they were the Dark Ages at their darkest. Recent historiography in the last forty years has produced a more nuanced view on the time between c. 500–1000 [8, 9]. Indeed, we find a lot of learned men who both want to preserve the canon of knowledge as well as to broaden it [9]. To broaden the canon of accepted knowledge, however, was no mean feat in that time. Contemporaries believed that all knowledge a man could possibly attain already existed in the world – God had revealed it in his word and through the actions and teachings of *patres* and Saints [9].

Still, when Alcuin reportedly rediscovered the text *Decem categoriae,* a Latin paraphrasing of Aristotle's *Categories* he sought to spread that text in the Frankish empire.[2] But why transform that text into a dialogue? As Nadja Rädler-Bohn says, almost half of *De dialectica* is a verbatim quotation of *Decem categoriae* [11]. Couldn't he just as well just let that text be copied? Why did he intersperse the *Decem categoriae* with quotes from other authors like Isidor of Sevilla or Virgil?

These questions I will only answer insofar they pertain to the question asked above: how is new knowledge added to existing curriculum?

2 De Dialectica

The text *de dialectica* is a fictitious dialogue between Alcuin and Charlemagne about philosophy. It is a paraphrase of the text *Decem categoriae,* which itself is a paraphrase of Aristotle's work *Categories.* Additionally, Alcuin interjected parts of other works, most notably Isidor's *Etymologiae* as well as quotes from Cicero, Virgil and the Bible among others. Alcuin supposedly found *Decem categoriae* "coming from south-eastern Europe" (*deveniens e partibus Scythiae*) and translated them from Greek [15, col. 951 B].

The text *de dialectica* was composed in the late 8th century, possibly in 784/6 [11]. There are about thirty manuscripts extant. Most of them were written during the 9th century, though some were written later somewhere between 900 and 1200. Additionally, there are four or five printed editions between 1529 and 1863.

[2] The history of *Decem Categoriae* is more intricate. For more details see [11].

It is meant to be a didactic work (*opus didascalium*), so Alcuin didn't intent for it to replace *Decem categoriae*, and *Decem categoriae* was circulated parallel to *De Dialectica.*

3 Early Medieval Didactics

The goal of monastic education in the Early Middle Ages was to enable the future monk to strive for wisdom – *sapientia* – themselves [7]. However, in the understanding of the Early Middle Ages, it wasn't through the teacher that a student learned anything, it was through God. The teacher's job way merely to present the subject matter and to sharpen a student's mind for that. The student was to repeat the teachings (together with the teacher) until he understood [7]. Also, a teacher should be good example, at least in regard to behavior [7].

Concerning theoretical knowledge, Alcuin favors a dialogue between the students and the teacher, as not least E. Ann Matter has pointed out [12]. The students ask question which the teacher answers, prompting the students to more questions regarding unexplained items in the teacher's answer. This explains why most of Alcuin's didactic work takes the form of a dialogue (the only exception being *de orthographia*). In fact, some scholars go so far as to say that by stylizing his didactics as dialogues, Alcuin sketched how an ideal lecture should look like [13]. It seems this would be one answer to why Alcuin elected to convert the text into a dialogue.

Knowledge in the early middle ages meant mostly theological knowledge. But it couldn't be discovered, rather it was already revealed in the word of God as well as in the works, words and deeds of Saints and Church Fathers such as Augustine and Hieronymus. Thus, it was important for an early medieval scholar to show, that his opinions were actually the opinions of respected authorities [7, 10].

4 Variants in the Manuscripts *De Dialectica*

It is normal that manuscripts differ from one another. These variants may be just another word order, different spelling as well as mistakes, missing words or even whole lines and pages and other corruptions. While a hinderance concerning the reconstruction of the *urtext*, the textual archetype, these mistakes are vital when it comes to tracking the *stemma* – the genealogy of the manuscripts, which manuscript was paragon for which other manuscripts. Additionally, the state of the copies sometimes allows us to draw conclusions about the writers.

Some of the copies, for instance, appear to have been written by people who didn't quite understand the text as indicated by their unintelligible orthography of Greek words, shaky grammar and uneasy use of rare Latin words – sometimes just replacing them with more common, albeit contextually nonsensical words. All of these can exemplarily be seen in the manuscript 9581–95[3] of the Bruxelles Bibliothèque Royale:

1. on fol. 80v in line 9 instead of *polyonyma* the hand wrote *phylionyma*

[3] Digitized Version here: https://uurl.kbr.be/1558116.

2. on fol. 80v in line 10 the hand repeated the word *pluriuoca*
3. on fol. 80v in line 17 is an example of unsure grammar: *de boues* instead of *de boue*
4. on fol. 81r in line 34 the hand wrote *amari* ("to be loved") instead of *armari* ("to be equipped"). Although it is imaginable that the hand just missed the *r*, and indeed there are a few orthographic mistakes even in the Latin words, but in this case that seems unlikely, not least because the hand does these mix-ups later on as well, like on fol. 81r line 19 where the hand mistakenly writes *adorando* ("pleaing") instead of *odorando* ("smelling").

It would be tempting to ascribe copies with such apparent mistakes to the work of novices and other students. However, even if we accept that teaching was sometimes done by letting a student copy a text, this had to be was exceedingly rare. This could be concluded merely on the fact, that parchment was rather expensive which alone rules out mass-copying. But likewise, the evidence speaks against it. We know of thirty manuscripts of *De dialectica.* Even accounting for some losses during the centuries (like for instance 132 (262) in Chartres, which was destroyed in 1944) thirty manuscripts is not nearly enough for that kind of practice to have occurred frequently. However, if teaching through copying was done frequently, we would expect hundreds of surviving copies.

Interestingly, the aforementioned 9581–95 seems to be a manuscript, where the writer apparently didn't understand the lessons of the text. At the end of chapter III (*de categoriis*) the lessons learned of this lecture are repeated [15, col. 957 A-C], in this case how a human being could be defined using (Aristotle's) categories. Here Alcuin uses the opposite pair of all - single (*in omne – in solo*) to sharpen both the definition as well as to clarify and exemplify the recent lesson. So, in the following paragraph the words *solus* and *omnis* occur often. But the hand of 9581–95 at times writes *homo* ("man, human being") instead of *omnis* (fol. 81r), thus revealing that he may have understood what the paragraph was generally about (defining a human being) but he did not quite grasp the specific way this was supposed to be done. It is, of course, also possible, that the writer of 9581–95 merely copied these mistakes from another manuscript. Sometimes a writer or a reader would mark a section he is unsure about with a marginal R-abbreviation (for *requiro* – "I research"). This specific section, however, lacks any marginal markings. If the writer of 9581–95 was unsure about this part he did not mark it. Either way, we can conclude that the writer of 9581–95 did not understand the teachings of the lesson.

At first glance this conclusion might seem a bit extreme. Thus, I want to reinforce it with a maybe even clearer example.

In chapter I, Alcuin says the following [15, col 952 C f]:

Alcuinus: Theologica est, quae latine inspectiua dicitur, qua supergressi uisibilia de diuinis et caelestibus aliquid mente solum contemplamur. Nam in has quoque duas partes philosophia uera diuiditur, id est, inspectiuam et actualem.

Alcuin: Theology is, what in Latin is called "contemplation", through which we think about the divine and the celestial with our mind alone, after we have surpassed the visible things. Because the true philosophy is divided in these two parts, i.e. the contemplative and actual.

This paragraph is a bit puzzling. First, theology does not mean "contemplation" in Latin, neither is it part of philosophy, even less one of its fundamental parts. This passage becomes clearer when we consider that instead of "theology" there should be written "theory", because θεωρία means "contemplation" and it could reasonably be considered a fundamental part of philosophy. But every single manuscript writes "theologia" here. This part becomes even stranger, when we consider that this section is almost verbatim a quote from Isidor of Sevilla's *Etymologiae* – where Isidor of course writes "theoria" [22, II,XXIV].

It is difficult to assess what exactly happened here. One explanation would be that he inserted his own interpretation into the text and masking it as a quote from Isidor [7]. But this seems unlikely, because of the blatant discrepancies shown above.

The most likely explanation seems to be that Alcuin's copy of Isidor already read "theologica" here, and since theology could be described as "thinking about the divine and celestial" he thought it fitting. Additionally, since philosophy and theology were closely related in the early middle ages [7], he might have found it appropriate to think of theology as interlinked. But even considering this, he should have wondered why theology is considered a part of philosophy. But if Alcuin indeed had such thought he trusted the authority Isidor more than himself.

But either way, this example strengthens my point, that the writers sometimes didn't quite understand what they were writing – and that their Greek was at best shaky. Interestingly, even the later manuscripts, which tend to have better Greek, get this wrong, too – which further strengthens my point.

De dialectica frequently mentions the original Greek names of important terms, such as οὐσία for *substantia.* It is at this point difficult to say why these Greek words were included in the text, especially, since it seems that the writers, not even Alcuin himself, knew Greek.

As an example, in chapter IV (*de quantitate*) Charlemagne asks, whether quantity itself can be ascribed the attribute more or less, i.e. can something be more or less two-legged [15, col. 958]:

Carolus:	Si maius aut minus, quod Greci malon cetoton dicunt quantitas habere potest?
Alcuinus:	Non potest. …
Charlemagne:	Can quantity have the property of more or less, which the Greeks call malon cetoton?
Alcuin:	It cannot. … [20, p. 13][4]

A few things are interesting here. First, it is actually Charlemagne, who appears to know Greek here. Second, *malon cetoton* is not Greek. Or rather it is heavily corrupted Greek. If correct Greek was used here, Charlemagne would have said μᾶλλον καὶ τὸ ἧττον (mâllon kai to hêtton). It has to be stressed here, that the manuscripts and print editions vary wildly in this place. I use a manuscript here that allows for a clearer backtracking of the original Greek words. In Mignes edition this part reads.

"*malon, ceuton* [Mss., *ceceton.*; al. μάλλον et ἐλάττον]" [15, col. 958]

[4] Digitized version here: http://www.e-codices.unifr.ch/de/csg/0270/13/0/Sequence-454.

And these aren't even all variants of those Greek words. Almost each manuscript has its own distinct variant.

This indicates that hardly anyone of the (first) writer knew Greek. They just copied what they could read. Notable exception is the aforementioned manuscript B.IV.6, and BPL 84 in Leiden university, both of which are of significantly later date (early 12th century and 11th century), a time when knowledge of the Greek language was not quite as uncommon as in the 9th century. Thus, the writer of these later manuscripts could correct the corrupt Greek words. Notable as well is the manuscript Reg.lat.1209 of the Vatican library, where a later hand corrected corrupt Greek, albeit in other parts of the text.

On one hand, we can clearly see the effects of Alcuin's teaching, especially, when Alcuin actually uses the lectures even in different contexts. In a letter to an unknown female noble about the adoptionist controversy, Alcuin uses the exact same analogy as in *De dialectica*: comparing a painted human to an actual human [16, p. 339] [15, col 955].

On the other hand, based on the evidence of corrupt Greek and shaky understanding of Latin in context and overall teaching (as in 9581–95) we can conclude that some writers did not fully understand what they wrote. And since we can be reasonably sure, that not even Alcuin himself knew Greek, this pertains to him as well at least partially.

On the other hand, there are some manuscripts which appear to be written for and by lecturers, for instance in manuscript B.IV.6 of the Durham University Library.[5] Here the hand frequently inserted synonyms and additional definitions after he words apparently deemed requiring further explanation. Following here is an example from that manuscript fol. 159v. In most other manuscripts, there is just a list of Aristotle's categories in this place; the additions in B.IV.6 of this hand are in italics:

> … Quantitas, *ut magnus paruus*, ad aliquid, *ut dominus seruus*, qualitas, *id est albedo nigritudo*, facere – *lego*, pati – *legor*, situs – *iacere*, ubi – *locus*, quando – *tempus*, habere – *habitus*. His nouem *accidentibus* iunctis ad substantiam, quam greci usian uocant, fiunt decem categoriae.
>
> … Quantity, *like big and small*, relative, *like lord and serf*, quality, *i.e. whiteness and blackness*, doing – *I chose*, suffer – *I am chosen*, attitude – to lie, where – *place*, when – *time*, to have – *disposition*. The ten categories are made from these nine *accidents* plus the substance, which the Greeks call οὐσία.

It seems the hand of B.IV.6 added here some more fodder. So, in case the students didn't immediately understand, what any of these categories denoted, a teacher could supply them with additional information and definitions to help them understand it. At other instances, this manuscript provides cross references to earlier parts, as if to recall things learned earlier (e.g. on fol. 160v, lines 10ff). The writer of B.IV.6 clearly engaged with the text's subject.

But even in this manuscript, some of the more peculiar mistakes are present, thus thy point still stands: sometimes the writers didn't know, what they were writing about. It can

[5] Digitized version here: https://iiif.durham.ac.uk/index.html?manifest=t1mzs25x866g&canvas=t1t6t053k49f.

be speculated, that they trusted so much in the authority of their source that they copied it, in the hopes that someone might eventually understand, even though, they themselves couldn't. The writers' trust in their sources exceeded their trust in their knowledge.

5 Why a Dialogue?

There are two easy ways of answering this question. And while they are valid, they neither fully satisfactory. One answer is, that since Plato there has been a long, rich and fecund history and tradition of educational, indeed scientific, dialogues. By emulating this style, Alcuin was putting himself in that tradition. Another reason would be, that Socratic dialogues were the preferred mode of teaching in the Early Middle Ages – no doubt at least partially because of that said rich tradition.

However, these answers, while applicable, don't quite cut it. As E. Ann Matter has pointed out, these texts cannot neatly be qualified as dialogues, but better as question-and-answer text [12]. One could argue, that they are somewhat less than that, since sometimes Alcuin's fictitious dialogue partners are relegated to ask rather peculiar questions like, what introduction means [15, col. 953 B]. Scholar E. Ann Matter even goes so far as to call Charlemagne in this dialogue "dull" [12].

The question, why Alcuin chose the dialogic form even for *De dialectica* becomes stranger still, when one considers that Alcuin made the underlying text, *decem categoriae*, a present to Charlemagne when he rediscovered it. That act alone would have probably been enough to encourage his contemporaries to use that text also in educational circumstances, because it meant that the most influential learned man in the entire Latin world endorsed that text – and with him Latin Christianity's most powerful ruler.

The conversion into a dialogue may have hastened the text's adoption into curriculum – converting it into a familiar form – but it did not facilitate it. However, the conversion into a dialogue, indeed, the act of transforming it at all, gave Alcuin the opportunity to play with the text – and insert other material into it. Most notably he inserted passages of Isidor's *Etymologiae* as well as quotes by Virgil and Cicero. This shall be demonstrated in the following passage. Passages quoted verbatim from *Decem categoriae* are broad, passages from Isidor's *Etymologiae* are in italics.

Albinus: **Quantitas, ad aliquid, qualitas, facere, pati, situs, ubi, quando habere** [...]. His nouem iunctis ad substantiam, **quam greci usian uocant**, fiunt decem categoriae. *Et inter haec decem uerba quicquid homo loquitur ineuitabiliter inuenitur.* Sed priusquam de his singulis disputare incipiamus *omonymorum legem quae categoriarum sunt instrumenta uideamus.* [15, col. 955 A]

I think this happened, to intersperse the new material with familiar material, so as to make the new material more palatable. But not only in a pedagogic-didactic way, also in as a means to justify it. As scholars like Sita Steckel or Marc-Aeilko Aris have shown, Carolingian scholars used the method of compiling and quoting from forebearers to both justify their opinions as well as to show that even new interpretation of older material were actually already revealed [9, 10].

In fact, the latter point may be the reason why he elected to include a lot of Greek words, even though they are heavily corrupted. The inclusion of the Greek words may have elevated the surrounding text through their polyglot appeal. Additionally, it gave Alcuin the chance to showcase Charlemagne's learnedness by letting him speak Greek words. It has to be stressed, however, that as of now it cannot be said with certainty, why Alcuin chose to include Greek at all in his text, especially, since he himself does not appear to know Greek.

It's a way of rhetoric, just as we today use jargon and references as a way of convincing our fellow scholars that our opinions may be taken seriously [17].

6 Conclusion

We can see with the example of Alcuin's *De dialectica* how one introduces new material into curriculum. The new material is:

1. poured into a familiar form. Here a dialogue (or dialogue-like)
2. interspersed, and thus interlinked with old, familiar knowledge.

In general, this is still the way how modern societies keep their curricula up to date. The only major difference between Alcuin and us today is the relationship between the old material and the new. We are used to the idea, that the new material expands on old material, supersedes it and indeed sometimes replaces the old material. In Alcuin's time, however, the new material wasn't viewed as really new, merely a new way of looking onto familiar objects. Hence, he stresses in an accompanying poem, that "old genius may be praised as wonderful" (*ingenium veterum mirabile laudet)* [15, col. 951 B].

The example of epigenetics could teach us the benefits of this medieval approach. Lamarck's theory of epigenetics was discarded in favor of Darwin's theory of evolution (and Mendel's laws of inheritance). But recently, biologists have discovered that some phenomena can better be described using epigenetics [23].

Thus, the medieval approach could show us the benefits of looking favorably onto old approaches and appreciate the things they got right. It could demonstrate the power of dialectical thinking.

References

1. Rohmann, D.: Christianity, Book-Burning and Censorship in Late Antiquity. De Gruyter, Boston (2016). https://doi.org/10.1515/9783110486070
2. Müller, C.D.G., Weiß, H.-F.: Alexandrien. Theologische Realenzyklopädie Online. De Gruyter, New York (2010). https://www.degruyter.com/document/database/TRE/entry/tre.02_248_12/html. Accessed 2021-03-02
3. Russel, B.: Philosophie des Abendlandes: ihr Zusammenhang mit der politischen und der sozialen Entwicklung. Europaverlag, München/Wien (2000)
4. van Renswoude, I., Raaijmakers, J.: The ruler as referee in theological debates: reccared and charlemagne. In: Meens, R., et al. (eds.) Religious Franks. Religion and Power in the Frankish Kingdoms: Studies in Honour of Mayke de Jong, pp. 50–70. Manchester University Press (2016)

5. Kramer, R.: Adopt, adapt and improve. Dealing with the Adoptionist controversy at the court of Charlemagne. In: Meens, R., et al. (eds.), Religious Franks. Religion and Power in the Frankish Kingdoms: Studies in Honour of Mayke de Jong, pp. 32–50. Manchester University Press (2016)
6. Dales, D.: Alcuin: his life and legacy. James Clarke, Cambridge, UK(2012)
7. Steckel, S.: Kulturen des Lehrens im Früh- und Hochmittelalter: Autorität. Wissenkonzepte und Netzwerke von Gelehrten. Böhlau Verlag, Köln/Weimar/Wismar (2011)
8. Maurus, H.: Lehrer, Abt und Bischof; [Symposion d. Akad. d. Wiss. u.d. Literatur, Mainz, 18. - 20. September 1980]/Akademie der Wissenschaften und der Literatur, Mainz. Hrsg. von Raymund Kottje u. Harald Zimmermann. Steiner, Wiesbaden (1982)
9. Steckel, S.: Von Buchstaben und Geist: Pragmatische und symbolische Dimensionen der Autorensiglen (nomina auctorum) bei Hrabanus Maurus. In: Becker, J., Licht, T., Weinfurter, S. (eds.) Karolingische Klöster, pp. 89–130. De Gruyter, Berlin (et al.) (2015)
10. Aris, M.-A.: Nostrum est citare testes. Anmerkungen zum Wissenschaftsverständnis des Hrabanus Maurus. In: Schrimpf, G. (ed.) Kloster Fulda in der Welt der Karolinger und Ottonen, Verlag Josef Knecht, Frnkfurt a. M. (1996)
11. Rädler-Bohn, E.: Re-dating Alcuin's De dialectica: Or, did Alcuin teach at Lorsch? Anglo-Saxon England **45**, 71–104 (2016). https://doi.org/10.1017/S0263675100080236
12. Matter, E.:. Alcuin's question-and-answer texts. Rivista Di Storia Della Filosofia (1984-), 45(4), 645–656 (1990). http://www.jstor.org/stable/44022670. Accessed 2 Mar 2021
13. Tremp, E., Schmuki, K., Flury, T.: Karl der Grosse und seine Gelehrten. Zum 1200. Todestag Alkuins († 804) , Verlag am Klosterhof, St. Gallen (2004)
14. Bullough, D.A.: Alcuin: Achievement and Reputation: Being Part of the Ford Lectures Delivered in Oxford in Hilary Term 1980. Brill. Leiden (2004)
15. Patrologiae Latinae vol. 101, Beatus Flaccus Albinus seu Alcuinus ed. J.-P- Migne, Paris (1863)
16. Monumenta Germaniae Historica (MGH), Epistolarum tomus IV (Epp. 4), ed. E. Dümmler, Weidmann, Berlin (1895)
17. McCloskey, D.N.: The Rhetoric of Economics: Vol, 2nd edn. University of Wisconsin Press, Madison (1998)
18. Manuscript: Bruxelles Bibliothèque Royale 9581–95, fol. 79r–87r. https://uurl.kbr.be/1558116
19. Manuscript: Durham University Library B.IV.6, fol. 158r–169r. https://iiif.durham.ac.uk/index.html?manifest=t1mzs25x866g&canvas=t1t6t053k49f
20. Manuscript: St. Gallen, Stiftsbibliothek, Cod. Sang. 270, pp. 4–38. http://www.e-codices.unifr.ch/de/csg/0270/13/0/Sequence-454
21. Manuscript: Vatican Library Reg.lat.1209, fol. 1r – 49v. https://digi.vatlib.it/view/MSS_Reg.lat.1209
22. Isidor of Sevilla: Etymologiae
23. https://www.spektrum.de/thema/epigenetik/1191602. Accessed 06 Jun 2021

Approaching Sustainability in Interdisciplinary Teams

Rebecca Fischer(✉) and Hauke Hasenknopf

Masterstudiengang ZukunftsDesign, Hochschule Coburg|Coburg University of Applied Sciences, Friedrich-Streib-Straße 2, 96450 Coburg, Germany
{Coburgrebecca.fischer,hauke.hasenknopf}@hs-coburg.de

Abstract. This paper discusses the question of how interdisciplinary teamwork on sustainable projects influences the perception of the concept of sustainability by the team members. The target groups are two projects of the master program "ZukunftsDesign" (Roughly translated as designing the future). The research is carried out by conducting explorative interviews. For a methodical triangulation observation of the group and a survey were used. Finally, the paper summarizes how the perception of the concept of sustainability may or may not change through an interdisciplinary environment to give hints on how sustainable concepts can be communicated more efficiently.

Keywords: Sustainability · Interdisciplinary · Teamwork · Project work

1 Introduction

Sustainability is an inevitable necessity. Developing a sustainable society and economy that will preserve itself as well as the environmental parameters and diversity needed for enabling flourishing human life, is a task that is passed on from generation to generation. Since the topic of sustainability came to larger transnational attention and more and more object of international cooperation in the 1980s, various approaches developed during the decades. Another crucial development in our knowledge-based and heavily industrialized society is the necessity of interdisciplinary approaches to research and development as well as business activities due to structures and tasks that get more and more complex in any field. Technological advancement and individualization are exemplary drivers of complexity in society and the economy.

The interdisciplinary master course ZukunftsDesign of the University of Applied Sciences Coburg in Germany aims to foster regional social and economic development. A huge part of designing a future that is identified by individuals as a scenario worth living in is the aspect of sustainability. The main didactic tool of ZukunftsDesign is interdisciplinary teamwork on real projects initiated by local companies, institutions, or students themselves.

C. Leitner et al. (Eds.): AHFE 2021, LNNS 266, pp. 126–131, 2021.
https://doi.org/10.1007/978-3-030-80840-2_13

2 Interdisciplinary Teamwork at the Master Program "ZukunftsDesign" in the Context of Sustainability

In the winter term of 2020/2021 different projects of the Master Program ZukunftsDesign targeted sustainable goals. Two of the projects were "MitweltHeld Kronach" and "Planetary Health" (later renamed "Challenge4Future"). ZukunftsDesign itself is organized as a part-time study that can be carried out while fully employed. Therefore, students bring various experience from different branches and educational backstories to the table.

Due to the COVID 19 pandemic, most of the project work was carried out online. This was an additional challenge since both projects targeted an audience either from the town Kronach or its future campus of Coburg University.

Both projects align with the SDGs (Sustainable Development Goals) of the United Nations. Especially goal 3 "Ensure healthy lives and promote well-being for all at all ages", goal 9 "Build resilient infrastructure, promote inclusive and sustainable industrialization and foster innovation", goal 11 "Make cities and human settlements inclusive, safe, resilient and sustainable", goal 12 "Ensure sustainable consumption and production patterns" and goal 13 "Take urgent action to combat climate change and its impacts" [1].

2.1 MitweltHeld Kronach

The project MitweltHeld Kronach is centred around a sustainable and holistic approach called "Mitwelt" by German physicist and philosopher Klaus Michael Meyer-Abich. The basic idea of this concept is an integration of the surrounding nature and social environment against other concepts that see the environment as a separate sector besides human existence.

The concept of the project came from a student. The initial idea was to find different tools and approaches to foster the concept of "Mitwelt" in the region of Kronach. Therefore, the name "Mitwelt Held" (Mitwelt hero) is derived. Starting as a relatively open concept, the project evolved around the idea of establishing an office and different services based on the idea of "Mitwelt" at the upcoming Lucas-Cranach-Campus.[1] The professions represented by the students in the team were Design, Teaching, Consultancy, Tourism Management, Business & Administration, Marketing, Social Work and Psychology.

2.2 Planetary Health|Challenge4Future

The project Planetary Health, later renamed to Challenge4Future, was also initiated by a student. The starting point of the project was the concept of "Planetary Health" which

[1] The Master Program ZukunftsDesign is mainly carried out in the town of Kronach, where a new joined Campus (Lucas-Cranach-Campus) of Coburg University of Applied Sciences and Hof University of Applied Sciences is established. Coburg as well as Kronach are both towns in Upper Franconia, Germany.

is an understanding that human health and the health of the environment are inseparable intertwined.

In the beginning, the idea and therefore direction of the project was open. The first basic decision was to develop an improvement for the upcoming Lucas-Cranach-Campus. In the end, the students decided to realize a competitive concept called "Challenge4Future" at the campus which connects an online platform and on-sight infrastructure to give the students of the campus access to different challenges centred around sustainability and even create their own challenges. The professions, represented through the students in the team, were Social Work, Coaching, Consultancy, Change Management, Business & Administration, Engineering, Teaching and Governmental Administration.

3 Methodological Approach

The methodological approach was based on a methodological triangulation of in-depth explorative interviews, participant observation and a survey. The observation was based on the project work conducted online and in physical presence. The focus was on the interaction between different disciplines in the context of sustainability. Through the survey, basic parameters of the individual perception of sustainability and intertwined life experiences were determined. Additionally, intersections of the overall concept of sustainability and the project aims were identified. The interviews focussed on the definition of Sustainability, the intersections of the project and Sustainability, the changes in the personal understanding of sustainable concepts through the project work, the benefits of interdisciplinary teamwork and potentials for improvement of the project work. The interviews and survey were carried at the end of the project work. The observation was carried out through the whole project work.

4 Individual Concepts and Definitions of Sustainability

Sustainable development was defined by the World Commission on Environment and Development in the late 1980s: "Sustainable development is development that meets the needs of the present without compromising the ability of future generations to meet their own needs." [2].

Sustainability in general can be understood as a concept that guarantees parameters in which a system can sustain itself. Sustainability can also be defined by three pillars: Society, Environment and Economy. While there are various approaches to the topic in different disciplines, each person has a uniquely individual approach to the topic as well.

Through the interviews which were carried out with the members of the two project groups, it became clear that in around 60% of the cases the individual perception of Sustainability and therefore its definition is determined by activities, certain behaviour or objects that are identified as "sustainable". Therefore, the concept of what Sustainability is in general, is less dominant in perception than different of its practical applications. In nearly all cases it seemed easier for the participants to name a sustainable activity or behaviour than to give a general and abstract definition of the concept of Sustainability. However, since roughly 40% of the participants could name a more abstract definition of the concept and the other participants were also able to do so in the progression of the interview, Sustainability itself seemed a known concept, also in abstract terms, in the participant group.

In summary, the participant group were already sensitized to the topic, either from work experiences or personal life. A common definition of the concept of Sustainability was the basis of the interdisciplinary teamwork on projects with sustainable aims.

5 Sustainability in Relation to Profession, Personal Life, and the Semester Project of ZukunftsDesign

The survey and the interviews conducted showed that there is a close relationship between the importance of Sustainability in personal life and the importance of Sustainability in the semester project and the professional setting. Often ratings of importance correlated in the sections of semester project and personal life. However, the rating of importance of Sustainability in the profession of the participants often differed from the other two sectors.

The difference of the importance rating of Sustainability between the personal life and semester project on the one side and the importance rating in the profession was further discussed in the explorative interviews. The main pattern that various that different participants shared was that the importance of Sustainability in the profession was rated lower as the importance in personal life or semester project. There were two main reasons why participants did not rate the importance of Sustainability in their profession as high as in personal life or the semester project. The first reason was that sustainable aspects affecting the work of the participant were not in the influence or competence of the interviewee. In detail, this means that the interviewee is not able in his daily profession to have a sustainable impact on the company or institution he or she is working in. This can be through limitations of competencies or restricted work frames. The second reason was that the interviewee is forced through profit-oriented requirements of their employer to neglect sustainable aspects. These can be through pressure from the market or competitors. Implementing a different business model which integrates sustainable aspects and allows profit as well was not an option according to the interviewees.

The relation in rating between the importance of Sustainability in personal life and semester project was in nearly all cases congruent. Since interviewees had control over most aspects of personal life and the semester project, they choose to integrate sustainable aspects as far as possible. However, most interviewees didn't give themselves the highest rating in both sectors, even if through the interview it became clear that participants, in general, had high ambitions in integrating Sustainability into their daily lives and further. This was always explained that there is still potential for improvements.

In summary, it can be stated that where participants or interviewees had the unrestricted freedom to choose sustainable alternatives or integrate sustainable aspects, they did on a high level. Still, the necessity for profit-oriented decision making and a restriction of competencies of employees remained the two major hindrances for establishing sustainable approaches.

6 The Impact of Interdisciplinary Project Work with Sustainable Aims on the Perception of Sustainability

The participants were asked to describe if the interdisciplinary teamwork on sustainable projects affected their perception of Sustainability and if yes in which way. Nearly all participants agreed that the work on the project changed their perception of Sustainability. The main argument was that different approaches through different professional disciplines represented by the team members came to the attention of the interviewees. Also, the requirement of doing further research during the project work led to a broader understanding of the topic. Even though nearly all participants were affected by a shift in perception the change itself was mostly gradually. Most interviewees already researched the topic and tried to integrate it into their personal life.

7 The Characteristics of Interdisciplinary Teamwork

A main part of the explorative interviews was the discussion on the characteristics of interdisciplinary teamwork. The most stated trait of interdisciplinary teamwork was the need for a certain kind of translation due to different technical language and varying standards according to different disciplines. Further, it was stated that people from certain disciplines also often share certain traits. For example, do engineers focus on technical and rational aspects while artistic professions can have a different focus or prefer holistic approaches. These stereotypes may not be applying in any case, but it was clear that different professions often result in a different type of persons. Therefore, trying to understand the view of another person becomes more important than in teams with only one profession represented. All in all, interviewees mostly argued that the necessity to try to understand one another is per se a positive aspect since it forces oneself to leave established and static thinking patterns.

A further major aspect of interdisciplinary teamwork according to the interviewees are the different viewpoints on one topic. This fact was identified as a huge benefit from interdisciplinary teams by the interviewees. It was stated that therefore interdisciplinary projects can integrate more aspects into the process and the results as teams consisting of only one or similar professions.

However, these benefits of interdisciplinary teamwork seem to come with possible challenges [3, 4]. For example, often a good moderation and keeping track of different arguments are required to ensure a productive environment [5]. Also, according to the participants, interdisciplinary teamwork can tend to stronger friction amongst the team members resulting from different social approaches based partially on professional background.

In summary, interdisciplinary teamwork comes with benefits but requires stronger attention to interpersonal communication related to translation and moderation.

8 Interdisciplinary Teamwork in the Context of Sustainability

Another important outcome of the conducted interviews was the view of the participants on interdisciplinary teamwork in the context of sustainability. The statement that interdisciplinary teamwork is especially useful in a sustainable context was provided by most of the interviewees. It was argued that Sustainability is a topic that is of importance in any part of society and economy and therefore can only be fully processed by a combination of different disciplines. One interviewee also stated that truly sustainable solutions can only be developed by interdisciplinary teams. Asked for improvements for the teamwork in the context of sustainability a few interviewees also stated that a more diverse team that would represent even more disciplines would be even more effective.

9 Summary

Sustainability is a necessity for the future existence of our society. Therefore, ZukunftsDesign as a master course needs to integrate this aspect. The project groups of "MitweltHeld Kronach" and "Challenge4Future" were examined through a methodical triangulation of survey, interviews, and observation. The focus was to investigate how the personal perception of Sustainability can change through interdisciplinary teamwork with sustainable goals. Gradually changes in the perception of the participants were discovered which were related to the accessibility of other points of view through different disciplines represented in the project team. Interdisciplinary teamwork was identified as a working mode with various benefits but also a demand for advanced interpersonal communication like translation and moderation. Especially sustainable projects benefit from interdisciplinary teams since Sustainability itself is a holistic approach. Potential lies in the further diversification of project teams.

References

1. The United Nations: The 17 Goals. https://sdgs.un.org/goals. Access 15 Nov 2020
2. World Commission on Environment and Development: Our common future. Oxford University Press. Oxford (1987)
3. Nancarrow, S.A., Booth, A., Ariss, S., et al.: Ten principles of good interdisciplinary team work. Hum Resour Health **11**, 19 (2013). https://doi.org/10.1186/1478-4491-11-19
4. Baldwin, D.C., Tsukuda, R.A.W.: Interdisciplinary Teams. In: Cassel, C.K., Walsh, J.R. (eds.) Geriatric Medicine. Springer, New York (1984). https://doi.org/10.1007/978-1-4612-5232-0_30
5. Corney, R.H.: Collaboration or conflict: factors affecting interdisciplinary teamwork. In: Hippius, H., Lauter, H., Ploog, D., Bieber, H., van Hout, L. (eds.) Rehabilitation in der Psychiatrie. Springer, Berlin, Heidelberg (1989). https://doi.org/10.1007/978-3-642-72705-4_12

The Impact of the Master's Program ZukunftsDesign on Student Creativity

Lena Pieper[1(✉)], Annika Linsner[1], and Franziska Leutner[2]

[1] Coburg University of Applied Sciences and Arts, Coburg, Germany
{Lena.Pieper,Annika.Linsner}@hs-coburg.de
[2] Goldsmiths College, University of London, London, UK
f.leutner@ucl.ac.uk

Abstract. This research examined the impact of the Master's program ZukunftsDesign on student creativity. The subjective and objective creativity of 42 students were evaluated by means of an online questionnaire and external personality assessment. Findings confirmed the direct link between subjective and objective creativity. The study also revealed promising results for the Master's program and its students. More in-depth research is required to manifest the identified trends and provide detailed practical and theoretical implications.

Keywords: Creativity assessment · Creative self-image · Teaching innovation

1 Introduction

Students need certain competencies to generate innovations that will endure the environment of a rapidly changing VUCA (versatile, uncertain, complex, ambiguous) world [1]. Creativity is an important precondition for innovation [2], as it includes the ability to think of new, unexpected, and appropriate solutions and products [3]. One of the goals of the Master's program ZukunftsDesign (Eng. FutureDesign) is to teach and promote creativity. Previous qualitative studies have examined specific methods of promoting creativity and their effect on students and their creative self-efficacy [4, 5]. The overall aim of the current study was to examine whether the Master's program ZukunftsDesign influences student creativity. In particular, this research was designed as a pilot-study to identify the impact of the experiences gained within the Master's program on student creativity and investigate the development of student creativity over the course of the degree.

2 Theoretical Background

2.1 Definition of Creativity

In science and practice the perceived importance of creativity for individual, societal, and also organizational success is growing, driving research on the personal, social, and

C. Leitner et al. (Eds.): AHFE 2021, LNNS 266, pp. 132–140, 2021.
https://doi.org/10.1007/978-3-030-80840-2_14

situational conditions of creative performance (cf. [2]). Therefore, a variety of definitions of creativity exist, which emphasize its different dimensions. Some of them only address the quality of the product and the process, such as unique, useful [6], novel-original and adaptive to reality [7, 8]. Other definitions focus on the creative person(s) individually or within a small group of individuals [3], or highlight the aspect of interaction of personal, social, and situational conditions [9].

2.2 Master's Program ZukunftsDesign

The Master's Program ZukunftsDesign (ZD) is an extra-occupational degree for innovation. ZD aims at collaborative cooperation and wants to create interfaces of scientific disciplines to "systematically tackle changes, uncertainty, and complexity" [10]. Throughout the program students work on their own ideas and project topics set by regional companies and institutions in a unique inter- and trans-disciplinary environment. This opens students' minds for new ideas and promotes a "pro-entrepreneurship attitude" [10]. According to the ZD teaching concept, the conscious confrontation with opposing positions and contradictions delivers impulses for new ways of thinking. The diverse contents of the academic program range from innovation technologies and ethical questions to aspects of project group dynamics and their communication structures to teach the students how to deal with uncertainty (i.e., *VUCA thinking*). To achieve this, ZD breaks through the traditional hierarchy between professors and students. Teachers guide through projects as coaches and the curriculum is aligned with the situational needs of the students and relies on "learning by doing" to foster forward-looking, innovation-oriented skill development and change management [10].

The various elements of the program can be assigned to Rhodes' 4p model, *person, process, product,* and *press* [11]. The setting of interdisciplinary project teams of very different age groups and experience horizons, thrown together by lottery, form the *press*. Aspects of the *process* are covered in Innovation Techniques and Theory, taught and accompanied in application by coaches, and supported by a digital method toolbox for agile project work. Project results (e.g., prototypes) are the *products*. The *person* is the focus of the subjects communication and mediation, team dynamics, experience of one's own limits, and leadership.

2.3 Assessment of Creativity

According to Rhodes the *creative person* is one of the basic elements of creativity [11]. One approach to assessing creativity at the personal level is through the survey of creative self-image. A person's self-image reflects the extent to which they view themselves as creative [12]. This self-assessment serves to predict the motivation to be creative, acts as a guiding goal in creative activities, and can also shape self-selection in professional (choice) processes [13]. When considering the creative person, both cognitive aspects and aspects of personality can be examined. Since there are many components associated with creativity, this study focuses on *cognitive flexibility, curiosity* and *openness to experience*:

- *Curiosity*: the recognition, pursuit, and intense desire to explore novel and uncertain events [14].
- *Openness to experience*: a Big Five personality trait, which is considered a promoter of creativity [7, 15, 16].
- *Cognitive flexibility*: the ability to switch cognitive sets to adapt to changing environmental stimuli [17]. In the literature, *flexibility* and *adaptation* are often used interchangeably. The importance of adaptation is emphasized here: "Creative thought or behavior must be both novel-original and useful adaptive" [7] and "if a response is to be called original [...] it must be to some extent adaptive to reality" [8]. In particular, the aspect of the ability to adapt seems to be crucial for the ability to innovate as the fit into the market is indispensable for a successful innovation.

2.4 Image Based Tests

Interpersonal differences are primarily measured by means of psychometric tests in verbal form [18]. However, psychometric tests are criticized for their inability to engage the test taker [19], the ease of faking responses [20], and adverse impact [21]. Recently, interest in alternative forms of assessment has been growing [22]. In particular, the use of image-based testing to assess individual differences has increased substantially. Supporters argue that they offer a more engaging alternative to text-based psychometric tests [23]. One possibility to explain this is that images are more vivid than texts. However, studies based on this hypothesis have not been able to prove a "vividness effect". Rooted in social psychology, this effect describes information as "vivid" if it is emotionally stimulating, tangible and challenging (imagery-provoking), and is close in sensory, temporal or spatial terms [24].

Leutner et al. developed an image-based creativity test, which is based on the following three measures [23]: 1. *Curiosity and Exploration Inventory-II* (CEI-II; [14]), measures two traits: stretching (e.g., 'I actively seek as much information as I can in new situations') and embracing (e.g., 'I am the type of person who really enjoys the uncertainty of everyday life'). 2. *Cognitive Flexibility Inventory* (CFI; [25]), a self-report measure of adaptive thinking in stressful situations, which assesses behaviours related to alternatives (e.g., 'I consider multiple options before making a decision'), behaviours related to control (e.g., 'When I encounter difficult situations, I feel like I am losing control'), and 3. *Openness to experience* [26] (e.g., 'I enjoy hearing new ideas'). This image-based test is part of the Red Bull Wingfinder personality assessment (WF). The WF is based on psychological research and focuses on components important to employability and career success for knowledge-based jobs. Its four core components are *drive, creativity, connection* and *thinking* [27].

The WF includes many dimensions that are interesting against the background of the study program and provides an unconventional and engaging approach compared to an ordinary and uninspiring student survey. Thus, pairing this form of assessment with the current study was deemed suitable to investigate the creativity of the ZD students. Examining the influence of ZD on students' self-perceived creativity is relevant to laying the groundwork for further research in the field of teaching innovation.

3 Methods

At the time this research was carried out a total of 137 students were enrolled in the Master's program ZukunftsDesign at the Coburg University of Applied Sciences and Arts in Germany. All students were asked to participate in the study by means of an online questionnaire. In addition to demographics and subjective questions relating to the Master's program as well as self-rated creativity, the creativity of participants was measured objectively as part of the WF. The WF results range from 1–100 and are adjusted to the RedBull global norming group. Due to the focus of this research only the creativity component of the WF was included in the analysis. Creativity is a summary measure of the sub-components *cognitive flexibility*, *curiosity* and *openness to experience*, which showed good scale reliability (Cronbach's $\alpha = .76$).

The creative self-image was measured by including questions about the participants' self-rated creativity (e.g., 'how creative are you', 'how creative are you compared to your fellow students'). Participants assessed these items on a 5-point Likert scale (1 = *not creative at all*; 5 = *very creative*). Scores on items were combined to form an overall measure of students' creative self-image. The resulting scale was internally consistent with a Cronbach's α of .77.

Responses were analyzed using SPSS to investigate any trends, underlying structures and relationships between variables, in particular between subjective and objective creativity in view of ZD. Only fully completed surveys were included, entries that showed consistent missing values were deleted. Correlations were used to examine associations between variables and Analysis of Variance (ANOVA) was calculated to identify existing differences between mean scores.

3.1 Participants

Forty-two students fully completed the questionnaire (25 females, 17 males). Due to multiple inconsistent responses one female participant was removed from the data set ($N = 41$). The majority of participants were within the 21–29 years of age category (20; 48.8%), 11 (26.8%) were aged 30–39, eight (19.5%) were aged 40–49 and two participants were aged 50–59 (4.9%). Thirteen (31.7%) participants were currently studying in their first semester. The highest academic degree of most students was a Bachelor's degree (26; 63.4%). Most participants (26; 63.4%) stated that they were currently employed with weekly working hours exceeding 35; only four (9.7%) participants were unemployed.

The majority of students (90.2%) agreed that ZD encourages creativity, only four (9.8%) were undecided. Further, 34 (82.9%) participants stated that the program has a positive effect on their individual creativity; none reported negative effects. Of the 41 students 68.3% agreed with the results of the WF, while only four (9.7%) disagreed.

4 Results

The ANOVA results indicated several significant differences between female and male participants. Females ($M = 4.17$; $SD = 0.64$) found themselves to be more creative compared to their private sphere than males, $p = .04$. Additionally, females provided significantly higher scores when asked about the importance of creativity for their current job (females: $M = 3.83$, $SD = 1.34$; males: $M = 2.82$, $SD = 1.29$; $p = .02$).

Another difference was detected for the WF creativity component, where female students performed significantly better than their male counterparts (females: $M = 63.79$, $SD = 16.81$; males: $M = 49.24$, $SD = 27.25$; $p = .04$). Closer examination of the three sub-components of creativity identified the cause of this difference. Females ($M = 69.75$, $SD = 16.37$) achieved significantly higher scores for the sub-component *adaptable* than males ($M = 45.82$, $SD = 30.60$; $p = .002$).

For further inspection of the differences between the genders, the data set was split into those who found that ZD encourages creativity (37) and those who were undecided (4). Calculations were continued with $n = 37$. The significant differences described strengthened for this sample (see Table 1 for descriptive statistics). Further differences emerged for the importance of creativity for the current employer (females: $M = 3.52$, $SD = 1.08$; males: $M = 2.71$, $SD = 0.91$; $p = .02$). The significant differences also increased for the WF creativity component, with *open to experience* being the only sub-component without notable differences between genders (see Table 2).

Table 1. Descriptive statistics for split sample ($n = 37$).

		N	Mean	SD	Min	Max
How creative are you compared to your private sphere?	Female	23	4.17	.650	3	5
	Male	14	3.57	1.016	2	5
	Total	37	3.95	.848	2	5
How important is creativity for your current job?	Female	23	3.52	1.082	2	5
	Male	14	2.71	.914	1	4
	Total	37	3.22	1.084	1	5
How important is creativity for your current employer?	Female	23	3.96	1.224	1	5
	Male	14	2.64	1.216	1	5
	Total	37	3.46	1.366	1	5
Creativity (Wingfinder)	Female	23	62.91	16.61	36	94
	Male	14	43.29	25.566	7	84
	Total	37	55.49	22.31	7	94
Adaptable (Wingfinder)	Female	23	70.00	16.687	34	99
	Male	14	40.21	29.689	5	92
	Total	37	58.73	26.513	5	99
Innovative (Wingfinder)	Female	23	61.96	20.851	17	98
	Male	14	43.00	32.824	1	96
	Total	37	54.78	27.233	1	98

Table 2. Significant differences between male and female participants for split sample ($n = 37$).

One-way ANOVA		Sum of Squares	df	Mean Square	F	Sig.
How creative are you compared to your private sphere?	Between Groups	3.159	1	3.159	4.864	.034
	Within Groups	22.733	35	.650		
	Total	25.892	36			
How important is creativity for your current job?	Between Groups	15.018	1	15.018	10.075	.003
	Within Groups	52.171	35	1.491		
	Total	67.189	36			
How important is creativity for your current employer?	Between Groups	5.674	1	5.674	5.427	.026
	Within Groups	36.596	35	1.046		
	Total	42.270	36			
Creativity (Wingfinder)	Between Groups	3352.560	1	3352.560	8.055	.008
	Within Groups	14566.683	35	416.191		
	Total	17919.243	36			
Adaptable (Wingfinder)	Between Groups	7720.940	1	7720.940	15.368	.000
	Within Groups	17584.357	35	502.410		
	Total	25305.297	36			
Innovative (Wingfinder)	Between Groups	3127.314	1	3127.314	4.644	.038
	Within Groups	23570.957	35	673.456		
	Total	26698.270	36			

Further statistically significant differences between groups as determined by one-way ANOVAs were detected between students currently studying in semesters 1–4 (31) and those above (10) with regards to creativity ($F(10, 30) = 2.41$, $p = .03$). Students in higher semesters are more creative (semesters 1–4: $M = 3.29$, $SD = 0.81$; higher semesters: $M = 4.00$, $SD = 0.87$).

The Spearman's rank order correlation coefficient (i.e., Spearman's rho) was performed to identify relationships between variables. The Spearman's rho revealed a statistically significant positive relationship between the participants' self-rated creativity and the WF creativity score ($rs(41) = .38$, $p = .014$). Closer inspection of the relationships of the participants' self-rated creativity with the creativity sub-components showed a low significant relationship for *innovative* ($rs(41) = .31$, $p = .05$) and a highly significant relationship for *open to experience* ($rs(41) = .46$, $p = .003$). Additionally, positive relationships were identified for the effect of the ZD Master's program on the participants' creativity and participants' creativity compared to their private sphere ($rs(41) = .38$, $p = .016$).

5 Discussion

Overall, the findings showed that the higher students rated their creativity (subjective creativity), the higher their score for the WF creativity component (objective creativity). This confirms a direct link between subjective and objective creativity. The study also produced promising results for the Master's program ZukunftsDesign. Descriptive statistics established the students' positive attitude towards the program with regards to creativity. Only a small number of students were unsure whether the program promoted their creativity. A possible reason for this could be the professional background of those students. If they worked in creative jobs or had significant creative education before starting ZD, they may already hold a high level of creativity. Since this was not enquired as part of this study, this hypothesis remains to be tested. Future studies will include questions regarding the educational and professional background of the students to determine their level of creativity before commencing ZD.

One result that confirms the development of subjective creativity throughout the course of ZD is that students in higher semesters find themselves more creative compared to their fellow students in contrast to students in the first four semesters. This is an indication for an increase in perceived self-creativity throughout the Master's program and confirms the effectiveness of the creative learnings. Additionally, students who felt that ZD has a positive effect on their creativity also found themselves to be more creative compared to their private sphere. This provides further evidence for the individual creative development and learnings that are part of the Master's program.

Additional findings indicated that women tend to hold jobs where creativity is more important and they also achieved higher scores for the WF creativity component compared to men. However, when self-rating their creativity, there were no significant gender differences. This suggests that even though the objective creativity of women is significantly higher, they underestimate their subjective creativity while men overestimate the latter. The literature expresses caution when it comes to gender differences for creativity, as the field is very complex and the influencing factors are multiple [28]. Possible factors include barriers to promotion and career advancement, different domains, socialization, roles, and context (e.g., early childhood environments such as family, school, community, and college experiences) [29]. "Relative equality" between genders is assumed [28, 30]. However, according to a meta-analysis, there is a modest tendency for the female gender to score higher in creativity, particularly in *open to experience*, *fluidity*, *cognitive flexibility/adaptation*, and *divergent thinking* [28]. From the 67 studies included in this analysis, 30 did not find any statistically significant differences between the genders. Eight studies found higher creativity for females and four for males. Thirty studies showed no difference between genders on absolute scores but did reveal variations in their patterns. This trend also fits with the results of the current study, where female students performed significantly better than their male counterparts in the objective creativity assessment but similar gender differences could not be identified for all questions relating to creativity. Further research is required to explore this topic more deeply and include background information such as socialization and context.

5.1 Limitations

Although this study was only introduced as a pilot-study some limitations have to be noted. Due to the short time frame to collect data and a naturally limited pool of participants the sample size was comparatively small. Future research should allow more time for data collection and also include students outside of ZD to enable detailed comparative analysis with a random student sample. This will help to manifest the identified trends and provide opportunities for further practical and theoretical development.

Pairing the study with the WF also represented limitations as this naturally increased the length of the survey, while only the WF creativity component was used for the subsequent analysis. Despite the aforementioned advantages of image-based tests, questions and possible answers are limited to those that can be represented visually. Some images may not suit the respondents, so that the participants choose any arbitrary option and, thus, distort the result. Further, several questions of the WF were not relevant in times of a global pandemic due to the external constraints.

6 Conclusion

Findings of this pilot-study indicated that the Master's program ZukunftsDesign has a positive effect on students' individual creativity and does indeed promote creative talents. More in-depth research is required to manifest the identified trends and provide further practical and theoretical implications, which are important for the Master's program itself but also for any other similar degree. Further, results will indicate whether adaptations to the program may help to facilitate and improve the individual creativity of students. Overall, this pilot-study provides a baseline for future accompanying research and longitudinal comparative studies to examine the development of creativity within the educational landscape.

References

1. Mack, O., Khare, A., Krämer, A., Burgartz, T. (eds.): Managing in a VUCA World. Springer, Cham (2016). https://doi.org/10.1007/978-3-319-16889-0
2. Palmer, C.: Berufsbezogene Kreativitätsdiagnostik. Springer Fachmedien, Wiesbaden (2016)
3. Amabile, T.M.: A model of creativity and innovation in organizations. Res. Organ. Behav. **10**(1), 123–167 (1988)
4. Pieper, L.: Association Memory: Prototype and evaluation of a new creativity method in the context of divergent and convergent approaches for application in business as well as a teaching concept. In: Naples Forum on Service 2019, Naples, Italy (2019)
5. Pieper, L., Fischer, R., Hasenknopf, H.: "Innovation? Yes, I Can"–individually perceived creative self-efficacy as an effect of vividness targeting creativity methods. In: Spohrer, J., Leitner, C. (eds.) AHFE 2020. AISC, vol. 1208, pp. 17–21. Springer, Cham (2020). https://doi.org/10.1007/978-3-030-51057-2_3
6. Runco, M.A., Jaeger, G.J.: The standard definition of creativity. Creativity Res. J. **24**(1), 92–96 (2012)
7. Feist, G.J.: A meta-analysis of personality in scientific and artistic creativity. Pers. Soc. Psychol. Rev. **2**(4), 290–309 (1998)

8. Barron, F.: The disposition toward originality. J. Abnorm. Soc. Psychol. **51**(3), 478–485 (1955)
9. Plucker, J.A., Beghetto, R.A., Dow, G.T.: Why isn't creativity more important to educational psychologists? Potentials, pitfalls, and future directions in creativity research. Educ. Psychol. **39**(2), 83–96 (2004)
10. Futuredesign: Coburg University of Applied Sciences and Arts (2021). https://www.coburg-university.de/academic-studies/degree-programs-in-german/interdisciplinary/futuredesign.html
11. Rhodes, M.: An analysis of creativity. Phi Delta Kappa **42**, 305–310 (1961)
12. Hughes, D.J., Furnham, A., Batey, M.: The structure and personality predictors of self-rated creativity. Thinking Skills Creativity **9**, 76–84 (2013)
13. Schuler, H. Görlich, Y.: Kreativität. Ursachen, Messung, Förderung und Umsetzung in Innovation. Hogrefe, Göttingen (2007)
14. Kashdan, T.B., Silvia, P.J.: Curiosity and interest: the benefits of thriving on novelty and challenge. Oxford Handb. Positive Psychol. **2**, 367–374 (2009)
15. Furnham, A., Bachtiar, V.: Personality and intelligence as predictors of creativity. Pers. Individ. Differ. **45**(7), 613–617 (2008)
16. Martindale, C.: Personality, situation, and creativity. In: Glover, J.A., Ronning, R.R., Reynolds, C. (eds.) Handbook of Creativity, pp. 211–233. Plenum Press, New York, London (1989)
17. Scott, W.A.: Cognitive complexity and cognitive flexibility. Sociometry 405–414 (1962)
18. Ahmetoglu, G., Chamorro-Premuzic, T.: Personality 101. Springer, NewYork (2013)
19. Krosnick, J.: Response strategies for coping with the cognitive demands of attitude measures in surveys. Appl. Cogn. Psychol. **5**, 213–236 (1991)
20. Morgenson, F.P., Campion, M.A., Dipboye, R.L., Hollenbeck, J.R., Murphy, K., Schmitt, N.: Reconsidering the use of personality tests in personnel selection contexts. Pers. Psychol. **60**(3), 683–729 (2007)
21. Hough, L.M., Oswald, F.L., Ployhart, R.E.: Determinants, detection and amelioration of adverse impact in personnel selection procedures: Issues, evidence and lessons learned. Int. J. Sel. Assess. **9**(1–2), 152–194 (2001)
22. Attali, Y., Arieli-Attali, M.: Gamification in assessment: do points affect test performance? Comput. Educ. **83**, 57–63 (2015)
23. Leutner, F., Yearsley, A., Codreanu, S.-C., Borenstein, Y., Ahmetoglu, G.: From Likert scales to images: Validating a novel creativity measure with image based response scales. Pers. Individ. Differ. **106**, 36–40 (2017)
24. Nisbett, R.E., Ross, L.: Human Inference: Strategies and Shortcomings of Social Judgment, p. 45. Prentice-Hall, Englewood Cliffs (1980)
25. Dennis, J.P., VanderWal, J.S.: The cognitive flexibility inventory: Instrument development and estimates of reliability and validity. Cogn. Therapy Res. **34**(3), 241–253 (2010)
26. Goldberg, L.R.: A broad-bandwidth, public domain, personality inventory measuring the lower-level facets of several five-factor models. Pers. Psychol. Europe **7**(1), 7–28 (1999)
27. RedBull Wingfinder. https://www.redbull.com/int-en/wingfinder
28. Baer, J., Kaufman, J.C.: Gender differences in creativity. J. Creative Behav. **42**(2), 75–105 (2008)
29. Henderson, S.J.: The correlates of inventor motivation, creativity and achievement. Unpublished doctoral dissertation, Stanford University (2003)
30. Kogan, N.: Creativity and sex differences. J. Creative Behav. **8**, 1–14 (1974)

A Human-Centered Approach to Designing a Blended Learning Environment for an Interdisciplinary Audience

Matthia Leyendecker(✉) and Christian Zagel

Coburg University of Applied Sciences, Friedrich-Streib-Str. 2, 96450 Coburg, Germany
{Matthia.Leyendecker,Christian.Zagel}@hs-coburg.de

Abstract. This paper intends to provide insight into a project recently initiated and funded by the European Social Fund (ESF) which deals with the design of a blended learning environment for an interdisciplinary and cross-industry audience. The professional diversity of the participants as well as the different levels of expertise they contribute to the project is what makes it unique. The focus within this project is to investigate how human-centered design applied in a blended learning environment can narrow the gap between traditional ways of face-to-face teaching and e-learning. This human-centered approach shall optimize learning experiences to attain the greatest possible benefit for the diverse audience.

Keywords: Human-centered design (HCD) · Blended learning · Micro-learning · Interdisciplinary learning · Sustainability

1 Introduction

Companies in industrialized regions are increasingly exposed to strong public criticism in relation to how they deal with the topic of sustainability. This is due to multiple reasons: enormous energy consumption, pollution of air and water, waste that is difficult to recycle, and many more. Basically, many of the companies face the same - or at least similar – challenges, independent of their specific sector. The need for a changed mindset is recognized by companies, yet it is often only implemented partially. Regarding sustainability efforts, there is a lack of basic knowledge of the opportunities and, above all, of successful examples from the industry.

Creating cross-industry collaborations and fostering knowledge transfer by implementing innovative learning solutions appear to be a suitable approach to tackle these challenges. However, as learners' preferences in business contexts and their potential learning settings differ from typical educational settings, specific solutions are required. Therefore, the goal of the research project "Sustainability Cluster Upper Franconia" is to develop a blended learning environment for a business context which is based on a human-centered design approach in order to focus on the learners' individual needs.

After summarizing and discussing the underlying theory at the beginning, the project and its background will be introduced briefly. Lastly, the vision and the intended human-centered approach will be elucidated.

C. Leitner et al. (Eds.): AHFE 2021, LNNS 266, pp. 141–148, 2021.
https://doi.org/10.1007/978-3-030-80840-2_15

2 Theoretical Background

2.1 Technology Enhanced Learning (E-Learning)

With the emergence of the Internet starting in the mid 90's, the use of Information and Communication Technologies (ICT) in education has gained increasing importance within the last two decades. The so-called "Technology Enhanced Learning" (TEL) enabled a new way of learning. However, in the beginning the same content which had been used within classrooms was simply provided on a new platform, the Internet. But with the progress of technology, the boundaries of learning have increasingly dissolved. Learning can take place everywhere. The expansion of broadband Internet access and the development of technological devices have transformed us into a digital society. Today, TEL (or e-learning in general) offers a wide range of tools, including video streaming, quizzes, digital pinboards and many more.

Benefits of current forms of e-learning include providing an unlimited access to learning materials and information, flexibility and comfort, as well as individual pace and time for reflection, as online learning environments use to be asynchronous [1, 2]. However, virtual classrooms cannot provide instant clarification of the content being covered, nor do group discussions or activities that are normally conducted in personal modalities unless they are held synchronously. Blended learning concepts try to pick up on that by combining physical presence learning and online-learning.

2.2 Blended Learning

The objective of blended learning formats is to compensate the shortcomings of both concepts, f-2-f-classrooms and online learning, and combine "the best of both worlds" [2, 3]. Empirical studies have shown that blended learning has a positive impact on learners' engagement as well as learning outcomes [4, 5] and satisfaction [6]. Furthermore, it facilitates learner-learner interactions, as well as learner-instructor interactions [7, 8] and the development of learner autonomy [9, 10].

Blended learning is a quite ambiguous term as there are multiple definitions and interpretations as for example demonstrated by Oliver and Trigwell [11] or Driscoll [12]. A very common definition was coined by Graham, who defines blended learning formats as learning systems that "combine face-to-face instruction with computer mediated instruction" [13]. Another definition has been given by Garrison and Kanuka, who see blended learning as "the thoughtful integration of classroom face-to-face learning experiences with online-learning experiences" [14]. This definition highlights the effort needed for the design of a blended learning environment and implies that it involves more than merely combining two formats. For example, having computer-based trainings in labs or having f-2-f lessons and only providing content online (e.g., for homework) would be a rather simple combination.

Having examined different blended learning approaches, such as the "flipped classroom", intensive learning or blended Massive Open Online Courses (MOOCs), it appears they all share the same problem: the *thoughtful integration* or, in other words, ensuring the right balance of face-to-face (f-2-f) and online modalities and activities in order to achieve optimum learning outcomes and meet learners' needs [13, 15, 16]. While the balance within the flipped classroom model is relatively equal, the intensive learning framework predominantly focusses on online modalities by having only a few f-2-f units. Blended MOOCs combine f-2-f workshops with a closed circle of participants with a wide offer of online content which is made available to a public audience [17]. Accordingly, a major challenge is to establish a well-balanced ratio of f-2-f teaching and online modalities. However, there is more to consider. Further challenges include technological issues, an appropriate choice of media components, maintaining ongoing communication and motivation on the part of the audience and most importantly, meeting learners' expectations and needs. Most of the current blended learning formats seem to lack personalization in relation to content, learning behavior and learning experience. However, gaining a deep understanding of learners needs, abilities, and limitations helps to create an intuitive and seamless learning experience which makes e-learning easier and more inspiring.

Consequently, offering a blended learning format does not necessarily mean creating a successful learning environment. The balance and blend of the various components have to be chosen and planned carefully to achieve effective learning outcomes and satisfy learners' needs. Even when combining the benefits of both formats, they need to be suitable to the context, the content and the audience. Learning concepts cannot be seen as universal and eternal because contextual framings differ and technological modalities develop continuously. Hence, a specific framework can only serve as a foundation but needs to be adjusted to and designed for the context of the particular learning environment.

2.3 Personalized Learning

Personalization gains more and more importance, whether it is peoples' desire for individuality, or to distance from mass production. Within the educational field, the term aims at personalized learning methods and materials in order to enhance the learning outcome [18]. Personalized learning can be defined as "tailoring learning for each student's strengths, needs and interests—including enabling students' voices and choices in what, how, when and where they learn—to provide flexibility and supports to ensure mastery of the highest standards possible" [19]. Hence, in contrast to most frameworks which earmark identical types of instruction, assignments and assessments for all, personalized learning considers learners' individual differences and needs. An individualized instruction aims to enable students to set their own target, to learn at their own pace and to ensure a flexible learning environment [20].

However, individualized instruction or personalized learning cannot be seen as the overall solution. Besides many benefits, there are also drawbacks. Designing personalized learning activities also means authoring and providing a variety of materials which requires extra preparation time and money [21]. Another problem is the lack of collaboration within this concept. Collaboration plays a crucial role regarding learning outcomes as social processes are part of the educational and learning psychology and may affect the learning process in a positive way.

All of the approaches introduced above have their respective benefits and drawbacks. It is obvious that there is no one-size-fits-all instructional model which satisfies all learners as there are individual disparities such as prior knowledge or cognitive style

[22]. In the context of the project, we will examine whether a human-centered approach can bridge the gap between the different education models. However, the aim is not to define an all-embracing solution but rather to establish human-centered design as a key principle for designing learning systems in general.

3 Case Study: Blended Learning for an Interdisciplinary Audience

3.1 Background

The project "Sustainability Cluster Upper Franconia – Future viability through sustainable learning and business" is funded by the European Social Fund (ESF) and the Virtual University of Bavaria (vhb). As a project sponsor of the ESF, the vhb coordinates and supervises university projects from a wide range of disciplines which are primarily aimed at small and medium-sized enterprises (SMEs) in parts of Bavaria that are particularly affected by demographic change. The goal is to sustain these enterprises by imparting new types of methods and state-of-the-art knowledge. The aim is to positively affect the companies' ability to innovate. In addition, the projects create important networks and partnerships between universities and the participating companies.

The project specifically addresses the region of Upper Franconia (Bavaria, Germany). Although it is one of the regions with the highest density of industrialization in Europe, it faces a strong negative demographic development. Against this background, the part-time master's degree program "ZukunftsDesign" (Future Design) was created in 2016. Its goal is to maintain intensive and sustainable contact with the regional economy and organizations in a special way. ZukunftsDesign is part of a large regional innovation network, consisting of more than 70 companies from a great variety of sectors (manufacturing industry, service providers, etc.). This heterogeneity is also reflected within the companies participating in the "Sustainability Cluster Upper Franconia" project. The experience from numerous workshops and cooperation meetings with these companies has shown that there are ongoing social and political discussions on the subject of sustainability. Sustainability is of great interest and represents a common challenge for the upcoming years, regardless of the respective field of business. The topics of e-learning and blended learning are also seen as a suitable way of reaching employees in a simple and innovative way, and to provide sustainable on-the-job trainings. There is a great willingness to intensify the existing networks, but also to cooperate with regional universities.

3.2 Vision

The central goal of the project is to strengthen and to sustain the innovative power of the Upper Franconian region by promoting the local sustainability management and offering economic advantages at the same time. Therefore, an important aspect is not only to address the companies' needs and challenges in relation to sustainability, but also to identify best practices among the companies. Some of them have already started implementing excellent approaches to solve particular aspects in relation to sustainability management. Hence, the idea is not to have a unilateral flow of information and knowledge but rather to create an environment where participants can learn from one another and engage in cross-industry networks. Consequently, the companies can rely on an interdisciplinary knowledge transfer resulting in synergetic effects which emerge by exchanging experiences and implementing goals as a team.

3.3 A Human-Centered Approach to E-Learning

In order to focus on the interaction and exchange of experiences, the focus within the blended learning environment is set on the participants. Therefore, human-centered design principles and activities will be considered. The involvement of all actors has a significant role for obtaining a profound understanding of what they need, what they already know and what they aspire. A central aspect of getting a better understanding of the users is their learning context and learning behavior. The different ways learners use to perceive, to grasp and to conceptualize information must be considered when designing the content. Furthermore, the fact that all participants are full-time employed demands for content and synchronous lessons which fit into their individual work situation.

After having examined the user needs via workshops, interviews and online-surveys, the content and course structure are drawn up to fit the identified needs. Therefore, a central learning platform is designed which will then serve as the technological foundation. It has the purpose of providing the particular content to the participants in different ways such as video lectures, working sheets or quizzes. The content, however, should not only be based on the theoretical knowledge with regard to sustainability and the Sustainable Development Goals (SDGs) elaborated by the United Nations (UN), but also on the best-practices which have been identified within presence workshops and can directly be integrated into the e-learning content. This can be done in form of interviews, experience reports, illustrative examples or case studies. The approaches and best-case examples of other participating companies should stimulate discussions and network activities among the participants. The goal is to create content that enables participants to learn from each other. The interaction and exchange of experiences between the participants can result in synergetic effects and deeper learning experiences. Hence, focusing on participants as well as their interaction and collaboration is essential. Collaboration in general means working with others. In the educational field, working can also include learning with others. However, the collaboration should not be reduced to learning *with* others only; it should rather be understood as a resource to learn *from* others. The objective is to seek active learner-learner as well as learner-content-interaction. Therefore, the central platform requires certain tools and possibilities that allow an ongoing exchange of information, experiences, and knowledge in addition to the provision of an adequate learning material. While exchanging information and discussing certain topics, participants might realize that even though they work in completely different industry sectors, they have common issues and challenges regarding sustainability management. Within f-2-f workshops they are able to continue and deepen discussions of greater importance. Furthermore, these workshops serve as a practical framework in which participants have the possibility to generate innovative ideas and find solution approaches to specific overlapping challenges.

Within this concept, the idea of linking digital content to presence workshops is of great importance. A previous study already presented an innovative method to combine physically available information with digital content as so called "Method Cards" [23]. In the form of traditional playing cards, individual thematic aspects are presented as a brief overview. Each participating company will receive a set of method cards which can be used as a working tool within workshops, either to retrieve and deepen special knowledge or to generate ideas. The 17 SDGs serve as the basis for thematic content, especially those which are more relevant within entrepreneurial contexts. In addition, the cards illustrate a variety of different methods such as brainstorming techniques, "Six thinking hats" or the "6-3-5 method". For a more in-depth discussion, the cards refer to further content on the online platform (via QR code or NFC tag). In particular, micro-learning content in the form of detailed short videos or quizzes is presented. According to perspectives in teaching and learning psychology [24], the usage of cards in learning situations can significantly enhance the learning process. In addition to that, the playful aspect of the card set might strengthen the social aspect.

Considering human-centered design principles, iterations play an essential role. HCD is an inherently iterative approach to problem solving as its core idea is obtaining feedback from the target group and incorporating it into any subsequent steps of development. By continually iterating, refining and improving, the design solution can be validated before it is finally adopted and embraced. Regarding the project, this means that a core focus will be set on exploring, failing, adjusting, testing, refining in order to have a solution that finally fits the participants needs. Precisely, there will be usability-tests of the platform which reveal if learners can reach their goals in an effective, efficient and satisfying way. The findings serve as an indication for what must be improved and what already works well.

Hence, by applying a human-centered approach we aspire the *thoughtful integration* of classroom f-2-f learning experiences with online-learning experiences.

4 Conclusion and Future Prospects

Designing an e-learning environment is naturally challenging as learning methods and communication differ in many ways from traditional teaching and learning. With blended learning concepts organizations and educational institutions pick up on that by combining online learning materials and interactions with traditional classroom-based lessons. A major challenge, however, is to establish a well-balanced ratio of f-2-f teaching and online modalities. It appears that there still is a research gap of how online and f-2-f modalities can be thoughtfully combined. Although a lot of research is done within the field of blended learning and exploring the benefits of combining f-2-f and online modalities, less is known about how to ideally blend these two formats. Even though there are many approaches and frameworks such as the flipped classroom, there is no one-size-fits-all solution. This is because most of them either focus on how to combine both modalities or how to use the rising number of technological functionalities than on considering the concrete learners' expectations and needs. Having a deep understanding of learners' needs, abilities and limitations helps to create an intuitive and seamless learning experience. Learning environments with a focus on the learner and their learning behavior enable them to achieve their goals in a more efficient and satisfying way. Furthermore, a *learner-centered* approach enhances the overall learning experience which affects aspects such as motivation and learning progress.

By means of the project "Sustainability Cluster Upper Franconia" we want to create a new blended learning approach by applying human-centered design processes and activities. One main goal of the project is to examine how HCD can enhance the learning environment with all its components. The core idea is not to simply see the technological infrastructure as a delivery tool, but rather as a cognitive partner remaining the focus on the learners, their interaction and their learning experience.

References

1. Cundell, A., Sheepy, E.: Student perceptions of the most effective and engaging online learning activities in a blended graduate seminar. OLJ 22 (2018)
2. Hall, S., Villareal, D.: The hybrid advantage: graduate student perspectives of hybrid education courses. Int. J. Teach. Learn. Higher Educ. **27**, 69–80 (2015)
3. Baragash, R.S., Al-Samarraie, H.: An empirical study of the impact of multiple modes of delivery on student learning in a blended course. Ref. Librarian. **59**, 149–162 (2018)
4. Dziuban, C., Hartman, J.L., Cavanagh, T., Moskal, P.: Blended courses as drivers of institutional transformation. Blended Learning across Disciplines: Models for Implementation 17–37 (2011)
5. Means, B., Toyama, Y., Murphy, R., Bakia, M., Jones, K.: Department of Education (ED), O. of P., Evaluation and Policy Development, SRI International: Evaluation of Evidence-Based Practices in Online Learning: A Meta-Analysis and Review of Online Learning Studies. US Department of Education (2009)
6. Martínez-Caro, E., Bolarín, F.: Factors affecting students' satisfaction in engineering disciplines: traditional vs. blended approaches. Eur. J. Eng. Educ. 36, 473–483 (2011)
7. Aspden, L., Helm, P.: Making the connection in a blended learning environment. Educ. Media Int. **41**, 245–252 (2004)
8. So, H.-J., Brush, T.A.: Student perceptions of collaborative learning, social presence and satisfaction in a blended learning environment: relationships and critical factors. Comput. Educ. **51**, 318–336 (2008)
9. De George-Walker, L., Keeffe, M.: Self-determined blended learning: a case study of blended learning design. Higher Educ. Res. Dev. **29**, 1–13 (2010)
10. Snodin, N.S.: The effects of blended learning with a CMS on the development of autonomous learning: a case study of different degrees of autonomy achieved by individual learners. Comput. Educ. **61**, 209–216 (2013)
11. Oliver, M., Trigwell, K.: Can "Blended Learning" be redeemed? E-Learning. **2**, 17–26 (2005)
12. Driscoll, M.: Blended learning: let's get beyond the hype. E-learning 3 (2002)
13. Graham, C.: Blended learning systems: definition, current trends, and future directions. Presented at the January 1 (2006)
14. Garrison, D.R., Kanuka, H.: Blended learning: Uncovering its transformative potential in higher education. Internet Higher Educ. **7**, 95–105 (2004)
15. Bliuc, A.-M., Goodyear, P., Ellis, R.A.: Research focus and methodological choices in studies into students' experiences of blended learning in higher education. Internet Higher Educ. **10**, 231–244 (2007)
16. Torrisi-Steele, G., Drew, S.: The literature landscape of blended learning in higher education: the need for better understanding of academic blended practice. Int. J. Acad. Dev. **18**, 371–383 (2013)
17. McKenna, K., Gupta, K., Kaiser, L., Lopes, T., Zarestky, J.: Blended learning: balancing the best of both worlds for adult learners. Adult Learn. **31**, 139–139–149 (2020)

18. Inayat, I., Amin, R. ul, Inayat, Z., Salim, S.S.: Effects of collaborative web based vocational education and training (VET) on learning outcomes. Comput. Educ. **68**, 153–166 (2013)
19. Patrick, S., Kennedy, K., Powell, A.: International Association for K-12 Online Learning: Mean What You Say: Defining and Integrating Personalized, Blended and Competency Education. International Association for K-12 Online Learning (2013)
20. Campbell, R.J., Robinson, W., Neelands, J., Hewston, R., Mazzoli Smith, L.: Personalised learning: ambiguities in theory and practice. Br. J. Educ. Stud. **55**, 135–154 (2007)
21. O'Donnell, E., Sharp, M., Wade, V., O'Donnell, L.: Challenges encountered in creating personalised learning activities to suit students learning preferences. Learn. Manage. Syst. Instruct. Des. Best Pract. Online Educ. 263–287 (2013)
22. Chen, S.Y., Macredie, R.: Web-based interaction: a review of three important human factors. Int. J. Inf. Manage. **30**, 379–387 (2010)
23. Zagel, C., Grimm, L., Luo, X.: Method cards – a new concept for teaching in academia and to innovate in SMEs. In: Ahram, T.Z. (ed.) AHFE 2018. AISC, vol. 787, pp. 230–241. Springer, Cham (2019). https://doi.org/10.1007/978-3-319-94229-2_22
24. Mietzel, G.: Pädagogische Psychologie des Lernens und Lehrens. Hogrefe, Göttingen (2017)

Interdisciplinary Service Development for Construction Suppliers with the Help of an Innovation Support System

Matthias Hille(✉), Melissa Dilek Izlisu, and Freimut Bodendorf

Information Systems – Services, Processes, and Intelligence, Friedrich-Alexander-University Erlangen-Nuremberg, Nuremberg, Germany
{matthias.hille,dilek.izlisu,freimut.bodendorf}@fau.de

Abstract. In the construction supply industry, there is increasing price pressure on semi-finished products. This price pressure is strongly influenced by the commoditization of the product business. One way of differentiating from competitors in product-heavy industries such as the construction industry is to develop and offer digital services. First, this paper presents the status quo of service development processes within construction suppliers based on deep-dive interviews with long-term industry experts. Second, a process model to enhance service development at construction suppliers is presented. The proposed methodology recombines existing phases and actions while adding additional process steps and decision gateways. Third, an innovation support system is outlined which supports the interdisciplinary development process for smart services. The support system enables the sharing of specialized knowledge among broad and interdisciplinary user groups.

Keywords: Smart services · Digitalization · Development processes · Decision support system · Digital transformation · Business models · Construction industry · Digital services

1 Challenges for Suppliers in the Construction Industry

Up to now, various supplier industries have been dependent on the manufacture of semi-finished products and products with comparatively low added value [1]. For many products, technical standardization is increasing while the number of variants remains high. As a result, product-side differentiation between individual competitors is shifting towards a price war. Due to a tense price situation, decreasing margins are increasingly worsening the competitive situation of existing companies [2]. Similar to mechanical engineering and manufacturers of finished products, manufacturers of semi-finished products are also under increasing pressure from the market [3]. One strategy to counteract declining margins is to offer complementary smart services (also called digital services). By offering complementary services or stand-alone digital solutions, companies can open up new business areas and thus ensure their competitiveness [4, 5]. For this paper, smart services are defined as solutions that collect, process, and interpret information and are marketed as value-added digital solutions through digital distribution

C. Leitner et al. (Eds.): AHFE 2021, LNNS 266, pp. 149–156, 2021.
https://doi.org/10.1007/978-3-030-80840-2_16

channels [6–8]. This paper focuses on the supplier sector of the construction industry. Subcontracting companies in the construction industry often lack experts in the design, evaluation, and technical development of software and digital services [9, 10]. The objective is to facilitate the development of intelligent service portfolios from strategic, organizational and technical perspectives using a decision support system (DSS). The system intervenes after idea generation for digital services and leads through a structured process to technical conception. The final evaluation gives an outlook on which advantages and influences the system represents in business practice. Testing in practice shows how the development of digital services can be improved in the environment of manufacturing companies.

2 Status Quo of Development Processes for Smart Services in the Construction Industry

Notable examples of development processes of digital services are in particular the linear models according to [11, 12] and[13] as well as the iterative ones according to [14, 15]. At the moment, a status quo prevails for the development process in supplying companies. In order to provide a better description of the current situation within construction suppliers, expert interviews are used. The interviews are performed with 16 experts within nine companies from the construction industry. The interviews are built upon questionnaires within semi-structured qualitative interviews according to [16]. All 16 interviews are embedded into a multiple case study design by [17]. The results obtained from the survey are combined with the results from the literature research about service development processes and presented below.

Although there are various similarities with the linear phase model and the classic view as a stage-gate process, there exist also other views in the sense of agile working during the development of digital solutions. As the focus of current development processes within construction suppliers is most of the time within one activity and not across the board; the interviews highlight a combination of linear phase model and agile working [18]. The current development process is thus composed of the phases *idea generation*, *technical realization* and *market launch*, the activities involved in each with a corresponding illustration, can be found in Fig. 1.

During the idea generation, the customer's needs are first analyzed, and solutions are proposed, then resources and a cost-benefit factor are considered; these are subject to various uncertainties in this early phase; a decision is made on the basis of the resulting recommendation. This is worked out in more detail in the technical implementation phase, where the focus of the implementation is typically outsourced to external service providers; in-house developments are usually an exception. Once the MVP has been developed, the aim is generally to achieve feedback from customers and sales colleagues with minimal functionalities. The further development and improvement are addressed afterwards. Possible interface problems and architectural challenges are only addressed in a few cases in early phases, although this can lead to considerable additional costs at a later stage. The next logical step is the market launch phase, in which feedback from the customer and further adaptation of the solution strategies take place.

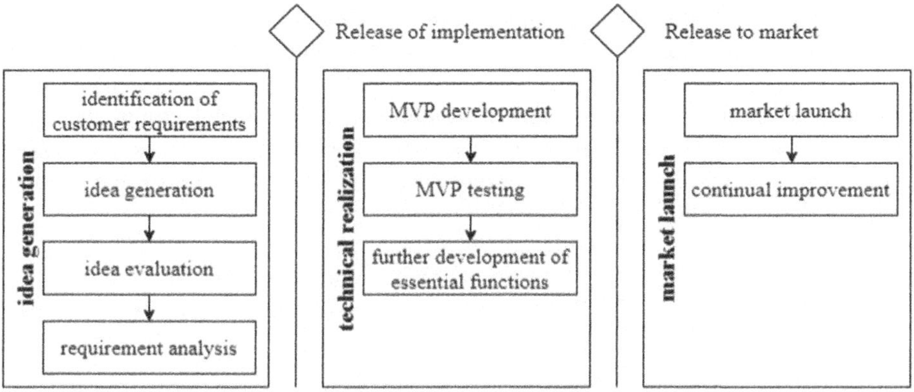

Fig. 1. Status quo of the smart service development process in construction suppliers

This current situation is accompanied by a number of problems, which will be briefly outlined here. Example problem areas include customer needs, requirements analysis and market introduction. For customer needs, it is above all the difficulty of obtaining homogeneous feedback from the various target groups as well as the increasing complexity of possible solutions for possible needs. The requirements analysis, on the other hand, shows difficulties during the requirement specification and especially for technical requirements. Often requirements are not known during the conception phase and the interviewed companies are not aware of the technical efforts needed. In the case of market introduction, the overall duration of development is seen critically, as this usually results in immense costs. Longer than planned developments are often caused by inflexible, entrenched development structures, but fundament on missing information during the requirement analysis and the technical realization phase. With these problems in mind, the following proposals for solutions will be elaborated accordingly.

3 Modern Development Processes for Service Engineering

Since a structured development process for digital services is necessary for said industries, the status quo requires revision. The new variant corresponds roughly to the stage-gate process [11], but is significantly more detailed, as it comprises five phases with a total of 21 activities. The proposed process is based on service modularization and includes multiple iterations and allows switching to previous phases while the process is ongoing. A simplified graphical representation can be found in Fig. 2.

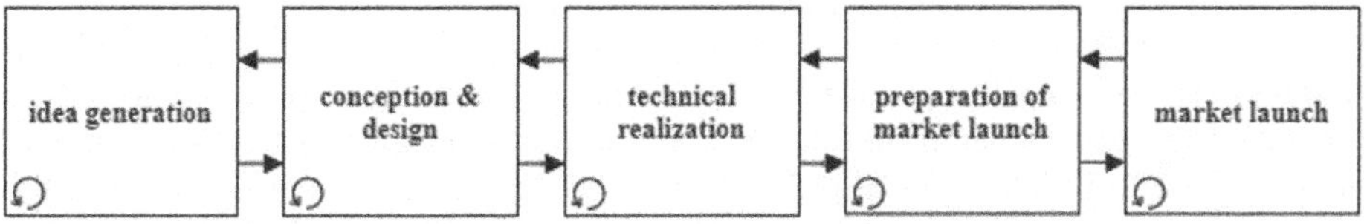

Fig. 2. Proposed process model for the development of smart service in supplying companies

In addition to the previously mentioned phases, the phases "conception & design" and "preparation for Market launch" are now added. The idea generation sets itself again to the goal of finding and evaluating appropriate solutions for analyzed problems of the customers, whereby it can concern both existing and future target groups. From the evaluation, parameters are obtained, which are to be regularly re-evaluated and revised if necessary. In order to elaborate the requirements profile and the necessary technical modules for the new digital service, a DSS is used.

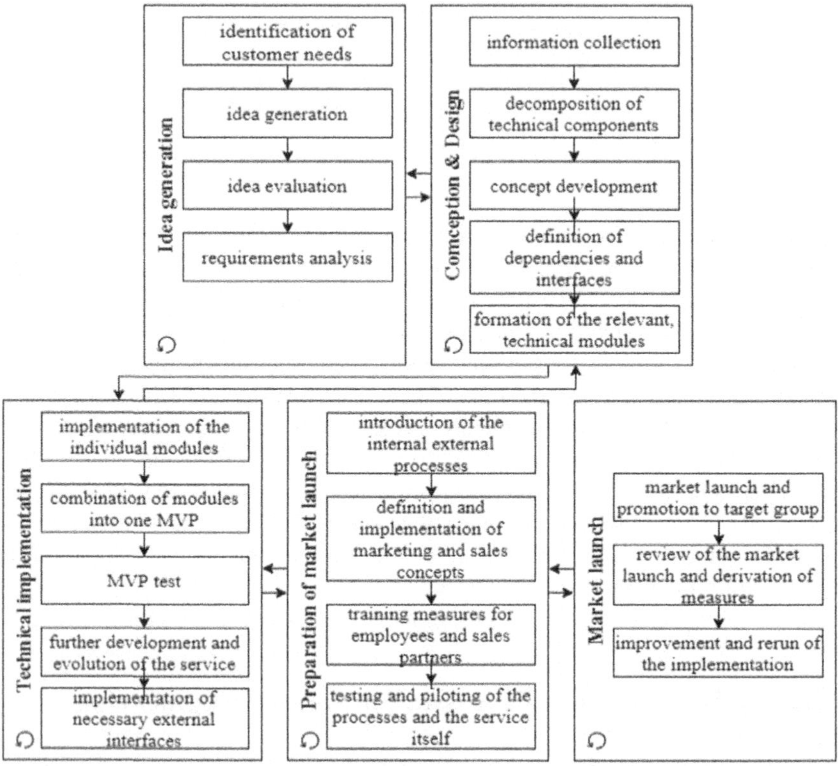

Fig. 3. Proposed process model for the development of smart service in supplying companies, in detail

Conception & design deals with the modular structure of the service to be developed on the basis of the previously compiled data; the required modules, as well as their dependence on each other and external interfaces are worked out. The resulting structure serves as the basis for further development, whereby the application of a DSS generates the greatest possible benefit here; a significant extension of [19] and [14] can be found. The technical implementation takes into account the implementation of the elaborated approach, whereby the elaborated modules can be developed individually but simultaneously. Modules from already existing applications are also integrated into this process, this composition enables the implementation of the Smart Service. According to the elaborated MVP, initially only minimum requirements and basic functions are implemented, so that sufficient functionality as well as marketability can be ensured. A continuous feedback collection ensures constant refinement and expansion of the functional scope. This phase takes particular account of the concept of [15], recommending the use of an agile stage-gate process for development [11].

In order to highlight the innovations of the process flow described here in comparison to the status quo, a detailed graphical representation is provided in Fig. 3.

Before the actual market entry takes place, it is prepared. This includes technical aspects and test runs, as well as training and marketing or sales concepts. The conclusion of this phase begins with a test run in live operation; any problems that arise are re-evaluated and rectified; the length and diligence of the previous phases has a decisive influence on the quality of the system now in use. This is followed by the market launch, where the experiences and solution approaches of the previous test phases have a significant influence on the further development and application. The service created is now used for service delivery; feedback to be collected regularly is used to define improvement measures and to revise the technical implementation phase again in order to achieve an increase in performance as well as a wider range of functions. This continuous improvement is based on [15].

4 Decision Support Systems for Service Innovation Within the Construction Supplier Industry

To implement the in Sect. 3 proposed smart service development process a DSS is developed [20, 21]. This system focuses especially on the phases *conception & design* and *technical implementation*. Therefore, the system already starts with the activity requirements analysis within *idea generation* and influences the various activities until the market launch. The main problems which are outlined in Sect. 2 are based on missing employee competencies and a lack of existing experience with smart service development in supplying companies [22, 23]. The DSS has the aim to get the requirements for a smart service from an unexperienced user and transform them into structured and machine-readable requirements in the background. With these transformed requirements the DSS evaluates which functionalities and technical dependencies the described service has and creates a list of technical modules [24, 25]. These modules are a part of a list of 67 technical modules for smart services in the construction industry. This database is the result of extensive market research with 468 analyzed smart services from 297 companies in the construction industry in eight different countries [26, 27]. The DSS evaluates which functionality depends on which technical module and which module depends itself on other modules. At the end a list of all needed modules is created and shown to the user of the DSS. All modules are enriched with an estimation of the development effort for this module and relevant external factors which have to be kept in mind. The user interface to the user is as simple as possible and consists of initial questions to collect all requirements and the full list of recommended modules [28, 29]. The amount of questions depends on the answers from the user and the shown questions are created with structured analysis and design technique (SADT) and displayed dynamically to present only relevant questions to the user [30, 31]. Figure 4 presents the user journey of a regular user within the DSS [32]. The user journey describes the key functionality of the given DSS, further functionality like management of datasets, evaluation of ratings and user action are also implemented.

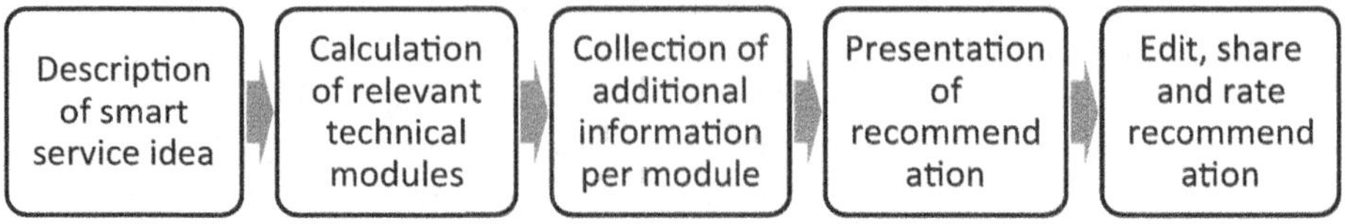

Fig. 4. User journey within the DSS

The user is normally part of the project team for the service development team and start to use the system within the requirements analysis activity of the idea generation phase. Afterwards the DSS enables the project team to perform the conception & design phase faster and with a better information base. With this information base the project team can interact directly with external service providers and internal IT departments. The conception & design phase can be executed more structured and relevant external factors as well as dependencies, estimated effort and necessary interfaces are highlighted in an early phase of the project. As described in Sects. 2 and 3 especially the interfaces, dependencies and internal requirements are a key problem with current development processes. Therefore, the DSS supports the earlier technical conception of a new service as well as the possibility to identify necessary interfaces and internal dependencies. The users of this innovation support system are set in a position where the knowledge given by the DSS helps to include more stakeholders and company departments earlier to gain higher quality, sunken costs and improved schedules during the development process. These improvements are proofed by [27] with an extensive practical evaluation in more than ten corporates in Europe.

5 Discussion and Summary

This paper describes the status quo of the current situation of smart service development within construction suppliers. The interviews gain deep insights into practical approaches which can be found within these companies nowadays, but they also highlight multiple problems, e.g., employee know-how, existing structures, previous experience. Therefore, a new approach for a development process based on modularization is proposed in Sect. 3. This approach is tested and evaluated with a described DSS in Sect. 4 and focuses on the main conception phase of the smart service development process. The interviews show the benefit of such an innovation support tool and its integration into a suitable, modern and agile development process. Future research should focus on a comparison of practical development projects in supplying companies with and without the proposed process and innovation support tools. The authors want to continue to accompany projects in multiple companies and measure relevant KPIs to get objective and non-biased indicators on the effectiveness of the proposed processes and tools.

References

1. Lusch, R.F., Nambisan, S.: Service innovation: a service-dominant logic perspective. MIS Q. **39**, 155–176 (2015)
2. Jacob, F., Ulaga, W.: The transition from product to service in business markets: an agenda for academic inquiry. Ind. Mark. Manage. **37**, 247–253 (2008)
3. Moeller, S.: Characteristics of services - a new approach uncovers their value. J. Serv. Mark. **24**, 359–368 (2010)
4. Gebauer, H., Gustafsson, A., Witell, L.: Competitive advantage through service differentiation by manufacturing companies. J. Bus. Res. **64**, 1270–1280 (2011)
5. Agarwal, N., et al.: Nutzergenerierte Dienstleistungssysteme zur digitalen Transformation von Organisationen. In: Stich, V., Schumann, J.H., Beverungen, D., Gudergan, G., Jussen, P. (eds.) Digitale Dienstleistungsinnovationen, pp. 281–306. Springer, Heidelberg (2019). https://doi.org/10.1007/978-3-662-59517-6_15
6. Pöppelbuß, J.: Smart service. https://www.enzyklopaedie-der-wirtschaftsinformatik.de/lexikon/informationssysteme/Sektorspezifische-Anwendungssysteme/smart-service
7. Beverungen, D., Breidbach, C.F., Poeppelbuss, J., Tuunainen, V.K.: Smart service systems: an interdisciplinary perspective. Inf. Syst. J. **29**, 1201–1206 (2019)
8. Kagermann, H., et al.: Smart service welt. Umsetzungsempfehlungen für das Zukunftsprojekt Internetbasierte Dienste für die Wirtschaft (Abschlussbericht) (2015)
9. Oesterreich, T.D., Teuteberg, F.: Understanding the implications of digitisation and automation in the context of Industry 4.0: a triangulation approach and elements of a research agenda for the construction industry. Comput. Ind. **83**, 121–139 (2016)
10. Boton, C., Rivest, L., Forgues, D., Jupp, J.: Comparing PLM and BIM from the product structure standpoint. In: Harik, R., Rivest, L., Bernard, A., Eynard, B., Bouras, A. (eds.) PLM 2016. IAICT, vol. 492, pp. 443–453. Springer, Cham (2016). https://doi.org/10.1007/978-3-319-54660-5_40
11. Cooper, R.G.: Stage-gate systems: a new tool for managing new products. Bus. Horiz. **33**, 44–54 (1990)
12. DIN Deutsches Institut für Normung e.V.: Service Engineering. Entwicklungsbegleitende Normung für Dienstleistungen. Deutsches Institut für Normung, Berlin (1998)
13. Edvardsson, B., Olsson, J.: Key concepts for new service development. Serv. Ind. J. **16**, 140–164 (1996)
14. Kingman-Brundage, J., Shostack, L.G.: How to design a service. In: Congram, C.A., Friedman, M.L. (eds.) The AMA Handbook of Marketing for the Service Industries, pp. 243–261. AMA, New York (1991)
15. Scheuing, E.E., Johnson, E.M.: A proposed model for new service development. J. Serv. Market. (1989)
16. Lamnek, S.: Qualitative Sozialforschung: Lehrbuch. Beltz (2005)
17. Yin, R.K.: Case Study Research and Applications. Sage, Thousand Oaks (2018)
18. Cooper, R.G.: Agile–Stage-Gate hybrids: the next stage for product development blending Agile and Stage-Gate methods can provide flexibility, speed, and improved communication in new-product development. Res. Technol. Manag. **59**, 21–29 (2016)
19. Anderl, R., et al.: VDI 2221 Entwicklung technischer Produkte und Systeme - Gestaltung individueller Produktentwicklungsprozesse Blatt 1 Entwurf - Inhaltsverzeichnis (2018)
20. Sprague Jr., R.H.: A framework for the development of decision support systems. MIS Q. 1–26 (1980)
21. Kampker, A., Schuh, G., Schittny, B.: Unternehmensstruktur. In: Schuh, G., Kampker, A. (eds.) Strategie und Management produziernder Unternehmen: Handbuch Produktion und Management 1, pp. 133–229. Springer, Berlin (2011)

22. Goger, G., Piskernik, M., Urban, H.: Studie: Potenziale der Digitalisierung im Bauwesen (2017). http://www.forschung-bau.at/media/1369/201802_studie-potenziale-der-digitalisierung.pdf
23. Rieder, A., Bröckl, A.: Für die sinnvolle Umsetzung von digitalen Tools müssen sich Geschäftsmodelle und Strukturen ändern: Anton Rieder, Geschäftsführer des Bauunternehmens RIEDERBAU über Potenziale bei der Digitalisierung in der Baubranche und die Auswirkungen der neuen Technologien auf die Arbeitswelt. AMS Info 2018 (2018)
24. Eissens-Van der Laan, M., Broekhuis, M., van Offenbeek, M., Ahaus, K.: Service decomposition: a conceptual analysis of modularizing services. Int. J. Oper. Prod. Manag. (2016)
25. Schilling, M.A.: Toward a general modular systems theory and its application to interfirm product modularity. Acad. Manag. Rev. **25**, 312–334 (2000)
26. Hille, M., Bodendorf, F.: Smart Services in der Baubranche. In: Automatisierung und Personalisierung von Dienstleistungen, pp. 417–438. Springer, Wiesbaden (2020). https://doi.org/10.1007/978-3-658-30166-8_17
27. Schaub, C., Hille, M., Schulze, J., Bodendorf, F.: Implementierung eines Entscheidungsunterstützungssystems für die Entwicklung von digitalen Dienstleistungen bei Zulieferunternehmen der Bauindustrie. In: Proceedings of the International Conference on Wirtschaftsinformatik (2021)
28. Sommerville, I.: Software Engineering. Pearson Education Limited, Harlow (2016)
29. Vogel, O., et al.: Architektur-Anforderungen (WARUM). In: Software-Architektur: Grundlagen — Konzepte — Praxis, pp. 103–124. Spektrum Akademischer Verlag, Heidelberg (2009). https://doi.org/10.1007/978-3-8274-2267-5_5
30. Ross, D.T.: Structured analysis (SA): a language for communicating ideas. IEEE Trans. Softw. Eng.16–34 (1977)
31. Marca, D.A., McGowan, C.L.: SADT: Structured Analysis and Design Technique. McGraw-Hill Inc., New York (1987)
32. Balzert, H.: Lehrbuch der Softwaretechnik: Basiskonzepte und requirements engineering. Spektrum Akademischer Verlag, Heidelberg (2009). https://doi.org/10.1007/978-3-8274-2247-7

Detecting Effective Impulses for the Socio-ecological Change - Presentation of the Interdisciplinary Research Concept Behind Game of Rain

Matthias Scheibe(✉), Hauke Hasenknopf, and Oliver Frenzel

Coburg University of Applied Sciences and Arts, Master "ZukunftsDesign", Friedrich Streib Straße 2, 96450 Coburg, Germany
Matthias.scheibe@hs-coburg.de

Abstract. The waste of valuable fresh water is one of the major environmental threats for a future worth living in. Our goal is to make people, especially children, aware of the subject of virtual water and encourage their civic involvement to reduce the water consumption in the production of everyday goods. In the next three years we intend to research if and how such a socio-ecological change of individuals' and groups' consciousness could be facilitated through game-based mechanics. For this endeavor, it is crucial to identify and research specific change agents. To ensure this, the use of qualitative and quantitative methods at several measuring points is planned. In addition, reactive and non-reactive research methods are combined and the results are supplemented by surveys in the environment of the studential participants about any behavioral changes.

Keyword: Virtual water · Civic engagement · Mixed-methods · Gamification

1 Introduction

The insight that geological raw materials and fossil fuels in particular are just as limited as the planet's ability to absorb waste, including in the form of emissions [1], coincides with the realization that the social challenges of the global south stem from lifestyle of the global north. Climate change, exorbitant resource consumption and drinking water pollution affect everyone and endanger the future of humanity. Private decisions and individual changes in behavior have only a negligible influence on this development. Laws from influential political bodies are required. In democracies, popular topics have the potential to influence the outcome of elections, since law makers are inclined to model their decisions after public opinion. Thus, they can appeal to the electorate and subsequently win elections. Here lies the chance to influence global developments with civil society involvement and to protect the planet.

C. Leitner et al. (Eds.): AHFE 2021, LNNS 266, pp. 157–162, 2021.
https://doi.org/10.1007/978-3-030-80840-2_17

Following these considerations, we[1] ask ourselves the question: Which conditions favor impulses that make people willing to engage in social and ecological engagement? Since this question cannot be answered in a comprehensive way in a single research project, we focus in our planned study on the topic of the consumption of virtual water and how this topic can be relevant for children between 8 and 13 years and their families. This gives us a target demographic of nine million children in Germany. We favor a participatory approach and plan on actively involving the target group in the research process.

In the study we use a narrative gamification approach. The involved children follow a red line along the Theme we titled Game of Rain. In the research process we want to combine quantitative with qualitative survey methods and use both reactive and non-reactive social science methods to analyze the effect mechanisms of action of various combinations of offers for reflection specially developed for the question in two different settings. The personal attitudes of the participants shall be documented at five measuring points and a potential change will be made visible.

2 Paper Preparation

We divide the fundamental question in two sub-questions. On the one hand, we take for granted that people are different and thus should be addressed in individual ways. That is why we developed four different offers for reflection and want to find out how the amount and combination of these influences the willingness to get involved socially and ecologically. To this end, we divide two groups: one group who gets access to every possible offer for reflection, and another who will get staggered access to these same offers. On the other side we also expect differences, depending on whether the participants use the options for reflection in the Systemwelt in their Lebenswelt. We use these terms based on Jürgen Habermas [2]. The interlinking of the two sub-questions creates four groups (Fig. 1).

		Research setting	
		Systemwelt	Lebenswelt
Offer intensity	increasing	Group A	Group C
	completley	Group B	Group D

Fig. 1. Crossed dimensions of research setting and offer intensity

[1] We are students from different disciplines got together for the master's "ZukunftsDesign" the Coburg University of Applied Sciences and Arts and our names are Oliver Frenzel, Elias Germeroth, Julian Göller, Hauke Hasenknopf (coach), Jana Lenhard, Matthias Scheibe, Stefan Stammberger.

3 Offers for Reflection

The central component of our research approach are the so-called offers for reflection. We understand these as social tools with which people can be sensitized to a socio-ecological topic and new impulses for action can be set. We define the concept of action based on Alfred Schütz as a conscious decision for an alternative. This can also be an active omission [3].

Our four offers for reflection are:

Homepage - Here one gets easy access to information about the game, the characters and the topic.
Card game - This playfully ignites interest for the challenge of sustainable use of resources.
Storybook - It provides more content, deeper information and wider context through an engaging narrative.
Forum/Community - Here players and people interested in the subject can exchange their experiences, develop collaboratively new approaches to deal with this challenge, discuss possible implementations and potential trials (Fig. 2).

		type of medium	
		digital	analog
Mode of activity	Turned invard	Homepage	Storybook
	Turned outward	Forum / Community	Card game

Fig. 2. Crossed dimensions of activity and type of medium

The four offers of reflection are deliberately designed differently in order to appeal to children with different preferences and different possibilities and abilities. In the activities, two aim to be active, while the other is supposed to thought provoking. Furthermore, the media presentation differs: one half is digital and the other half is analog.

4 Research Design

This chapter describes the similarities and differences between the research procedure in the columns Systemwelt and Lebenswelt. In both cases, participants must first register on the Game of Rain homepage. Subsequently, they are offered a standardized online questionnaire at five defined measuring points. As stated above, there are nine million people in our target demographic. Since we are expecting that a lot of them will participate we aim for a sample size of 500 children. These are distributed equally across the four groups. The data from this quantitative method is used to trace how the social-ecological commitment of the participants from the four groups changes in the course of the study. In addition, two groups are implemented for each column, one already receives all the offers for reflection at the second measuring point, the other starts with the card game and then continuously receives further ones.

The participatory development of the individual offers for reflection is based on the conditions of the different settings. Different qualitative processes are used here for the system and living environment. The attractiveness of the individual offers for reflection for the target group (game mechanics of the card game, graphic design of the game characters and scientific models, emotional framing of the story, barrier-free interaction and moderation in the forum) is determined at each measuring point by a specific survey instrument and the related adapted evaluation method analyzed. The findings from all four groups are triangulated and the respective options for reflection are revised. The further developed variant is then made available to the participants for the next measurement interval.

4.1 Offer Intensity in the Setting Systemwelt

The system world is the externally structured, institutionalized and bureaucratized area of human life. Through purposeful, strategically success-oriented action, those involved are treated as if they're merely affected. A prominent example for this in the epistemic world of our target demographic is school. The offers for reflection are understood here as teaching material and lessons in school are conceivable as a possible setting.

While the participants use/perceive the individual offers for reflection, they are observed by the researchers and then asked about the mechanics of the card game, the graphic design of the characters and scientific models, the emotional framing to the story, barrier-free interaction and moderation in the forum.

4.2 Offer Intensity in the Setting Lebenswelt

The Lebenswelt is the place with equal, mutual help in the social environment, i.e. the neighborhood or family. This area can neither be planned, nor is it reliable, since the order is created by all those involved through their respective actions. This part of the investigation is integrated into the everyday life of the participants, because they structure the setting in which they try out the various offers for reflection themselves. These are understood here as objects of distraction and are part of the leisure time with befriended children and adolescents as well as family members.

The data collection should also take place here via the homepage, with the participants keeping video game diaries between the individual measuring points, for example, participating in narrative and design competitions and discussing ideas for improvement in the forum. The data should be evaluated promptly, based on the qualitative content analysis [4]. This part of the research is organized as a running process with more individual responsibility of the involved children, that is reason while we work here with measurement intervals instead of points (Table 1).

Table 1. Roadmap for the research in the setting Lebenswelt

	Group C	Group D
First measurement interval	Material: Homepage	
	Objective: To inform people about the topic and to win them over to the research project	
	Method: Self-assessment form	
	Sample: Different children (ages 8–13, of various ethnicity, gender, disability, social status…)	
	Self-assessment sheet with indicators related to socio-ecological commitment, as well as knowledge of virtual water	
Second measurement interval	Test card game	Test card game
	Material: Card game	Material: Card game, storybook, forum/community
	Goal: Improvement of game mechanics, graphics, etc.	Goal: Improvement of game mechanics, graphics, etc.
	Method: Individual game diary book/self-assessment questionnaire	Method: Collaborative game diary book/self-assessment questionnaire
Third measurement interval	Test a combination of card game and storybook	Test a combination of card game and storybook
	Material: Card game, storybook	Material: Card game, storybook, online forum/community
	Goal: Analyze the interaction between the card game and the storybook, test the attractiveness of the story, the fit of the narrative and information, further develop the graphic design, etc.	Goal: To analyze the interaction of the card game and the storybook, to test the attractiveness of the story, the fit of the narrative and information, to further develop the graphic design, etc.
	Method: Competitions/self-assessment questionnaire	Method: Competitions/self-assessment questionnaire
Fourth measurement interval	Test a combination of card game, storybook and forum/community	Test a combination of card game, storybook and forum/community
	Material: Revised card game, story book, forum	Material: Revised card game, story book, forum
	Goal: Analyze the interaction of the card game, storybook and forum, collectively further develop the idea	Goal: Analyze the interaction of the card game, storybook and forum, collectively further develop the idea
	Method: Forum communication/self-assessment questionnaire	Method: Forum communication/self-assessment questionnaire
Fifth measurement interval	Socio-ecological commitment	Socio-ecological commitment
	Material: revised card game, revised story book, forum	Material: Revised card game, revised story book, forum
	Goal: Are the participants still active on the topic?	Goal: Are the participants still active on the topic?
	Method: Forum communication/self-assessment questionnaire	Method: Forum communication/self-assessment questionnaire

5 Conclusion

There is widespread consensus that humanity should hastily concern itself with a responsible, sustainable handling of Earth's resources. With Game of Rain, we respond to social lethargy by looking for impulses for socio-ecological engagement in our research approach. We involve people in the process, as we closely interlink science and practice, and not only describe the effects, but also partially produce them ourselves.

In the next step, the four presented options for reflection will be made accessible to a broad public. For this purpose, the digital ones will be activated for all interested parties and the analogue ones will be available in toy stores. As well as downloadable content. In this phase, we evaluate our knowledge. In conclusion we would like to state that our project offers many connection options for further research activities.

References

1. Sukhdev, P., Sharma, K.: Eine neue Ökonomie. In: Möllers, N., Schwägerl, C., Trischler, H. (eds.) Willkommen im Anthropozän. Unsere Verantwortung für die Zukunft der Erde, pp. 74–78. Deutsches Museum, München (2015)
2. Habermas, J.: Theorie der kommunikativen Handelns. Zur Kritik der funktionalistischen Vernunft. Suhrkamp, Berlin (2011)
3. Schütz, A., Luckmann, T.: Strukturen der Lebenswelt. UTB, Konstanz (2017)
4. Mayring, P.: Qualitative Inhaltsanalyse: Grundlagen und Techniken. Beltz, Weinheim, Basel (2015)

Resilient Design for Service-Oriented Value Creation

How Higher Education Institutions and Technology Innovation Centres Can Support Organisations Through IoT Digital Transformations

Maximilian Ge[1(✉)], Lena Ahner[2], Yasmin Bayrak[2], Imogen Cleaver[1], Jens Neuhüttler[2], and Florian Urmetzer[1]

[1] Institute for Manufacturing, Department of Engineering, University of Cambridge, 17 Charles Babbage Road, Cambridge CB3 0FS, UK
{lg304,ibc21,ftu20}@cam.ac.uk

[2] Fraunhofer Institute for Industrial Engineering IAO, Nobelstr. 12, 70569 Stuttgart, Germany
{Lena.Ahner,Yasemin.Bayrak,Jens.Neuhuettler}@iao.fraunhofer.de

Abstract. Digital technologies in Industry 4.0, such as, the Internet of Things (IoT), are enabling transformation of organisations across industry sectors at increasing scale. Organisations are predicted to benefit from both economic efficiencies and the adoption of novel business models, enabled by smart technologies. However, there is reported to be low human capability within industry to achieve IoT transformations. We present an empirically derived model of organisations' educational learning needs at three stages along a digital transformation journey, focusing on the combination of business and technology knowledge transfer in executive education programmes. Our work contributes to the adoption of new digital technologies in industry by improving the offerings of knowledge transfer providers in this area.

Keywords: Digital transformation · Executive education · Industry 4.0 · Internet of Things · Knowledge transfer

1 Introduction

The fourth industrial revolution (4IR), or Industry 4.0, offers great potential to transform industry and society through smart products and services that create added value [1]. The Internet of Things (IoT) brings together enhanced sensing and network capabilities, cloud computing, simulation, big data and analytics in a ubiquitous, digitally interconnected world, and is reshaping competition and industry boundaries [2]. Technology and the awareness of its potentials and pitfalls are therefore playing an increasing role in strategic decision making [3]. Organisations face diverse challenges with IoT including security and privacy, data management, interoperability, scalability and real-time access [4]. Predicted benefits include efficiency gains in existing processes and the adoption of novel business models enabled by the new technologies [5]. However, there is reported

C. Leitner et al. (Eds.): AHFE 2021, LNNS 266, pp. 165–172, 2021.
https://doi.org/10.1007/978-3-030-80840-2_18

to be low human capability within industry [6], for example, to design digital platforms or generate monetary value from them.

Education and training are considered crucial for organisations transitioning towards Industry 4.0 [7–9]. Executive Development Programmes (EDPs) [10, 11] are relatively short courses intended to develop capabilities of an organisation's leaders. Here we are concerned with EDPs run by publicly funded external knowledge providers, such as, higher education institutions (HEIs) and technology innovation centres (TICs). Success in this new economy critically depends on how we can transfer knowledge between research institutions and industry [12]. HEIs and TICs in turn are facing an increasing demand from organisations and funding institutions to demonstrate the impact of their EDPs aimed at improving such capabilities [13].

In this paper, we review the role of executive education and the integration of strategy and technology education. Based on empirical data from the UK and Germany, we present organisations' learning needs, as knowledge transfer receivers, at three stages along a digital transformation journey. We seek to provide a more fine-grained understanding at the interface between knowledge providers and knowledge receivers, to support more effective transfer of knowledge for organisations adapting to Industry 4.0.

2 Literature Review

EDPs fulfil a variety of roles. First, they provide managers the opportunity to adapt to the challenges of changing work environments and develop their skills and competencies [10]. Second, they are a method for organisations to build competencies. Third, in the leadership field, Conger and Xin noted a shift from skills-centric "abridged MBAs" that served as a reward and preparation for senior management, to EDPs used as a tool facilitating strategic transitions by aligning organisations to new directions [14]. They argued EDP content should be tailored to an organisation's specific learning needs, but noted ever-greater customisation risked cementing existing ways of working. They anticipated organisational and technological change would continue to be a central theme of learning needs, so EDP content should address ecosystems and there would be a role for technology experts in delivering EDPs.

A more recent study of 197 UK-based EDPs between 2000 and 2012 in the area of supply chain management [15] classified programmes into three main types: knowledge gap/sheep dip; specialist/technical; and strategic support programmes. Knowledge gap/sheep dip programmes cover a broad general knowledge overview with the objectives of introducing the state of technology and strategy, providing a unified framework across the organisation, and creating a common language to communicate about it. Courses had 18–30 delegates and ran over 3–5 days. Specialist/technical programmes have a narrower subject focus aimed at providing specialist skills to operational roles, 10–20 delegates and run shorter. Strategic support programmes provide a counterpoint to the classic competency based EDPs. In common with EDPs as tools of change, their aim is to bring about new strategy deployment. These programmes are highly customised and had a preparation time of 6–18 months before the courses ran. The longitudinal analysis shows a prevalence of competency focused programmes between 2000–2006 and a shift towards strategic support programmes between 2006–2012.

The combination of technology and strategy content in the knowledge gap EDPs has also been noted in the service science, management and engineering, or SSME, literature. The 2016 IBM Conference on SSME – Education for the 21st Century described an urgent need for graduate and executive education combining management and technology education [12]. SSME critically depends on knowledge transfer between science and industry and lifelong learning for people of all ages. Subsequently, Master's programmes in SSME were created [16, 17]. We build on the supply chain management and SSME literatures on EDPs with an empirical study of the learning needs of organisations adjusting to Industry 4.0. Existing literature addresses the consequences of 4IR for jobs [6], the preparedness of educators to address such changes [18] and forecasting of skills required for implementing IoT technologies [8]. While there is some reference to a role for lifelong learning in providing skills for working on the technologies, there is a dearth of research on a broader role for EDPs to introduce Industry 4.0 concepts.

3 Methodology

This study is based on data from projects in the UK and Germany. In both countries the research teams were developing new EDPs to transfer relevant knowledge from the HEI/TIC to industry. The focus of this paper was on how HEIs and TICs can support organisations through smart digital transformations, so an exploratory, qualitative method was appropriate.

Qualitative data was gathered from 24 people in the UK. In advance of developing the pilot course, we conducted 5 semi-structured interviews online with senior managers with experience of digital programmes in firms in various sectors; and two virtual focus groups with a further 8 managers from the manufacturing sector. After piloting the course, interviews were conducted remotely with 11 participants from two runs of the course. The UK EDP of five 2-h modules over one week broadly fits the knowledge gap type [15], with content focused on IoT technology, platform business and ecosystem strategy. The organisations were drawn from the private and public sectors.

Our semi-structured interviews were designed to investigate organisations' learning needs, allowing the interviewer to be sensitive to what the interviewees considered important [19]. Expert interviews and focus groups included an overview of our proposed course; pilot participant interviews included reflection on the course materials they had experienced. Interviews were recorded and transcribed. Atlas.ti aided organisation, exploration, interrogation and interpretation of the data [20]. Finally, to minimise any bias towards verification, the emerging findings were shared with the authors based in Germany for their criticism based on data collected from developing similar interventions. The German EDP ran over several months and was similar to the strategic support type [15] but with a technology implementation emphasis.

4 Data and Analysis

As with many qualitative studies, it is challenging to share the depth of the data behind the findings in an article format. We give examples in Table 1 of some supporting data behind one finding, before summarising the full findings and analysis.

Table 1. Sample data.

Stage of adoption of an IoT investment	EDP participant	Organisation learning need	EDP learning content	Desired organisational impact
Generating awareness: "we won't be actually building anything for a couple of years" D14, Senior project manager, Manufacturing	*Project managers and engineers:* "senior to mid-level management [in] engineering … also… the programme management side" D15, Engineer, Energy sector	*Increased awareness of IoT when specifying planned infrastructure investment:* "it would just be … on their radar about what to consider about a project… spec" D11, Strategy and deployment manager, Transport sector	*Technical overview:* "connectivity, sensors, data, it's all very relevant to how we design and execute this" D14 *Strategy overview:* "case studies… very useful" D14	*Improved efficiency:* "there's an opportunity there to build in some IoT technology to kind of increase the efficiency of how we use the site" D15

Our study supports the SSME call for combined strategy and technology education. As noted above, this combination of content becomes critical for an organisation when the new technology enables new business models in their ecosystem, as is anticipated with IoT [5]. We have identified various pathways for HEIs/TICs to use EDPs to help organisations face this challenge, at three stages of adoption: generating awareness, deciding to invest in the technology and implementation of the digital change programme (Fig. 1).

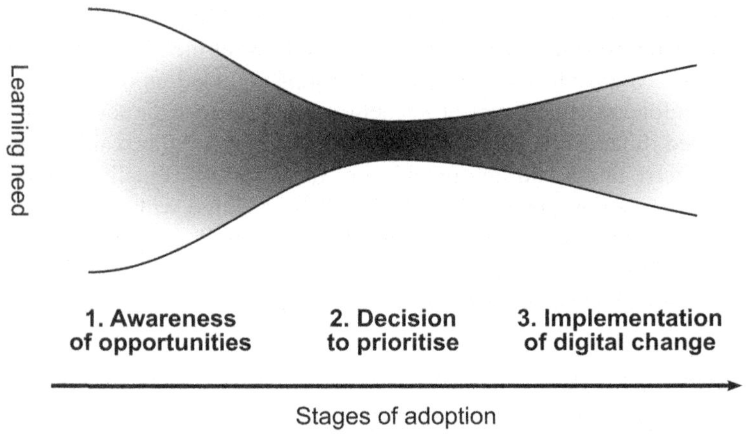

Fig. 1. Illustration of learning need for combined strategy/technology external EDPs by stage of adoption.

Overall, we found an hourglass shape of broad need for EDPs in the first stage, a much narrower need at stage two and then a different broad need during implementation. We focus on EDPs to encourage IoT adoption; depending on the organisation, competency-building EDPs in, for example, change leadership, may also be relevant. While this analysis is not exhaustive of all pathways, we provide insight for HEIs and TICs.

4.1 Generating Awareness: A Wide Range of EDP Participants

In this first stage we identified calls for project managers and engineers, a wide range of domain specialists, consumer interface functions and executives to receive forms of EDPs.

As the interface of the internet with the 'real' world, *IoT involves things* and, therefore, investment in physical kit. Senior project managers and engineers are in functions which can make an impact in three ways: by future-proofing large, planned infrastructure investments which may operate for 15 to 30 years; by introducing piecemeal IoT when specifying solutions to immediate problems; and by proactively suggesting potential efficiency gains. EDP learning content needs to generate understanding of the opportunities IoT brings, otherwise "people will just do it the way they've always done the project" (D14) or "get too stuck on the initial problem itself" (D16). As well as an overview of functional topics, such as, sensors, connectivity, edge computing and data, they found case studies teaching strategies for adoption/new business models relevant as well. However, courses involving strategy were considered most suitable for middle and senior managers, to achieve an impact inside the organisation.

The fact that *IoT involves things* also drives the second classification of participants at this stage: domain specialists. For organisations with a wide range of specialists, for example, hospitals, there was perceived benefit in increasing awareness of the technology among as many people as possible: "the opportunities in occupational therapy... will be very different to in dietetics" (D1). This was not seen as relevant to organisations which mass produce one product. By increasing awareness organisations hoped to generate ideas to lead to efficiency gains. Conger and Xin reported the cascading of EDPs within organisations to engage the collective in change programmes. We see EDPs as a potential *tool to combine the domain expertise of the collective with knowledge in new concepts to achieve idea generation.* Again, the learning content needs to explain the technology and strategy concepts, but not provide deep technical training or develop a new organisation strategy. However, caution was suggested regarding the cost of taking such an initiative below the managers of each function.

Just as IoT involves things, *IoT involves digital technology*. There was a perceived demand for IT to meet the business side need for an understanding of IoT, but also to understand how that would fit with organisational strategy. Again, there were perceived opportunities to introduce piecemeal IoT, it being within their remit and not requiring a change of focus of the business. Even though the UK EDP was advertised for non-technical executives, it attracted several IT participants (or executives delegated it to IT managers). More junior IT employees had specific questions about challenges they were facing on implementing specific projects, for example, MQTT (Message Queuing Telemetry Transport) use in a ventilation installation. In Germany, during the initiative, the participants had access to individual feedback outside the group workshops on up to five occasions. *Junior IT employees may benefit from longer-term action-learning method EDPs, more focused on building skills*, rather than a short course introducing concepts. While the strategy case studies were considered useful for understanding how third parties might benefit from/pay for an IoT project and for improving communication with the business side, more junior IT employees reported a lack of access to people in the organisation who would be able to change the focus of the strategy, for example, to take advantage of the opportunities of automated supply chains which might involve increased sharing of commercial data with third party organisations. There was also a perceived difference between IIoT (Industrial Internet of Things) as internally focused and IoT as more externally focused, therefore requiring EDPs with different learning content for each. *Learning content focused on IoT, rather than IIoT, was suggested as particularly valuable for functions such as commercial managers or consumer product managers, to promote development of products to improve the understanding of customers or to enhance the consumer interface.*

Finally, while there was a need for the executive level to have improved awareness of trends in business and technology, there was recognition that executives are less likely to generate organisational impact through idea generation. There was perceived potential for an introductory IoT EDP to prompt messaging from the top which would disseminate through the organisation, to empower a wider group of employees to engage with Industry 4.0 and a reported need for business sponsors to be aware of how strategies might need to change to optimise investment. However, a critical perceived impact of the executive is at stage two: decision-making.

4.2 Deciding to Invest: EDPs for a Small Group of Gatekeepers

Senior managers/executives with investment decision responsibility are the gatekeepers of adoption of new technologies. The size and scale of the investment determines the seniority of those involved. The impact of this group is on identifying priorities for the organisation, both between competing potential investments and within existing projects. The learning need is to have a basic level of understanding of what the benefits and problems might be, to evaluate opportunities. Regarding learning content, technology experts suggested it is critical for executives to understand their ecosystem to develop their IoT strategy, for example, to understand if they should govern a platform or participate in a third-party one. Even for an intra-organisation project the directors need to "get their mindset around spending this money on this system" (D5). Within a given major project, the question might be "how much of a state-of-the-art plant do we want" (D14). In contrast to Conger and Xin's (2000) reporting of a shift towards EDPs being used for developing social-behavioural skills, we suggest *Industry 4.0 will see HEI/TICs increase their focus on cognitive teaching methods for functional knowledge, combined with experiential methods for supporting strategy development*. However, there was caution against asking executives to, for example, try programming a Raspberry Pi for two hours. The learning content should provide enough understanding of the technology to support evaluation of competing investment opportunities/strategic plans. We suggest providers promote the strategy element to avoid non-technical executives perceiving the EDP as more relevant to IT specialists.

4.3 Implementation: EDPs as a Tool for Digital Change Programmes

In common with Conger and Xin we found a learning need for EDPs which can create engagement with a change programme. *In addition, we propose the organisation impacts sought from these EDPs beyond engagement are: to optimise development collaboration; and using the technology to its full potential.* From a learning content perspective, a high-level overview of the opportunities of IoT technologies and strategies can provide a third-party endorsement to support strategy set by the executive. Second, new technologies which are prototypes require collaboration with developers to maximise the opportunity. EDPs which educate plant managers in the concepts could support communication and optimisation. Change programmes require stamina over months/years and we see a role here for the *cascading integrating mechanisms* reported by Conger and Xin, as technology is rolled out throughout an organisation.

As well as the learning needs of participants identified in Sects. 4.1, 4.2 and 4.3 above, we have two cross-cutting themes. In contrast to Conger and Xin's description of bespoke EDPs focused on an organisation's sector, we found a demand for case studies from other sectors as well as their own and for opportunities to develop ideas with people from other sectors. Second, several participants on the pilot courses perceived personal benefit from attending, for example, gaining an understanding of how IoT might solve problems in their personal lives, such as, using Tile technology to track children's school bags (D7); competence building for a business they are developing themselves (D9); and improving the strength of their annual appraisal (D9). While these benefits will not accrue directly to the employer purchasing the course, the employer may still benefit indirectly from the increased motivation of the individual to engage fully with the course. Finally, while most interviewees saw a need for the course in their organisation, a few exceptions were suggested, for example, "within the realm of homeless medicine" (D7).

5 Conclusion

For impactful knowledge transfer from HEIs/TICs to organisations adapting to Industry 4.0, we recommend developing EDPs around the learning needs identified. We do not seek to suggest this is an exhaustive list of organisations' learning needs in Industry 4.0, nor do we wish to suggest there will be a single change programme to adapt to it. Rather we anticipate organisations may be implementing small IoT adoptions at the same time as seeking to optimise the opportunities of large infrastructure projects. We have sought to provide insights on which participants should attend, the learning need of those participants and the emphasis of the course content to meet their needs at a given stage. Future research should focus on both developing tools for knowledge providers and receivers to evaluate the impact of these EDPs; and identifying the most effective pedagogies for such programmes.

Acknowledgements. The UK project was sponsored by Research England's Connecting Capability Fund award CCF18-7157 – Promoting the Internet of Things via Collaboration between HEIs and Industry (Pitch-In). The German project Business Innovation Engineering Center (BIEC) is funded by the Ministry of Economics, Labour and Housing Baden-Württemberg.

References

1. Kagermann, H.: Chancen von Industrie 4.0 nutzen. In: Vogel-Heuser, B., Bauernhansl, T., ten Hompel, M. (eds.) Handbuch Industrie 4.0 Bd.4. SRT, pp. 235–246. Springer, Heidelberg (2017). https://doi.org/10.1007/978-3-662-53254-6_12
2. Porter, M.E., Heppelmann, J.E.: How smart, connected products are transforming competition. Harv. Bus. Rev. **92**(11), 64–88 (2014)
3. Warner, K.S.R., Wäger, M.: Building dynamic capabilities for digital transformation: an ongoing process of strategic renewal. Long Range Plann. **52**, 326–349 (2019). https://doi.org/10.1016/j.lrp.2018.12.001
4. Čolaković, A., Hadžialić, M.: Internet of Things (IoT): a review of enabling technologies, challenges, and open research issues. Comput. Netw. **144**, 17–39 (2018). https://doi.org/10.1016/j.comnet.2018.07.017
5. Roblek, V., Meško, M., Krapež, A.: A complex view of Industry 4.0. SAGE Open **6**, 215824401665398 (2016). https://doi.org/10.1177/2158244016653987
6. Bonekamp, L., Sure, M.: Consequences of Industry 4.0 on human labour and work organisation. J. Bus. Media Psychol. **6**, 33–40 (2015)
7. Hecklau, F., Galeitzke, M., Flachs, S., Kohl, H.: Holistic approach for human resource management in Industry 4.0. Procedia CIRP **54**, 1–6 (2016). https://doi.org/10.1016/j.procir.2016.05.102
8. Benešová, A., Tupa, J.: Requirements for education and qualification of people in Industry 4.0. Procedia Manuf. **11**, 2195–2202 (2017). https://doi.org/10.1016/j.promfg.2017.07.366
9. Müller, J.M., Buliga, O., Voigt, K.I.: Fortune favors the prepared: how SMEs approach business model innovations in Industry 4.0. Technol. Forecast. Soc. Change. **132**, 2–17 (2018). https://doi.org/10.1016/j.techfore.2017.12.019
10. Wuestewald, T.: Adult learning in executive development programs. Adult Learn. **27**, 68–75 (2016). https://doi.org/10.1177/1045159515602256
11. Hura, G.: A new model for executive education. J. Exec. Educ. **2**, 1 (2013)
12. Hefley, B., Murphy, W.: Service science, management and engineering: education for the 21st century. Springer, Boston (2008). https://doi.org/10.1007/978-0-387-76578-5
13. Murray, S.: How can we measure the impact of executive education? (2019). https://www.ft.com/content/3c9b985a-76f6-11e9-b0ec-7dff87b9a4a2. Accessed 1 Feb 2021
14. Conger, J.A., Xin, K.: Executive education in the 21st century. J. Manag. Educ. **24**, 73–101 (2000). https://doi.org/10.1177/105256290002400106
15. Bernon, M., Mena, C.: The evolution of customised executive education in supply chain management. Supply Chain Manag. **18**, 440–453 (2013). https://doi.org/10.1108/SCM-07-2012-0262
16. Saviano, M., Polese, F., Caputo, F., Walletzký, L.: A T-shaped model for rethinking higher education programs (2016)
17. Choudaha, R.: Competency-based curriculum for a Master's program in Service Science, Management and Engineering (SSME): an online Delphi study (2008)
18. Flynn, J., Dance, S., Schaefer, D.: Industry 4.0 and its potential impact on employment demographics in the UK. In: Advances in Transdisciplinary Engineering, pp. 239–244. IOS Press BV (2017). https://doi.org/10.3233/978-1-61499-792-4-239
19. Bryman, A.: Social Research Methods. Oxford University Press, Oxford (2016)
20. Silver, C., Lewins, A.: Using Software in Qualitative Research: A Step-By-Step Guide. Sage, London (2014)

Impacts of Professional Education Measures on the Digital Transformation in Organisations

Lena Ahner[1(✉)], Yasemin Bayrak[1], Imogen Cleaver[2], Maximilian L. Ge[2], Jens Neuhüttler[1], and Florian Urmetzer[2]

[1] Fraunhofer Institute for Industrial Engineering IAO, Nobelstr. 12, 70569 Stuttgart, Germany
{Lena.Ahner,Yasemin.Bayrak,Jens.Neuhuettler}@iao.fraunhofer.de
[2] Institute for Manufacturing, Department of Engineering, University of Cambridge, 17 Charles Babbage Road, Cambridge CB3 0FS, UK
{ibc21,lg304,ftu20}@cam.ac.uk

Abstract. Digitalisation in all sectors and industries has shortened innovation cycles and puts companies under pressure to retain their competitive abilities. The need to acquire new knowledge and capabilities has led to a rising demand for cooperation of industry with universities and technology innovation centres. These organisations now need to provide suitable measures for knowledge and technology transfer and ultimately account for effectivity and impact of their measures. Based on research on knowledge transfer as well as information technology, we suggest a model for evaluating impact of professional education measures that looks at individual and organisational impacts separately. This paper helps to better understand how teaching interventions can contribute to the capability of business organisations to implement digital technologies.

Keywords: Transfer research · Internet of Things · Service engineering · Digital transformation · Professional education · Knowledge transfer · University-industry cooperation · Information system success model

1 Introduction

With ongoing emergence of digital technologies, such as the Internet of Things (IoT), companies can no longer solely rely on their own expertise but have to integrate technological knowledge and expertise from external institutions [1]. Moreover, companies have to deal with rather strategic questions, e.g. the effects of new technologies on existing processes and business models or how new solutions can be developed across company and industry boundaries [2]. In these rapidly changing times, resilience of value creation also implies for companies to evaluate the potentials of digital transformation continuously and to build up relevant knowledge and capabilities as well as to continuously expand them to shortened technology innovation cycles. In order to exploit the manifold potentials, organisations with rather limited resources are particularly challenged by these dynamic and complex developments.

C. Leitner et al. (Eds.): AHFE 2021, LNNS 266, pp. 173–180, 2021.
https://doi.org/10.1007/978-3-030-80840-2_19

Research institutions are predestined to make a significant contribution to building relevant capabilities in companies by transferring current research findings in target group-oriented formats [3]. However, above-mentioned challenges require research institutions also to rethink the way and content that is transferred in order to achieve a sustainable impact in companies. On the one hand, the knowledge transfer (KT) providers' professional education programs designed to help organisations with picking up new technologies and implementing them in their businesses must adapt to the KT receivers in terms of previous knowledge and absorptive capacity, but also other intrinsic and extrinsic influence factors such as culture, technology and organisational factors [4]. On the other hand, shortened innovation cycles of technologies demand building up dynamic capabilities to continuously adapt to change and make use of new potentials [1]. With growing demand for relationships between public sector education institutions and the private sector, transfer providers face an increasing need to determine the impact of transfer measures [5].

Following this paper, we will refer to measures of professional education with the aim of knowledge and technology transfer by the term of teaching intervention. The theoretical basis of this work is the current state of inter-organisational knowledge and technology transfer, focusing on impact. This will serve as a basis to conceptualize evaluation questions, which can be used to adequately assess the impact of teaching interventions. The evaluation methods will be applied on two exemplary cases of professional education measures from the Fraunhofer Institute for Industrial Engineering (IAO) and the Institute for Manufacturing (IfM) at the University of Cambridge. The qualitative research consists of interviews and online-questionnaires among course participants both directly after the program, focusing on instant feedback, course design etc., as well as after a longer period of time, focusing on long-term impact. The findings of this paper can be used to assess how teaching interventions can contribute to the dissemination of knowledge into companies. The overall aim is to provide insights of the impact of these interventions in terms of enhancing an organisation's readiness to implement digital solutions and secondly to derive recommendations for other research institutions and KT providers on how to design similar support programs.

2 Theoretical Foundation

In order to assess impacts of professional education measures on the digital transformation in organisations, one important aspect is to understand how impact can be measured. The OECD defines impact as "the extent to which the intervention has generated or is expected to generate significant positive or negative, intended or unintended, higher-level effects" [6]. For this paper, we focus on intended positive effects of teaching interventions. A variety of existing academic literature investigates impact as a result of KT or university-industry cooperation [7, 8]. Particularly, KT between public institutions and the private sector is a widely studied field of research, which was previously analysed with regard to different dimensions [1, 3, 9]. In the field of KT, researchers have examined different aspects affecting the relationship between providers and receivers of information. In particular, the willingness to participate in technology transfer activities [1] and the benefits and impediments of university-industry cooperation [3] have been studied. Nevertheless, while research in the field of KT acknowledges impact as a result of performance [4], measures and characteristics of impact are insufficiently incorporated in KT processes.

In order to find a theoretical approach that investigates and operationalises impact while at the same time including transfer providers and receivers' perspectives, we looked at the field of information technology and based the fundaments of our approach on the Information System Success (ISS) Model developed by Delone and McLean [10, 11]. According to this model, success of information systems depends on different variables including the quality of information, systems and services along with the intention to use and satisfaction of the receiver. On a processual level, the model suggests that users are exposed to an Information System (IS) of varying quality, experiencing varying levels of satisfaction depending on the perception of these qualities while using the IS. Based on these experiences, the user will develop an attitude towards the IS (intention of use) and subsequently choose if to initiate behavioural changes [11]. The impact of IS is measured on individual and organisational levels, which the authors first describe with the influence of the system on management decisions and organisational performance [10] while they later acknowledge much broader implications of impact [11] up to societal levels. This general openness and adaptability of the model led previous researchers to apply the ISS Model on various areas including the readiness to use knowledge management systems [12], the empirical validation of a Knowledge Management Success Model [13] and the success of KT systems in e-learning systems [14]. However, all these investigate specific systems as a research object, such as software or technological solutions. Building upon existing research, we extend the application of the model by closer examining how teaching interventions can contribute to the capability of business organisations to implement digital technologies.

Investigation of KT between public institutions and the private sector shows that in order to build successful partnerships, motivational aspects on both sides have to be met [1, 3, 9], which means that both actors want certain goals to be achieved, if to engage in partnerships. An organisation's motivation for engaging in partnerships could lie in competitive advantages, access to research expertise or enhanced reputation [1, 3, 7]. In order to be able to evaluate the impact of transfer measures as holistically as possible, a structuring of impact based on the jobs-to-be-done framework can be helpful. The underlying assumption of the jobs-to-be-done theory is that people hire products or services in order to get a certain job done [15]. Instead of focusing on demographics or psychographic factors of consumption, jobs focus on what people are trying to achieve or accomplish in a certain context [16]. Transferred to our case, this means that participants take part in the courses in order to (better) fulfil a specific task for themselves or for their company during work. The specific task can vary for different participants as well as context factors (e.g. depending on the position within a company or cultural and organisational company context etc.). While one participant, for example, may have the task of implementing smart technologies in the companies' products, another would like to obtain general information about the potential of technologies and future effects on competition, since he is responsible for business development. The interesting thing about jobs-to-be-done theory for our purpose is that jobs are characterized as multi-layered and complex constructs: Besides a core functional dimension, they also possess social and emotional dimensions [17]: Emotional jobs define how customers want to feel or avoid feeling during executing the functional job, while social jobs state how a customer wants to be perceived by others [18]. Following this logic, the value of transfer measures not solely refers to its support to fulfil a certain task within the working environment, but also to how well it contributes to emotional and social aspects, such as security of decision making or perception of competence among colleagues. Although the jobs-to-be-done-theory is mainly used to identify in-depth needs and requirements [19], transferring the social and emotional dimensions could also add to a holistic evaluation of transfer measure impact.

3 Suggested Model for Evaluating Impact of Professional Education Measures

Based on the theoretical foundation, we suggest a model to evaluate both the perceived quality and thus satisfaction of the participants of a transfer measure and its impact on the participants personally or on their organisations in the sense of an adapted ISS model [10, 11] (Fig. 1). This core model helps to structure the general process of impact evaluation as it features all relevant elements: the information source, which in our case is described by teaching interventions while it originally evaluates information systems, the transfer receiver including perceptive as well as behavioural components, and finally the resulting impacts. For operationalising these elements, however, other disciplines as described above have been added in order to achieve a more refined view on impact of teaching interventions.

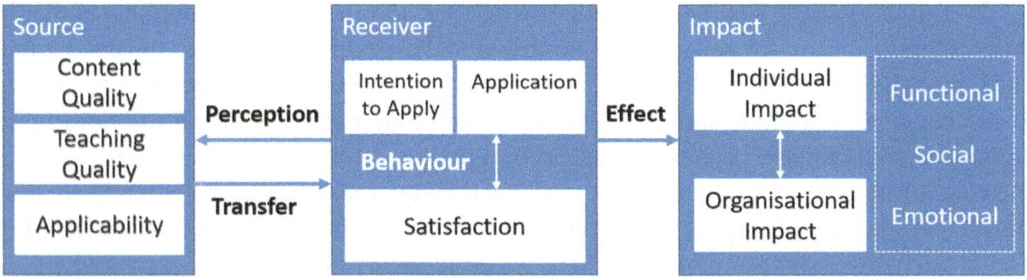

Fig. 1. Suggested model for evaluating impact of professional education measures.

Quality assessment of teaching interventions is vital for identifying the impact made on organisations as well as eliminating potential mistakes in the practice of KT [4]. We therefore ask participants to evaluate the content quality (e.g. appropriate use of methods and examples, comprehensibility, good fit of content with the organisation's demand), teaching quality (e.g. experts competencies, response to questions) and applicability (e.g. benefits of participation, intention of use).

The ISS model shows strong links between satisfaction and behaviour, which is described by 'use' [11]. While the ISS model examines the intention to use and the use of an IS, for this model we suggest that satisfaction with a teaching intervention will affect the application of the transferred knowledge. Therefore, participants were asked in the evaluation questionnaires after finishing the interventions, whether they meant to apply the gained knowledge.

Even though DeLone and McLean decided to switch from a taxonomy of impact to the broader single category 'net benefits' [11], this paper maintains individual and organisational impact as separate categories, in order to better understand interrelationships and characteristics of both. Process-wise, the ISS Model states that individual impacts collectively lead to organisational impact [11], a position that is also represented in research of change processes in organisations [20]. While validated measures for individual impact in terms of enterprise systems success, such as learning and decision effectiveness, could potentially be transferred to individual impact of transfer measures, organisational impact measures such as cost reduction or business process change [21] seemed harder to objectively be traced back to participation in a transfer measure. As discussed before, in order to successfully engage in private-public KT, an organisation must be intrinsically or extrinsically motivated. Therefore, the assumption was made that participants perceived the teaching interventions like a service used to solve a certain problem. This assumption led to the introduction of the Jobs-to-be-done-theory [17] to help define interview questions as well as create a suitable framework for analysis.

As a result, interview partners were asked whether the interventions have led to individual or organisational impact on a functional, emotional, or social level. Translated into interview questions, interviewees were asked whether they have been able to apply gained insights into their personal work practice or business processes (functional), whether they felt safer with decision making (emotional), or if their individual or the companies' reputation have been positively impacted (social). In addition, Neuhüttler et al. [16] emphasise the context in which a job has to be fulfilled. Therefore, questions regarding prior knowledge and company- or industry-specific characteristics have been added to the interview guideline.

4 Case Studies and Results

In order to find out what impact transfer measures of research institutions have on companies, selected teaching interventions of IAO and the IfM were examined with the help of questionnaires and participant interviews. The results from the collected data were then shared between the two transfer providing parties. Using research method triangulation [22], coherent as well as contradictory findings will be pointed out and analysed. Triangulation in this context refers to comparing results from different evaluation methods (e.g. questionnaires, observations during transfer measures or qualitative interviews) and angles, in order to identify and cross-check results.

Both teaching interventions present a multi-day course which addresses business applications for the digital technologies with a focus on IoT. Furthermore, both formats address KT receivers on an executive level, aiming towards facilitating informed decision making rather than in depth training on specific IoT applications and thus support companies in the context of digital transformation.

At IAO, two teaching interventions addressing development of smart services and one teaching intervention addressing development of digital business model innovations formed the basis of the studies. All three interventions ran over the period of about

6 months and were completed either in the summer of 2019 or 2020. The processes included monthly group workshops with homework to be prepared for each month.

After each intervention, participants filled out evaluation questionnaires. These served the purpose of evaluating the experts, the quality of content and perceived applicability. Since it is hardly possible to make statements about the impact of the transfer measures at this point, the participant interviews were conducted at least six months after the teaching intervention was completed.

At IfM, one teaching intervention with the aim of introducing participants to IoT strategy and technology for adopting these technologies in their respective organisations was investigated. It was a one-week intensive program. As the intervention was finished at the end of year in 2020, the 11 subsequent interviews were conducted few weeks after. For this reason, the majority of the following results on impact on an organisational level can be traced back to data collected from IAO, as for IfM the time of investigation was too early for participants to determine large scale impact.

For both individual and organisational impact, interview partners could state a large variety of examples especially for functional aspects. On an individual level, all interview partners could confirm knowledge acquisition especially with regard to methodology and structure of developing digital business models, products or services. This in many cases led to improved decision and work processes, which ultimately resulted in higher individual efficiency and productivity levels.

It is important to mention at this point, that while participants confirmed to have made progress with their digitalisation projects, none has finalised implementation yet, which made it difficult for many to specify organisational impact. However, many interviewees stated that the final implementation of the planned digitalisation project would ultimately lead to improved competitiveness, as well as improved profitability and efficiency. Many also state that the teaching intervention had been decisive in their efforts to concretise digitalisation projects and lead business models into the right direction. These more tangible aspects of organisational impact were mainly reported by participants with large decision-making power, such as smaller companies with less hierarchic company structures, or participants who were members of executive management. Some interview partners have incorporated the methodology conveyed in the teaching interventions into their business processes.

Emotional and social aspects of impact were mostly notable on an individual level. Emotionally, interview partners reported to feel secure in their decisions and actions due to broader information level. Many feel confirmed with their general ideas, and are now more convinced with the direction they are heading, even when changes of plans occurred along the way. In addition, the application of the suggested methodologies gave participants guidance when planning next steps. Some now feel empowered to talk to persons with a technological background at eye level. On a social level, few participants could report changes in perception from other people, such as colleagues, employees or supervisors. While some did receive respect and appreciation for pursuing innovative approaches of digital transformation, others assume that in their positions or industries, these kinds of efforts were expected.

The results also suggest that the context in which the participants make use of the teaching interventions does affect impact. Especially in larger companies, participants could not identify organisational impact. Participants on a higher hierarchy level assumed, that decision power had a great effect on the success of the project. In addition, especially service providers made clear that digitalisation have changed their industries in a way, that change or adaption was a mandatory condition to remain in business.

It is worth adding that some interview partners very positively mentioned cross-sectoral exchange within the teaching interventions. The variety of industries, problems and respective differences in approaching these problems, for many interview partners led to ideas they state would not have gotten if they stayed within their field of expertise.

5 Conclusions and Limitations

Using the ISS model as a framework for evaluating impact of teaching interventions, some insights and limitations were found: Almost all reported characteristics of organisational impact could be assigned to functional categories, while adding emotional and social aspects proved especially valuable for differentiating individual impact. Generally, it was harder for interview partners to draw conclusions on organisational impact, especially when asked for impact on overall corporate goals. We suggest to ask for specific metrics such as competitiveness, culture, profitability, innovative capabilities, productivity, efficiency etc. Especially very small (1–10 employees) and large (>10.000 employees) companies had difficulties to make assumptions towards organisational impact. While especially for self-employed participants personal and organisational impact was the same, employees of larger companies tended to negate impact on a larger organisational scale. This suggests that the introduced model is most informative if applied to small and medium-sized enterprises and participants in leadership roles.

For KT providers who aim at supporting business organisations to implement digital solutions, these findings could further suggest that participants should hold a higher ranked position within the company in order to initiate the changes necessary to observe organisational change. Additionally, the reported impact was most tangible, when the initial idea or project within the company was already in place. This suggests that transfer providers announce this aspect as a requirement for a faster pace of development. Finally, cross-industry participation has been very positively remembered due to a variety of different points of view and application areas. In this context, diverse industry backgrounds can be seen as a success factor for enhancing the participants' perspectives.

While the used approach proved valuable for programs that aim at more intensive supervision and support, we do not suggest that this model is equally suitable for other teaching concepts that aim to convey more general information. Finally, this model is merely a suggestion on how to conduct and operationalize impact research for transfer measures. For conclusions on causality of variables or validity of measures, more data needs to be collected.

Acknowledgements. The UK project was sponsored by Research England's Connecting Capability Fund award CCF18-7157 - Promoting the Internet of Things via Collaboration between HEIs and Industry (Pitch-In). The Business Innovation Engineering Center is funded by the Ministry of Economics, Labour and Housing Baden-Württemberg.

References

1. Lai, W.H.: Willingness-to-engage in technology transfer in industry–university collaborations. J. Bus. Res. **64**(11), 1218–1223 (2011)
2. Neuhüttler, J., Fischer, R., Ganz, W., Urmetzer, F.: Perceived quality of artificial intelligence in smart service systems: a structured approach. In: Shepperd, M., Brito e Abreu, F., Rodrigues da Silva, A., Pérez-Castillo, R. (eds.) QUATIC 2020. CCIS, vol. 1266, pp. 3–16. Springer, Cham (2020). https://doi.org/10.1007/978-3-030-58793-2_1
3. Valentín, E.M.M.: University—industry cooperation: a framework of benefits and obstacles. Ind. High. Educ. **14**(3), 165–172 (2000)
4. Liyanage, C., Elhag, T., Ballal, T., Li, Q.: Knowledge communication and translation–a knowledge transfer model. J. Knowl. Manag. **13**(3), 118–131 (2009)
5. Hepburn, N., Wolfe, D.A.: Technology and Innovation Centres: Lessons from Germany, the UK and the USA (2014)
6. Organisation for Economic Co-operation and Development (OECD): Better Criteria for Better Evaluation. Revised Evaluation Criteria Definitions and Principles for Use, OECD/DAC Network on Development Evaluation, November 2019
7. Ankrah, S.N., Burgess, T.F., Grimshaw, P., Shaw, N.E.: Asking both university and industry actors about their engagement in knowledge transfer: what single-group studies of motives omit. Technovation **33**(2–3), 50–65 (2013)
8. Kochenkova, A., Grimaldi, R., Munari, F.: Public policy measures in support of knowledge transfer activities: a review of academic literature. J. Technol. Transf. **41**(3), 407–429 (2015). https://doi.org/10.1007/s10961-015-9416-9
9. Oliver, C.: Determinants of interorganizational relationships: integration and future directions. Acad. Manag. Rev. **15**(2), 241–265 (1990)
10. DeLone, W.H., McLean, E.R.: Information systems success: the quest for the dependent variable. Inf. Syst. Res. **3**(1), 60–95 (1992)
11. DeLone, W.H., McLean, E.R.: The DeLone and McLean model of information systems success: a ten-year update. J. Manag. Inf. Syst. **19**(4), 9–30 (2003)
12. Clay, P.F., Dennis, A.R., Ko, D.G.: Factors affecting the loyal use of knowledge management systems. In: Proceedings of the 38th Annual Hawaii International Conference on System Sciences, pp. 251c–251c. IEEE (2005)
13. Kulkarni, U.R., Ravindran, S., Freeze, R.: A knowledge management success model: theoretical development and empirical validation. J. Manag. Inf. Syst. **23**(3), 309–347 (2006)
14. Holsapple, C.W., Lee-Post, A.: Defining, assessing, and promoting e-learning success: an information systems perspective. Decis. Sci. J. Innov. Educ. **4**(1), 67–85 (2006)
15. Christensen, C.M., Cook, S., Hall, T.: Marketing malpractice: the cause and the cure. Harv. Bus. Rev. **2**, 2–13 (2005)
16. Neuhüttler, J., Woyke, I., Ganz, W., Spath, D.: An approach for a quality-based test of industrial smart service concepts. In: Ahram, T.Z. (ed.) AHFE 2018. AISC, vol. 787, pp. 171–182. Springer, Cham (2019). https://doi.org/10.1007/978-3-319-94229-2_17
17. Christensen, C.M., Hall, T., Dillon, K., Duncan, D.S.: Know your customers' jobs to be done. Harv. Bus. Rev. **94**(9), 54–62 (2016)
18. Ulwick, A.W.: https://jobs-to-be-done.com/outcome-driven-innovation-odi-is-jobs-to-be-done-theory-in-practice-2944c6ebc40e
19. Kalmbach, J.: The Jobs to Be Done Playbook. Two Waves Books, New York (2020)
20. Balogun, J., Hailey, V.H.: Exploring Strategic Change, 3rd edn. Prentice-Hall, London (2008)
21. Sedara V., Gable G.: A factor and structural equation analysis of the enterprise systems success measurement model. In: Proceedings of the 25th International Conference on Information Systems,Washington D.C., pp. 449–463 (2004)
22. Wilson, C.E.: Triangulation: the explicit use of multiple methods, measures, and approaches for determining core issues in product development. Interactions **13**(6), 46–63 (2006)

Transformation Towards Smart Result-Oriented Product-Service Systems

Victor Naumann[1(✉)] and Alexander Pflaum[2]

[1] Fraunhofer Institute for Integrated Circuits, Nordostpark 84, 90411 Nuremberg, Germany
victor.naumann@iis.fraunhofer.de
[2] Otto-Friedrich-University Bamberg, Feldkirchenstraße 21, 96052 Bamberg, Germany

Abstract. Product-Service Systems (PSS) are an established field of research. In the era of digitalization, data of smart products becomes a valuable resource and can be used to provide smart result-oriented PSS. In an exploratory study in the machinery and plant industry, the socio-technical changes from a provider's perspective was researched. This paper outlines the socio-technical transformation frame developed on basis of the insights and the challenges identified with the transformation.

Keyword: Product-service systems · Digital transformation · Socio-technical system · Smart products · Servitization · Manufacturing and plant industry

1 Product-Service Systems in the Digital Era

As part of the continuous trend of servitization [1], customers increasingly ask for the solution of problems, instead of the possibility to purchase products [2]. A proposition representing this paradigm is the concept of result-product-service systems (PSS) [3]. PSS can be - exemplifying a variety of definitional approaches - defined as "tangible products and intangible services that are designed and combined to jointly meet specific customer needs" [3, 4]. The most service-oriented category of PSS is that of result-oriented PSS, whereby the provider undertakes all activities required to achieve an agreed result in the customers' value creation [3]. However, even result-oriented PSS were prophesied to be the dominant industrial business model in the year 2020 [5], until today only few examples exist.

Besides the trend of servitization, the megatrend digitalization is considered to be of great importance for future innovations [6]. The conceptual understanding of digitalization sees the implementation of novel products and associated services through the use of data from smart products at the forefront of a world of networked actors and objects [7–11]. Converging the two trends, smart result-oriented PSS can be described as fee-based service-oriented offers in which providers take responsibility for achieving measurable results in the value creation of customers and achieve them through the systemic combination of smart products and smart services, taking into account the respective environment. The smart products are part of an IoT solution and the use of the digital data and information generated by the IoT solution is an essential basis for

C. Leitner et al. (Eds.): AHFE 2021, LNNS 266, pp. 181–188, 2021.
https://doi.org/10.1007/978-3-030-80840-2_20

modular, smart services that generate added value for users along the life cycles of smart products [12]. As existing PSS publications recommend the explicit research on smart result-oriented PSS [13] as well as the transformation towards smart result-oriented PSS [14], the following research question was addressed: *What socio-technical changes and preconditions are required for the realisation of smart result-oriented PSS from the perspective of machine and plant manufacturers and what challenges are experienced?*

The following chapters outline the research, the insights as well as an outlook.

2 Socio-technical Systems as Theoretical Framing for an Explorative Research Approach

The field of service research is relating to a number of theoretical concepts, not all being grand theories. PSS concepts can be considered as a form of socio-technical systems (STS). Also the relevance of the concept of STS in regards to the issues of change towards smart PSS has been identified [15]. In the context of the IoT, the realization of PSS requires the right configuration of people, technology, organization and information on the side of the provider as well as the customer. This comprises new interactions between people and machines, as encompassed by STS. A STS consists of complex interaction between social humans and technical systems. The social subsystem consists of the organizational structure, which encompasses i.e. authority structures, knowledge and skills as well as attitudes and values. Because the STS is an open system and is susceptible to external inputs, it will be affected by the complex operating environment. The technical system consists of the technical and the tasks that have to be done, so that the STS can generate value as a whole. Therefore, the theoretical perspective chosen to answer the research question is the one of STS, making technology, activities, actors, the organization as well as the complex environment the dimensions of analysis.

To answer the outlined research question, 20 explorative interviews in the machinery and plant industry were conducted. All experts must either have realized, being in the process of piloting or currently developing results-oriented PSS. Their industrial products must be smart products. The interviews lasted between 116 and 64 min, with an average of 81 min. All interviews were transcribed and pseudonymized, resulting in a total of 410 pages of documentation. The coding process was structured and documented using the MaxQDA Software. The first order data coding was done deductively, following the structure of the elements of the STS. The second and higher order codes were derived inductively from the information provided by the experts.

3 Transformation Towards Smart Result-Oriented Product-Service Systems

The explorative study outlined various possibly necessary transformations in the socio-technical elements: organization, technology, actors, activities and environment. The degrees of these transformations are influenced by the characteristics of the PSS as well as the initial setting of the provider.

3.1 Transformational Framework

From an *organizational* perspective there is a need to adapt processes. Transparency in vertical and horizontal processes of involved departments has to be increased. Silos must be dissolved to enable an interdisciplinary project work in the provider's organization, which encompasses all necessary processes to reach an agreed result. This includes the determination of clear responsibilities and rules for conflicts in regard of allocating resources between a smart result-oriented PSS and the traditional business. Furthermore, an adaptation of the organizational structure can be seen as necessary. Various design options exist, ranging from integrated units, joint ventures as well as spin-offs. Such organizational structures are needed to frame the new processes and involved actors. Related to the actors, providers of smart result-oriented PSS need to expand their network of partners to an ecosystem. Traditional linear relationships must be changed to partnerships, which responsibly contribute to reaching the agreed results. In the ecosystem, partners must be connected and information exchange needs to be established in organizational design. Furthermore, contracts and directives must be adapted and expanded. Rules and agreements related to the smart result-oriented PSS must be included in traditional contracts. Especially concrete agreements on the handling of data in the ecosystem are needed. The dealing with intellectual property resulting from the activities in the PSS must also be defined. As contracts for smart result-oriented PSS can be comprehensive, a modular structure was recommended by the experts. One of the largest transformations required, is the adaption of the culture in the provider's organization. Providers of smart result-oriented PSS act as solutions providers and therefore the culture must be customer-focused. It is necessary to establish an understanding of the importance of the life cycle for all actors in the ecosystem and of the necessity to cooperate intensively in ecosystems. All communication processes – internal and external – must be aligned to contribute to the cultural change.

Regarding the *technological* transformation, there are changes related to the IoT-solution as well as the physical products. Products need to have technical capabilities, as sensors, to generate relevant data. Such capabilities can be included in new products or have to be retrofitted. Furthermore, the software of smart products needs to be compatible to other systems and able to transfer data, by relying on standard for data. In order to enable the transfer of knowledge who to operate smart products in different result-oriented PSS, products and their components need to become modularized and standardized. As the providers take over the life cycle responsibility in result-oriented PSS, the durability of products should also be lengthened.

The IoT-solution, which can be seen as the technology stack [7], requires the transformation of local, standalone servers into an IoT-Architecture. This IoT-Architecture encompasses global server, platforms and databases. Additionally, the implementation of high-performance data processing capacity and micro-services at different levels of the architecture is a required change to existing IT-architectures. Existing systems must be connected by the implementation of interfaces and the resources to collect and aggregate their data must be set up. Additionally, a technological capability for data processing and analysis needs to be increased. Software applications must be installed, which enable the necessary data analytics functions as well as user interfaces. Algorithms to support the decision making in activities must be implemented, whereby a high degree of automation should be aspired. Regarding the infrastructure and connectivity to transfer data from smart products, a bi-directional exchange of data must be enabled. Furthermore, security concepts must be expanded to ecosystem-wide IoT-security solutions. The last identified change in regards to the IoT-solution is the transformation of IT policies. They must be converted to a data governance, which includes the definition of technical standards, data usage, software codings and semantics of data.

The explorative study also outlined the need for transformation regarding the *activities* of providers. The development of products services and IoT-solution need to be transformed from separate to integrated activities. These must be guided by the aim to develop solutions that solve customers' problems and add value compared to other forms of PSS. This requires a change from focusing on technical specifications of products towards the analysis of processes and activities of customers. Managing software and infrastructure becomes more important in smart PSS. Ensuring of compatibilities and of modularity are seen as essential activities. Additionally, the provision of infrastructure for the transfer of data is not a traditional activity of industrial firms, but necessary in smart result-oriented PSS. As data is seen as a key resource in smart PSS, the management of data need to become a conscious and extended activity across the provider's activities and in the ecosystem. Instead of singular activities, data management must follow an explicit data strategy. Ensuring the purposeful aggregability of data across a PSS is a necessary new task. Intensifying data analytics from a descriptive activity to an activity ensuring the identification of patterns as a foundation of prescriptive applications is seen as a necessary change. Data Analytics has to be intensified to systemize implicit knowledge through data and thereby making knowledge available in smart functionalities in PSS. Traditional activities as purchasing need to adapt to the requirement of short notice procurement of partners and smart services that are needed for the smart result-oriented PSS. Existing product management activities must internalize the need of a wider scope and a coordination of all the necessary activities to achieve an agreed result for customers. Product management must cover the need to warrant a life cycle perspective and systemic view on all activities. Therefore, one of the foci, must be the development of strategies and activities for the end of PSS agreements and end-of life of smart products. Previous marketing activities must be transformed from their transaction orientation towards marketing smart solutions. Building trust and enforcing partnerships have to be essentials of marketing smart-result-oriented PSS. Sales need to transform into a proactive selling of smart solutions instead of selling products. Payments per agreed result along the duration of PSS agreements require an adoption of incentivizing sales with customer-lifetime value concepts. Data must be used in smart result oriented PSS as a source of sales leads. The traditional service activities also need to be changed from a reactive to a proactive service proposition. Using and analyzing data to ensure the achievement of the agreed results, must be internalized in all service activities. Besides service for physical products, service activities for the IoT-solution must be set-up at a high level of availability. Another need for redesign exists in relation to the financial accounting activities. They must be more flexible to realize variable pricings and must be based on the data of the smart result-oriented PSS. The provider's internal and external communication also needs to be transformed to multilateral, ecosystem-wide activities, using communication platforms to support the required level of information availability across a smart result-oriented PSS. Due to the larger scope of responsibility for the provider and the availability of data, risk management must be expanded to a continuous, data-based activity. Existing claim management principles should also be based on data, to clarify responsibilities and liabilities in an automated way.

More needs to transformation exist in regards to internal and external *actors*. The different roles can be filled by internal or external actors, depending on the PSS and the ecosystem. Compared to traditional service activities, the competences of service employees must be extended by knowledge about up- and downstreams value creation activities to achieve the results. Furthermore, they must be able to use available data and service IoT-solutions. Sales employees need to understand customers' processes and problems and the potential of data and smart functionalities in PSS in addition to the knowledge of technical product specifications. Traditional engineers need to rethink their objectives to an integrated development of products, services and IoT-solutions and extend their competences in the development of software. Legal experts must acquire a deep understanding of the functioning of smart result-oriented PSS in order to design contracts for the responsibilities of the results. As building up ecosystems for smart PSS is not usual for procurement experts, they must build up the respective competences. Financial partners and investors also need to understand the financial consequences of result-oriented pricing models and internalize them into their operations. Very important for the successful transformation towards result-oriented PSS seem to be changes in regards to potential customers. Potential customers must develop an openness to integrate providers into their value creation, share data with them and accept them as partners instead of suppliers. New roles that are not common in the machinery and plat infrastructure are employees responsible for the success of a smart-result-oriented PSS. Besides the PSS-internal coordination of the successful provision of agreed results, they represent roles of evangelists for the smart PSS across the ecosystem. In order to be able to improve the value creation of customers, consultants are needed. They analyse operations and give impulses to a more effective and efficient achievement of results in PSS. For the IoT-solution, actors setting up and managing the IoT-infrastructure as well as IoT-developers are needed. They bring the competences into the ecosystem, that are needed to provide data as key resource in smart result-oriented PSS. The importance of data analytics requires the establishment of data scientists, who are able to apply advanced analytics.

Further necessary changes, not directly in the influence of providers are the related to the *complex environment* for smart result-oriented PSS. Macro-economic conditions must motivate the need for a change of business models for providers and of the value creation of potential customers. Technological progress in regards to the performance and reliability of IoT-solutions must also evolve. Furthermore, legal foundations must be extended by international agreements, role models, product and data standards as well as data governance directives. In the social context, an attitudinal change towards an openness for non-ownership and the enhanced use of the IoT in a business context is needed (Fig. 1).

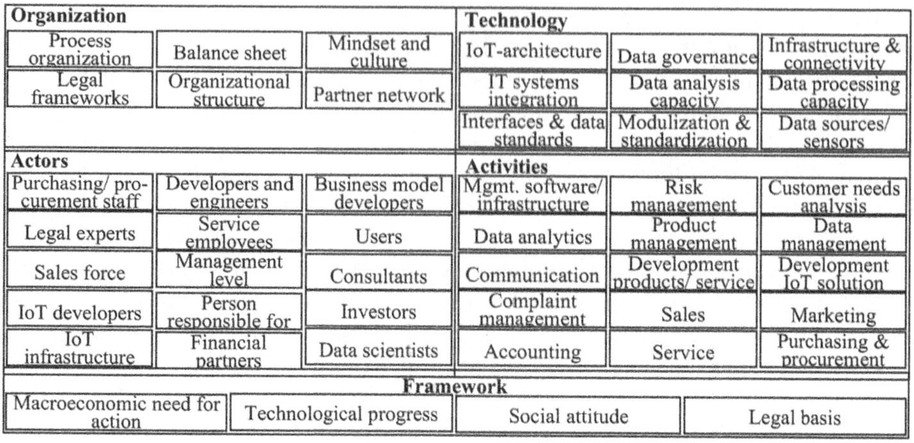

Fig. 1. Socio-technical transformation frame for smart result-oriented PSS

3.2 Challenges Related to the Transformation

In conducting the transformations, various challenges and barriers were observed.

Looking at *organizational* challenges, the lack of competences and capabilities to realize smart result-oriented PSS is hindering. Insufficient forms of organization are barriers for i.e. work in interdisciplinary teams. Many firms face cultural barriers due to a lack of openness and a missing link between the digital transformation and the firm's strategy. The impact on the balance sheet and the financial accounting exceeds the agility of many providers. Also the supply chains structures and the providers position are often seen as disadvantageous, due to different actors between the provider and a potential customer. Contracts for smart result-oriented PSS are seen to be far more complex than traditional contracts. The large partition of export sales increases this complexity even further, due to foreign regulations. From a *technological* perspective, the high degree of individuality of the IoT-solutions required is challenging. Existing IT-systems are also a limitation for realizing the IoT-paradigm. Innovative IoT-technologies appear to lack the reliability and performance for industrial applications. Additionally, the security requirements are perceived to be high. There is little experience with advanced analytics and insufficient standards to handle data. The market for digital technologies is seen as dynamic, thus hard to assess. The existing product portfolio in the machinery and plant industry is diverse, many products have individual specifications and are due to their size immobile and elaborate to set up. Regarding the challenges in relation to the transformation of activities, unsuitable practice of development with silo-like working of product, service and IoT-development was observed. The change in marketing and sales comes is extensive and requires a large effort. The number of qualified employees, able to realize smart result-oriented PSS is seen to be very limited. At the same time, there is intensive organization of service activities needed to achieve the results in such PSS. The complexity of product management activities increases and the management of the IoT-solution is laborious. Additionally, the effort necessary to conduct a sufficient data management is seen to be exhaustive. Providers have to manage risks; of which many are in transparent. The management of the required transformation is also seen to be complex and the strategic importance not clear for all actors. Regarding the transformation of actors, qualified actors are seen to be a very scarce resource. The network of actors of the provider grows, making it more intense and complicated. Potential customers' heterogenic requirements and suitability as customers are further barriers. By taking over responsibility in smart result-oriented PSS, providers expose themselves to additional competitive relations to other industrial actors.

4 Discussion and Contribution

The outlined insights show the extensive scope of the necessary transformations towards smart result-oriented PSS, thereby emphasizing the urgency to research transformation in a holistic way. PSS as an extensively researched field are facing additional complexity through the convergence with the context of digitalization. This context is in different ways related to all transformation needs. The need to take a systemic perspective along the lifecycle is substantiated in the complexity of industrial value creation, the responsibility of the PSS-provider and the integrated handling of products, services and IoT-solutions. The various transformation aspects related to data, its handling and management, show the importance of considering data as a key resource, when realizing smart result-oriented PSS. The identified challenges show how demanding the transformation is. The technological challenges are not perceived as the biggest challenges of the transformation.

This research contributes to the scientific community by approaching the complex and under-researched field of result-oriented PSS and digital transformation of PSS. The research additionally contributes to the necessary convergence of PSS- and IoT-research. It outlines the scope of the transformation behind the trend of servitization in a digital era. Furthermore, by applying the theoretical framing of STS, the research enhances the theoretical foundation in PSS research. Practitioners can benefit from the aggregation of the various transformation experiences for their own orientation. The theoretically based frame of possibly necessary transformation can be a support to assess and structure the own servitization activities. The description of the barriers enables the sufficient planning of realizations in order to prevent future providers from the same pitfalls.

5 Limitations and Further Research

There are some limitations to the research. The explorative design limits the findings to the experience of the sample. Further, interviews might have brought additional insights. Although STS are a suitable theoretical framing, they are not a grand theory and the interrelatedness of elements was not analysed. The underlying concept of smart result-oriented PSS needs further conceptual enhancement. The analyzed examples show, there is not "the" single form of smart-result oriented PSS.

Further research should intensify the structural analysis of smart PSS in a taxonomy or new categorization that includes economical and sustainable effects of smart PSS, i.e. on circular economy solutions. Additionally, PSS should be categorized PSS not from a providers view but in relation to the interaction in ecosystems and the solution's scope for customers. Research approaching the procedures and methods of how the transformation can be successfully conducted is a promising field of research. Further research is required in order to explore the concrete use of data in result-oriented PSS and the determination of value for the different actors in the ecosystem.

References

1. Vandermerwe, S., Rada, J.: Servitization of business: adding value by adding services. Eur. Manag. J. **6**, 314–324 (1988)
2. Abramovici, M., Filos, E.: Industrial integration of ICT: opportunities for international research cooperation under the IMS scheme. J. Intell. Manuf **22**, 717–724 (2011)
3. Tukker, A.: Eight types of product–service system: eight ways to sustainability? Experiences from SusProNet. BSE **13**, 246–260 (2004)
4. Kambanou, M.L., Lindahl, M.: A literature review of life cycle costing in the product-service system context. Procedia CIRP **47**, 186–191 (2016)
5. Meier, H., Roy, R., Seliger, G.: Industrial Product-Service Systems—IPS 2. CIRP Ann. **59**, 607–627 (2010)
6. Hamidian, K., Kraijo, C.: DigITalisierung – Status quo. In: Keuper, F., Hamidian, K., Verwaayen, E., Kalinowski, T., Kraijo, C. (eds.) Digitalisierung und Innovation: Planung – Entstehung – Entwicklungsperspektiven. Springer Gabler, Wiesbaden (2013). https://doi.org/10.1007/978-3-658-00371-5_1
7. Porter, M., Heppelmann, J.E.: How smart, connected products are transforming competition. Harv. Bus. Rev. **92**, 64–88 (2014)
8. Yoo, Y., Lyytinen, K.J., Boland, R.J., Berente, N.: The next wave of digital innovation: opportunities and challenges: a report on the research workshop 'digital challenges in innovation'. SSRN Electr. J. (2010)
9. Tilson, D., Lyytinen, K., Sørensen, C.: The missing is research agenda. Inf. Syst. Res. **21**, 748–759 (2010)
10. Hehenberger, P., Vogel-Heuser, B., Bradley, D., Eynard, B., Tomiyama, T., Achiche, S.: Design, modelling, simulation and integration of cyber physical systems: methods and applications. Comput. Ind. **82**, 273–289 (2016)
11. Whitmoore, A., Agarwal, A., Da Xu, L.: The Internet of Things - a survey of topics and trends. Inf. Sys. Front. **17**, 261–274 (2015)
12. Naumann, V.: Digitale Transformation von Produkt-Service Systemen, (Arbeitstitel). In: Pflaum, A. (Hrsg.): Schriftenreihe Logistik und Informationstechnologien, Fraunhofer Verlag, Stuttgart (planned 2021)
13. Alcayaga, A., Wiener, M., Hansen, E.G.: Towards a framework of smart-circular systems: an integrative literature review. J. Clean. Prod. **221**, 622–634 (2019)
14. Mittag, T., Rabe, M., Gradert, T., Kühn, A., Dumitrescu, R.: Building blocks for planning and implementation of smart services based on existing products. Procedia CIRP **73**, 102–107 (2018)
15. Kuhlenkötter, B., et al.: New perspectives for generating smart PSS solutions – life cycle methodologies and transformation. Procedia CIRP **64**, 217–222 (2017)

Towards a Capability Based Approach to Strengthen the Strategic Decision Making Process for Developing Smart Products and Services in SMEs

Sandra Frings(✉) and Holger Kett

Fraunhofer IAO, Nobelstrasse 12, 70569 Stuttgart, Germany
{Sandra.Frings,Holger.Kett}@iao.fraunhofer.de

Abstract. For manufacturing companies, especially small and medium enterprises (SMEs), the digital transformation is often a large step – creating digital and "smart" products even more. In advance, SMEs need to know how well they are prepared to smartify their products. We are developing a maturity, capability and measures based approach which is used to strengthen the strategic decision making process for developing smart products and services in SMEs. The methodical tool kit is designed to be the foundation to structurally analyse an SME and derive measures to initially smartify its product offers.

Keywords: Smart product · Smart service · Digitalization · Decision making · Maturity · Capability · Smartification · Measures

1 Digitalization and Smartification: Introduction and Challenges

Years ago, the digital world changed the lives of many people and the situation of many companies – large, medium or small – a little or very significantly. There are many advantages for small and medium-sized enterprises (SMEs) generally through digitalization and digital transformation like reaching competitiveness and foster value creation [1], specifically through the Internet of Things (IoT) and Industry 4.0, e.g. using approaches like smart engineering [2], through artificial intelligence like cost and effort reduction through optimised and automated processes [3], and by producing and offering smart products and services to satisfy customers expectations [4].

However, for SMEs, the steps from "analogue to digital" – whether in the context of products, services, processes, technologies or business models [5] – are mostly not trivial, not affordable, or not necessarily predictable [6]. The steps needed to be able to change a non-digital to a digital, smart, or intelligent product require expertise in different domains, which is often not available, or – more often – not recognized as a lack of expertise. Frequently, SMEs are not familiar with this situation; nevertheless new developments are spontaneously risked which do lead to a positive outcome if needed capabilities are somehow available. On the other hand, SME hesitate, often too long, and lose touch with digitalization [7, 8].

C. Leitner et al. (Eds.): AHFE 2021, LNNS 266, pp. 189–196, 2021.
https://doi.org/10.1007/978-3-030-80840-2_21

Companies – that manufacture analogue, non-digital physical products and have already proven that they can be successful on the market – have one option (of many) to jump on the bandwagon of digitalization by pursuing the aim of turning an analogue product into a digital product or, in other words, to "transform" them. The physical product is digitally refined or otherwise called smartified. Smartification here refers to equipping analogue products with various components so that they have the ability to identify and save data via a network, to record their environment with sensors, to influence them as actuators, and/or to process the data electronically in order to be able to use this as decision support [7]. Many success stories exist of such a smartification, e.g. smart sun visors, chopsticks, horse saddle belts, and strollers [6, 9, 10].

Within a research project in Germany a simple non-digital insect trap was smartified [11]. The reason why this project ended up very successful in a new founded SME is because the needed expertise was physically brought together in numerous meetings and discussions by experts in persons: one SME brought in the expertise for insect trap development, one SME adding expertise in software development and thirdly adding structured requirements analysis and project management expertise by the research institute. Agile software development was used. The SME producing insect traps did not go through digital transformation or built up software development expertise. This is one way to integrate missing capabilities, but not all companies have the possibility to do that or want to bring in long-lasting external expertise.

Successful SME's CEOs certainly do have strategic knowhow and are supported by some IT expertise within the SME. But the method of identifying – in an efficient and effective way – the digitalization and smartification aims as well as determining how to reach them – without external help and on their own – is not immediately available or does obviously not belong to the CEOs everyday routine.

2 Related Work

Before starting the smartification activities, SMEs need to make strategic decisions regarding the following: Which processes, technologies, business models, etc. have to be adapted and which competencies need to be built up so that the SME can follow a resilient design for a new smart product and its related services to create the most possible value for the company itself as well as for their customers?

Studies have shown that projects are more likely to be successful if the measures are planned well, structured and needs-based [12]. There are numerous standalone methods (e. g. capability models, maturity models, readiness checks, business potential analysis) to analyse and identify an organisation's strategic aims [13, 14] as well as the ability (through maturity or capability) to perform in the information technology domain where smartification of non-smart products is mostly relevant. The term "maturity" is defined as a way to evaluate "the state of being complete, perfect, or ready" [15] whereas the term "capability" refers to "quality or state of being capable" [16]. The terms are used according to their intent.

Maturity models reach back to the late 1980s. The Software Engineering Institute (SEI) developed the first capability and maturity model (CMM) "which can be used by any software organization to assess its own capabilities and identify the most important areas for improvement [17]." The framework defines five levels of process maturity, different subdomains and indicators to be fulfilled for a certain level. The intent is to determine the current software process maturity and the company's capability to be able to build up data to statistically identify where the process needs to be refined. From this measures can be defined to move up from the current level of maturity to a targeted level [18] 2010, the SEI CMM was replaced by the SEI Capability Maturity Model® Integration which now exists in Version 1.3, whereby the CMMI for Development is most relevant for our research [19]. "The use of CMMI-DEV is intended to ensure that the activities of an organization are directly linked to the business goals, the overview of the business activities is improved and it ensures that products or services actually meet the wishes of the customers" [20].

Today, maturity and capability models exist to the hundreds including guidelines in the development phase of the model [21] even for SMEs [22] – all having their specific reason to exist [23, 24]. In general, maturity and capability models are ideal for subject areas that are difficult to survey due to their ability to view complex objects broken down into their components.

When developing a new maturity or capability model, to be able to include the own focus, also many published procedures and guidelines exist [25]; the identification of the relevant domains, the definition of the maturity levels, as well as their respective indicators is a tedious task since just the right level of detail is important for later application of the model [26].

Once the model is applied in an organisation, the assessed maturity and/or capability levels in each domain are always a snapshot of the current situation. In some cases the models include how to define the targeted level; very few models propose exact measures to be taken to reach that target [27].

Next to CMM, there are many methods, frameworks, action lists, sets of advice etc. – mostly in the field of Industry 4.0 since this is the era which started the comprehensive digitalization of the industrial production [28] – which deal with smart product and service development to some extend and detail [29–32].

SMEs – developing non-smart products – usually do not have the capability to build up competencies to apply above mentioned methods and therefore need a compact "tool kit" to support them in their smartification vision and strategic decision support. According to this papers authors' intense research, as of today, such a usable methodical tool kit –for the purpose of smartification – is not publicly available to support SMEs in their smartification thrive.

3 A Smartification Tool Kit for SME

Literature shows that a methodical tool kit is not publicly available for SMEs to be able to straightforwardly find out, where and how they have to change processes as well as IT, and which capabilities to build up to be able to design smart products (from non-smart existing products) and services in the future.

Our hypothesis is: SMEs need a tool kit to methodically support them in their strategic decision for developing smart products and service starting with their successful but non-smart products. This approach needs to be backed up by documented guided advice, instructions and recommendations to be able to be applied successfully. The following sections give an overview of the current status of search of the tool kit, its components and their most important functional as well as non-functional requirements.

3.1 Capability, Maturity and Measures Model

The methodical tool kit, as mentioned above, is intended to be made up of a self developed underlying capability, maturity and measures base model for supporting in strategic decision making when following the aim of smartifying a non-smart product in SME. In domains like strategy, processes, smart products and services, IT, people (employees, customers), and data, the model seeks to split up these domains in process areas, and there within defines indicators on maturity and capability levels, as well as measures to be taken moving from one level to the next.

The following requirements are in discussion for the capability, maturity and measures model:

- The base model must be context relevant: There are numerous CMMs for many specific domains (like software development, quality management, project management, business process management, knowledge management, IT management) but not specifically for smartification of non-smart products. The focus of the model must be put on the smartification of existing non-smart products and services, which means that the domains and process areas of the model need to comply to that focus.
- The base model must contain the "right" number of maturity and/or capability levels: By definition, all CMMs have defined maturity and/or capability levels. All have level 1 to 3 or 4, many have a level 5, and some have the level 0. Having a CMM for a relatively new area of interest like smart product development, the model must start with level 0 to be able to declare a "does not exist"/"no capability available/present" level.
- The base model must define indicators for each maturity/capability level and include a comparison function: The research intent is to define and verbalize the indicators (within the process areas) in such a detailed way, that the comparison of any two levels delivers the measures to be taken to reach that next level.
- The base model must fulfil the combination of applicable degree of detail and usability: Many existing CMMs are very detailed which makes them hard to use (consultants must be at hand or user-centred tool comes with it). Many CMMs have very little detail which makes them usable but not really applicable to derive measures. The model must have the most ideal degree of detail in combination with the ease of use of the model. Criteria for determining this requirements-duo need to be aligned with the intended user group and organisation type as well as size.
- The base model must include two types of assessment tools:
  (1) Questionnaire: Based on the model, a questionnaire will be used to assess the strategic aims. It should contain easy to answer questions, regarding the strategic direction when the initial intent to smartify a products exists. There will be a kind of "translation" and "mapping" from questions as well as possible answers to indicators of the model. E.g. the question to the CEO is "why do you want to smartify your cloths hangers?". The CEO would state that the knowledge about customers needs

should be enhanced through movement sensors on the hanger in a store to be able to trace how many times a particular garment was tried on by customers. In the tool, there will be a translation from this answer to a model indicator in the process area "customers needs" as well as "IT infrastructure" and "market analysis".
(2) Indicator fulfilment: Using the model itself, the current maturity/capability levels will be identified through looking at each indicator defined for a process area and declaring which criteria are fulfilled. This gives direct information on the ability regarding a topic.

- The base model must include translation and mapping functions, to be able to translate the issues stated by the SME into structured aspects of the base model. See "methodical blocks" I, II and III in Fig. 1 and Sect. 4.

3.2 Methodology Framework

In addition to the underlying capability, maturity and measures base model, the tool kit will contain a methodology framework. Three different phases are foreseen using different methods to be applied:

(A) Identifying the presence: the methodology will comprise how to use different techniques to assess the current strategic aims as well as the respective current status of maturity and capability towards smartifying products.
(B) Defining the future: the methodology will propose techniques to support in defining the desired strategic aims as well as desired maturity and capability levels to be able to smartify a product.
(C) Deriving the steps from presence to future: The methodology will include the derivation of tangible measures to be taken which uses the comparison function within the capability, maturity and measures model.

4 Applying the Methodical Tool Kit for SMEs

The application of the methodical tool kit is demonstrated in the Fig. 1. On the left side, the capability, maturity and measures model resembles the underlying reference model, as mentioned above. In the following steps, the SME is involved.

(1) The questionnaire will be filled out by the SME's CEO (or representative who is responsible for the strategic development of the SME) to identify and rate the strategic orientation of smartifying an existing non-smart product in the future as well as to identify relevant further respondents in case the first respondents is not able to answer all questions in detail. In addition, interviews could be performed, if external support is available.
(2) The results of the questionnaire are input for (a) – the methodical block I – to be able to perform the current assessment according to the base model as well as input for (b) – the methodical block II – to be able to define the target situation.
(3) Using the capability, maturity and measures model, the respondents or other SME staff would help to identify, for each relevant process area and indicator, which maturity and capability level the SME currently has.

(4) In the next step, the targeted capability situation of the SME would be discussed, analysed and documented again using the capability, maturity and measures model and its indicators.
(5) Due to the structure of the base model, there will a pretty straight forward way to determine the delta between current and targeted criteria for each indicator.
(6) The methodical block III "mapping and translation from this delta to measures" is used to generate a list of tangible measures to be performed so that the SME can reach the defined target by smartifying a product in the future.
(7) With this information, the CEO could perform further activities like a cost-benefit analysis to be able to decide which concrete measures are to be followed or if external support is needed for the smartification.

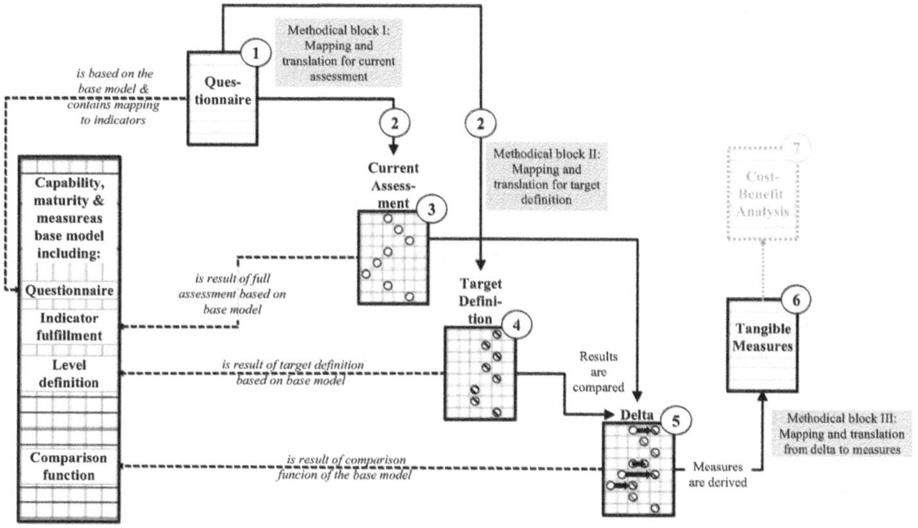

Fig. 1. Process of applying the Smartification Tool Kit

5 Summary and Conclusion

Summarized, the capability based approach – in form of the smartification took kit for SMEs to strengthen the strategic decision making process for developing smart products and services in SMEs – is designed to be the foundation to structurally analyse an SMEs ability and derive measures to follow the aim of smartifying a product for the first time. Once the SME has successfully used such a tool kit and implemented the measures, it will be more resilient to future changes since the change management process has been set up and/or improved.

In order to advance the approach described above, the processes of answers translated into content of structured formats such as current situation and future situation, as well as derived measures (methodical blocks) are addressed in the next steps of research.

References

1. Vukšić, V.B., Ivančić, L., Vugec, D.S.: A preliminary literature review of digital transformation case studies. In: Conference: 20th International Conference on Managing Information Technology (2018)
2. Abramovici, M., Acatech: Engineering smarter Produkte und Services Plattform Industrie 4.0 STUDIE. https://www.acatech.de/publikation/engineering-smarter-produkte-und-services/
3. Lichtenthaler, U.: Beyond artificial intelligence: why companies need to go the extra step | Emerald Insight, https://www.emerald.com/insight/content/doi/10.1108/JBS-05-2018-0086/full/html
4. Valencia, A., Schoorman, J.P., Mugge, R.: An Exploration of Smart Product-Service System Design. Delft University of Technology (2017)
5. Neuhüttler, J., Kett, H., Frings, S., Falkner, J., Ganz, W., Urmetzer, F.: Artificial intelligence as driver for business model innovation in smart service systems. In: Spohrer, J., Leitner, C. (eds.) AHFE 2020. AISC, vol. 1208, pp. 212–219. Springer, Cham (2020). https://doi.org/10.1007/978-3-030-51057-2_30
6. Oberli, W.: Die digitale Veredelung physischer Produkte (Teil 1/2). https://blog.carpathia.ch/2016/07/09/die-digitale-veredelung-physischer-produkte-teil-1-von-2/
7. Lindner, D., Leyh, C.: Digitalisierung von KMU – Fragestellungen, Handlungsempfehlungen sowie Implikationen für IT-Organisation und IT-Servicemanagement. HMD **56**, 402–418 (2019)
8. Tomiyama, T., Lutters, E., Stark, R., Abramovici, M.: Development capabilities for smart products. CIRP Ann. **68**, 727–750 (2019)
9. Bosch Media Service: Freie Sicht kann Leben retten: Bosch erfindet die Sonnenblende neu. https://www.bosch-presse.de/pressportal/de/de/freie-sicht-kann-leben-retten-bosch-erfindet-die-sonnenblende-neu-204928.html
10. Porter, M.E. and Heppelmann, J.E.: How Smart, Connected Products Are Transforming Competition. https://hbr.org/2014/11/how-smart-connected-products-are-transforming-competition
11. Frings, S., Domzig, S., König, S., Krieger, M., Kutzias, D.: Insekten-Cloud - Digitale Cloud-basierte Lösung zur Ermittlung, Lokalisierung und zum Monitoring von Schadinsektenbefall. https://cloud-mall-bw.de/wp-content/uploads/2019/12/5db1bc8e2969c191023-CMBW-PP01-Insekten-Cloud-Transferdoku-final.pdf
12. Ulrich, P., Rieg, R.: Agilität in Projektmanagement und Projektcontrolling – Ergebnisse einer empirischen Studie. Die Unternehmung **74**, 187–215 (2020)
13. Winter, R.: Modelle, Techniken und Werkzeuge im Business Engineering. In: Baumöl, U., Österle, H., Winter, R., Österle, H., Winter, R. (eds.) Business Engineering, pp. 87–118. Springer, Berlin Heidelberg, Berlin, Heidelberg (2003)
14. Österle, H., Blessing, D.: Business engineering modell. In: Baumöl, U., Österle, H., Winter, R., Österle, H., Winter, R. (eds.) Business Engineering, pp. 65–85. Springer, Berlin Heidelberg, Berlin, Heidelberg (2003)
15. Oxford University Press: Oxford English Dictionary (2021)
16. Merriam Webster, Incorporated: Merriam Webster Dictionary. https://www.merriam-webster.com/dictionary
17. Humphrey, W.S.: Characterizing the software process: a maturity framework. IEEE Softw. **5**, 73–79 (1988)
18. Dymond, K.M.: CMM® Handbuch. Das Capability Maturity Model® für Software. Springer, Berlin, Heidelberg (2002)
19. Software Engineering Institute: Capability Maturity Model® Integration for Development (CMMI-DEV), Version 1.3. Improving processes for developing better Improving processes for developing better products and services (Technical Report) (2010)

20. Wagner, K.-P.: Ermittlung des Reifegrads von Informationstechnologie in KMU. BWV Berliner Wissenschafts-Verlag, s.l. (2015)
21. Becker, J., Knackstedt, R., Pöppelbuß, J.: Developing maturity models for IT management. Bus. Inf. Syst. Eng. **1**, 213–222 (2009)
22. Williams, C.A., Lang, K.: Digital maturity models for small and medium-sized enterprises - a systematic literature review. In: Proceedings ISPIM Innovation Conference (2019)
23. Winkler, T.: Systematische Entwicklung von Smart Products – Konzeption eines Reifegradmodells. Studienarbeit am Institut für Arbeitswissenschaften und Technologiemanagement (IAT) der Universität Stuttgart (2020)
24. Freeze, R., Kulkarni, U.: Knowledge management capability assessment: validating a knowledge assets measurement instrument. In: Proceedings of the 38th Annual Hawaii International Conference on System Sciences, 251a-251a. IEEE (2005)
25. Berger, S., Bitzer, M., Häckel, B., Voit, C.: Approaching digital Transformation - Development of a muli-dimentional maturity model. Association for Information Systems Association for Information Systems/ECIS 2020 Proceedings (2020)
26. de Bruin, T., Freeze, R., Kaulkarni, U., Rosemann, M.: Understanding the main phases of developing a maturity assessment model. In: Proceedings of the Australasian Conference on Information Systems (ACIS) (2005)
27. Becker, J., Knackstedt, R., Pöppelbuß, J.: Dokumentationsqualität von Reifegradmodellentwicklungen. Arbeitsberichte des Instituts für Wirtschaftsinformatik, Westfälische Wilhelms-Universität Münster (2009)
28. Anderl, R., Fleischer, J.: VDMA: Leitfaden Industrie 4.0. Orientierungshilfe zur Einführung in den Mittelstand
29. Koldewey, C., Gausemeier, J., Fischer, S., Kage, M.: Entwicklung von Smart Service Strategien. Konferenz: Symposium für Vorausschau und Technologieplanung (2019)
30. Barata, J., da Cunha, P.R.: In search for a viable smart product model. In: Abramowicz, W., Corchuelo, R. (eds.) BIS 2019. LNBIP, vol. 373, pp. 101–112. Springer, Cham (2019). https://doi.org/10.1007/978-3-030-36691-9_9
31. Deuter, A., Otte, A., Ebert, M.: Extending the sliced v-model to smart product. In: Production Engineering and Management Proceedings 7th International Conference, pp. 115–125 (2017)
32. Falkner, J.: Leitfaden zur Entwicklung von intelligenten Diensten und Produkten. Mittelstand 4.0-Kompetenzzentrum Stuttgart (2020)

A People Centered Innovation Methodology for Its Application in Digital Transformation – Service Innovation Blocks

Gerhard Gudergan[1(✉)] and Yassi Moghaddam[2]

[1] FIR Institute for Industrial Management at RWTH Aachen University, Campus Boulevard 55, 52074 Aachen, Germany
gerhard.gudergan@fir.rwth-aachen.de
[2] Yassi Moghaddam, ISSIP International Society of Service Innovation Professionals, San Jose, CA, USA
yassi@issip.org

Abstract. Digitalization offers enormous opportunities not only to optimize operational processes, but also to redefine creative processes, e.g., in the area of innovation. This is becoming increasingly important in light of the fact that innovation is increasingly taking place in ecosystems, which means that an enormous amount of collaboration must be enabled in distributed and interdisciplinary teams. To be successful in this, innovation teams need easy access to the multitude of methods and assistance in selecting the appropriate method for the specific task. To this end, we propose a classification framework that structures methods from innovation management and service design based on higher-level task areas. The framework was developed and evaluated together with several companies. Results were implemented in the form of a playbook that won the red dot design award.

Keywords: Innovation · Methodology · People centricity · Innovation blocks

1 Introduction

Digitalization offers enormous opportunities not only to optimize operational processes, but also to redefine creative tasks. Transactions and value generation processes are changing [1, 2]. The good-centered logic, which was focused on tangible resources, value of the product, and single market transactions, changed to new perspectives with a focus on intangible resources and the co-creation of customers and service providers. This shift is characterized by Vargo and Lush as the Service-Dominant Logic of our economy [3]. For companies, besides changing toward a Service-Dominant Logic, the pure speed of innovating is becoming crucial to make use of upcoming opportunities. In order to achieve this, companies do not only have to rethink their approaches to innovation in general, but innovation teams need easy access to the multitude of methods and assistance in selecting the appropriate method for the specific task [4, 5].

C. Leitner et al. (Eds.): AHFE 2021, LNNS 266, pp. 197–203, 2021.
https://doi.org/10.1007/978-3-030-80840-2_22

2 Background and Relevance

This work is a result of longtime study with innovation experts from multinational companies. We build upon the basic concept proposed by Gudergan et al. [5] and discussed this before the context of the teamwork and collaboration needs. The innovation success of this teams depends on the one hand on an easy and quick access to tools and methods they need and on the other side on their capability to develop a more people centered perspective. The latter in particular accounts for innovation teams in more technology-focused organization. Thus, our methodology is intend to support innovation teams to get efficient access to the tools they need and to stimulate a more people – customer and other stakeholder – centered perspective. Methods and tools should be provided to team members according to the specific use-case situations, but users should not be required to strictly follow an overall sequence of methods and tools as typically proposed in traditional guidelines. However, there should be an order of methods and tools, which provide guidance at a high-level and correspond to typical innovation phases. We follow Chew in his perspective, that service innovation needs cross-disciplinary, holistic, and end-to-end approaches [6]. At the same time, innovation processes in the area of digital services are highly networked and knowledge-intensive processes and as such, they require specific attention in terms of organization and methods [7]. With respect to the set of methods presented in this paper, we build on results from Gudergan et al. [5, 8].

3 Objectives

From the perspective of innovation and development teams, our framework is intended to help increase the effectiveness and efficiency of collaboration by creating a collaboratively usable method base. Finding the methods that make sense for each individual user should be made easier. To achieve this goal in practice, the development of a classification approach is necessary from a research perspective. In addition, this should be presented in the context of user-centeredness in the sense of "people-centricity". The related research questions are listed below:

a) What are the typical tasks of innovation in collaboration teams and how can methods applied be classified according distinct phases?

Further detailing these phases leads to another question:

b) How can methods and tools be selected before the background of their applicability in the different phases?

The desired application of the classification and methods according to use-cases and contexts leads to the last research question:

c) How are use-case situations and Methods or tools related and can typical patterns of sequences be identified which collaboration teams should follow?

4 Research Approach

As presented in [5], we have developed and evaluated our basic framework together with companies. In doing so, we collaborated with experts from innovation teams and, in particular, with experts from the field of service innovation. We chose design science as our research methodology to meet the requirements of a pragmatic research approach. Design science as a research methodology is very well discussed by March and Smith [9]. In addition, Design science has been demonstrated as a suitable research methodology in other but similar use cases [10]. March and Smith view design science as a methodology used to create results for human use or application. Thus, design science is in contrast to the natural and social sciences, whose goal is to better understand reality [11]. The outcomes of design science research methodology consist of 1. constructs, 2. models, 3. methods, and 4. research activities. These four categories of results in turn include 1. building, 2. evaluating, 3. theorizing, and 4. justifying artifacts [11].

Constructs or concepts are a conceptualization used to describe problems within a domain. A model is a set of statements about relationships between constructs. A method is a sequence of steps in the sense of an algorithm or guideline used to solve a problem or task.

The operationalization of constructs, models, and methods in their specific environment constitutes instantiation. According to March and Smith, the activities of build and evaluation constitute the two essential activities of design science. According to Design Science, the construction of constructs, models, methods and artifacts belongs to the activity build. The first goal is always to show that constructs, models, methods, and artifacts can be developed in principle. The development of criteria as well as the assessment of the result when applying the criteria is summarized under the activity evaluation. Osterwalder describes this as follows [10]: "Constructs, models, methods, and artifacts are built to accomplish a specific task. These outputs then become the object of study that must be scientifically evaluated. They must be evaluated to determine if progress has been made. To do this, we need to develop metrics and measure the results against those metric" [10]. In the work presented here, the focus is on the 1. build and 2. evaluation activities. The focus is further on 1. constructs and 2. models as the outcomes. However, we can demonstrate that 3. methods and 4. instantiations can be derived from our classification.

5 Results

Based on our research, we developed the schema presented below, using the dimensions of exploration, evaluation, development, and realization for a top-level classification. According to the design science approach, these four dimensions represent the highest level of the constructs we used. In this context, the following chapter refers to the build phase in design science.

The starting point for our design is the work of March 1991, who presents activities of innovation in the tension between the competencies of exploitation and exploration [12]. Both competencies require combining knowledge into new things, but against the background of different objectives. According to Wadhwa and Kotha 2006, exploitation in particular requires efficiency and convergent thinking to bring offerings to market and continuously improve them [13]. In contrast, exploration includes activities such as searching, varying, or experimenting, with the objective of developing entirely new knowledge [12]. Exploration represents a major source of innovation and results from a significant deviation of an organization from its stated goals [14]. In our research, we understand the term exploration as the "pursuit of new knowledge about things that could become known" [15]. We include under it those activities whose goal it is to find entirely new opportunities outside the existing range of activities. Exploitation is best described by concepts such as selection, improvement, implementation, or efficiency [12]. We use this differentiation between exploration and exploitation, and for our purposes, we further divide the activities related to exploitation into the three activities of evaluating, developing, and realizing.

The activity of evaluation includes those activities that are part of assessing an idea and estimating its market potential. This creates the starting point for further detailing [16]. The development activity includes all the steps necessary to create a prototype [17]. The realization phase starts with a functional prototype and concludes with a service offering that could be successfully offered and scaled in the market [18]. In the following, we show in which way the constructs of exploration, evaluation, development and realization could have been put in a logical dependency to each other in different sequences and how patterns in terms of different innovation strategies result from this [5].

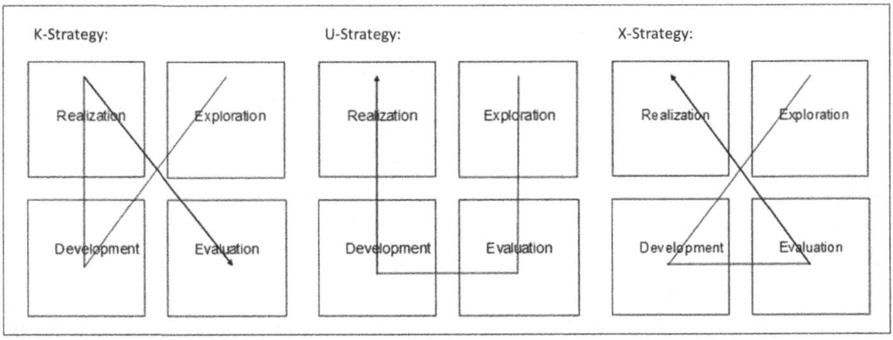

Fig. 1. Service Innovation Blocks and innovations strategies (K-U-X) (Own illustration)

The underlying framework is called "Service Innovation Blocks" [5]. As we will show below, the sequence in which the individual phases are run through is not fixed and depends on the specific context. However, it is characteristic for innovation processes that all phases are passed through (see Fig. 1). Each innovation team starts with the exploration phase and a selection of the methods contained therein. The start of an innovation is always to use different methods in the sense of an ideation and to get the best possible picture of the needs of a target person. This ensures explicit "people centricity", regardless of whether the framework is used, for example, in more technology-oriented teams or companies. However, the expert interviews we conducted clearly showed that the order in which the four phases are run through can vary greatly. The sequence then stands for a specific innovation strategy, the application of which in turn depends on the specific context. There are three strategies. The "K strategy" describes the pattern of a "rapid prototyping" approach. Here, a minimally functional product is developed from an idea immediately after the exploration phase. Based on this, the final product is developed and then launched on the market. The final evaluation is based on the market success achieved. The goal of the K strategy is the fastest possible realization of a minimally functional product. The U strategy basically corresponds to a "traditional innovation approach". This strategy has largely characterized the practice of innovation [19]. Idea evaluation immediately follows the exploration phase with the aim of reducing development costs by testing ideas too late. Only ideas that have been positively evaluated on the basis of suitable criteria are further detailed and finally realized. Although the advantage of this approach is understandable, it is too slow in view of today's development speeds and, above all, this traditional approach means that promising ideas are filtered out too early and thus do not reach prototype testing or implementation in the form of a minimally functional product. The U strategy tends to be applicable more for cost-intensive products or systems.

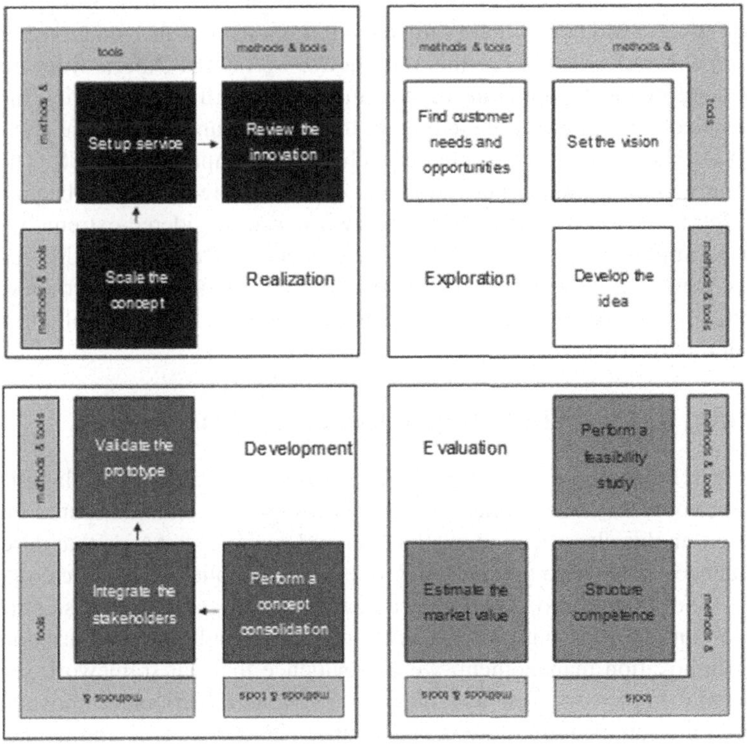

Fig. 2. Service Innovation Blocks with guidelines (level 2) and methods (level 3) (based on [5])

The X strategy is best described by the terms "trial and error". The development of an idea is immediately followed by the design of a concept. This concept is then quickly turned into a "minimum viable product", which leads to a very rapid evaluation in the market. This makes it possible to obtain very early feedback, which in turn avoids high investments in a fully detailed product that is ultimately tested on the market by far too late. This reduces the overall risk of an innovation.

The presented framework leads innovation teams step by step through the innovation process and offers also very interdisciplinary teams a suitable methodical framework for a cooperation. based on our interviews, the framework can serve as a joint and shared mental framework for innovation teams which in turn supports team identity and collaboration effectiveness at the same time. Each of the four blocks is subdivided into methods – "guidelines" - and these are in turn assigned suitable tools. This results in a hierarchical structure with a total of three levels (see Fig. 2).

6 Evaluation

The framework and its applicability has been evaluated in a two step approach. In a first step, we evaluated the consistency of our framework and in particular the applicability of the whole framework in the context of innovation teams. We conducted interviews and our interview partners clearly pointed out the high value of the framework in the specific use situations. They further evaluated our framework with a focus on consistency and gave a positive feedback. For this interviews, we invited experts from innovation teams from internationally operating organizations out of different countries. In a second step, we applied the framework including all the methods and tools in the context of an innovation hackathon program at a major technology company in the US. The goal of this program was to stimulate innovation and enable different teams to become part in a worldwide innovation challenge. The framework was considered extremely useful in particular, to implement a more people centered perspective within traditionally more technology centered innovation perspectives. At the same time, the framework was considered to build the joint mental framework, which is needed facilitate and ease collaboration in distributed teams.

7 Summary – contribution to Theory and Practice

From a research perspective, we applied design science to develop a classification of methods and tools for their application in interdisciplinary and distributed innovation teams. Based on this classification, we had derived specific strategies that describe different sequences of how the methods and tools can be applied in specific contexts. For this paper, we evaluated and tested the framework and in particular these strategies in the specific context of an innovation hackathon. As a result and a contribution to the practice of innovation management, we got evidence that our framework serves as a suitable mental framework to better integrate divers and distributed innovation teams and to develop a significantly more people centric perspective at team members and in the overall team process in particular in technology oriented organizations.

References

1. Keuper, F.: Digitalisierung und Innovation. Planung, Entstehung, Entwicklungsperspektiven. 1. Bearingpoint, Wiesbaden, s.l. (2013)

2. Hungerland, F., et al.: The digital economy. Strategy 2030 - Wealth and Life in the Next Generation, No. 21e. Hrsg.: Hamburg Institute of International Economics (HWWI) (2015)
3. Vargo, S.L., Lusch, R.F.: Evolving to a new dominant logic for marketing. J. Mark. **68**(1), 1–17 (2004)
4. Bitner, M.J., Ostrom, A.L., Morgan, F.N.: Service blueprinting: a practical technique for service innovation. Calif. Manag. Rev. **50**(3), 66 – 94 (2008)
5. Gudergan, G., Heikenfeld, H., Hessler, A.: Service innovation blocks. In: 2018 IEEE International Conference on Engineering, Technology and Innovation (ICE/ITMC) (2018)
6. Chew, E.K.: Service innovation through an integrative design framework. In: Agarwal, R., Selen, W., Roos, G., Green, R. (eds.) The Handbook of Service Innovation, pp. 481–500. Springer, London (2015). https://doi.org/10.1007/978-1-4471-6590-3_22
7. Khatibian, N., Hasan gholoi pour, T., Abedi Jafari, H.: Measurement of knowledge management maturity level within organizations. Bus. Strategy Ser. **11**(1), 54 – 70 (2010)
8. Gerhard Gudergan (Hrsg.): Framework and Model for Successful Service Solution Design - An Organisational Design / Dynamic Capabilities Perspective (2009)
9. March, S.T., Smith, G.F.: Design and natural science research on information technology. Decis. Support Syst. **15**(4), 251–266 (1995)
10. Osterwalder, A.: The business model ontology. A proposition in a design science approach, Lausanne Universite de Lausanne, Dissertation (2004)
11. Au, Y.A.: Design science I: the R role of design science in electronic commerce research. Commun. Assoc. Inform. Syst. **7**(1), 1–15 (2001)
12. March, J.G.: Exploration and exploitation in organizational learning. Organ. Sci. **2**(1), 71–87 (1991)
13. Wadhwa, A., Kotha, S.: Knowledge creation through external venturing. Evidence from the telecommunications equipment manufacturing industry. Acad. Manag. J. **49**(4), 819 – 835 (2006)
14. Nooteboom, B.: Innovation, learning and industrial organisation. Camb. J. Econ. **23**(2), 127–150 (1999)
15. Rothaermel, F.T.: Incumbent's advantage through exploiting complementary assets via interfirm cooperation. Strateg. Manag. J. **22**(6–7), 687–699 (2001)
16. Buczkowski, A.: Innovative Dienstleistungen gestalten. Eine Studie zum Einsatz von Best Practice Methodent. Igel Verlag RWS, Hamburg (2014)
17. Alam, I., Perry, C.: A customer-oriented new service development process. J. Serv. Mark. **16**(6), 515–534 (2002)
18. Bullinger, H.J., Scheer, A.W.: Service engineering — entwicklung und gestaltung innovativer dienstleistungen. In: Bullinger, H.J., Scheer, A.W. (eds) Service Engineering. Springer, Berlin, Heidelberg (2006). https://doi.org/10.1007/3-540-29473-2_1
19. Eversheim, W. (Hrsg.): Innovation management for technical products. Systematic and integrated product development and production planning. RWTH edition. Springer, Berlin, Heidelberg (2009)

Towards a Continuous Process Model for Data Science Projects

Damian Kutzias(✉), Claudia Dukino, and Holger Kett

Fraunhofer IAO, Fraunhofer Institute for Industrial Engineering IAO, Nobelstrasse 12, 70569 Stuttgart, Germany
{Damian.Kutzias,Claudia.Dukino,Holger.Kett}@iao.fraunhofer.de

Abstract. Process models can assist in structuring and managing projects. For typical IT-projects, there are plenty process models which evolved over the last decades. Compared to them, data science process models focus on the specific challenges and aspects of data-based projects. They started evolving just before the turn of the millennium. This paper evaluates contents which could and should be included in data science process models to be useful for enterprises, especially when they are small and medium-sized or do not have their core competences in data science or IT. Regarding these contents, some existing models are analysed, providing an overview of their focus. Concluding, a vision for a continuous data science process model is given, which not only addresses the previously discussed contents, but also fulfils additional aspects to be useful in practise.

Keywords: Data science · Process model · Data analytics · Project management · Methodology · Data mining · Economics · Industrial science · Transformation process · Business analytics · Artificial intelligence

1 Introduction

The current digitisation shifts requirements and potentials in today's economy in an extent which justifies calling it a second digital revolution [1]. It is often the data that matters instead of the service or product: the data can be used for developing new business models or services, which may be called «Smart Services» [2]. While such smart services can be developed without the usage of artificial intelligence (AI), our society has reached a stage at which it cannot develop independent from AI anymore [3]. Finding useful information (or patterns) in data is the core task of artificial intelligence [4], which is known by many names such as «Data Mining», «Knowledge Extraction», «Information Discovery» and many more [5]. One of the currently prominent terms is «Data Science» - it is now the leading term being favoured over data mining [6]. It has a rather long history started in 1974 [7] and has many different definitions and related terms [7, 8]. Especially some of the scientific definitions underline the part «Science» meaning the scientific approach, when working as a data scientist. However, data science has become a trend term in practice and is used by popular science, bloggers and enterprises alike. As an example, the data scientist is depicted as the sexiest job of the 21st century with

C. Leitner et al. (Eds.): AHFE 2021, LNNS 266, pp. 204–210, 2021.
https://doi.org/10.1007/978-3-030-80840-2_23

programming as his core skill [9]. The Google trend analytics show that «Data Science» was used more than two times as often as «Data Analytics» and more than four times as often as «Data Mining» over one year, checked at the 18th of November 2020 [10]. We explicitly want practitioners of data-driven projects from the industry to be addressed, which might be more interested in the results than a scientific approach, and therefore choose a simple and open definition, not necessarily including the scientific approach, but respecting the usage of the term in practice. Our definition is:

Data Science is the nontrivial acquisition of knowledge from data.

Also note, that this definition does not require machine learning or AI to be applied within a data science project. The realisation of data science projects is a multidimensional challenge requiring specialised tools, processes and methodologies [11]. To cope with this challenge, data science process models can assist in the successful realisation of data science projects. Whereas process models are common in more general areas such as software development, they are a rather new and evolving field for data science. Within this paper, we investigate data science process models regarding covered contents as well as gaps and weaknesses concerning industry usability and derive a vision in terms of characteristics of a next generation data science process model.

2 Existing Data Science Process Models

Within this section, some data science process models are briefly presented with their focuses and main phases. This list is not complete, for a broader overview, we refer to [12, 13], which provide graphical overviews of different data science process models and the dependencies within their evolution.

- **Knowledge Discovery in Databases (KDD)**. The authors define the KDD process as «The nontrivial process of identifying valid, novel, potentially useful, and ultimately understandable patterns in data». KDD process in more detail is described as an interactive and iterative process involving nine steps: 1) Learning the application domain, 2) Creating a target dataset, 3) Data cleaning and pre-processing, 4) Data reduction and projection, 5) Choosing the function of data mining, 6) Choosing the data mining algorithm(s), 7) Data mining, 8) Interpretation and 9) Using discovered knowledge [5]. KDD was the first process model to establish all the steps to be taken to develop a data mining project [13] and has severely influenced many of the following data science process models [12, 13].
- **CRoss-Industry Standard Process for Data Mining (CRISP-DM).** CRISP-DM was developed by an industry consortium and is based on the practical real-world experience of how people do data mining projects instead of a scientific approach. It includes six main phases which include going back and forth in an iterative manner: 1) Business understanding, 2) Data understanding, 3) Data preparation, 4) Modelling, 5) Evaluation and 6) Deployment. CRISP-DM comes with a step-by-step guide breaking down phases to generic tasks, generic tasks to specialised tasks, and specialised tasks to process instances. It includes many check lists or step lists for different tasks to be performed [14].

Two polls about the used process model in data science, which were conducted by KDnuggets, have shown that CRISP-DM was used by 42 percent of the participants in 2007 and 43 percent of the attendees in 2014 whereat in 2014 no other process model exceeded 10 percent [15]. It can still be considered the de-facto standard [6].

- **Analytics Solutions Unified Method (ASUM).** ASUM introduces itself as «The Analytics Solutions Unified Method (ASUM) is a step-by-step guide to conduct a complete implementation lifecycle for IBM Analytics solutions». It contains five defined phases: 1) Analyse, 2) Design, 3) Configure & Build, 4) Deploy, 5) Operate & Optimize and in addition lists «Project Management» as a sixth phase emphasising agile project management in conjunction with established models such as the V-model (a process model for software development including several test phases after the development) as an integrated part with each phase of a data science project [16]. During the analysis, ASUM was only available as a white paper with three pages. The given webpage and the given contact address from the white paper were dead as of 9. October 2020, therefore only the limited version was considered.
- **The lightweight IBM Cloud Garage Method for data science (ILG).** ILG introduces itself as «A process model to map individual technology components to the reference architecture». It includes no tasks for setting the objective and analysing the domain because it is intended to be used for projects where this is already done. It comes with seven phases which also support going back and forth: 1) Initial Data Exploration, 2) Extract, Transform, Load, 3) Feature Creation, 4) Model Definition, 5) Model Training, 6) Model Evaluation and 7) Deploy Model [17]. The process model is complemented by architectural decisions guidelines for data science giving an abstract reference architecture and propositions how the components can be selected in a concrete project [18].
- **Engineering Data-Driven Applications (EDDA).** EDDA assumes that the development of machine learning or AI components is part of a larger development process and thus proposes two complementary processes, a software engineering process and the EDDA process. The software engineering process consists of 1) Requirements Engineering, 2) Specification and Design, 3) Implementation, 4) System Test and 5) Operation. The EDDA process consists of 1) Is Machine Learning suitable?, 2) Data Exploration, 3) Model Requirements, 4) Model Development, 5) Model Integration and 6) Operation. EDDA emphasises roles representing responsibilities and specialisations and describes them in addition to the phases [11].
- **Team Data Science Process (TDSP).** TDSP is introduced as an agile, iterative data science methodology to deliver predictive analytics solutions and intelligent applications efficiently with a focus on team collaboration. It comes with four key components: 1) A data science lifecycle definition, 2) A standardized project structure, 3) Infrastructure and resources recommended for data science projects and 4) Tools and utilities recommended for project execution. The data science lifecycle can be compared to the other process models, since it gives major phases (stages), also explained as iteratively executed: 1) Business Understanding, 2) Data Acquisition and Understanding, 3) Modelling and 4) Deployment. These major phases come with several sub-elements and the process is ended by customer acceptance after the deployment phase [19].

- **Data Science Process Model (DASC-PM).** DASC-PM describes a data science process model, which is embedded in the domain (of the project) and is enclosed and permeated by a scientific approach. It starts by a project order and has four phases which are based on the relevant IT-infrastructure: 1) Data provisioning, 2) Analysis, 3) Utilisation and 4) Usage. It describes competence profiles and roles in detail and gives quick information for steps and key areas within the model about the required roles and competences. DASC-PM is licensed under CC BY-NC-ND 4.0, thus the usage, especially in commercial context, is restricted [20].

3 Contents of Data Science Projects

The survey of existing data science process models showed that technical aspects are often treated in detail and extensively. Data and algorithms and their handling are usually the core of the models. The iterative nature of data science projects is covered by most models, supporting adaptions in the applied approaches. This is meaningful due to the uncertainty of the right tools and the results based on unobvious information in the data. Four of seven models make concrete propositions on project management, e.g. favouring agile development [11, 14, 16, 19]. Also, some models include a focus on roles that are more precise than that of a data scientist, especially the newer ones [11, 14, 19, 20]. Related pre- and post-project tasks are often omitted or presented in a very reduced form. Even though some of them are the same as in any other projects, we believe that especially in data science projects the tasks: Requirements analysis, process adaptation, user view and integration into the business context, are particularly important and behave differently than in standard software projects. Especially because it is about the interaction of humans and AI as a new technology in broad enterprise applications. It is precisely the close integration between people, technology and process that brings a decisive advantage in the development of new applications with artificial intelligence, machine learning aspects to create acceptance in the corporate context [21]. Through automation and augmentation, there is a fear of substitution [22], which should be taken away from the human early in the process and acceptance created through integration into the process. Also, in technical parts, some emerging issues such as explainable AI or model robustness to data changes or attacks on the data plane can have a great impact on data science projects but are rarely addressed in detail.

4 Vision

Our vision is an up-to-date data science process model, which has a high usability for practitioners. To have the potential to fulfil this vision, we suggest the following characteristics for such a model:

- **Continuity** – By continuity we understand that the model includes the previously discussed contents, i.e. starting with (business) objectives and state analysis and ends with the utilisation within enterprises including changes for the affected people and processes, since specialities of data science projects may change these compared to classic projects.

- **Enterprise size** – The model should be able to assist enterprises independent of their size, i.e. especially small or medium-sized enterprises (SME) should be addressed. SME have urgent need for high quality assistance, since they usually have limited budget and a less complex IT-landscape [23].
- **Business section** – The model should be independent of the business section. The effects of the current digitisation are strong and not restricted to business sections [24], this also matches the beliefs of 63 percent of the CIOs according to a poll of Capgemini [25].
- **Based on the experience of practitioners** – CRISP-DM states in its preface, that pure scientific developments have only failed in the past and argues its usability based on the non-scientific, pragmatic approach from practitioners [14]. The success in terms of adaption and usage is a strong indicator for the correctness of this statement [6, 15].
- **Unrestricted usability** – The usability in terms of licensing. Enterprises should be allowed to freely use the model and everyone should be allowed to extend it. A published process model with a restricted license might be allowed for learning but can cause legal problems when it comes to the usage in consultancy and cooperation on the business level, yielding a reason not to use it.
- **Neutrality of vendors** – While concrete integrated tools from an ecosystem of a single vendor might unleash the highest potential for customers of this vendor, such models might be of little use to others. Also, the concrete vendor support in a model results in vendor specific learning invest and can be seen as a special kind of vendor lock-in, which even might be a hindrance for possible customers [26].
- **Tools** – A concrete tool to face challenges can provide more value than a simple explanation of what to solve or how to solve it on an abstract level. Even if the proposed concrete tool is not the right choice for all cases, it can support the selection since humans learn well from examples especially in early skill acquirement [27].

5 Conclusion

We briefly presented several data science process models including KDD as the first, CRISP-DM as the industry standard and young models up to 2020, describing their phases and focal points. Based on these models and additional research, we identified relevant contents for data science process models rarely addressed today. Finally, we described our vision for a continuous data science process model based on the relevant contents including additional characteristics.

To reach the described vision, further research will be necessary on each of the contents, especially to respect the current business requirements appropriately and to provide useful tools. An important aspect is the involvement of practitioners. To address this, we already started an interview series addressing a broad spectrum of different roles and positions from junior analyst to CEO.

References

1. Rindfleisch, A.: The second digital revolution. Mark. Lett. **31**(1), 13–17 (2019). https://doi.org/10.1007/s11002-019-09509-4

2. Marquardt, K.: Smart services – characteristics, challenges, opportunities and business models. In: Proceedings of the International Conference on Business Excellence, vol. 11, no. 1, pp. 789–801 (2017)
3. Bovenschulte, M., Stubbe, J.: Einleitung: „Intelligenz ist nicht das Privileg von Auserwählten.". In: Wittpahl, V. (ed.) Künstliche Intelligenz, pp. 215–220. Springer, Heidelberg (2019). https://doi.org/10.1007/978-3-662-58042-4_13
4. High-Level Expert Group on Artificial Intelligence: A definition of AI: main capabilities and disciplines (2019)
5. Fayyad, U., Piatetsky-Shapiro, G., Smyth, P.: The KDD process for extracting useful knowledge from volumes of data. Commun. ACM **39**(11), 27–34 (1996). https://doi.org/10.1145/240455.240464
6. Martinez-Plumed, F., et al.: CRISP-DM twenty years later: from data mining processes to data science trajectories. IEEE Trans. Knowl. Data Eng. 1 (2020)
7. Cao, L.: Data science: a comprehensive overview. ACM Comput. Surv. **50**(3), 1–42 (2017)
8. Dhar, V.: Data science and prediction. Commun. ACM **56**(12), 64–73 (2013)
9. Davenport, T.H., Patil, D.J.: Data scientist: the sexiest job of the 21st century (2012)
10. Google: Google Trends for "Data Science", "Data Mining" and "Data Analytics" (2020)
11. Hesenius, M., Schwenzfeier, N., Meyer, O., Koop, W., Gruhn, V.: Towards a software engineering process for developing data-driven applications. In: 2019 IEEE/ACM 7th International Workshop on Realizing Artificial Intelligence Synergies in Software Engineering (RAISE), pp. 35–41. IEEE (2019)
12. Mariscal, G., Marbán, Ó., Fernández, C.: A survey of data mining and knowledge discovery process models and methodologies. Knowl. Eng. Rev. **25**(2), 137–166 (2010)
13. Martínez-Plumed, F., et al.: CASP-DM: context aware standard process for data mining (2017)
14. Chapman, P., et al.: CRISP-DM 1.0: step-by-step data mining guide (2000)
15. Piatetsky, G.: CRISP-DM, still the top methodology for analytics, data mining, or data science projects (2014)
16. IBM Corporation: Analytics solutions unified method: implementations with agile principles (2016)
17. Kienzler, R.: The lightweight IBM Cloud Garage Method for data science: a process model to map individual technology components to the reference architecture (2019)
18. Kienzler, R.: Architectural decisions guidelines: an architectural decisions guide for data science (2019)
19. Microsoft: Team Data Science Process Documentation (2020)
20. Schulz, M., et al.: DASC-PM v1.0 - Ein Vorgehensmodell für Data-Science-Projekte (2020)
21. Tombeil, A.-S., Kremer, D., Neuhüttler, J., Dukino, C., Ganz, W.: Potenziale von Künstlicher Intelligenz in der Dienstleistungsarbeit. In: Bruhn, M., Hadwich K. (eds.) Automatisierung und Personalisierung von Dienstleistungen, Forum Dienstleistungsmanagement, pp. 135–154. Springer, Wiesbaden (2020).https://doi.org/10.1007/978-3-658-30168-2_5
22. Frey, C.B., Osborne, M.A.: The future of employment: how susceptible are jobs to computerisation? (2013)
23. Lindner, D., Leyh, C.: Digitalisierung von KMU – Fragestellungen, Handlungsempfehlungen sowie Implikationen für IT-Organisation und IT-Servicemanagement. HMD Praxis der Wirtschaftsinformatik **56**(2), 402–418 (2019). https://doi.org/10.1365/s40702-019-00502-z
24. Leyh, C., Bley, K.: Digitalisierung: Chance oder Risiko für den deutschen Mittelstand? – Eine Studie ausgewählter Unternehmen. HMD Praxis der Wirtschaftsinformatik **53**(1), 29–41 (2015). https://doi.org/10.1365/s40702-015-0197-2
25. Capgemini: Studie IT-Trends 2018: Digitalisierung: Aus Ideen werden Ergebnisse (2018)

26. Opara-Martins, J., Sahandi, R., Tian, F.: Critical review of vendor lock-in and its impact on adoption of cloud computing. In: International Conference on Information Society (iSociety 2014), pp. 92–97. IEEE (2014)
27. Renkl, A., Atkinson, R.K., Maier, U.H.: From example study to problem solving: smooth transitions help learning (2000)

Emerging Research Innovations in AI, User Experience and Design

Value Co-creation Through Collaborative World-Building and Cosplay: QwörkSpace

Joshua Ian Robles(✉)

The Red Mage: World-Building in Design, 1636 Sargent Court, Los Angeles, CA 90026, USA
http://theredmagepodcast.com/

Abstract. Cosplayers are content creators who are symbolic of the universal struggle of micro-entrepreneurs in being fairly compensated for the value they generate. As industries gained exposure and increased sales, they have historically repaid cosplayers who act as brand ambassadors, community leaders, and content creators with symbolic victories. The core issue is a deeply rooted social and cultural learned behavior that causes cosplayers, much like the micro-entrepreneurs they represent, to fail to recognize their value. To transition cosplayers from celebrated hobbyists to paid professionals, this research identifies the potential of game-based learning, world-building, and cosplay as an opportunity for value generation for the contributing cosplayers.

Keywords: World-building · Game design · Game-based learning · Cosplay · Community building · Value Co-creation · Behavior change · Value proposition

1 Introduction

Cosplayers are content creators who are symbolic of the universal struggle of mi cro-entrepreneurs in pursuit of fair compensation. As industries gained exposure and increased sales, they have historically repaid cosplayers who act as brand ambassadors, community leaders, and content creators with symbolic victories. The core issue is a deeply rooted social and cultural learned behavior that causes cosplayers, much like the micro-entrepreneurs they represent, to fail to recognize their value. This research identifies the potential of game-based learning, world-building, and cosplay as an opportunity for value generation for the contributing cosplayers to transition cosplayers from celebrated hobbyists to paid professionals.

2 Methodology

A world-building methodology explores the value co-creation of a business model for creative micro-entrepreneurship. A world-building methodology is optimal as it explores the solutions in virtual and actual space with designs integrated for cross reality solutions. Cross reality solutions are systems that transition between physical and virtual interfaces to bridge local and remote users and provide various touchpoints for inclusion and

C. Leitner et al. (Eds.): AHFE 2021, LNNS 266, pp. 213–218, 2021.
https://doi.org/10.1007/978-3-030-80840-2_24

accessibility. The methods for this research will include field visits, field activities, ethnographic interviews with cosplayers, literature review, popular media searches, and exploration of the business model canvas. The research will drive the design of a saleable MVP that identifies opportunities for curving behaviors, game-based learning, and value co-creation with the cosplay community.

Ethnographic interviews with cosplayers will provide insight into the experience of cosplay, pain points of cosplayers, and key elements involved that define cosplay as a system. Interviews with cosplayers will provide insight into the meaning and purpose of cosplay as well as delineate the aspirations the drive cosplayers to pursue a career in cosplay. Questions derived from popular media searches and literature reviews to challenge assumptions, receive community input and guide an informed social narrative. To understand the world that cosplayers inhabit, field visits and field activities were conducted. Attendance at conventions and participation in panels and events hosted at these conventions provide first-hand experience in understanding operations and systems. Mixed-methods research defines the narrative of cosplayers and reveals avenues to co-create solutions for cosplanuership.

3 What Will be Discussed

Creatives are invaluable content creators and value generators for companies. Creatives are critical to developing virtual communities, propel brand engagements, and act as ambassadors for a product, company, or service. Connecting with these communities and partnering with these brand ambassadors allows for value co-creation. Value co-creation ensures resilience, relevance, and success of the iterative release of products, services, and platforms. The complication that arises is that only a select few have had the opportunity to establish a sustainable career through being a community leader and brand ambassador. In regards to cosplay, the economics of a cosplaneur, a professional cosplayer seeking to establish a career or use cosplay as a vehicle for a long term career, is punishing. Obstacles that cosplaneurs face is the inundation of talented cosplayers, difficulty negotiating for financial compensation, communicating the value they generate, social and cultural concepts of value, and creating a sustainable business model. The solution provided must allow for autonomy, skill mastery, knowledge generation, and behavioral change to ensure a sustainable transition from hobbyist to professional. These objectives present an opportunity for implementing a game-based learning solution.

Definitions

Cosplay is the practice of "fans donning costumes and performing characters from popular media text such as comics, animated or live action films and television, games and other popular cultural media including music videos" [1].

DM is an abbreviation for Dungeon Master. In table-top role playing games such as Dungeons and Dragons, a DM is responsible for guiding players through adventures, overseeing battles, and informing players of the results of interactions with non-player characters, environments, or objects [2].

Cosplaneur is a user segment within the cosplay community that aspires to establish a career as a professional cosplayer, use cosplay as a vehicle for promoting their business, or use cosplay as an entryway to a creative career.

Magic Circle is described by Johans Huizinga as a dedicated space where players suspend their disbelief of reality. In doing so, players abide by a series of rules while voluntarily partaking in a game for a given duration of time. This space is separate from the real world [3].

Shared Fantasy as described by Gary Allan Fine in his book Shared Fantasy to be a state in which players fall into a state of *Folie Á Deux* or shared madness. Active participants are able to distinguish fantasy from reality but agree to abide by a set of rules and behaviors while immersed in the gameplay [2].

Value Co-creation is the collaboration between businesses and communities to develop an efficient, viable, and resilient value proposition that addresses the needs of the community. Co-creating with the community reveals what community members value, what they have access to, and elements that impact their decision making.

4 Popular Media Search

Popular media searches reveal that cosplay is a part of a system that is inter-woven with the gaming industry, fashion industry, and entrepreneurial pursuits of creatives. Fans of videogames attend conventions while dawning the personas of video game characters. This celebration of gaming culture not only generates organic promotion of a series, it is an experience that builds brand community through the process of co-creation. Blizzard is aware of the importance that the cosplay community plays in the promotion of their brand and integrates cosplayers into their marketing system. Blizzard does so through highlighting their social media images, providing details on characters to help cosplayers achieve accurate character depictions, hiring professional cosplayers for promotion, and co-creating with famous personas [4]. The marketability of these characters or avatars has not gone unnoticed in the fashion industry. High fashion brands such as Louis Vuitton incorporated Final Fantasy XIII characters into their promotional material in 2016 [5]. This trend has also influenced street fashion and fueled micro-entrepreneurial pursuits in creating clothing that celebrates gaming culture. When exploring cosplay as a system, it becomes apparent that cosplay as a celebration of gaming culture promotes the prosperity of the gaming industry which then creates opportunities for the fashion industry to market their products to a new user segment [1].

Cosplayers attending conventions and gatherings also contribute to local economics. As recorded by San Diego, the regional impact of San Diego Comic Con in 2019 was $149 million dollars with an attendance of 135,000 people over the course of four days [6]. Similarly, in 2019 Anime Expo made an estimated regional impact of over $100 million dollars with 350,000 fans at-tending in person and virtually [7]. 12 Insights into cosplay demographics from Anthony Thomas Marketing reveals that 60% of cosplayers are between the ages of 23–30 with 64% of cosplayers being female [8]. When attending conventions 70% of cosplay makers spend $100 or more at conventions they attend. In regards to the cost of cosplay materials 31.2% of cosplayers spend $100–$200 on their cosplay while 70% spend $100–$600, and few even spend $1000 or more for advanced cosplays [8]. In regards to the number of cosplays and conventions attended per year, 32.1% of cosplayers make 3–4 cosplays per year and 64% attend 3 or more events per year [8]. As cosplays are worked on prior to attending conventions, the investments into materials for 3–4 cosplayers per year provides additional economic impact [8]. The challenge that many cosplayers endure is the lack of economic support and time in creating their cosplays. As mentioned in an article by BTR Today titled *The Punishing Economics of Cosplay*, it becomes apparent that cosplayers making six figure salaries are incredibly rare [9]. The majority of cosplayers pursuing cosplay as a full time career are able to build large followings and community support, but have difficulty monetizing that following. Challenges in negotiating for compensation, communicating the value of one's services, and establishing a sustainable business model, and the inundation of cosplayers makes the transition from cosplay as a hobby to a profession extremely arduous [9].

5 Literature Review

Literature exploring cosplay reveals that cosplay is a critical practice that extends beyond the dawning of a persona for entertainment. In *Planet Cosplay: Costume Play, Identity, and Global Fandom* cosplay is discussed from the perspective of détournement [1]. This perspective frames cosplay as a system in which cosplayers and their audience require a minimum threshold of familiarity with fandom and storyworld be met in order to engage a community of play. Cosplayers transcend the limitation of fans as they garner critical faculty to franchises they reference through recontextualizing sources and creative expression [1]. Avenues as such involve music videos, fanfiction, and character re-designs expressed through cosplay [1]. Cosplayers do have an internal struggle within the community when it comes to the monetization of cosplay. The di-vide in the community regarding monetization stems from the concern that 'hyper-commercialism of many cons and the media sensationalization and public trivialization of fan cultures so decried by many committed cosplayers" [1]. Concerns of commodification culture in the cosplay regard the re-appropriation of the predominately female DIY culture, distortion of messages regarding body image, derailing the critical practice of cosplay, and deterioration of elements such as identity and empowerment for financial gain [1].

6 Creative Business Models and Opportunities for Game-Based Learning

Commercialization demonstrates the potential to disenfranchise communities when implementing a business model that focuses solely on metrics and financial gains [1]. This concern is validated when communities may potentially be stripped of their agency, their spaces, and a sense of belonging. Industries may lose opportunities to expand their key target audience if their efforts are regarded as opportunistic.

Factors which contribute to this potential loss in the infinite game in establishing a relationship between one's brand and the cosplay community arise from barriers of cosplanuership. Cosplanuership is on the rise, and it would behoove industries to engage in value co-creation to sustain a relationship with cosplayers in their marketing strategies. Cosplayers are part of the infinite game when it comes to engaging with brands, platforms, and services.

Currently cosplayers that have been able to create a career out of cosplay such as Yaya Han and Kamui Cosplay. Scrutinization of these business models solely in regards to entrepreneurship demonstrates the need to engage in a multi-channel business model. *Yaya Han's World of Cosplay* illuminates the need to have a diversification of services to build capital, multiple social channels for communication and community building, branding, and partnerships [10]. The most important aspect of a cosplay business model is community building and contribution to the community that supports one's brand [10]. This business model conveys a double bottom line where the monetization of cosplay supports the content creator who in turn invests in the cosplay community. Investments include education and knowledge sharing, providing community resources, and community empowerment.

To achieve this goal, it is necessary to provide a resource that educates and empowers through self-discovery. Interviews with cosplayers during field visits shows that while many aspire to transition from the realm of hobbyist to professional it is difficult to establish a sustainable model for building a professional network, communicating the worth of their services, and illustrating the value they generate. Many of the cosplayers interviewed express the frustrations of finding paid work, but did not mention key target audiences or express familiarity with a value proposition. However, cosplayers interviewed did express a love of videogames and/or table top games such as dungeons and dragons. This revealed an opportunity to leverage game-based learning to teach cosplaneurs about building a value proposition and communicating their value.

The MVP (minimal viable product) would leverage role-playing and a series of quests scaffolded to introduce elements of a value proposition. The MVP will use web XR to ensure accessibility and incorporate elements such as voice, chat, and emotes to ensure inclusivity. Future iterations would expand upon the MVP to incorporate a business model canvas and a triple bottom line. A value proposition is necessary to start as popular media searches and cosplayer interviews identified communicating the value of one's services, professional skills, and negotiating payment as the largest obstacles when transitioning from a hobbyist to a professional. Using games as a critical technology will build intrinsic motivation to learn, allow for full autonomy granting players agency, and allow for mastery of subjects introduced during gameplay. The fantasy role-playing game will assign 1–5 players to a DM who would guide players through quest lines as they navigate through 3D environments learning about value propositions. Each quest is scaffolded to teacher players about value propositions and build upon their previous knowledge. Sessions would be scheduled in advance and carried out in 90 min intervals.

7 Outcomes

The MVP that came to be known as QwörkSpace shed light on the potential of game based learning to teach cosplayers how to construct a value proposition. Interviews with play testers revealed that while more development was needed on the first iteration of QwörkSpace, the RPG was a viable option for teaching cosplayers how to build a value proposition. Allowing full autonomy in how players reached certain milestones immersed players in the experience. Additionally, interdisciplinary teams of players demonstrated knowledge sharing and collaboration to resolve problems. Various playthroughs conveyed player collaboration to negotiate for community building and potential for networking with future iterations. Lastly, follow up interviews with play testers revealed core behaviors that act as impediments to career growth and financial compensation for the services they provide. Reflections on actions taken in the game demonstrate an aversion to risk when negotiating for fair compensation. An importance is placed upon financial security, when the compensation for one's services are grossly undercut. This stems from an uncertainty in communicating the value of services or professional skills being provided. There is potential to implement interactions, choices, and narrative elements that curve these behaviors.

QwörkSpace benefits industries influenced by cosplayers such as the gaming or fashion industries by paving the way for value co-creation with content creators, integrating entrepreneurial ventures into a larger system that can mitigate pangs regarding sustainability, and collaborating with cosplaneurs to empower and engage communities through campaigns, events, and workshops. Future iterations of QwörkSpace will integrate a business model canvas that emphasizes a triple bottom line and can promote the adoption of cosplaneurs advocating for a triple bottom line. These cosplaneurs are the future innovators, community leaders, and brand ambassadors that can mobilize a community driven, inclusive, and accessible future.

References

1. Mountfort, P., Peirson-Smith, A., Geczy, A.: Planet Cosplay: Costume Play, Identity and Global Fandom. Intellect Books Ltd. Kindle Edition, Bristol
2. Fine, G.A.: Shared Fantasy: Role-Playing Games as Social Worlds. University of Chicago Press, Chicago (2002)
3. Huizinga, J.: Homo Ludens. Beacon Press. Kindle Edition, Boston
4. Debost, Constance: "How Blizzard Is Successfully Using Cosplayers to Promote Its Brand." Medium, Cube, 18 Feb 2017. medium.com/the-cube/how-blizzard-is-successfully-using-cosplayers-to-promote-its-brand-c24f79502c52
5. Ap, T.: "Model of the Future? Louis Vuitton's Newest Face Is a Video Game Character." CNN, Cable News Network, 29 Dec 2015. www.cnn.com/style/article/louis-vuitton-fi-nal-fantasy-model/index.html
6. Staff: "Breaking Down Comic-Con 2019 by the Numbers." San Diego Convention Center, 11 July 2019. scene/13-breaking-down-comic-con-2019-by-the-numbers?Itemid=101
7. Staff, Anime Expo: "28th Annual Anime Expo Delights More Than 350,000 Fans of Japanese Pop Culture." Anime Expo, 18 July 2019. www.anime-expo.org/2019/07/11/28th-an-nual-anime-expo-delights-350k-fans-japanese-pop-culture/
8. Decker, I.: Cosplay Makers: How Much They Spend on Costumes (and Why) (2017). blog.anthonythomas.com/cosplay-makers-how-much-they-spend-on-costumes-and-why
9. Handlin, T.: "The Punishing Economics of Professional Cosplay." BTRtoday // Emerging Culture. Curated Daily (2017). www.btrtoday.com/read/featured/the-punishing-economics-of-professional-cosplay/
10. Han, Y.: Yaya Han's World of Cosplay: A Guide to Fandom Costume Culture. Sterling (2020)
11. Flanagan, M.: Critical Play: Radical Game Design. MIT Press, Cambridge (2013)
12. Stampler, L.: "Welcome to The New Era of High Fashion and Video Game Collabora tions: Inside Moschino and The Sims Partnership." Fortune, 7 June 2019. fortune-com/2019/04/19/moschino-sims-high-fashion-video-games/
13. Clark, M.: "Why Anime Is Menswear's Next Big Obsession." Esquire, Esquire, 29 May 2018. www.esquire.com/uk/style/a20882017/anime-menswear-next-big-thing-trend/

Evaluating Innovation Strategies in Online Education in Higher Education

Debra Satterfield(✉), Jose Rivera-Chang, David Teubner, Tom Tredway, and Wesley Woelfel

California State University, Long Beach, 1250 Bellflower Blvd., Long Beach, CA 90840, USA
{debra.satterfield,jose.rivera,david.teubner,tom.tredway, wesley.woelfel}@csulb.edu

Abstract. The impact of rapid changes to online course content delivery via platforms such as Zoom and the new demands placed on learning management systems such as Canvas and Blackboard to maintain a semblance of order, were staggering for many educators. However; out of the rapidly changing higher education landscape, there were successes that emerged and we witnessed a burgeoning of online education communities for sharing what can be best described as DIY teaching survival strategies. This research describes a process for sharing success strategies between educators using a mixed method data collection survey and an online participatory group brainstorming protocol using Zoom, an online whiteboard, and an observation method to identify and evaluate teaching strategies in face to face, hybrid and online formats. This online education evaluation strategy, "Innovation in Teaching (I-Teach) Event," will be discussed with regard to the tools and the techniques of documenting, comparing, and evaluating teaching strategies and technologies.

Keywords: Online education · User participatory design · Design thinking · Teaching evaluation · Higher education · Hybrid teaching · I-Teach events

1 Introduction

In a study on factors influencing academic success in design students, it was noted that creative people tend to be both emotional and analytical in how they think and solve problems. This unique method of problem solving is often referred to as "design thinking." With regard to complex problem solving, research in neuroscience says that cognition actually includes thinking, reasoning, intelligence and emotion. Therefore, cognition and emotion can both be considered unconscious processes and are equal in an all-encompassing view of the mind [1]. Therefore, to be most effective in reaching our design students through our educational strategies it is important to keep our course content challenging and engaging in both analytical and emotional ways. However, methods which worked to achieve this balance in a face-to-face (f2f) classroom did not automatically have the same impact and engagement when taken into a hybrid or online format. Therefore, there is a critical need for methods of evaluating teaching effectiveness

C. Leitner et al. (Eds.): AHFE 2021, LNNS 266, pp. 219–225, 2021.
https://doi.org/10.1007/978-3-030-80840-2_25

and sharing successful strategies among faculty with regard to online, hybrid and f2f modalities. In addition, as faculty themselves were less likely to have opportunities to share teaching ideas and strategies in day-to-day interactions such as meeting each other in the hallways or at coffee shops on campus, there was an emerging need to reimagine an online way of sharing optimal teaching experiences.

2 Design Thinking as a Framework for Online Teaching

With limited toolsets, instruction within the online environment can seem repetitive and stale. The familiarity of tools is outshined by the monotony of breakout rooms, muted presenters, and screen sharing. New methods of instruction often require facilitators to find software on the fringes or navigate a maze of screens and applications. With the spirit of improving engagement and instruction effectiveness, in what ways might we innovate in this area?

The utilization of creative collaborative strategies emerged in the 1960s with the works of Osborn and others. It later evolved into methods such as Design Thinking in the late 1990s with firms like IDEO. The tools, processes, and exercises of Design Thinking, CPS, and other frameworks, foster the creative development of ideas. These creative processes are commonly linked to innovation, and can serve as a framework to foster innovation within the online environment. In the online event, a standard recipe of these tools and exercises were developed using Miro to frame out the collaboration and idea generation of groups of participants. These tools require finesse, and the experience typically requires interaction, multi-sensory outputs, and curation from experienced facilitators.

In a 2015 case study, the process of using new technologies, such as online 3D printing could be added to the content of other design courses thereby creating a model for hybrid or online pedagogy [2]. In another class, it was observed that when working in groups using innovative communication strategies and technology to collaborate on creating a crowdfunding presentation to a specific audience, the students demonstrated better understanding of design, business models and an increased awareness about funding issues using these techniques [3]. In a survey of graduate and undergraduate university students [4], it was found that students had preferences with regard to course content delivery methods with regard to online versus traditional face to face (f2f) teaching methods specifically when given the choice of tests, written papers, live presentations of research, designing models or prototypes or lab experiments. However, preferences for teaching strategies of faculty in online versus face-to-face modalities and best practices is unclear. In addition, as faculty taught remotely their ability to share successes in teaching strategies and the use of technologies to create novel or optimal online teaching experiences with their peers was disrupted. Therefore, there is a critical need for faculty to have a mechanism to share teaching and technology success strategies in a way that allows for peer-to-peer mentoring.

3 I-Teach Event Methodology

The methodology of the research utilized a mixed method data collection strategy by incorporating both a Qualtrics survey and online focus groups (IRB ref #21–141,

1668510–2 Innovation Strategies in Higher Education, Approved January 20, 2021 by the CSULB IRB). A series of three focus group case study events were conducted to evaluate the I-Teach Event tools and focus group protocol with regard to their impact on the quality of the participant engagement and their ability to gather information on best practices about online teaching tools and strategies.

Each participant first completed an online Qualtrics survey. This survey uses self-evaluation of perceived effectiveness in the teaching modalities of face-to-face, hybrid, and online course content delivery. It ended with questions about teaching strategies and successes to prime participants for the I-Teach whiteboard brainstorming event.

3.1 Research Strategies and Survey Instruments

Three I-Teach Event survey instruments were designed for gathering demographic and disciplinary specific data before and during the online event; 1) a mixed-method Qualtrics survey administered prior to the event, 2) a participatory scripted online group event held in a whiteboard environment that has been customized into a pre-event set up to facilitate the specific topics and participants, and 3) an observational data collection tool using the Connectivity Model ([5] to collect, document, and code the social, emotional, motivational and behavioral verbal and non-verbal information displayed by the participants during the live event.

3.2 Qualtrics Mixed Method Survey

The Qualtrics Mixed Method Survey is sent out as a link via email with the informed consent and invitation to participate in the research information. The survey collects demographic data including the discipline of the respondents; years teaching online; type of educational institution affiliations; self-reported effectiveness of teaching in all modalities; information about trainings for online technology and educational strategies; perceived successes and challenges in technology; and perceived successes and challenges in online content delivery. The last question on successes and challenges is used to prime participants for the online group participation event.

3.3 Scripted Online Participatory I-Teach Event

To coordinate the I-Teach Event, participants are emailed a Zoom ID with the date and time to meet online. Upon gathering in Zoom, the Em Cee uses a scripted agenda to greet the participants and introduce them to the team members. After the first phase introductions, a link to the online whiteboard is pasted into the chat window. The whiteboard is pre-customized with the necessary sticky notes and graphical elements and printed instructions for the I-Teach Event. The structure of the digital whiteboard is designed to directly support the phases of the discussion and activities in the event. The focus of the event is based on a design thinking methodology that promotes an initial plethora of divergent ideas, an assessment of those ideas to group synergistic ideas into nodes and connect related elements with arrows or hand drawn associations, and then using verbal discussions to identify three main themes for use in the processing phase. The processing uses a strategy called the "POWER Tool." This phase asked participants to assess and refine their chosen idea by addressing the **P**ositive aspects of it, identify **O**bjections, expand it with a **W**hat else questioning, add **E**nhancements, and apply **R**emedies to make the idea fully functional. The outcomes of the POWER Tool are then written into a summary statement of the vision for this best practice teaching strategy or technology. The actionable data is collected in the form of color-coded responses by individual participants, labels applied to the concept buckets, identification of all aspects of the idea and solution and a final summary of the group data.

3.4 Connectivity Model Observation Sheets

A Google doc was used to record event times stamps and the flow of the activities. It was also used to record verbal interactions and exchanges of information, non-verbal communication and cues and social gestures. Categories also recorded the emotional tone of the events, behaviors associated with the activity, and motivating conditions or interactions.

4 I-Teach Event Implementation Strategies

The I-Teach Event tools and setup included a system of 1) three pre-assigned roles delegated to team members to manage the time and sequence of event activities, 2) a Zoom link for the initial phase of the meeting, 3) a pre-configured custom whiteboard set up in Miro for the event brainstorming activity, and 4) Connectivity Model Observation Sheets in a Google doc set up for recording verbal and non-verbal data during the live event. The implementation strategy involves three specific team member roles, an agenda outline and a coding strategy.

The online setup includes a whiteboard in Miro.com designed to integrate activities and directives that supported the verbal script. (See Fig. 1) The beginning space was visually indicated via a large arrow, and also visually displayed reminders of tools (1). The activities were broken down into four main sequenced areas. Each area was numbered, contained reminders of directives and large numbers to assist with navigating to the correct area. The following board set up was used:

Area 01: Rapid divergent ideation framed via "technologies" and "teaching ideas" (See Fig. 1); **Area 02:** Discuss and cluster ideas (See Fig. 2); **Area 03:** Converge to three concepts; and **Area 04:** Convergent tool and written summary (See Fig. 3). To avoid confusion, the board locked all framework content, while keeping the active content (such as stickies and drawing tools) accessible. After each area was completed, a screen shot was taken to document the content generated.

Fig. 1. (Left) I-Teach Event from third screen shot and four areas. (Right) Area 1.

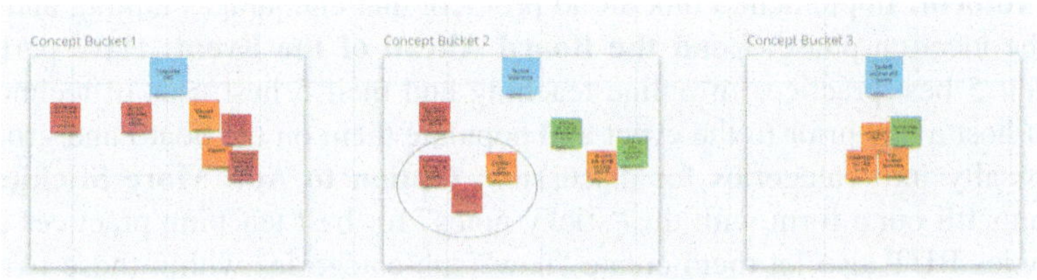

Fig. 2. Case study 2 concept buckets.

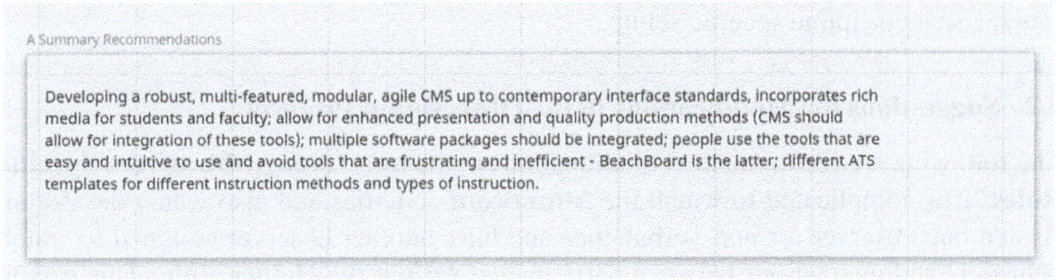

Fig. 3. Case Study 2 participant summary statement.

5 Team Facilitator Roles and Agenda

The I-Teach events are facilitated by three team members in the following roles: **Em Cee**- They used a script to give verbal instructions, time prompts, and verbally responded to questions; **Observers-** They recorded the event in terms of sequence of steps, start and stop times, verbal and written responses of the participants, and non-verbal data in the areas of social situations, emotional tone, behavioral tendencies and an overall summary of the event; and **Helper(s)**- They assisted the other participants by helping people who seem confused by the task or need assistance with entering data on the whiteboard. The event used this agenda: **Introduce:** Explain the quest and meet participants; **5 MIN**; **Demonstrate:** Online tools and environment; **5 MIN** (Screen Shot 1); **Explore:** Talk, add content and discuss; **20 MIN** (Screen Shot 2); **Coding Data:** Combine ideas and add labels; **10 MIN** (Screen Shot 3); and **Wrap up:** Discuss Summaries, Outcomes, and Findings; **10 MIN.**

6 I-Teach Case Study Event Evaluations

Three case studies were conducted with a total of 10 participants (3–4 participants per event) who were all full-time educators with backgrounds from a variety of disciplines and all with experience in f2f university level education and currently teaching in online environments for two semesters. They were conducted via Zoom and Miro.

6.1 Findings for Modifications to the Online Environment

The following changes are recommended based on the case study observation data: **Talk Aloud Protocol:** Implement a talk aloud protocol that encourages talking and sharing during the ideation phase; **Load the Board Ahead of the Event:** Have participants email their 5 best practices in online teaching and their 5 best uses of technology to the event host a day prior to the event and populate them on the board and group them synergistically into categories for discussion; **Option to Add More Stickies:** Have participants fill out a form with the "sticky notes" for best teaching practices and best technologies BUT also let them create "new" sub-categories within those two during the event; **Simplify the Board:** The whiteboard set up is too complicated and the first two phases and should be populated on the board beforehand; and **Adjust the Board According to the Audience:** Not all audiences may see the same categories or flow the same. Use a discipline specific set up.

6.2 Suggestions for Modifications to the Observation Protocol

The following are recommended modifications to the roles: **Assign Observers Specific Roles:** It is complicated to watch the Miro board activities and also watch the Zoom. Assign one observer for non-verbal cues and have another observer assigned to watch behaviors and interactions between participants; **Adjust the Helper Role:** This person should run the technology, links, add stickies for participants and take screenshots; and **Reduce Presence of Observers:** Turn off cameras during the event to reduce the "fishbowl" element.

7 Conclusions and Areas for Future Research

I-Teach was designed to be a robust teaching evaluation platform developed to support educators and administrators working in f2f, hybrid, and online teaching modalities by identifying the most effective and innovative strategies in their teaching methods. Specifically, the data collected by the I-Teach platform allows the faculty participants to code and summarizes their own information and thereby identify the best technologies and strategies as identified in their collaborative discussions. The data is directly based on faculty input, processed in a clear and understandable way, and summarized into actionable information. The inclusion criteria and selection of participants should be done with the input of participants and should combine people who are comfortable working together and have meaningful synergies in their teaching content or styles. The data is designed to illuminate meaningful practices and facilitate sharing in ways that has been inhibited when faculty are not working on campus.

References

1. Satterfield, D.: Factors influencing academic success for design students: a study of curricular expectations and ethical issues. In: Freund, L.E., Cellary, W. (eds.) AHFE 2017. AISC, vol. 601, pp. 153–161. Springer, Cham (2018). https://doi.org/10.1007/978-3-319-60486-2_14
2. Rivera-Chang, J.: Case study: use of online tools in the classroom and their impact on industrial DESIGN PEDAGOGY. Procedia Manuf. **3**, 2275–2280 (2015). https://doi.org/10.1016/j.promfg.2015.07.372
3. Rivera-Chang, J.: Case study: how design students tailor their crowdfunding presentation to different audiences. In: Di Bucchianico, G., Shin, C.S., Shim, S., Fukuda, S., Montagna, G., Carvalho, C. (eds.) AHFE 2020. AISC, vol. 1202, pp. 927–931. Springer, Cham (2020). https://doi.org/10.1007/978-3-030-51194-4_120
4. Satterfield, D., Lepage, C., Ladjahasan, N.: Preferences for online course delivery methods in higher education for students with autism spectrum disorders. Procedia Manuf. **3**, 3651–3656 (2015). https://doi.org/10.1016/j.promfg.2015.07.758
5. Satterfield, D.: Ethics in service design for children with autism and cognitive disabilities. In: Advances in Intelligent Systems and Computing, pp. 63–72 (2016). https://doi.org/10.1007/978-3-319-41947-3_7

Color and Flavor Perception

Sunghyun R. Kang(✉), Carol Faber, Nora Ladjahasan, and Andrea Quam

College of Design, Iowa State University, 715 Bisell Road, Ames, IA 50011-3092, USA
{shrkang,chfaber.aquam,nading}@iastate.edu

Abstract. Color is a critical element in how people perceive products and can affect the decision-making process, and it is also known to be an element that influences the perception of taste and flavor. People associate colors with flavors; therefore, natural color can be an indicator of food flavors. Our experiments began several years ago with children and adults within the context of snack packaging. In this study, we removed the context of food packaging and children as audience members, focusing only on color itself within an adult population.

This study investigates five colors used in the market with computer-generated color palettes to review relationships between: color and flavor, preferred color to eat, perception of healthier color, and whether favorite colors influence individual's eating selections. In this study, four questions were given, with five colors options. The following questions were asked: 1) Rank the color you prefer to eat, 2) Identify flavors that you associate with the color, 3) Rank the color you think is healthier to eat, and 4) Rank your favorite colors.

As a research method, an online survey was distributed to a university community. The data analyzed 804 adults' responses. Respondents ages 18 to 65+ participated in the online survey. Red is the most selected color as the first choice to eat, followed by blue, green, yellow, and pink. Green is selected most often as the healthiest color expectation, followed by yellow, red, blue, and pink. Blue was chosen as the overall preferred color, followed by green, yellow, pink, and red. This study will additionally report the color association with flavors and correlations among the age group, gender, and ethnicity.

Keywords: Color · Flavor · Perception · Preference

1 Introduction

Color is closely associated with our daily life in various ways and is a critical element in how people perceive products. Perception of color can be influenced by many factors, such as its context, culture, age, and gender. A study shows that favorite colors are predetermined by gender but does not relate to age and education [1].

Color can be a tool to enhance brand loyalty and recognition of a product; simultaneously, color is often taken for granted in our surroundings, e.g., green trees, yellow bananas. Color association with flavor certainly happens in food and beverage [2]. The previous study with children ages 9 to 13 shows that most children's answers related to flavor were associated with a fruit's name [3]. However, that study was conducted with the context of snack packaging.

C. Leitner et al. (Eds.): AHFE 2021, LNNS 266, pp. 226–232, 2021.
https://doi.org/10.1007/978-3-030-80840-2_26

Previous color studies related to 'blue' and 'number seven' have been conducted by several researchers [4–6]. The studies show that blue is most selected when subjects were asked to write down a color's name and it also indicated gender differences [4, 5].

This study seeks answers to relationships between color and flavor, the preferred color to eat, and whether favorite colors influence individuals' eating selections. The previous research with children finds that green was perceived as the healthiest color within the context of snack packaging [3].

2 Research Methods

The study presented here investigates five colors used in current market places, conducted with computer-generated color palettes. Figure 1 shows the five colors used in snack products. It was determined in the previous snack packaging study that blue was associated with blueberries, green with apples, pink as strawberry, red as cherry, and yellow as a lemon [3].

Fig. 1. Color for the experimentation, blue, green, pink, red, and yellow respectively.

For this study, four questions were given as follows: 1) Rank the color you prefer to eat, 2) Identify flavors that you associate with the color, 3) Rank the color you think is healthiest to eat, and 4) Rank your favorite colors. Also, demographic information such as age, gender, occupation, ethnicity, income, and educational background was collected in the study.

The survey is approved by Iowa State University's Internal Review Board and the survey link was emailed to students, staff, and faculty at Iowa State University. A total of 804 individuals participated in the study and the data was processed using SPSS statistical software.

3 Data Analysis and Discussion

Among 804 participants, some participants did not answer demographic questions and therefore data is analyzed only based on the valid answers. Sixty-five percent were female and thirty-four percent were male, and one percent did not specify their gender. Sixty percent were 18–22 years old, followed by 23–28 years old (11%) and 29–34 years (6%), 35–40 (5%), 41–46 (5%), 47–52 (4%), 53–58 (4%), 69–74 (4%) and over 65 was less than one percent. The majority of participants (83%) were white followed by Asian/Pacific Islander (8%), Hispanic/Latino (5%), and rest of them were Africa American, Native American, or others. Due to the testing environment, a majority of the participants (89%) have a college degree or are enrolled in college. The rest of the participants had a high school or equivalent degree. Income was higher than expected as about fifty percent of the participants made under $50,000 a year, twenty-six percent made $50,000 to $99,999 and twenty four percent made over $100,000.

3.1 Color Preferred to Eat

On average, red was the color most preferred to eat at 31%, followed by yellow (23%), green (15.5%), pink (15%), and blue (14.5%). Table 1 shows the gender differences in colors selected to eat. Females selected red as the first choice, followed by yellow, pink, green, and blue, respectively. While males chose yellow as the first choice, followed by red, blue, and green as a tie, and pink respectively. However, the difference between red and yellow as a first choice is not significant in frequency.

Table 1. Color selected to prefer to eat by gender (variance = percentage), n = 740

Color	Blue		Green		Pink		Red		Yellow	
Gender	M	F	M	F	M	F	M	F	M	F
1st choice	15.2	13.7	15.2	16.1	13.4	17.7	27.6	32.4	28.6	20.1
2nd choice	19.4	14.9	24.0	23.7	12.9	19.4	30.4	24.9	13.4	17.0
3rd choice	20.7	21.6	21.2	20.4	18.9	18.7	21.2	19.2	18.0	20.1
4th choice	18.0	21.3	27.2	26.1	22.6	19.7	16.1	15.6	16.1	17.3
5th choice	26.7	28.5	12.4	13.7	32.3	24.5	4.6	7.9	24.0	25.4

Interestingly, the choice of younger participants ages 18 to 22 were different from those aged over 22 years. Younger participants' first choice was red (33.3%) followed by pink (19.9%), yellow (19.1%), blue (15.5%), and green (12.1%). The other group, over 23 years old, chose yellow (28.5%), followed by red (27.3%), green (21.3%), blue (12.3%), and pink (10.7%). Incomes and education show a similar trends seen within the age groups. Ethnicity does not show any differences compared to the total rankings.

In the previous study (with children in the context of snack packaging) the same numbers of participants selected blue, red, and green as the first choice [3]. The other study [7] also found red and green were strongest preference followed by yellow and orange among children. However, this study finds that blue was a preferred color for consumption by adults—both males and females.

3.2 Color and Flavor

Most participants perceived colors as a fruit. Blue is recognized as blueberry, raspberry, or grape. Green as a green apple, lime, or lemon, and some participants see it as vegetables such as lettuce, green beans, or a salad. Pink is recognized as a strawberry, watermelon, but some participants perceive it as bubble gum or cotton candy. Red is acknowledged as cherry, strawberry, apple, or raspberry and yellow as lemon, banana, orange, or pineapple. Even though the survey and its questions were focused on flavor, some participants identified color as taste. Some participants identified color as tastes, such as: salty, sour, sweet, spicy, or savory. Nine percent of participants said pink was sweet, seven percent of participants said red was spicy, and four percent of participants perceived yellow as salty and green as sour. Table 2 shows top 10 flavors of each color. This data shows that the first ranked flavor matched with the previous study with children, showing colors in the context of packaging [3].

Table 2. Top 10 flavors of blue, green, pink, red, and yellow (variance = percentage), n = 639

	Blue		Green		Pink		Red		Yellow	
	Flavor	%	Flavor	%	Flavor	%	Flavor	%	Flavor	%
1	Blueberry	48.6	Apple	37.6	Strawberry	35.9	Cherry	47.3	Lemon	31.2
2	Raspberry	11.5	Lime/ lemon	23.1	Watermelon	12.7	Strawberry	13.7	Banana	13.2
3	Grape	7.0	Vegetable	11.1	Sweet	9.2	Spicy	7.0	Orange	5.9
4	Blue raspberry	5.7	Sour	4.3	Bubble-gum	7.5	Apple	6.4	Salty	4.1
5	Berries	5.0	Mint	2.3	Lemonade	5.5	Raspberry	4.6	Mustard	3.3
6	Cool	2.4	Sour apple	1.8	Raspberry	4.4	Cinnamon	2.6	Cheese	3.0
7	Chocolate	2.3	Salad	1.7	Fruity	3.5	Fruit	2.0	Honey	2.0
8	Sweet	1.8	Kiwi	1.5	Cotton candy	4.4	Hot	1.8	Pineapple	2.3
9	Salty	1.7	Fresh	1.1	Candy	1.8	Fruit punch	1.7	Savory	2.1
10	Blue 20	1.4	Nuts	1.1	Berry	1.4	Savory	1.5	Corn	2.0

3.3 Healthiest Color to Eat

As expected, green is selected as the healthiest color to eat, followed by yellow, red, blue, and pink. Table 3 show the rankings by gender for each color.

Table 3. Healthier color choice by gender (variance = percentage), n = 638

	Blue		Green		Pink		Red		Yellow	
Gender	M	F	M	F	M	F	M	F	M	F
1st choice	2.7	8.0	71.7	67.3	1.8	2.9	10.5	8.7	13.2	13.1
2nd choice	20.1	18.9	12.3	15.3	12.3	7.5	23.3	29.8	32.0	28.6
3rd choice	29.2	21.1	6.4	8.7	19.6	20.1	28.3	24.2	16.4	25.9
4th choice	23.7	23.0	4.6	5.3	31.5	29.8	22.4	24.5	17.8	17.4
5th choice	24.2	29.1	5.0	3.4	34.7	39.7	15.5	12.8	20.5	15.0

For the healthiest color choice, there is no gender differences for the selections. Pink is ranked as the least healthy color among five colors from both genders. Also, there are no differences among ages, ethnicities, incomes, and educations.

3.4 Favorite Color

On average, blue (39.4%) was selected as the favorite color followed by green (17.5%), yellow (17.4%), pink (16.3%) and red (9.4%). However, there is a gender difference, and Table 4 shows the ordering of color choices broken down by gender differences. Both genders chose blue as the favorite color; however, more males preferred blue compared to females. Yellow is ranked second by males. Females chose pink as the second, but males selected it as the least favorite color. Red was the least favorite color chosen by males.

Table 4. Favorite color choice by gender (variance = percentage), n = 639

	Blue		Green		Pink		Red		Yellow	
Gender	M	F	M	F	M	F	M	F	M	F
1st choice	46.1	36.3	18.3	17.2	2.7	23.0	7.3	10.4	25.6	13.1
2nd choice	29.7	30.3	23.3	23.7	7.8	16.2	8.7	10.4	30.6	19.4
3rd choice	11.9	18.6	30.6	23.5	14.2	16.2	18.7	18.4	24.7	23.2
4th choice	8.7	8.7	17.4	22.0	26.0	17.7	35.2	25.4	12.8	26.2
5th choice	3.7	6.1	10.5	13.6	49.3	26.9	30.1	35.4	6.4	18.2

Blue is the favorite color across all demographic categories, and red was the fifth choice by all demographic categories, except the non-white group. Non-white groups' favorite color order was different from other categories. White participants' favorite color order was blue (39.3%), green (19.1%), yellow (16.3%), pink (16.9%), and red (8.5%), while non-white groups' favorite color order was blue (39.8%), yellow (23.2%), pink (13.9%), red (13.0%), and green (10.2%).

3.5 Correlations in Color Selections

The data was also analyzed to assess correlations among preferred color to eat, healthier color, and favorite color using Pearson correlation (1 as the first choice and 5 as the fifth choice).

Colors that are blue and green showed positive correlations among preferred color to eat, perceived as healthiest, and favorite colors. There was a stronger correlation between preferred color to eat and favorite color ($r(628) = .30$ for blue and $r(628) = .33$ for green), compared to other combinations. The data showed that participants who selected blue or green as their first choice to eat were more likely to choose the same colors as healthiest and favorite.

Pink also showed positive correlations between preferred color to eat with healthy and favorite. However, there was no relationship between pink as favorite and healthy color. Table 5 shows the Pearson correlation in five colors.

Table 5. Pearson's correlations in color selections

	Blue	Green	Pink	Red	Yellow
Prefer to eat/healthy color	.087*	.122**	.195**	.122**	0.000
Prefer to eat/favorite color	.303**	.333**	.195**	−0.073	−0.041
Favorite color/healthy color	.107**	.148**	0.039	−0.033	.080*

*. Correlation is significant at the 0.05 level (2-tailed).
**. Correlation is significant at the 0.01 level (2-tailed).

In comparison, red only showed a positive correlation between preferred color to eat and healthiest. Yellow, on the other hand, showed a positive correlation between favorite and healthiest. However, the value is not strong ($r(628) = .08$).

In summary, the data showed positive correlations in participants' preferred color to eat and their favorite colors as blue, green, and pink. However, red and yellow did not show correlation in those categories. Red showed positive correlations in preferred to eat and healthiest color choice, and yellow showed correlations in participants' favorite and healthiest.

4 Conclusion

This study investigates the relationships between color and flavor perception. Research in color and food has been studied for eighty years [8] but most studies done focus on food colors, specifically in relation to drinks [2, 8]. This study focuses on color related to flavor, and preferences on color rather than color and food tastes. In general, red was the favorite color to eat, and yellow was the second choice. Green, pink, and blue were very similar in frequency, and further study is needed to for conclusive answers. Also, this study confirms gender difference exists in color preference [1, 4, 5].

Blue is identified as blueberry, green as apple, pink as strawberry, red as a cherry, and yellow as a lemon. The flavor perception in the colors was the same as the previous study [3] and could be used to guide the food industry and packaging design.

This study found ethnicity a significant indicator in favorite color preferences. The first choice was blue, but the order after blue was different between white participants and non-white participants. However, further study is required with a larger, more diverse ethnic group.

This study finds correlations in preferred colors to eat, perceived heathiest colors, and favorite color choices, predominantly blue and green. However, red and yellow do not show correlations between preferred to eat and favorite colors.

This study shows that most participants perceived color as related to fruits. There was a consensus that blue is blueberries, green is apples, pink is strawberries, red is cherries, and yellow is bananas. Additionally, green is perceived as a healthiest color among most participants. This information can be used in marketing, package design, and food colors.

References

1. Sliburyte, L., Skeryte, I.: What we know about consumers' color perception. Procedia. Soc. Behav. Sci. **156**, 468–472 (2014)
2. Clydesdale, F.M.: Color as a factor in food choice. Crit. Rev. Food Sci. Nutr. **33**(1), 83–101 (1993)
3. Kang, S.R., Satterfield, D., Ladjahasan, N.: Color preference study on snack package design to children as a health communication strategy. In: 2016 AIC Conference, pp. 201–204. Organizer: Chile, Asociación Chilena del Color (2016)
4. Simon, W.E.: Number and color responses of some college students: preliminary evidence foe a "Blue Seven Phenomenon." Percept. Mot. Skills **33**, 373–374 (1971)
5. Simon, W.E., Primavera, L.H.: Investigation of the "Blue Seven Phenomenon" in elementary and junior high school children. Psychol. Rep. **31**, 128–130 (1972)
6. Benel, R.A: The "Blue-Seven Phenomenon" or the "Blue, Seven, ... Phenomena"?. J. Gen. Psychol. **94**(1), 145–146 (1976)
7. Walsh, L.M., Toma, R.B., Tuveson, R.V., Sondhi, L.: Color preference and food choice among children. J. Psychol. **124**(6), 645–653 (1990)
8. Spence, C., Zampini, M.: Does food color influence taste and flavor perception in human? Chemosens. Percept. **2010**(3), 68–84 (2010)

Disruptive Innovation: Designing a Shifting Pedagogy for Creative Disciplines in Higher Education Learning

Kimberly Mitchell(✉)

University of Tennessee – Knoxville, Knoxville, TN, USA
kmitch57@utk.edu

Abstract. The 2020 pandemic caused an unexpected digital shift in education and professional practices all throughout the world. For higher education faculty to effectively teach the second half of the semester, teaching strategies had to transfer from brick-and-mortar classrooms and studios to the virtual classroom. Transitioning from an entirely face-to-face (f2f) format to an entirely online synchronous format disrupted curriculum and required an immediate pedagogical redesign.

This ongoing study suggests that classroom shifts from entirely f2f to an online synchronous format can be successful in an undergraduate creative discipline if appropriate technologies are introduced and there is faculty presence. The findings from the surveys show that there are several disruptive technologies, that when used together, allowed for a seamless transition from f2f to online. In addition, the study suggests that students felt better prepared for remote work success after completing half of a semester or an entire semester entirely online.

Keywords: Pedagogy · Education · Course design · User experience · Synchronous · Disruptive technologies

1 Introduction

For many years, graphic designers have worked remotely with clients from all over the world. In fact, a lot of the design industry is now remote – when searching for a "graphic design" job on job search sites like indeed.com, 701 active remote jobs were displayed. Turning to the innovative tools that remote designers use to work in the industry helped in shifting traditionally face to face (f2f) graphic design classes to completely online formats during the 2020 pandemic.

Innovations in pedagogical strategy and technology have helped strengthen online learning so that it has become more similar in experience as that in a f2f environment. Video communication technologies allow professors to teach in the comfort of their own spaces, while still being able to communicate and converse with students in the classroom setting. Breakout rooms allow for small group discussions, similar to what would happen in round table discussions in a classroom. There are technologies that allow for student interaction and engagement while working together to brainstorm ideations,

C. Leitner et al. (Eds.): AHFE 2021, LNNS 266, pp. 233–239, 2021.
https://doi.org/10.1007/978-3-030-80840-2_27

just as it would be if they were in person using a whiteboard in a classroom. There are technologies that allow professors in creative disciplines to actively demonstrate drawing, printmaking, or painting. And there are technologies that allow for easy access for conversation and critique between students and the student and professor.

This ongoing study suggests that classroom shifts from entirely f2f to an online synchronous format can be successful in an undergraduate creative discipline if appropriate technologies are introduced and there is faculty presence. During the study, students were surveyed after the completion of the courses about the course contents, course structure, and the technologies used. The findings from the surveys show that there are several technologies, that when used together, allowed for a seamless transition from f2f to online. In addition, the study suggests that students felt better prepared for remote work success after completing half of a semester or an entire semester entirely online.

Using technologies created using the Disruptive Innovation Model, published by Clayton Christensen in 1995, students not only prepare themselves for working in the remote world, but they become more adaptable, more flexible, more engaged, and more innovative in their own ways of creating and doing. This paper will explain how several recent technologies were used in the online experience that have assisted the pandemic disruption of having traditionally f2f class formats into a successful online modality.

2 Disruptive Innovation Model

According to researcher and business consultant, Clayton Christensen, disruptive innovation is the process in which a smaller company, usually with fewer resources, is able to challenge an established business by entering at the bottom of the market and continuing to move up-market [1].

Disruptive innovations are not an attempt to bring better products to established customers in existing markets. Rather, they disrupt and redefine that trajectory by introducing products and services that are not as good as current available products [2]. Disruptive technologies offer other benefits, typically offering technologies that are more simple, more convenient, and less expensive and appeal to new or less-demanding customers [3].

Using Christensen's Disruptive Innovation Model, this paper will outline several disruptive technologies that were used in assisting a positive disruption of a traditionally face to face classroom format into an engaging, rich virtual presence.

2.1 Zoom

Zoom Video Communications, Inc., or "Zoom" is a technology company providing videotelephony and online chat features through a cloud-based software. Zoom was founded in 2011, and launched in 2012. Zoom's ease of use and reliability made it an instant attraction during the 2020 pandemic when there were major increases in remote work and distance education. Initially, Zoom had trouble finding investors because it was considered a highly saturated market with companies like FaceTime, SKYPE, Webex, and BlueJeans [4]. Founder Eric Yuan saw a demand for a video conferencing room with "multiple solutions: cloud based, good voice over IP quality, good video quality, and seamless single-click solution that works regardless of if you're on a phone or computer" [5].

Zoom allows for a human connection. Professors can actively lecture using the software as if class was in-person–only instead of students sitting in seats in a classroom, they're in virtual seats onlines. In-class discussions and critiques can be done in real time. Zoom allows for screen sharing while work is being discussed. Zoom's Breakout Room feature was helpful in relationship building, as students could be grouped into small groups every class period and talk with one another, while working together as if they were in an in-person setting working around small tables. Breakout rooms were essential to creating a sense of community in the classroom – students were able to "meet" with one another in small groups and get to know each other.

Another way Zoom was used in the classroom was to give feedback. Students were assigned peers that they were to critique outside of class time, and they were to upload their critique by the next class period. Students would login to Zoom, share their screen, press record and critique other student's work. After they were done recording, they would simply upload the critique to Microsoft Teams or a shared Google Drive folder for the classmate to review. Professors did the same for the students – and it was much less time consuming than writing on top of jpeg files or PDF files in Adobe Acrobat.

2.2 Microsoft Teams and Slack

Slack is a collaboration platform that brings people, information, and tools together [6]. Everything happens in a "channel," organized by projects and teams. There is also the option for users to video conference or call one another.

Slack is basically a chatroom for the entire team, or class. It is a platform that was created as a way for organizations to communicate both as a group and in personal one-on-one discussions. Users can communicate as a group through channels or privately via "direct messaging." With over 12 million users, many design agencies use Slack, so it was a great tool to introduce to the students, because it prepared them for the remote working world.

Microsoft Teams is similar to Slack in a lot of ways and it was easier to manage because it was integrated into the Canvas LMS, so after Spring 2020, our classes left Slack and started using Teams. Additionally, Microsoft Teams has a suite of subscription services in their Microsoft 365 line, which was free for all faculty and staff at our University.

"The hub for teamwork," is how Microsoft describes its Teams service. It "brings together everything a team needs: chat and threaded conversations, meetings & video conferencing, calling, content collaboration with the power of Microsoft 365 applications, and the ability to create and integrate apps and workflows that your business relies on" [7]. Microsoft Teams was used every class period for critiques, process, and discussions. Each student has a "channel" or threaded conversation, and in their channel, they post process of their projects throughout the semester. Under each submission, others can go in and given written feedback, or markup their work and submit an attachment. Because it is threaded, it is a great way to track participation and to give pretty immediate feedback. Nothing had to be downloaded and it is easy to click from one student to the next by selecting their channel or name. To see project development, all one had to do was simply scroll up or down in the channel to see the work produced, rather than have to go to multiple discussion prompts in an LMS.

A big competitor to Zoom now is Microsoft Teams. As of December 2020, there is the option for creating breakout rooms and a "Together mode" where students can open up a Whiteboard and sketch out their ideas. One drawback to using Microsoft Teams versus Slack for class lectures is that the maximum number of people that can be seen at once is 49 people, whereas in Teams, everyone can be viewed with a click of a button.

2.3 Mural and Miro

Mural and Miro are very similar web applications that have the same purpose: to serve as visual collaboration platforms as digital whiteboards, where you can write on sticky notes, facilitate group exercises and brainstorming sessions, and create presentations like Powerpoint or Keynote. Students would work in Zoom breakout rooms, upload sketches in Mural or Miro, vote on their top choices with sticky notes, and talk as a small group as if they were in class. Mural or Miro was used for mind mapping exercises in small groups and brainstorming sessions in larger group settings. Both applications took the concept of a physical whiteboard, where because of the pandemic most people did not have access to, and incorporate it into our new online world.

3 Research and Methods

Online surveys, observation and discussion with students were used as evaluation methods. Overall, results from the observation and discussion with students felt more valid because of the lack of response rate from the surveys.

3.1 Surveys

An anonymous online survey was created and distributed via email across 5 sections of undergraduate students, ranging from freshmen to seniors and recently graduated seniors (graduated Spring 2020), at two institutes of higher education. The survey was distributed at differing times – students who graduated in Spring 2020 received the survey 8 months after graduation and students currently enrolled in graphic design classes received it 3 weeks after completing their Fall semester.

The survey was opened by 46 people and completed by 16 individuals – with a response rate of 34.78%. 5 participants (29.4%) graduated in Spring 2020, and their classes abruptly shifted from face-to-face settings to an asynchronous online format. Results from the survey may be skewed because of the large return rate of the graduated students in comparison to the current students, because both the faculty and the students had little time to prepare for the abrupt shift in teaching modality.

Prior to the COVID-19 pandemic 9 participants (56.25%) had taken an online class, and 7 (43.75%) had not. We recorded data from 6 (35.29%) year-one graphic design students, 4 (23.53%) sophomores, 0 juniors, 2 (11.76%) seniors, and 5 (29.41%) recently graduated. Of the technologies used, besides Adobe, and the Universities' chosen LMS systems (Canvas and Sakai), 100% of the participants used Zoom, 12 participants used Microsoft Teams, 2 participants used Slack, 12 used Mural, and 4 used Mural.

Students found Microsoft Teams easy to learn and use (4.36 out of 5). They used Microsoft Teams to post their own work for critique and to view and critique other's work. Overall, they found that the format of this technology to be an acceptable experience (4.64/5). A majority of the participants saw it as a software that they'll use once they graduate (3.73/5).

The survey results found that students found the Zoom software's ease of use to be a little easier than Microsoft Teams (4.44/5). Participants didn't enjoy that format as much for giving critiques (4/5) in comparison to Microsoft Teams. A majority of the participants found that this is a software that they will use after graduation (4.44/5).

Mural was a bit more challenging in terms of ease of use from the student survey results (3.64/5). They didn't feel it to be as valuable for class discussions (2/5) or team building (2.73/5), and overall, they didn't see it as a software that will be used following their graduation (2.82/5) in comparison to Microsoft Teams or Zoom.

When looking at data from the currently enrolled students, a majority of the participants who completed the survey had 4+ classes online in the Fall semester (81.8%). Overall, students did prefer face-to-face classes and felt that they were more effective scholars in those classes (Fig. 1). 10 rated their effectiveness as students as "excellent" in f2f classes. 4 rated their effectiveness as "good," and 1 "average". 0 of the participants rated their ability to be effective in a f2f class as "poor" or "terrible". The effectiveness dropped for the "excellent" ranking for hybrid, online synchronous, and online asynchronous. 6 students ranked their effectiveness as a student as "excellent" for hybrid, 2 students ranked themselves as "excellent" for online synchronous, and 0 participants for online asynchronous. This seemed to be a consistent pattern–student effectiveness ratings dropped significantly the less f2f time the student had with his or her instructor.

The data was divided when asking if participants felt that they learned well in an online synchronous format. 8 participants (53%) felt that they learned well, and 7 participants (46.6%) felt that they did not learn well. 1 participant did not answer. When asked what difficulties students faced when taking online classes, common responses were: "*motivation/focus, ability communicating, distraction, time management, loss of connection with internet and humans.*" When asked what made online classes easier to manage, common responses were: "*more time to work together, no commuting, keeping a schedule, a flexible professor, and Microsoft Teams.*"

The last piece of data the survey evaluated was an interest in online classes following the pandemic. The likelihood of the participants in this survey to take an online class in the future are little. 8 participants (50%) said that they will never sign up for an online class again, 3 participants (18.75%) said that they are undecided, 3 participants (18.75%) said that they have graduated, and only 2 participants (12.5%) said that they enjoyed the flexibility online classes gave.

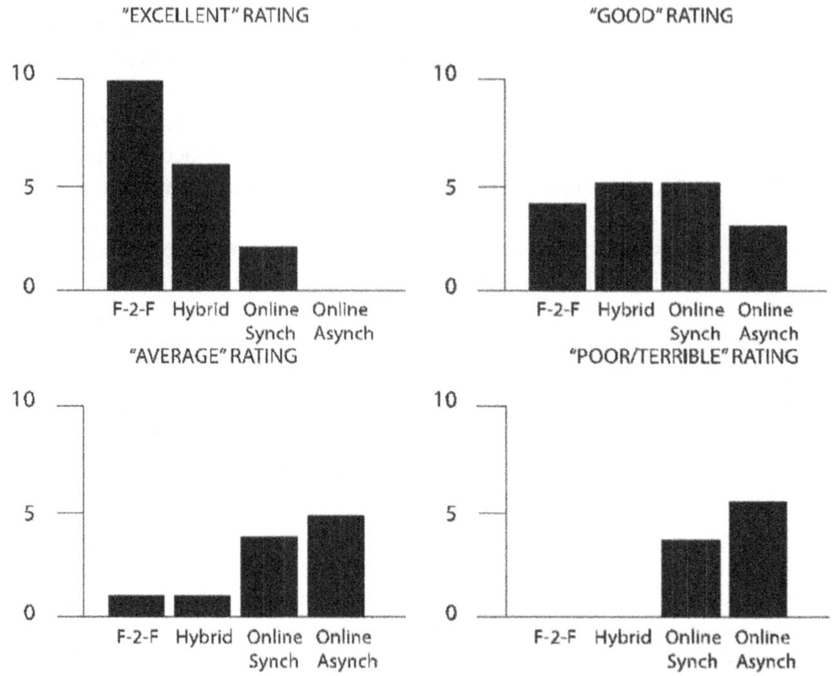

Fig. 1. Results from the survey. 15 out of 16 surveyed participants responded to these series of questions. The questions asked the participant to rate their "effectiveness as a student" in face-to-face (f2f), hybrid, online synchronous, and online asynchronous classroom settings. The rating scales were: *excellent, good, average, poor/terrible.*

3.2 Observation and Discussion as Evaluation Methods

Observation and discussion with students were also used as evaluation methods. The results from these methods differed from that of the survey. Through observation, faculty members noticed that students engaged in critique and conversation more in online formats because there were specific requirements built into the class where students were to give written or recorded feedback to one another as part of the project process.

Additionally, because content was posted online, students weren't confined to a specific class time, and so they had more time to look at their peer's work and digest what they saw before giving feedback. They conducting this feedback from their computer, so sometimes students were able to reference specific examples as part of their feedback and post it to the discussion forum. Faculty also found that they were able to give more detailed feedback because they weren't confined to a specific class time, which often resulted in speed critiques to get through everyone in the class time.

Students seemed to be more engaged with each other in smaller breakout rooms versus larger settings in Zoom and in-person classes. Typically, f2f freshmen design classes are quiet and critique is a forced part of the classroom experience, where faculty have to call on names to get students to engage. In the online synchronous format, faculty found that when students were in small breakout rooms students more likely to offer feedback to one another and engage in conversation.

Students felt that they were better prepared for remote work success when discussing the class and structure at the end of the Fall 2020 semester. Students said that they felt engaged, enjoyed Zoom breakout rooms, and still felt part of their classroom community, even though they were virtual versus in-person.

There were other benefits to being remote that students may not have seen. Students were able to join speaker series from other institutions. Portfolio review sessions, which were typically held 3 h away from campus on a school night were held via ZOOM.

4 Conclusion

Overall, there were many benefits of remote learning. While some of the survey data felt skewed because of the large return rate ratio of the graduated students (where both the faculty and the students had little time to prepare for the abrupt shift in teaching modality) in comparison to the current students. The observation and discussions with students served as another method of evaluation, and that showed that students seemed to adjust well as long as there was consistency and clear organization of the class. The shift to online allowed for more learning experiences outside of the classroom because of the shared resources across many different college and universities. Brick and mortar no longer define education and that there are many more innovative ways to think of the classroom experience. There has been great technology and software advancement that have helped supplement online learning so that students gain similar engaged experiences as if they were in an in-person classroom setting.

References

1. Larson, C.: Disruptive Innovation Theory: What It Is & 4 Key Concepts, Harvard Business School (2016). https://online.hbs.edu/blog/post/4-keys-to-understanding-clayton-christensens-theory-of-disruptive-innovation
2. Christensen, C., Bauman, H., Ruggles, R., Sadtler, T.: Disruptive innovation for social change. Harv. Bus. Rev. **84**, 94–101 (2006)
3. Christensen, C.: Disruptive Innovation, The Encyclopedia of Human-Computer Interaction, 2nd edn., Interaction Design Foundation. https://www.interaction-design.org/literature/book/the-encyclopedia-of-human-computer-interaction-2nd-ed/disruptive-innovation
4. Parth, B.: How Zoom Bested Google, Cisco, and Microsoft to Become the Poster Child of Video Conferencing Services (2020). https://www.appvoir.com/blog/how-did-zoom-slowly-conquered-an-over-saturated-market/
5. Ray, T.: Is There Room for Zoom Video to Continue Zooming Upward? The Street (2020). https://www.thestreet.com/investing/can-zoom-continue-zooming-upward
6. Slack. www.slack.com
7. Microsoft Teams. www.microsoft.com

Virtual Exhibit Design: The UX of Student BFA Design Shows in Social VR

Laura Huisinga(✉)

California State University – Fresno, Fresno, USA
lhuisinga@csufresno.edu

Abstract. Creating exhibitions in virtual reality offers unique user experience design challenges. This paper examines a case study of a past VR exhibit and discussions on future UX and usability research for VR exhibitions. Due to COVID19, student BFA shows across the country were canceled as most Universities went virtual. In an effort to create virtual student shows, many programs put together websites or zoom-style showcases. At Fresno State University, the Graphic Design BFA students put on a virtual show using Social Web VR. This paper examines the UX design of the VR show. Next, it examines the UX planning for the 2021 BFA show using social web-based VR. Finally, the paper concludes with plans for researching best practices for the UX of Virtual Exhibition Design moving forward. While social web-based VR offers great potential, designers must be careful to keep VR Exhibits human-centered and consider all participants' total user experience.

Keywords: Virtual Reality · User experience · Web-based VR · Social VR

1 Introduction

Virtual Reality [VR] technology is not new but still not used widely for daily use. Accessibility to VR technology, including hardware like Head Mounted Displays [HMD], has become more cost-effective yet is not commonly owned by the masses the way a smartphone is. Using web-based VR or VR accessed through a browser, anyone with a device and an internet connection has access. Social VR allows people to communicate in real-time through avatars that can interact in a virtual world. Web-based social VR facilitates communication, collaboration, and interaction between people in a virtual world when meeting in real-life is too difficult. Meeting in a virtual environment allows for the use of virtual space and interactions that might benefit from movement inside that space.

Much like a web page, one should consider the total user experience of social web-based VR when designing the virtual experience. How will people navigate through the virtual space? Can they find what they need and accomplish their goals? How can you encourage people to engage with each other and the space in the virtual environment? The accessibility of a web-based VR should consider what type of devices can access the virtual space and how can virtual spaces be accessed by someone with a vision impairment, hearing issue, or other impairments?

C. Leitner et al. (Eds.): AHFE 2021, LNNS 266, pp. 240–246, 2021.
https://doi.org/10.1007/978-3-030-80840-2_28

2 Background

While VR has increased in both popularity and accessibility, most people still do not use it regularly. However, more people are considering Interacting in virtual environments a viable option instead of meeting in person. The pandemics shutdowns worldwide have caused people to think about leveraging technology to replace in-person interactions. Not being able to meet in person, especially for large events, made the virtual environment an exciting alternative. Conferences, expos, and festivals turned to VR to interact, collaborate and meet up with people around the world.

Social VR also simulates the ability to have conversations with a few people, even in a large group. You can be in a public place with 30 people and have a conversation with just the three people closes to you in real life. If you are on a zoom with 30 people, you are having a conversation with 30 people simultaneously, unless you separate into breakout zoom groups—Social VR allows for the same organic nature of group conversations as real life. The avatars who are closest to you will hear you best, and those farther away will not be able to hear as well.

A virtual alternative to the BFA show was needed after the cancellation of all in-person events on campus. Many senior shows nationwide decided to put together websites for a virtual show presentation, sometimes accompanying live zoom events. A website did not seem like an adequate replacement since the graduating BFA graphic design seniors already used websites to accompany the physical show. The use of social-web-based VR became an exciting alternative. Social VR was essential to allow people to interact in real-time and discuss student work as well as have students interact with local professionals, family, and friends.

3 BFA Show 2020 Case Study

The use of web-based VR was essential to ensure that most people had access to the virtual environment. Web-based VR allowed for the use of cell phones, tablets, desktops, as well as head-mounted displays. Mozilla Hubs was chosen after assessing available options for social web-based VR. While Mozilla Hubs has some limitations, its benefits for our use case far outweigh these limits.

Benefits

- Allows most mobile devices and computers to access
- Allows for real-time interactions
- Allows groups to interact in a space with each other and the items in the space.

Limits

- Only 24 people could be in one room at a time.
- Need decent internet connection
- Headphones are needed if someone wants to use their microphone.

Mozilla hubs allowed students with no prior VR or 3D modeling experience to jump in and set up their gallery. It was important for students to learn how to customize their gallery quickly without needing code or complex programs. Mozilla offered the out-of-the-box functionality we needed with a low barrier of entry.

3.1 Results of the 2020 BFA Show

The 2020 show was a success. All the students were able to fill their virtual gallery with their design work. We had many guests, including friends, family, university faculty and staff, and professionals from the area. It was exciting that family members of students from other countries could attend (Fig. 1).

Fig. 1. Group talking verbally and via chat during the 2020 reception.

3.2 Pain-Points of the Virtual Exhibition

One major limitation to Mozilla Hubs is the number of people in a room. You are limited to 24 people in a virtual room at a time. Since we had over 30 students in the show, each student needed a room. I set up a central gallery as an entry point with information about the show, navigating the space, and portals to each of the student's galleries. A guest would spawn into the main gallery and navigate to a student's portrait to jump into their gallery. The connected galleries solved having too many people in one room but was isolating because the students were all in their separate rooms, making it difficult for them to interact with each other. Most people had not used Mozilla Hubs before or even VR in general. Figuring out the best way to navigate and how to interact was a learning curve for many guests. Some students wanted to have multiple videos and encountered issues with adding too many videos or getting videos to link. The last major hurdle was our award ceremony. We usually announce the student awards and have a little speech;

however, only 24 people could fit in one room. Most people had to watch from the lobby. Unfortunately, several of the students who won awards were stuck in the lobby and could not enter to receive their awards.

4 BFA Show 2021

In Based on last year's show and feedback from the students we plan to address as many of the pain points as possible. Some areas of improvement include Student interactions, the learning curve for students creating content, the learning curve for guests navigating the show, and loading issues of content in rooms.

The rooms will again be divided so each student will have their own room. This allows students more space to showcase their work and allows for 24 people per room but it can isolate the students from each other. To mitigate this discord will be incorporated into Mozilla hubs. There will be a public channel that all guests and students can communicate on as well as a private channel for all of the students. Discord will allow for students in different rooms to communicate with each other without leaving their room. Also, it is recommended that all students create a short intro video for their show in case they want to leave their room and visit other galleries during the reception.

To improve student workflow a two-hour workshop will introduce students to Mozilla hubs and showcase tips and tricks for lining up content straight on the walls. A reference guide will also be created for the students with an interactive tutorial posted in Hubs Room connected to the main gallery during the development phase. This will also cover how to reduce loading issues.

Improved signage and guides for the guests as they enter will inform them how to navigate and interact with the show. Short videos will introduce the space in addition to signage that imparts navigation information as visitors enter.

After the show, an optional survey for guests and students will provide feedback on the experience. This survey will collect both quantitative and qualitative data (Fig. 2).

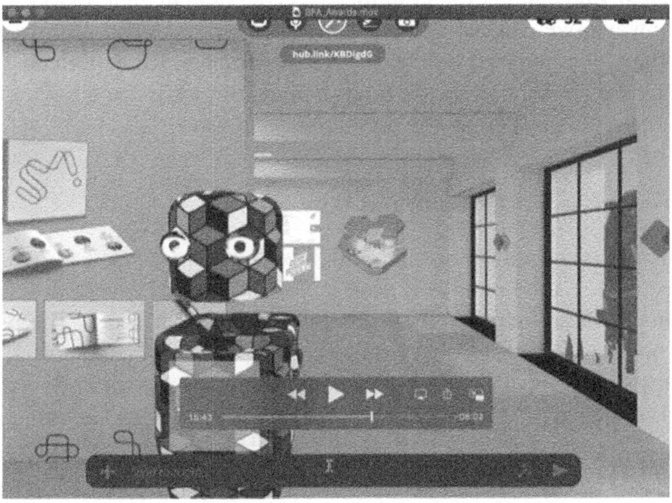

Fig. 2. Inside student gallery space from 2020.

5 Researching Best Practices for VR Exhibitions

5.1 Why Do We Need VR Exhibitions?

Virtual Exhibits allow for work to be shown without the physical limitations of a gallery space. A virtual exhibit can be viewed by any guest around the world with a device and access to the internet. The limitations of both time and space regarding a physical exhibition are taken away with a virtual exhibit. In fact, many museums around the world are interested in creating virtual experiences to supplement their physical content [1]. While existing virtual museums exist, they have not always been as appealing as attending a museum in person [2]. By leveraging virtual web-based technology exhibits accessibility increases. Instead of focusing on the use of virtual exhibits as a replacement or proxy for a physical exhibit how can they be leveraged to evolve into something even better?

5.2 Benefits of Remote Access

The last year has accelerated both interests and the need for virtual spaces to interact. Many people who previously were uncomfortable or uninterested in virtual spaces are now willing to try them [3]. When interacting in a physical space is no longer an option the use of a virtual environment offers features beyond phone or video conferencing. Virtual environments allow users to immerse themselves in a space. Being able to move through and around the space allows us to interact with others, objects, and the space itself in a natural intuitive way.

5.3 Using VR Exhibits in the Classroom as a Teaching Tool

The use of VR in the classroom can be used to bring students together when they cannot physically interact. Before the pandemic, a classroom of students could be gathered together physically all using VR and immersed separately in their VR learning environment. Now however we can see how students who are separated physically can all be brought together and immersed in a shared VR experience. VR experiences do not have to be bound by the physicals world's limitation or time or space allowing for learning experiences that would be impossible in the physical world.

5.4 Who Would Use VR Exhibits/and Why

There are already many museums and galleries that have created virtual experiences of their collections. As the pandemic shut down and limited access to museums more people sought out these increasingly promoted virtual experiences. The use of VR for exhibitions should go beyond a proxy replacement of visiting in person. By leveraging VR technology new levels of experiences can engage more people. By stripping away the physical limitations of an exhibit new levels of interaction and learning are possible.

5.5 Pain Points

The main hurdles or pain points to designing the UX of virtual exhibits include the limitations of the hardware needed to access the virtual content, access to fast internet connections, the creation of quality content for the virtual experiences, and making sure that VR exhibitions are accessible to all interested visitors regardless of any physical limitations including mobility, sight, or hearing.

5.6 Usability Testing

Usability testing for VR experiences is of the utmost importance. Iterative testing through the design and development process of a virtual exhibit ensures not only that people can interact with the information in a meaningful way but also that we don't create unintended hurdles. Testing can help prevent issues with VR sickness, eye strain, and repetitive movement. Looking at the following measures: Learnability, Ease of Use, Flexibility, and Attitude will help assess and improve Usability [4, 5]. The Systems Usability Scale (SUS) widely used for websites and desktop applications is a reliable quick way to test VR Experiences as well [6, 7]. The Questionnaires for User Interaction Satisfaction (QUIS) can also be used to assess VR Experiences [5].

6 Conclusion

Moving forward Virtual Exhibitions can complement or expand physical exhibits. VR exhibitions do not need to replace physical exhibits but they can be leveraged to allow exhibits that could not or would not otherwise exist due to financial, physical, or time constraints. The student BFA show has previously been constrained by the number of students, the physical space of the gallery, access to monitors, tablets, power outlets, availability of gallery guards, and timing of what other shows need to share the space.

With virtual exhibits, students can each have a full gallery instead of a 4 x 8 foot wall space. They can put up videos without needing to mount monitors on the wall and there is no need to worry about someone stealing or damaging equipment. Guest can visit the gallery anytime from any location. Someone in a different time zone could view an exhibit when it is convenient for them and not be bound by the operating hours of the gallery. The social VR component still allows for all kinds of interactions including live guided tours, receptions, and even spontaneous gallery conversation. Moving forward a virtual component should be part of all future BFA senior shows at Fresno State even once we are holding in person student shows. The many advantages to a virtual exhibition and peoples' willingness to use the technology after the shut downs of 2020 make VR a critical component of future exhibits.

References

1. Wojciechowski, R., Walczak, K., White, M., Cellary, W.: Building virtual and augmented reality museum exhibitions. In: Proceedings of the Ninth International Conference on 3D Web Technology, pp. 135–144, April 2004

2. Wolf, K., Reinhardt, J., Funk, M.: Virtual exhibitions: what do we win and what do we lose?. In: Electronic Visualisation and the Arts, pp. 79–86 (2018)
3. Amorim, J.P., Teixeira, L.M.L.: Art in the Digital during and after Covid: aura and apparatus of online exhibitions. Rupkatha J. Interdiscip. Stud. Humanit. **12**(5), 1–8 (2021)
4. Preece, J.: A Guide to Usability – Human Factors in Computing. Addison-Wesley Longman Publishing Co. Inc., Boston (1993)
5. Fang, Y.M., Lin, C.: The usability testing of VR interface for tourism apps. Appl. Sci. **9**(16), 3215 (2019)
6. Brooke, J.: SUS: A "quick and dirty" usability scale. In: Jordan, P.W., Thomas, B., Werdmeester, B.A., McClelland, A.L. (eds.) Usability Evaluation in Industry. Taylor and Francis, London (1996)
7. Bangor, A., Kortum, P.T., Miller, J.A.: An empirical evaluation of the System Usability Scale (SUS). Int. J. Hum. Comput. Interact. **24**, 574–594 (2008)

Human Centered Service Design

Human Centered Service Design (HCSD): Why HCSD Needs a Multi-level Architectural View

Markus Warg[1,2](✉) and Ulrike Deetjen[3]

[1] Institut Für Service Design, Hamburg, Germany
markus.warg@ifsd.hamburg
[2] University of Applied Sciences, Wedel, Germany
[3] McKinsey and Company, Inc., Stuttgart, Germany

Abstract. The motivation of this work is to examine and describe the relevance of architecture for Human Centered Service Design (HCSD). HCSD is the agile practice of improving service exchange (value co-creation and human well-being) in multi-level actor-to-actor networks. In this agile practice of HCSD, an architectural view means analyzing and describing the building blocks at micro, meso, and macro levels. All social and economic actors are embedded in broader social networks, coordinated by institutions and capable of acting on resources to mutually co-create value [1-3]. Since actions, institutions, and structure influence each other, this work considers the entire multi-level network in the understanding of a holistic service system ecology to investigate the relevance of architecture for HCSD [4, 5].

Keywords: Human Centered Service Design · Value co-creation · Holistic service system · Institutional arrangements · Service Dominant Architecture

1 Motivation

Human Centered Service Design (HCSD) is an approach that intents to co-create positively valued changes in human well-being by shaping service exchange in actor-to-actor networks. The intended and purposeful action of HCSD encounters given structures and resource constellations. This situation, which HCSD is facing on its way to creating valuable human centered solutions, was already summarized by Karl Marx who stated that "humans make their own history, but they do not make it of their own free will, not under self-chosen circumstances, but under circumstances that are directly found, given and handed down" [6].

Outcomes of HCSD depend largely on how the process of value co-creation is organized and this in the context of existing exogeneous structures and institutional arrangements. While the actor-to-actor nature of this exchange is increasingly recognized, the full extent of the interconnected, collaborative, polycentric and systemic nature of value (co) creation seems to be often underappreciated. The motivation of this work is to analyze the role of architecture as enabler of the process and outcomes of HCSD. The central research question therefore is "How can architecture support the process and

C. Leitner et al. (Eds.): AHFE 2021, LNNS 266, pp. 249–256, 2021.
https://doi.org/10.1007/978-3-030-80840-2_29

outcome of HCSD in multi-level networks?" For this purpose, HCSD is first defined and the core building blocks are brought together, and presented on the basis of their theoretical foundations. Based on this the relevance of architecture for HCSD is worked out and described.

2 Methodology

The research design of this paper is aligned to the research goal to demonstrate the impact of architecture for the process and outcome of HCSD in multi-level networks. A conceptual paper as methodology and within this methodology "theory synthesis" as approach are chosen [7]. "Theory synthesis" is understood as the conceptual integration across multiple theoretical perspectives. After outlining and summarizing the conceptual domain theories of HCSD the phenomenon of architecture within the process and outcome of HCSD is analyzed through new perspectives to increase the understanding and to extant the existing conceptual boundaries [8, 9]. This paper is based on the conceptual domain theories of service-dominant logic [10], service science [11] and institutional economics [3, 12]. Building on the domain concepts and relationships such as service, value co-creation, institutions, resource integration and multi-level networks the conceptual domain is displayed. In order to expand the boundaries of the domain theories for the research topic, new perspectives by Elinor Ostrom's Institutional Analysis and Development (IAD) framework [3] are elaborated and integrated.

3 Conceptual Domain Theories of HCSD

Based on a research of theories, and concepts of service-dominant logic, service science and institutional economics the core building blocks of HCSD [13, 14] are worked out. Service-dominant logic is a meta-theoretical framework for explaining value creation through service exchange among multiple resource-integrating actors forming institutionally coordinated service ecosystems [15]. Service systems and service-dominant logic are the key constructs of service science [16, 17]. The core principles of service science center on how value is created among entities (systems, service (eco-) systems), how interaction depends on access to resources and capabilities (institutions), and on how value-creation depends on the context of mutually agreed upon value propositions (intention) [2]. Actors are capable of acting on potential resources to cocreate value and service is defined as the application of resources for the benefit of another actor or oneself [15, 16, 18, 19].

Design Science interprets design as an "act of creating an explicitly applicable solution to an problem" that serves as a commonly accepted framework [20]. Service design is understood as explorative approach to creating novel forms of value cocreation [21]. In order to be human centered service design the creative act and its results must serve human needs [22, 23]. This knowledge and understanding of service, design and service design leads us to the following definition for HCSD: Human centered service design (HCSD) is an approach that intends to co-create value as explicitly applicable solution to a problem and as positively valuated change in human well-being through service exchange in actor-to-actor networks.

For the beneficiaries of HCSD the concepts of "holistic service systems" and "whole service" are of particular relevance. Holistic approaches which from an analysis of the parts lead to the comprehension of the "whole" are characterized by dynamic interaction and widespread value co-creation processes [24]. Holistic service systems provide "whole service" to the humans within them and allow them to survive for a period of time independently of all external service system interactions [25]. The latter is also characterized by the existence of sufficient rules and norms for the holistic service system to be independent and self-organizing. Therefore, institutional economics are relevant for the analysis of the systematic functioning of HCSD. Institutions are the rules of the game that coordinate actors. More formally, institutions are the humanly devised incentives and constraints that shape human interaction. In consequence they structure incentives in human exchange and shape the way new outcomes evolve [26, 27]. Taking into account the concepts of the domain theories the following building blocks of HCSD can be summarized (Table 1).

Table 1. Building blocks of Human Centered Service Design

HCSD	Service Science/Service-Dominant Logic	Ref
Actor	Actors are capable of acting on potential resources to cocreate value, either positively or negatively valanced; actor as carrier of operant and/or operant resources	[15, 28]
Holistic service (eco) system	A relatively self-contained, self-adjusting system of resource-integrating actors connected by shared institutional arrangements, mutual value creation and whole service through service exchange	[15, 25,17, 29]
Institutions	Institutions are the human-made rules, norms and beliefs that provide stability and meaning to social life by constraining and enabling collective action	[15, 30, 31]
Value co-creating	The process through which multiple actors, often unaware of each other, jointly contribute to an actor's wellbeing	[15]
Value proposition	Businesses, firms, and customers are hence all viewed as socio-economic actors who connect through value propositions within "complex service systems" and perform actions aimed at reaching desired outcomes	[1]

4 Previously Unconnected Conceptual Perspectives

Ostrom's Institutional Analysis and Development framework [32] contains a nested set of building blocks that can be used in efforts to understand human actions and outcomes across diverse settings. The purpose of the framework is to understand complex systems and to analyze how institutions affect the incentives of individuals, their choice and their resulting behavior [25, 32, 33].

4.1 Institutional Analysis and Development (IAD) Framework

An important strength of the IAD framework is structurally detailing the action situation relevant to actors as participants in specific situations and following a systematic set of rules. Thus helps to link rules to the action situation they constitute [32, 34]. As shown in Fig. 1 the IAD framework is a multi-tier conceptual map.

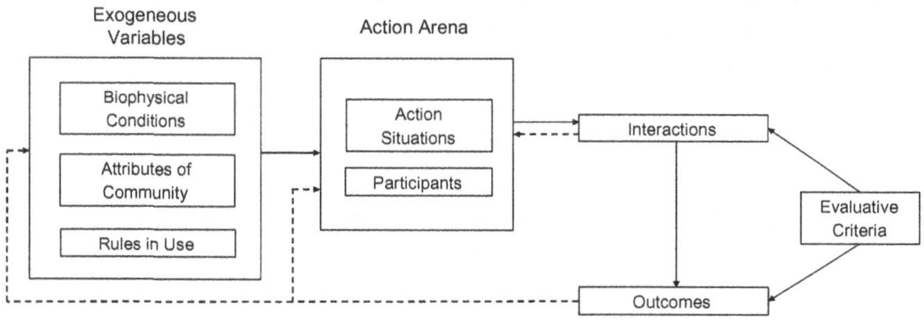

Fig. 1. Ostrom's Institutional Analysis and Development (IAD) framework [32, 35]

For further detailing, we structure the framework into the following three areas: "exogeneous variables", "action arena" and "interactions & outcomes". Starting on the left in the framework are the exogeneous variables that affect the participants, positions and actions of an action arena and by this its structure. The exogeneous variables include three clusters of variables [33]: 1. The attributes of states of the world that are acted upon in these arenas, e.g. the physical possibilities of actions, the producibility of outcomes and the linkage of outcomes to actions depend on the physical world and its transformations [35]. 2. The attributes of a community that affect the structure of an action arena are e.g. expected norms of behavior. 3. The third set of variables that effect the structure of an action arena relates to the rules "as prescriptions to define what actions (or outcomes) are required, prohibited [35]. From this point, the action arena is viewed as a set of variables dependent upon other factors. In the action situation individuals act on their own or as agents of organizations, observe information, select actions, engage in patterns of interaction, and realize outcomes from their interaction [36]. The action arena can be utilized to describe, analyze, predict, and explain behavior within institutional arrangements. The action arena is linked to the third area of the framework the "interactions & outcomes". Outcomes are generated by a given action situation, other closely related action situations, and exogenous influences that may not always be subject to effective control of human intervention [36]. Actors make choices based on their own preferences, the costs and benefits that they assign to alternative actions and outcomes, and strategic considerations (i.e., expectations of the behavior of others) [37].

5 Theory Synthesis and Relevance of Architecture

The synthesis of the domain theories and the IAD framework provides new perspectives to the building blocks of HCSD: Actors are defined in IAD framework more narrowly as humans whereas s-d logic and service science have a broader understanding of the term and include, for example, technology and combinations of humans and technology. On the other hand, the IAD framework complements the comprehension of the actor of the domain theories. It helps to link institutions to action situations and action situations to actors who then become participants. While s-d logic and service science focus on mutual value generation through service exchange and resource integration, the IAD allows to analyze the actor's decision-making situation and to show the impact on the outcome. Both in the domain concepts and from the perspective of the IAD framework the aggregated actions lead to pattern of interactions [37]. The framework complements the domain theories through a continuously evaluation and feedback process that the outcomes go through. Dynamically feedback to both the action arena and to the higher level exogeneous variables [37]. From a process perspective the IAD framework is characterized by an uni-directional production process. The process is designed like according to goods-dominant logic; the outcome as result is achieved via a production function that relates input and output. S-d logic and service science complement this perspective: The exchange of service is always the basis for economic exchange and the process of value cocreation is always (with the exception of self-service) bi- or multi-directional. As part of the action arena HCSD-actors and their decisions are nested in polycentric, multi-level and multi-tier institutional arrangements. Polycentric systems are themselves complex, adaptive systems without one central authority dominating all of the others [32, 38]. At the same time, the decision situation of the actors is influenced by multi-levels as operational choice (micro-), as collective choice (meso-) and as constitutional choice (macro-level) [37].

For the creation, robustness, autonomy and self-organization of HCSD solutions as holistic service systems architecture is particularly relevant: as process of institutionalization, as pattern and as (instantiated) structure as shown in Fig. 2 [39, 40, 25, 32].

As a process, architecture enables both the institutionalization of own norms or the adaptation to exogenous variables as well as the establishment of patterns for interaction and design patterns. Thus the dynamic process links as action the situation of the actors with the intended outcome.

As (static) pattern architecture establishes rules for the coordination of actors and the integration of resources. Interactions like service exchange are dependent upon design pattern as broadly defined institutions governing the flows of knowledge, materials, tools, energy, information, and people into and out of structures [25, 32, 37]. Finally, architecture enables holistic service systems as an (instantiated) structure, which again influences the formation of architectures through the described system. In a multi-tiered process, the suitability of the architecture as process, pattern and structure (outcome) is evaluated and analyzed again and again (Fig. 2) influencing the action arena and also taking into account the exogeneous variables. The actors then decide either to adapt and revise the pattern and outcome or to adopt it.

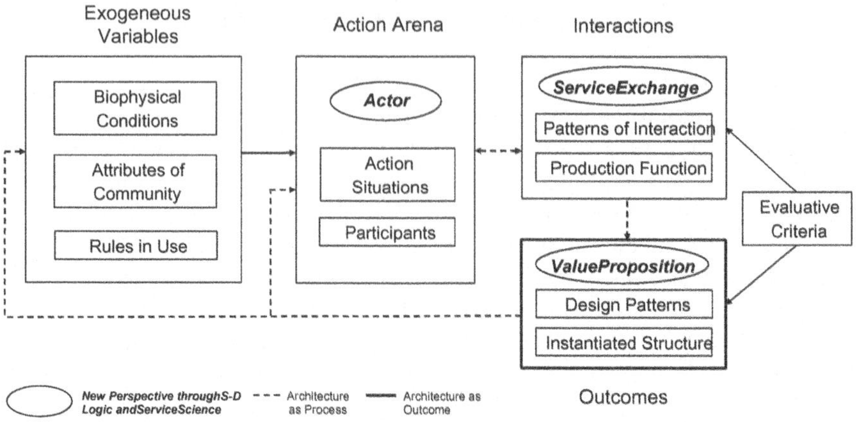

Fig. 2. Synthesis of Ostrom's IAD framework, S-D Logic, Service Science and Architecture; extended by Warg/Deetjen 2021

6 Implication for Practical Approaches

Brittle changes such as those caused by pandemics, climate change or digitization pose major challenges to the autonomy and self-organization of human centered solutions in polycentric networks. Architecture is particularly important for the mastering practical challenges by designing and instantiating holistic service systems. In this architecture has many roles, which can be integrated by the synthesis of this paper. As shown by Alexander (1977), patterns as solution parts enable the self-organization of the design and construction of e.g. houses [39]. Lusch and Nambisan (2015) emphasize in the context of service innovation the need to devise and implement an architecture of participation to coordinate actors and their service exchanges [41]. Spohrer, Piciocchi, Bassano (2012) introduce the term "entity architecture" as architectural approach for holistic service systems providing "whole service" to the humans inside and independence from external systems [25]. Warg (2015) introduced Service Dominant Architecture (SDA) derived of the concepts of service-dominant logic and service science. SDA enables the process of service exchange and mutual value creation. And in this way creating, holistic service systems and service innovation by orchestrating resource integration and value cocreation within polycentric service eco-systems, service platforms and institutions [42].

7 Implication for Theory

In the interplay with (holistic) service systems, architecture offers a variety of exciting fields of analysis for future theoretical research, especially from the point of view of self-organization, self-sufficiency and the greatest possible independence from external systems. The time of the pandemic in particular illustrates the great importance of independent value propositions and value chains [25, 43].

8 Observations and Outlook

Architecture facilitates the process of HCSD and its outcome as holistic human centered service system: as process of institutionalization pattern of interaction and design are institutionalized which link action to intended value proposition and outcome; as norm for coordinating actors, resource integration and service exchange and finally as (instantiated) outcome and holistic service system. Especially in times of pandemic and climate change, qualities of holistic service systems such as sustainability, self-organization or independence are becoming increasingly relevant.

Thus architecture enables the design and outcome of holistic human centered service systems that provision whole service to the humans inside them independently of all external service system interactions.

References

1. Wieland, H., Polese, F., Vargo, S.L., Lusch, R.F.: Toward a service (eco) systems perspective on value creation. Int. J. Serv. Sci. Manag. Eng. Technol. (IJSSMET) **3**, 12–25 (2012)
2. Maglio, P.P., Spohrer, J.: A service science perspective on business model innovation. Ind. Mark. Manag. **42**, 665–670 (2013)
3. Ostrom, E.: Governing the Commons: The Evolution of Institutions for Collective Action. Cambridge University Press, Cambridge (1990)
4. Spohrer, J.C., Vargo, S.L., Maglio, P.P.: The service system is the basic abstraction of service science. In: Proceedings of the 41st Hawaii International Conference on System Science, Big Island (2008)
5. Giddens, A.: The constitution of society: Outline of the theory of structuration (1984)
6. Marx, K.: Der achtzehnte Brumaire des Louis Bonaparte. BoD–Books on Demand (2016)
7. Gilson, L.L., Goldberg, C.B.: Editors' comment: So, what is a conceptual paper? SAGE Publications, Los Angeles (2015)
8. Jaakkola, E.: Designing conceptual articles: four approaches. AMS Rev. **10**(1–2), 18–26 (2020). https://doi.org/10.1007/s13162-020-00161-0
9. MacInnis, D.J.: A framework for conceptual contributions in marketing. J. Mark. **75**, 136–154 (2011)
10. Vargo, S.L., Lusch, R.F.: Evolving to a new dominant logic for marketing. J. Mark. **68** (2004)
11. Spohrer, J.C., Kieliszewski, C.A., Lyons, K., Maglio, P.P., Sawatani, Y., Patrício, L.: Handbook of Service Science. Springer, Heidelberg (2019)
12. Scott, W.R.: Institutions and Organizations: Ideas, Interests, and Identities. Sage Publication, Thousand Oaks (2013)
13. Wieringa, R.J.: Design Science Methodology for Information Systems and Software Engineering. Springer, Heidelberg (2014). https://doi.org/10.1007/978-3-662-43839-8
14. Imenda, S.: Is there a conceptual difference between theoretical and conceptual frameworks? J. Soc. Sci. **38**, 185–195 (2014)
15. Vargo, S.L., Lusch, R.F.: The SAGE Handbook of Service-Dominant Logic. SAGE Publications Limited, Thousand Oaks (2018)
16. Maglio, P.P., Spohrer, J.: Fundamentals of service science. J. Acad. Mark. Sci. **36**, 18–20 (2008). https://doi.org/10.1007/s11747-007-0058-9
17. Spohrer, J.C., Maglio, P.P.: Toward a science of service systems. In: Maglio, P., Kieliszewski, C., Spohrer, J. (eds.) Handbook of Service Science. Service Science: Research and Innovations in the Service Economy, pp. 157–194. Springer, Boston (2010). https://doi.org/10.1007/978-1-4419-1628-0_9

18. Storbacka, K., Brodie, R.J., Böhmann, T., Maglio, P.P., Nenonen, S.: Actor engagement as a microfoundation for value co-creation. J. Bus. Res. **69**, 3008–3017 (2016)
19. Ostrom, E.: Institutional analysis and development: Elements of the framework in historical perspective. In: Historical Developments and Theoretical Approaches in Sociology, vol. 2 (2010)
20. Peffers, K., Tuunanen, T., Rothenberger, M.A., Chatterjee, S.: A design science research methodology for information systems research. J. Manag. Inf. Syst. **24**, 45–77 (2008)
21. Vink, J., Edvardsson, B., Wetter-Edman, K., Tronvoll, B.: Reshaping mental models–enabling innovation through service design. J. Serv. Manag. **30**, 75–104 (2019)
22. Goodwin, K.: Designing for the Digital Age: How to Create Human-Centered Products and Services. Wiley, Hoboken (2011)
23. Warg, M.: Architecture and its multifaceted roles in enabling value co-creation in the context of human-centered service design. In: Spohrer, J., Leitner, C. (eds.) AHFE 2020. Advances in Intelligent Systems and Computing, vol. 1208, pp. 79–85. Springer, Cham (2020). https://doi.org/10.1007/978-3-030-51057-2_12
24. Golinelli, G.M., Barile, S., Spohrer, J., Bassano, C.: The evolving dynamics of service co-creation in a viable systems perspective. In: The 13th Toulon-Verona Conference Proceedings of the International Conference in Coimbra, Portugal
25. Spohrer, J., Piciocchi, P., Bassano, C.: Three frameworks for service research: exploring multilevel governance in nested, networked systems. Serv. Sci. **4**, 147–160 (2012)
26. North, D.C.: Institutions, Institutional Change and Economic, Cambridge (1991)
27. Williamson, O.E., Brousseau, E., Glachant, J.: The New Institutional Economics Guidebook. New Institutional Economics (2008)
28. Löbler, H.: Service-dominant networks. J. Serv. Manag. (2013)
29. Vargo, S.L., Lusch, R.F.: Institutions and axioms: an extension and update of service-dominant logic. J. Acad. Mark. Sci. **44**(1), 5–23 (2015). https://doi.org/10.1007/s11747-015-0456-3
30. Edvardsson, B., Kleinaltenkamp, M., Tronvoll, B., McHugh, P., Windahl, C.: Institutional logics matter when coordinating resource integration. Mark. Theory **14**, 291–309 (2014)
31. North, D.C.: Institutions, Institutional Change, and Economic Performance, New York (1990)
32. Ostrom, E.: Understanding Institutional Diversity. Princeton University Press, Princeton (2009)
33. Kiser, L., Ostrom, E.: The Three Worlds of Political Action. Sage, Berverly Hills (1982)
34. Li, R., Van Den Brink, M., Woltjer, J.: Rules for the governance of coastal and marine ecosystem services: an evaluative framework based on the IAD framework (2016)
35. Ostrom, E., Gardner, R., Walker, J., Walker, J.M., Walker, J.: Rules, Games, and Common-Pool Resources. University of Michigan Press, Ann Arbor (1994)
36. McGinnis, M.D.: Updated Guide to IAD and the Language of the Ostrom Workshop: A Simplified Overview of a Complex Framework for the Analysis of Institutions (2013)
37. Rudd, M.A.: An institutional framework for designing and monitoring ecosystem-based fisheries management policy experiments. Ecol. Econ. **48**, 109–124 (2004)
38. Polanyi, M.: The Logic of Liberty, 1951. Cited in Fuller, L. Collective bargain (1951)
39. Alexander, C.: A Pattern Language: Towns, Buildings, Construction. Oxford Press, Oxford (1977)
40. Gamma, E., Helm, R., Johnson, R., Vlissides, J.: Elements of Reusable Object-Oriented Software. Addison-Wesley, Reading (1995)
41. Lusch, R.F., Nambisan, S.: Service innovation: a service-dominant logic perspective. MIS Q. **39**, 155–171 (2015)
42. Markus, W., Peter, W., Ronald, E., Zolnowski, A.: Service Dominant Architecture based on SD logic for Mastering Digital Transformation: The Case of an Insurance Company. Tiziana Russo-Spenaand Cristina Mele, p. 807 (2016)
43. Maglio, P.P., Srinivasan, S., Kreulen, J.T., Spohrer, J.: Service systems, service scientists, SSME, and innovation. Commun. ACM **49**, 81–85 (2006)

Analyzing Actor Engagement in Data-Driven Business Models Innovation in the Context of Smart Cities by Creating a Common Understanding

Babett Kühne[1](✉) and Michel Muschkiet[2]

[1] IT-Management and -Consulting, University of Hamburg, Vogt-Kölln-Straße 30, 22527 Hamburg, Germany
babett.kuehne@uni-hamburg.de
[2] Business Informatics and Integrated Information Systems, University of Duisburg-Essen, Universitätsstraße 9, 45141 Essen, Germany
michel.muschkiet@icb.uni-due.de

Abstract. Since the advancing digitization, the amount of data increases exponentially and the potential to use this data to innovate data-driven business models is high. In the city, in particular, the new technologies generate large amounts of data that can be used to make a city smarter. However, the process of innovating new data-driven business models is complex and not trivial to create a shared understanding between the different involved actors. Our research shows that the link between data and the citizen's value proposition is not clear and requires specific steps in between. To support such processes, this work develops a graphical artifact for the Smart City area. Firstly, we supplement the existing knowledge of data-driven Smart City Services and lay the foundation for further concepts in this area. Second, the artifact can be used in practice to make their city smarter and to innovate data-driven business models.

Keywords: Data-driven services · Smart cities · Methodical artefact · Data-driven business models

1 Introduction

Innovative and affordable technologies enable cities to produce and analyze a large amount of data via various sources such as smartphones, sensors and commercial transactions [1, 2]. This data can generate new knowledge and drive innovation by developing new data-driven business models that can improve citizens' quality of life [3]. Data-driven business models can be defined as business models in which value is created through data as a key and necessary resource [4]. For several reasons, smart cities can achieve significant added value through service innovations and the development of data-driven business models. Firstly, smart cities are characterized by the overarching goal of sustainably improving their citizens' quality of life [5]. Citizens, therefore,

C. Leitner et al. (Eds.): AHFE 2021, LNNS 266, pp. 257–264, 2021.
https://doi.org/10.1007/978-3-030-80840-2_30

play a central role in urban planning and value creation [6]. Second, smart cities are increasingly instrumented and networked by using and exchanging data across multiple areas such as energy, mobility and health [7]. As a result, cities are increasingly gaining benefit from data-driven value creation [6]. Third, a key factor for smart city initiatives is the collaboration between various actors from different fields to identify and use data and conduct value creation activities by integrating the data, skills and knowledge of each actor about the citizen's needs [3].

Collaborative methods can support cities in designing data-driven business models and thus offer valuable assistance. Such methods are already known from other areas of innovation and are being used successfully there. For example, the Business Model Canvas [8] supports companies in innovating new business models. These methods can provide a common understanding of the involved actors about the most important business model innovation factors by limiting the complex system with few elementary aspects and presenting them in a clear graphic [9]. The literature lacks in such collaborative methods for the unique requirements of smart cities. Abella and colleagues [10] offer a methodology for improving smart city services' design as an approach to design smart cities by describing and evaluating services from the perspective of the citizen's experience and defining steps to improve them. Liu and colleagues [11] present a method for designing citizen-oriented services in smart cities. It describes the phases and related tasks of service design. While this work suggests overarching methods for a citizen-centered business model design, it does not provide methods for working together on the specific steps for identifying and describing business model ideas. Besides, they do not contain data as a key resource for creating value for citizens. As a consequence, this research aims to answer the following research question:

RQ: How could cities be supported to design data-driven business models with an easily applicable artifact?".

To answer this question, we derive a collaborative method designing of data-driven business models from the existing literature. In principle, the Business Model Canvas [8] can be used for generic business model development. Still, it was found that the step between the key resources data and the value proposition is too huge for data-driven business models and the involved actors to generate new innovative business models with the Business Model Canvas alone [12]. That is why we use a collaborative method Data Insight Generator in this research, which connects the key resources data and the value proposition for the customer [13]. In workshops in a smart city environment, it was found that a specific adaptation of such collaborative methods is necessary to map the specific characteristics of smart cities. This is why this research is developing a specific collaborative method for smart cities based on the Data Insight Generator using the Design Science Research method [14]. To evaluate our artifact, we conduct two workshops with six participants each and then present exemplary results in the collaborative method, which represent new ideas for data-driven smart city business models.

The following chapter provides an overview of current research in the field of smart cities and in connection with data-driven business models. The methodical approach on which the results of this article are based is then presented. Afterwards, we offer the Data Insight Generator adapted for smart cities. Finally, the implications for cities, which result from the present research work results, are examined.

2 Theoretical Background

Information and technology resources are increasingly recognized as an essential resource for the creation of services and especially new types of services-based business models driven by data, example "Data-as-a-Service" and "Analytics-as-a-Service" [15]. Data is a key resource in such business models [16]. However, when comparing traditional business models based on data with data-driven business models, there is no specific threshold of the amount of data for the dependence to distinguish these business models [4].

There are some methods available in the literature that support the innovation of data-driven business models. Kronsbein and Mueller present a method for developing data-driven ideas as tabular documentation of the entire design process [17]. Hunke and Schüritz developed a tool for designing analysis-based services [18]. In further studies, methods for developing key activities for data-driven business models [19], the Data Canvas [20] and a concept for the financial part of data-driven business models [21] were developed. However, this work does not meet smart cities' requirements, as it does not allow different actors in a network to merge their data and service ideas at different touchpoints to jointly design data-driven business models for a holistic citizen experience.

Over the past fifteen years, smart city initiatives have emerged as an evolution of previous efforts to modernize and digitize public services, infrastructure, urban planning, and more [22]. Although there is no standard definition for smart cities, theory and practice have at least one common basis: the sustainable improvement of citizens' quality of life through the use of information technology is seen as a smart city concept [23]. One aspect of improving the citizens' quality of life is taking the public value of services into account and not only considering their monetary value, but also acknowledging civic and democratic principles such as justice, freedom and participation [24]. Another important feature of smart cities is the role of technology which serves as an operational resource to reduce innovation boundaries and supports innovation activities in ecosystems composed of diverse interest groups [25]. Collaboration between stakeholders in identifying, collecting, generating and using data has been identified as a key factor for smart city initiatives [26]. In addition, smart cities also face challenges in practice when creating value from data [27]. As a consequence, there is a lack of such collaborative methods for the unique requirements of smart cities in the literature.

3 Methodical Approach

Our research design is based on the commonly used Design Science Research methodology by Peffers and colleagues [14] which is suitable for the design of artifacts, in order to achieve the goal of our study.

(1) In a first step, we derived the motivation for our research from the identified research problem, which is described in the introduction and the theoretical background of this paper. We also recognized the Data Insight Generator artifact's potential in an initial workshop of a smart city initiative in a German city. Many new ideas arose during this, but the need to adapt the tool to the characteristics of a smart city also became apparent.
(2) In a second step, we defined the need for related supportive methods and set the goal of this research as developing collaborative methods to facilitate the design of data-driven business models in smart cities.
(3) In step three, we adapted the Data Insight Generator to the smart city context. For this purpose, we derived design requirements from the literature to ensure that our methods master smart cities' described challenges in the design of data-driven business models.
(4) In step four, we discussed and evaluated the existing artifact within the group of four researchers to challenge and improve the Data Insight Generator for the smart city context. We presented the artifact to our colleagues and discussed their applicability which helped us in our further development.
(5) In step five, we evaluated and redesigned our artifact in two workshops, each with six participants with an experience background of the smart city context. introducing the research topic, we presented the Data Insight Generator adapted for smart cities principles and processes as guidelines applying our methods. The participants were then divided into two independent groups and asked to test the artifact together. We acted as passive observers during the workshops. To give the participants input for the application of the methods, and achieve comparable results, we introduced a fictitious person to all participants. We recorded the workshops on video, made notes and asked for detailed feedback in a final group discussion at the end of each workshop. All in all, we could identify that the adapted version of the Data Insight Generator for smart cities could be used to innovate new service ideas and business models in the field of smart cities. Further, the artefact served as a tool in order to create a common understanding between the participants with different backgrounds in the field of smart cities. As an additional finding, we could identify groups of actors which should be integrated into the innovation process of data-driven business models for smart cities to end the innovation successful as these groups have different perspectives in the field of smart cities and data usage in order to improve citizens life.

4 Findings

The collaborative method presented in this article and adapted to the smart city context is the Smart City Data Insight Generator which is based on [13]. Figure 1 shows our adapted version. Current research shows that the design of data-driven business models is not trivial, especially the connection between the value proposition for the citizen and the most important resource data is complex for the different actors in the innovation process [12]. Therefore, the Smart City Data Insight Generator was developed to provide a visual model for specifying this link and create a shared understanding between the different actors in a smart city. This artifact could be used in workshops with the different actors of a smart city.

The Smart City Data Insight Generator is designed in vertical and horizontal rows. The horizontal lines link the data and value proposition to the following fields: data quality, data combinations, pipes (infrastructure for exchanging data), algorithms and tools for data analysis, and their insights. In the field "data" the participants of the workshop could save all data that is available, or they think is relevant. The field "data quality" reflects if that data is complete and consistent. With the field "combination", the workshop participants can find useful combinations of this data and afterwards find how to link these data technically in the field "pipes". The field "analytics" reflects all relevant methods and tools needed to analyze the data. This analyzes should provide new insights that are not available in the pure data and could be noted in the field "insights". The field "value proposition" uses this insight to create a new value for the citizen. Adjustments compared to the original Data Insight Generator [13] where manly changes in the fields "data" and "value proposition".

Since the Smart City Data Insight Generator is an iterative method, the horizontal lines – "think", "validate", and "know" (facts) - offer the possibility of working with the artifact in iterations. First of all, all unlimited thoughts are stored ("think") in the first line. Then all items in the first row are sorted to the second (to be "validated") or third ("know" (facts)) row, depending on whether the fact is clear, or something needs to be validated at the next workshop.

The method is suitable so that the city can develop innovative new ideas for their city together with the citizens, which can ultimately be implemented in a business model. The city can start more from the perspective of the existing data and the citizens more from the perspective of the added value to be created for themselves. The order in which the individual fields are processed is not specified, and it is also possible to go back and forth as desired to fill in the particular fields. The Smart City Data Insight Generator serves as a supplement to the Business Model Canvas [8], with the interfaces data (key resources) and added value (value proposition), in order to then design a complete business model that can be implemented. No hurdles in the area of data and its analysis are needed to be overcome, as these have already been processed in the Smart City Data Insight Generator.

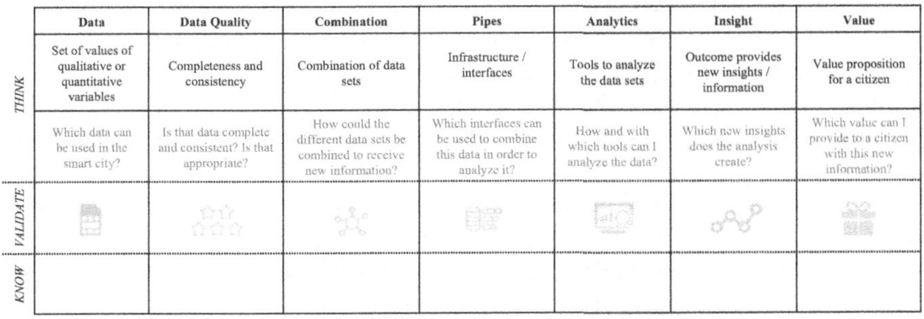

Fig. 1. Smart city data insight generator

The participants' analysis yields the following findings: (1) There need to have a participant who keeps the workshop flowing and acts like a moderator in the group. Such a person does not need to be a defined facilitator. One participant could develop this role during the workshop. (2) In order to utilize data-driven business model innovation in smart cities successfully, a mix of different roles is supporting the innovation process. We could identify the following roles: actors who know databases and data sets, actors who have a smart city background, actors who know the citizens, and actors who are a data scientist specialist. As the name of the roles could be different from organization to organization, we mentioned the different actors' knowledge.

5 Implications and Conclusion

Identifying the major steps and their characteristics for the description of data-driven business models in the context of smart cities is an essential step for a deeper understanding of this type of business model. The transfer to business model innovation is just as necessary. So far, there has been a lack of collaborative methods that address smart cities' specific characteristics. The Smart City Data Insight Generator contributes to closing this gap. In doing so, the artifact expands the knowledge in the area of business model innovations to include smart cities to increase citizens' quality of life.

The results presented can be directly applied in urban practice because the tool is self-explanatory and allows cities and citizens to innovate new ideas in the field of data-driven business models. Particularly in the early phases of business model innovation, it will enable actors to test whether an idea can be implemented considering the data and the technical conditions. It helps identify ideas at an early stage that can be implemented as a complete business model.

Besides, the collaborative method serves as a useful tool during business model development in the smart city area to identify the strengths and weaknesses of the concept of the data-driven business model. It evaluates which features the business model shares with other smart city initiatives and which features can serve as a unique selling point. In this way, potentials can be recognized, and improved ideas or possible extensions from other smart city projects can be added.

Another possible use of the presented collaborative method is the identification of new data-driven smart city business models. The development of smart city innovations is mostly still unstructured, and the potential of data is not yet fully used. The Smart City Data Insight Generator is used to use this potential of the data in the smart city ecosystem and increase the citizens' quality of life. Based on the predefined holistic characteristics, this collaborative method can be used to identify a theoretical design space for data-driven smart city business models at an early stage by examining various constellations of characteristics for their feasibility in terms of data and technical implementation. New developments and trends in the field of data acquisition (Internet of Things) or new data analysis methods made possible new directions by developments in programming (machine learning), for example, can be included in these considerations to identify application scenarios and potentials at an early stage.

In sum, the results presented here provide cities with valuable suggestions and methodological support to start developing data-driven smart city innovations in their city and thus be able to use the existing potential of data in the city.

References

1. Hashem, I.A.T., et al.: The role of big data in smart city. Int. J. Inf. Manage. **36**, 748–758 (2016)
2. Woetzel, J., et al.: Smart Cities: Digital Solutions for a More Livable Future, pp. 1–152 . McKinsey Global Institute, New York (2018)
3. Bifulco, F., Tregua, M.: Service innovation and smart cities: linking the perspectives. In: Russo-Spena, T, Mele, C., Nuutinen, M. (eds.) Innovating in Practice Perspectives and Experiences, pp. 261–287. Springer, Cham (2017). https://doi.org/10.1007/978-3-319-43380-6_12
4. Schüritz, R., Seebacher, S., Dorner, R.: Capturing value from data: revenue models for data-driven services. In: Proceedings of the 50th Hawaii International Conference on System Sciences, pp. 5348–5357 (2017)
5. Chourabi, H., et al.: Understanding smart cities: an integrative framework. In: 45th Hawaii International Conference on System Sciences, pp. 2289–2297. IEEE (2012)
6. Giourka, P., et al.: The smart city business model canvas—a smart city business modeling framework and practical tool. Energies **12**, 4798 (2019)
7. Cheng, B., Longo, S., Cirillo, F., Bauer, M., Kovacs, E.: Building a big data platform for smart cities: experience and lessons from santander. In: IEEE International Congress on Big Data, pp. 592–599. IEEE (2015)
8. Osterwalder, A., Pigneur, Y.: Business Model Generation: A Handbook for Visionaries, Game Changers, and Challengers. Wiley, New York (2010)
9. John, T., Szopinski, D.: Towards explaining the popularity of the business model canvas: a dual coding approach. In: Multikonferenz Wirtschaftsinformatik, pp. 1509–1515 (2018)
10. Abella, A., Ortiz-de-Urbina-Criado, M., De-Pablos-Heredero, C.: A methodology to design and redesign services in smart cities based on the citizen experience. Inf. Polity **24**, 183–197 (2019)
11. Liu, N., Gavino, A., Purao, S.: A method for designing value-infused citizen services in smart cities. In: Proceedings of the 15th Annual International Conference on Digital Government Research, pp. 34–43 (2014)
12. Kühne, B., Böhmann, T.: Data-driven business models - building the bridge between data and value. In: 27th European Conference on Information SystemsStockholm, Sweden (2019)
13. Kühne, B., Böhmann, T.: Formative evaluation of data-driven business models–the data insight generator. In: Proceedings of the 53rd Hawaii International Conference on System Sciences, pp. 427–436 (2020)
14. Peffers, K., Tuunanen, T., Rothenberger, M.A., Chatterjee, S.: A design science research methodology for information systems research. J. Manage. Inf. Syst. **24**, 45–77 (2007)
15. Chen, Y., Kreulen, J., Campbell, M., Abrams, C.: Analytics ecosystem transformation: a force for business model innovation. In: SRII Global Conference (SRII), pp. 11–20, IEEE (2011)
16. Hartmann, P.M., Zaki, M., Feldmann, N., Neely, A.: Big Data For Big Business? A Taxonomy of Data-Driven Business Models Used by Start-up Firms. Cambridge University (2014)
17. Kronsbein, T., Mueller, R.: Data thinking: a canvas for data-driven ideation workshops. In: Proceedings of the 52nd Hawaii International Conference on System Sciences, pp. 561–570 (2019)
18. Hunke, F., Schüritz, R.: Smartere Produkte durch analysebasierte Dienstleistungen-Ein methodisches Werkzeug zur strukturierten Entwicklung. HMD Praxis der Wirtschaftsinformatik **56**, 514–529 (2019)
19. Hunke, F., Wambsganß, T.: Turning data into value: towards an ideation tool for key activities of data-driven business models. In: 3rd Karlsruhe Service Summit Research Workshop (2017)

20. Mathis, K., Köbler, F.: Data-need fit–towards data-driven business model innovation. In: Service Design Geographies. Proceedings of the ServDes. 2016 Conference, pp. 458–467. Linköping University Electronic Press (2016)
21. Zolnowski, A., Anke, J., Gudat, J.: Towards a cost-benefit-analysis of data-driven business models. In: International Conference on Wirtschaftsinformatik, pp. 181–195, St. Gallen, Switzerland (2017)
22. Shelton, T., Zook, M., Wiig, A.: The 'Actually Existing Smart City.' Cambridge J. Regions, Econ. Soc. **8**, 13–25 (2015)
23. Alawadhi, S., et al.: Building understanding of smart city initiatives. In: Scholl, H.J., Janssen, M., Wimmer, M.A., Moe, C.E., Flak, L.S. (eds.) EGOV 2012. LNCS, vol. 7443, pp. 40–53. Springer, Heidelberg (2012). https://doi.org/10.1007/978-3-642-33489-4_4
24. Castelnovo, W., Misuraca, G., Savoldelli, A.: Smart cities governance: the need for a holistic approach to assessing urban participatory policy making. Soc. Sci. Comput. Rev. **34**, 724–739 (2016)
25. Lusch, R.F., Nambisan, S.: Service innovation: a service-dominant logic perspective. MIS Q. **39**, 155–175 (2015)
26. Paskaleva, K., Evans, J., Martin, C., Linjordet, T., Yang, D., Karvonen, A.: Data governance in the sustainable smart city. In: Informatics, p. 41. Multidisciplinary Digital Publishing Institute (2017)
27. Lim, C., Maglio, P.P.: Data-driven understanding of smart service systems through text mining. Serv. Sci. **10**, 154–180 (2018)

Fostering Human Centred Design in Public Service Development

Stephan Raimer[1](✉) and Peter Weiß[2]

[1] University of Applied Sciences for Administration and Service, Rehmkamp 10, 24161 Altenholz, Germany
raimer@fhvd-sh.de
[2] Pforzheim University, Tiefenbronnerstraße 65, 75175 Pforzheim, Germany
peter.weiss@hs-pforzheim.de

Abstract. The German Online Access Act presents a disruption in the way public services and service ecosystems are conceived, planned and developed. In a longitudinal case study approach, we aim to analyse, how the OOA is implemented so far and what can be learned towards conceptualizing service innovation with a regard and connection to human centred design. Our research focus is, how a Service-Dominant logic approach can contribute to innovate public service delivery through service systems design.

Keywords: Service innovation · Service logic · Value co-creation · Public service · Human-centered service design

1 Introduction

New emerging digital technologies are transforming all industries. More and more governments and public administrations are facing dramatic change. Citizens explore new ways to interact and communicate with government and public administrations as they are used from digital consumer offerings such as Amazon, AirBNB, Uber, etc. From this perspective, digital transformation redefines the role of public services and how they need to be delivered to citizens. Innovations in public service requires to systematically develop new service and offerings, through reconfigurations of re-sources that are linked together in new patterns and new logic [1]. This necessitates to elaborate a clear understanding of citizens needs and expectations. As consequence, human centered design plays a salient role in developing innovative public services. From a public service delivery point of view and in the light of service logic theory [2, 3], an important question is how the co-creation of (public) value is considered. Service-Dominant (S-D) logic [4] can offer guidance to concretize what capabilities will be required to respond to new behavioral patterns and expectations of actors (with focus on new preferences and behaviors of citizens). In this article, we offer a contribution in analyzing a large-scale case study through the service innovation "lense" presented by [4, 5]. Moreover, we want to contribute to the distinction, how the concepts of value creation and resource integration can be seen in a public service context compared to a for-profit context [2, 4,

C. Leitner et al. (Eds.): AHFE 2021, LNNS 266, pp. 265–272, 2021.
https://doi.org/10.1007/978-3-030-80840-2_31

5]. The paper is structured as follows: first, we present and explain our research methodology. Next, we briefly overview the theoretical foundations which are relevant and help us to set up and analyze our case, and finally discuss findings leading to our conclusions.

2 Methodology and Approach

The methodology and approach of this paper is to analyze the experienced gained in implementing the German Online Access Act from 2017. The vision of the OAA is that by the end of 2022 and over all levels of government (federal, state and local level) public administration will be easier to reach: central government portals will offer user-friendly digital forms to apply for (more than 7000) services and benefits. Private citizens and companies were assured that they can expect fast and convenient online services that reduce bureaucratic burdens and costs. Furthermore, there is also a European dimension with the goal to implement the EU Regulation to establish a Single Digital Gateway, hand-in-hand with the Online Access Act. Our research investigates based on a longitudinal case study approach how Online Access Act is being implemented so far using the example of one of the German federal states and the publications on the national level. Research approach and methodology is eclectic and combines Design Science Research (DSR) [6] and Information Systems Development [7]. In addition, we further apply a longitudinal, multi-case study approach [9: 65] to analyze and elicit lessons learned and observations from respective real-life Service Design projects in selected public administrations and agencies. A special focus is laid on the three elements service ecosystem, service platform and value cocreation together with proposed research implications by [6].

3 Public Service Innovation

The public sector and its organizations are currently facing dramatic change are subject to fundamental transformations [9–12]. In essence, public service organizations (PSOs) are claimed "[...] tend to function in an inward-oriented manner" [2: 787]. Often PSOs are perceived to be less "[...] user-focused and service-oriented as private service organisations" [2: 787]. [2: 787] argues further that "[...] although some financial investments probably are needed in the process, mostly it is a matter of having visionary people with an understanding of what it takes to be a service-oriented PSO and the courage and strength to implement their visions". To transform PSOs two fields of activities can be identified: (1) strengthening a service perspective to strive for service innovation and new value propositions and (2) building respective capabilities to execute digital strategies and offerings (building the foundation of execution) [13]. One key activity is to match organization's capabilities to future needs of envisioned strategic initiatives. However, it is difficult to make informed decisions, due to the uncertainties about future societal developments, and due to the dynamics and complexities in the organizations' environment, which are mainly fueled by the advancements of digital technologies. S-D logic offers explanations and insights concerning principles and puts emphasis on respective mechanisms at work. Hence, S-D logic allows to explore respective value creation processes and activities, including social mechanisms and service experience

[14]. Of particular relevance for PSOs are institutional arrangements [15, 16] which comprise regulative, normative and cultural-cognitive elements in the context of PSOs value cocreation activities, actors and resources [12: 56]. This includes as well published standards and regulations, which serve as prescriptive documents to which service systems need to comply. In addition, S-D logic is able to contribute to implement human-centered design approaches in PSOs and digital transformation activities, as it offers relevant concepts concerning a multilevel design perspective of service systems and value cocreation (macro level) as well as actors' engagement systems and service platforms (meso level) and actor engagement (micro). Meso level transformational mechanisms (micro-macro) allow to understand and design properties of service systems and to stabilize emerging resource integration patterns as result of innovative value proposition and new ways of value cocreation [12, 16].

4 The Case of the German Online Access Act

The publication the German Online Access Act in 2017 was intended to achieve great and ambitious goals until the end of 2022. Germany was and still is not be found in the top places of the eGovernment Benchmarks in an EU comparison [18]. Therefore, the Tallin Declaration on eGovernment from 2017 and following initiatives have acquired a special importance: "By 2020, public administrations and public institutions in the European Union should be open, efficient and inclusive, providing borderless, personalized, and user-friendly, end-to-end digital public services to all citizens and businesses in the EU. Innovative approaches are used to design and deliver better services in line with the needs and demands of citizens and businesses. Public administrations use the opportunities offered by the new digital environment to facilitate their interactions with stakeholders and with each other." (Vision of the eGovernment Action Plan 2016–2020) [18–22]. Two digitalization programs for implementing the Online Access Act (OAA) were consequently planned and started in Germany. The first focusing on all services and benefits provided at federal level (Federal Digitalization Program in Germany) and the second for services provided by the state and local governments (Digitalization Program Federal States in Germany). The strategic challenges for those two programs were (and still are) manifold. For example, digitalization only for the frontend. The OAA requires federal, state and local governments to provide their services for private citizens and companies also in digital form via their own portals by the end of 2022. Not within the scope of the digitalization programs are the administrative backends, i.e. organizational change and development, required new workflows or software solutions (see Fig. 1). There are further field of actions and problem areas that are available but will not presented in the scope of this paper. Implementation complexity and lack of standards for infrastructure, platforms and interfaces. The previously mentioned distributed responsibility is also evident in the IT solutions that are operated at the different federal levels. The target is set to integrate the services into central, customer facing platforms (portals for the federal, state as well as the municipal level) with user accounts. The implementation is still under development. In addition, a service standard was published in July 2020 in order to define holistic quality criteria and principles as a guiding framework. With regard to human-centred design, the involvement of end users was realised quite differently. Around 50 digitisation labs were conducted for high-priority OAA services in the thematic areas. The digitisation labs were geared towards simplifying the mostly complex processes and adapting them more to the needs of citizens, businesses and employees in the administrations.

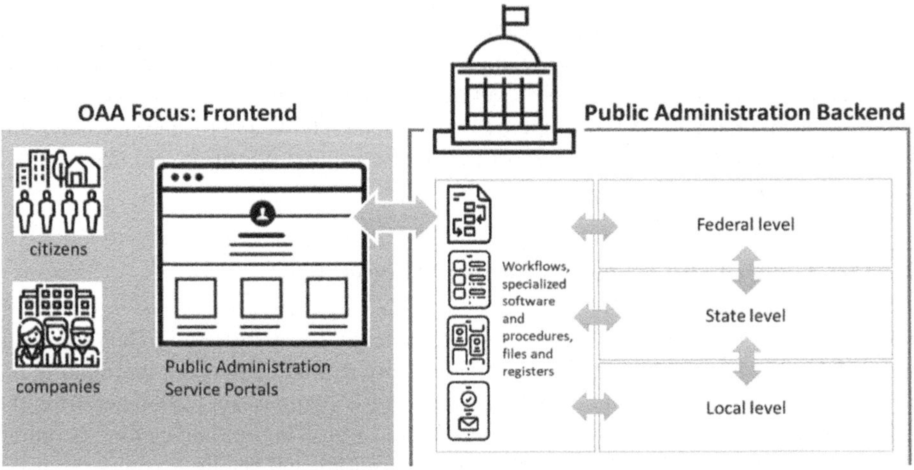

Fig. 1. Implementing the online access act (OAA) - frontend focus

An assessment of the situation at the end of 2020 for the OOA implementation in Schleswig-Holstein shows, that this large-scale digitisation project is also a change project. Realising the state service portal and first services (in December 2019) was pioneering work within the public service organisations. The progress is strongly dependent on the availability of the professional administrative contact persons and a high speed of implementation of the number of services has not yet been achieved.

5 Discussion and Findings

Looking at the efforts to implement the OOA in Germany –in a highly generalized way – it can be stated that a first step away from a "document-centered world" or the administrative equivalent of a G-D Logic (a "Fordist models of public service delivery" [9]) was taken. This is particularly evident in the structure of the OOA implementation catalogue, organizing services by life situations and business situations and not by logic and structure of the administrative bodies. Furthermore, it can be stated that the OOA is the first national administrative project that explicitly embraces and applies human-centered design activities (digitalization labs, prototyping and testing) and defines service standards with usability benchmarks. A clear limitation, however, is the sole focus on the front end for citizens and businesses. The OOA is creating great pressure for PSOs to reorganize their service provision as a result of the digital transformation. In application of [5] we can try to summarize the afore described project with the central theme and findings (Table 1). In order to approach the aspect of value cocreation, we follow the argumentation of [9] that PSOs only provide a public service offering. Resulting research design and framework is elaborated by adapting and extending the value creation sphere concept and approach of [3] for PSOs.

Table 1. Case analysis: conceptualization of service innovation applying the tripartite framework of [23].

Central theme	Case findings
Service Ecosystem (S-D Logic: Actor-to-Actor Network)	The Service Ecosystem which was realized based on the German OAA exists in two dimensions: On the horizontal layer, we have frontend web portal for citizens and business stakeholders with a coupled backend consisting of connected public service organizations (whose processes and procedures are surprisingly NOT in the scope of the OAA) The vertical layer is built through interconnected and interoperable portals adapted to the German federal structure with local, state and national level. Legal basis regulates the responsibility and thus level in which a service is made available Technical protocol standards for data transfer between platforms (OSCI and XTA) and the nature of cooperation often platforms are being further developed. So far standard elements such as user accounts (for citizens and businesses), a secure mailbox and payment interfaces have been defined for the service portals
Service Platform (S-D Logic: Resource liquefaction; resource density)	A newly developed and published web service platform for the state level of Schleswig-Holstein is at the core of the case described. Beforehand, public services could only be identified and accessed through on-site inquiries or appointments at the PSO, a central service number or a web-based authority finder service. In accordance with the implementation catalogue, services will be gradually developed and made available on the (frontend) service platform. The extent to which the interaction between frontend and backend is efficient, effective and satisfactory from the user's point of view cannot be assessed conclusively at the present time

(*continued*)

Table 1. (*continued*)

Central theme	Case findings
Value Cocreation (S-D Logic: Resource integration)	The possibility to use public services of administrative bodies mediated by IT (a web service platform) is in itself an innovation. Resource integration happens for users on the frontend side but up to now there are only few studies that further differentiate and research value cocreation in context of the OAA One can assume with some certainty, that user would mention similar aspects like in e-Business settings (ease-of-use, saving time)

For the consideration of the intersection, depicted as Joint Sphere, the value approach by [9] can be used: "It is the citizen and/or service user who creates the performance and value of a public service, with the PSO acting as a facilitator of this process. S/he does this by integrating the service offering of the PSO with their needs, personal abilities and experiences, and their societal context." If we want to pursue this actor perspective (i.e. a human-centered approach) in the design of service systems, we can draw very well on findings in the form of process models and evaluation models for the design of information systems (hardware/software). The international standard family ISO 9241 [24, 25] has been under development for about 30 years and addresses various aspects of ergonomics in human-computer interaction. Moreover, we can analyze the type and quality of interaction (value creation in interaction) through the lens of usability. ISO defines usability as "The extent to which a product can be used by specified users to achieve specified goals with effectiveness, efficiency, and satisfaction in a specified context of use." For the concrete investigation of usability, established usability test methods can still be used, which provide benchmarks such as error frequency, time requirement or overall user satisfaction.

6 Conclusion and Outlook

This paper investigates based on a longitudinal case study of PSOs in Germany how the Online Access Act is being implemented so far using the example of one of the German federal states and the publications on the national level. Public Service Innovation can be seen as a permanent task that has gained particular momentum in recent years for the EU, it is a race against the rapid development of new innovative digital technologies that determine citizens' everyday life (in the B2C context). There are strong efforts to apply innovations that are available in the economy (e.g. blockchain and the distributed ledgers, Artificial Intelligence, Big Data Analytics etc.) in order to take advantage of the benefits of the technologies such as scalability or 24/7 availability. This is ongoing research. At the current stage, the research design and theory development are prioritized to ensure that gained insights and lessons learned can be generalized. Hence, we elaborate on research questions and development of our conceptual framework [26: 37–38]. Our research design embraces the proposition of [5] by applying the tripartite framework in our case studies to conceptualize public service innovation [5]. We strive further for deeper understanding of social mechanism at play implementing service systems and value cocreation in public service by studying and exploring micro-macro-transitions [15, 16] and transformational mechanisms to implement and operate service systems [5] to engage actors.

References

1. Normann, R.: Reframing Business. When the Map Changes the Landscape. Wiley, Chichester (2001)
2. Grönroos, C.: Reforming public services: does service logic have anything to offer? Public Manag. Rev. **21**(5), 775–788 (2019)
3. Grönroos, C., Voima, P.: Critical service logic: making sense of value creation and co-creation. J. Acad. Mark. Sci. **41**(2), 133–150 (2013)
4. Lusch, S.L., Vargo, R.F.: Service-Dominant Logic – Premises, Perspectives, Possibilities. Cambridge University Press, Cambridge (2014)
5. Lusch, R.F., Nambisan, S.: Service innovation: a service-dominant logic perspective. MIS Q. **39**(1), 155–175 (2015)
6. Hevner, A.R., March, S.T., Park, J., Ram, S.: Design science in information systems research. MIS Q. **28**(1), 75–105 (2004)
7. Nunamaker, J.F., Chen, M., Purdin, T.D.M.: Systems development in information systems research. J. Manag. Inf. Syst. **7**(3), 89–106 (1990)
8. Yin, R.K.: Case Study Research: Design and Methods, 5th edn. SAGE Publications, Los Angeles (2014)
9. Osborne, S.P.: From public service-dominant logic to public service logic: are public service organizations capable of co-production and value co-creation? Public Manag. Rev. **20**(2), 225–231 (2018)
10. Price, R., Brodie, R.J.: Transforming a public service organization from inside out to outside in: the case of Auckland City, New Zealand. J. Serv. Res. **4**(1), 50–59 (2001)
11. Alford, J.: Co-production, Interdependence and publicness: extending public service-dominant logic. Public Manag. Rev. **18**(5), 673–691 (2016)
12. Petrescu, M.: From marketing to public value: towards a theory of public service ecosystems. Public Manag. Rev. **21**(11), 1733–1752 (2019)
13. Weiß, P., Zolnowski, A., Warg, M., Schuster, T.: Service dominant architecture: conceptualizing the foundation for execution of digital strategies based on S-D logic. In: Proceedings of the 51st Hawaii International Conference on System Sciences (HICSS), Kona, Hawaii, January, pp.1630–1639 (2018)
14. Prahalad, C.K., Ramaswamy, V.: Co-creation experiences: the next practice in value creation. J. Interact. Mark. **18**(3), 5–14 (2004)
15. Scott, W.R.: Institutions and organizations. Ideas, interests and identities. Fourth edition. Sage Los Angeles, London, New Delhi, Singapore, Washington DC (2014)
16. Hedström, P., Swedberg, R. (eds.): Social Mechanisms. An Analytical Approach to Social Theory. Transferred to digital printing. Cambridge Univ. Press (Studies in rationality and social change), Cambridge (2006)
17. European Commission, eGovernment Benchmark (2017). https://op.europa.eu/en/publication-detail/-/publication/7f1b4ecb-f9a7-11e7-b8f5-01aa75ed71a1/language-en. Accessed 29 Jan 2021

18. Federal Ministry of Justice and Consumer Protection, Germany, in German (Bundesministerium für Justiz und Verbraucherschutz). http://service.juris.de/ozg/BJNR313800017.html. Accessed 29 Jan 2021
19. Federal Ministry for Internal Affairs (Bundesministerium des Inneren, für Bau und Heimat), April 2018. https://www.it-planungsrat.de. Accessed 29 Jan 2021. (in German)
20. Federal Ministry for Internal Affairs (Bundesminsterium des Inneren, für Bau und Heimat), OZG-Information platform. https://informationsplattform.ozg-umsetzung.de/. Accessed 29 Jan 2021. (in German)
21. Federal Ministry for Internal Affairs (Bundesministerium des Inneren, für Bau und Heimat), Integration Guidelines (Integrationsleitfaden Bund (Version 1.5), December 2020. https://www.onlinezugangsgesetz.de/. Accessed 29 Jan 2021
22. Guidelines for OZG Implementation (Leitfaden OZG-Umsetzung). https://leitfaden.ozg-umsetzung.de/. Accessed 29 Jan 2021. (in German)
23. Federal Ministry for Internal Affairs: Maturity Model (Reifegradmodell). https://www.onlinezugangsgesetz.de/Webs/OZG/DE/grundlagen/info-ozg/info-reifegradmodell/info-reifegradmodell-node.html. Accessed 29 Jan 2021. (in German)
24. ISO 9241-1:1992 International Organization for Standardization, published 1992. https://www.iso.org/standard/16873.html. Accessed 29 Jan 2021
25. ISO 9241-11 International Organization for Standardization. https://www.iso.org/obp/ui/#iso:std:iso:9241:-11:ed-2:v1:en. Accessed 29 Jan 2021
26. Talin Ministerial Declaration on eGovernment, 6 October 2017. http://ec.europa.eu/newsroom/document.cfm?doc_id=47559. Accessed 29 Jan 2021

Digital Service Innovation and Actor Engagement: A Multilevel Design Perspective - Impacts from a Case Study of an Insurtech

Peter Weiß[1(✉)], Anselm Kronibus[2], Fabian Riedel[1], and Roman Rittweger[2]

[1] Pforzheim University, Tiefenbronnerstraße 65, 75175 Pforzheim, Germany
{Peter.Weiss,Fabian.Riedel}@hs-pforzheim.de
[2] Ottonova Services GmbH, Ottostraße 4, 80333 Munich, Germany
{Anselm.Kronibus,Roman.Rittweger}@ottonova.de

Abstract. In the following we present a multilevel design perspective to analyze the case of a fully digitized health insurance company in Germany which is striving to innovate by (re)designing practices to institutionalize new health services and solutions. To nurture digital service innovation companies are required to strengthen and expand design knowledge concerning actor engagement and related systems of engagement. Presented research strives for deeper understanding of the microfoundations of value co-creation, namely actor engagement, for the emergence of service innovations from a theoretical and practical point of view. This research is conducted in context of evolving Service Dominant Architecture (SDA) as implementable IT artifact and conceptual framework.

Keywords: Service innovation · Value co-creation · Resource orchestration · Service dominant architecture · Human-centered service design

1 Introduction

Digital service innovations are driven by universal connectivity [1: 5]. Hence, Systems of Engagement (SoE) can be seen as key element of S-D logic [2] informed digital transformation strategies leveraging actor engagement and value cocreation [3, 4]. Presented results are expected to yield deeper understanding and insights from a theoretical and practical point of view with regard to applied social mechanisms of value co-creation, in particular of actor engagement. Actor engagement needs to be analyzed applying a multilevel design perspective to study respective transformational social mechanisms nurturing emerging properties of service systems. Transformation mechanism are of interest because they are discussed in context of institutional arrangements and structural support leading to new actor behaviors, value constellations and market structures. Hence, we argue that design activities in the context of service systems require deeper understanding of how those mechanisms contribute to innovative configurations of actors and resources. In this context, emerging resource integration pattern need to be stabilized, reproducible, internalized and then supported by existing systems and structures

C. Leitner et al. (Eds.): AHFE 2021, LNNS 266, pp. 273–279, 2021.
https://doi.org/10.1007/978-3-030-80840-2_32

[5: 14]. This is seen as valuable and insightful concerning broader understanding of digital service innovation. In consequence, in the remainder we address digital transformation challenges in the health care sector. The paper is structured as follows: first, our research methodology and approach are described. Next, we briefly overview the theoretical foundations of actor engagement and present our research approach. Then, we describe results and insights yielding from analyzing a real-life case of an insurtech company in Germany. After a brief discussion of the results, we highlight major points in a summary and end with concluding remarks and an outline of the next research steps to be taken.

2 Methodology and Approach

As methodology this research uses a longitudinal single case study approach [6: 15] in combination with developing and applying a multilevel design framework for service systems [1, 5, 7]. This research addresses key challenges of implementing value cocreation concept in a real-life context by striving for deeper understanding of respective design elements and mechanisms at work [3, 8]. Collected practitioner feedback [6, 9, 10] based on use cases is fed back into the theorizing process of SDA [11, 12]. Followed research design and methodology is eclectic and grounds on various disciplines and related practices. Our research embraces IS development as research approach [13]. In addition, we incorporate elements and requirements of Design Science Research (DSR) which show significant relevance to achieve our research objectives [6]. Furthermore, we embrace a multi-level design perspective [5, 14] with a clear focus on the microfoundation perspective. Microfoundations are seen to bridge the existing gap between macro level research and micro level research on engagement of actors and individuals [7].

3 Digital Service Innovation and Actor Engagement

Subsequently, we argue that building systems of engagement are crucial to overcome the challenges of digital transformation [4]. They require systematic development and a systematic design of value propositions and service systems. Engagement properties and resource integration patterns can be identified as major design elements which needs to be reflected in the IT artifact, namely the service platform [7: 3014, 3, 8].

3.1 Microfoundations for Actor Engagement

Value cocreation involves generic actors integrating resources coordinated through service exchange within actor-generated institutions nested and interlocked in service ecosystems [7: 3008-9]. According to [15: 21–23] value cocreation and related social mechanisms (situational, action-formation and transformational) have to be analyzed and designed on three distinct levels, which reflects our research design. [16: 4–5] list nine systemic principles derived from synergetics which grant deeper conceptual understanding and can be applied for analysis and design. Resource integration patterns (meso level) are transformational mechanism at work and are described as combination of the elements: (1) engagement platform, and (2) engagement properties, and various activities [7: 3013]. Heteropathic resource integration patterns (mechanisms) [17: 3002] are of particular interest because they are able to explain and drive emergence [18: 25] of new actor behaviors and properties, value constellations and markets [7: 3015]. However, without supporting structure in place the effect is not reproducible and to be stabilized [17: 3002] (see Fig. 1). [19: 16] motivate to develop a better understanding how actors determine values upon which to build value propositions. In particular, the relationship to institutional arrangements and "institutional entrepreneurship" or deinstitutionalization processes is of relevance. Figure 1 outlines our research design and theoretical framework on microfoundations for actor engagement as foundation to discuss and evaluate the relevance and broader applicability of identified mechanism, mainly observations and lessons learned from presented case [6: 40–41]. Of particular interest are transformation mechanisms at the meso-level where "[…] the engagement behaviors of individuals linking in with the engagement practices of social collectives" [20: 178]. Hence, engagement platforms and related system properties need to be in focus of respective design activities and conceptualizations of actor engagement.

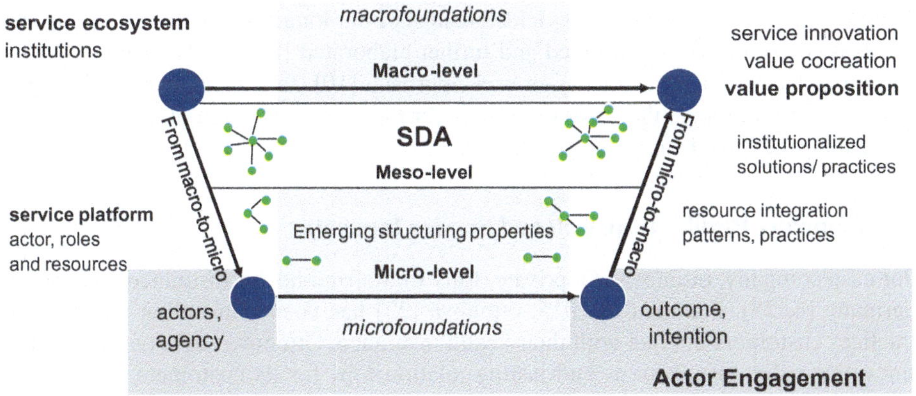

Fig. 1. SDA and institutions: microfoundations for actor engagement [3] (modified from [5, 7, 15]).

3.2 Building Engagement Platforms

Engagement platforms are defined as "multisided intermediaries that actors leverage to engage with other actors to integrate resources" [7: 3011] (see as well [21: 238, 22: 166–167]). [7: 3011-12] argue three particular areas of interest for further research: (1) platform modality, (2) roles of platforms vs. actors, and (3) platforms and engagement-related network effects. Engagement platforms are nested in service platforms [22: 166]. Design of service systems comprise to "develop the mechanism and configurations of actors and resources that leads to an effective resource integration pattern" [5: 14]. [5: 5] propose an iterative process model which is oriented towards requirements of Design Science Research [9] and consists of two cycles which link two design areas, namely (1) institutional set-up and (2) actor engagement [5: 14]. Each cycle foresees to support in systematic way (plan, build, operate, validate) relevant design activities and decisions. First cycle (1) institutional set-up and design activities aim at enabling appropriate arrangements and commitments of actors and resources to achieve effective resource integration patterns. The goal is to enable the required level of resource mobilization and density [5: 6]. During the second cycle (2) engagement design activities aim at building design artefacts such as engagement platform and engagement properties [5: 6]. At both levels design decisions concretize necessary actor and resource requirements and configurations (resource mobilization and resource density). Gained insights are used to shape and package appealing value propositions that achieve the expected level of intensity to stimulate emergence of resource integration patterns [19: 12–13, 5: 8]. In the next section, we present the identified and selected real life case of an insurtech company and how it applies value co-creation to develop service innovations by means of its service and engagement platform.

4 Case Study Insurtech

Subsequently, gained insights and yielded results of our longitudinal case study [6] with the insurtech ottonova is continued and further elaborated [8, 23, 24]. In particular, in the remainder we follow a "theory-in-use" approach [10] (to gather inputs and insights with regard to midrange theory development for the practical design of service systems in a real business environment).

4.1 Company Profile: Insurtech and Service Innovator

Our case company, ottonova is a private, fully digitalized health insurance company in Germany [8, 23]. Founded in 2015, ottonova [23] has been reinventing the way and practices customers interact with their health insurance. Ottonova strives for cocreating innovative value propositions and lasting relationships for its customers. IT plays an essential role in realizing respective co-creation environments and experiences [8: 10]. Ottonova designs fully digitally-enabled service systems using its technology platform with a user-friendly front-end mobile application. Hence, the company has gathered profound knowledge and competence how to engage with its customers and partners. Ottonova continuously expands and maintains its service ecosystem which enhances its capability to mobilize resources and to cocreate new resource linkages in given contexts for its customers. ottonova strives to shape and design service innovations for its customers building on value cocreation and actor engagement.

4.2 New Value Proposition: Corona Travel-App

The following use case "Corona Travel-App" is subject of further analysis. Based on the use case we underline the salient role of actor engagement platforms and understanding respective properties and patterns. Two particular design elements are in focus, namely resource integration patterns and actor engagement properties. We strive for building respective design knowledge concerning design and operation of service systems following a multi-level design perspective. In the remainder, our focus is on actor engagement systems and evaluating the applicability and relevance of available theory, with focus on practicability of selected frameworks and related concepts. In May 2020, the company initiated and kicked off a new service innovation development project (see Fig. 2). Due to the growing pandemic crisis, ottonova and its partners are thinking of ways to support their customers and the wider community through this difficult time. The idea of a Corona Travel App (see Fig. 2) is spearheaded by Peter Bauer, CIO of ottonova. He is bringing in two partners and is developing the concept around this solution.

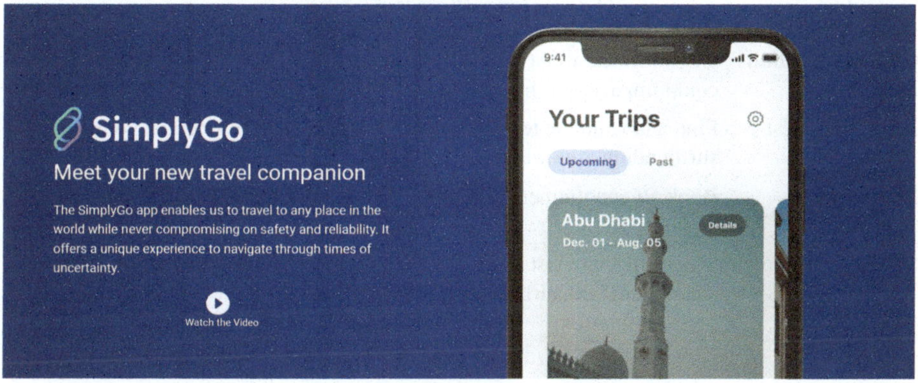

Fig. 2. Establishing a new digital service innovation: Corona Travel-App [24]

It follows a press release from ottonova, inHealth and Nortal that they are working on an app to enable traveling by providing information on entry requirements and a platform to upload test results. In June 2020, several employees of ottonova, Nortal and inHealth join the team to develop the concept and map out the different use cases. To facilitate actor engagement various options are validated, such as using the established health X Service [8]. The health X Service has been launched in 2020 and offers a reward system which engages with customers in an unprecedented way. Its central aim is to yield deeper customer insights concerning customers' preferences and needs.

5 Discussion and Findings

Digital transformation challenges incumbent organizations through "frame-breaking reconfigurations" [14: 242] which require incumbent organizations to rethink their value creation activities and strive for new value constellations. Engagement management capabilities are seen as vital to engage in "market-shaping" activities to generate market innovations [1: 7]. Firms which intend to shape markets through inventing new innovative service systems require a systemic view [16, 17] to be able to grasp opportunities for actor engagement [1: 7, 5]. The case study methodology has been chosen because we want to illuminate decisions made and strive for deeper understanding of the contextual setting for the design of service systems, in particular setting up engagement platforms.

In the scope of this paper we primarily focus on preparing and sharpening the research design [6: 29] through elaborating and explicating the theoretical framework. Applied method is expected to provide insights into mechanisms at play and to elicit generalizable findings and lessons learned concerning actor engagement concept [6: 40–41] (Table 1).

Table 1. Corona travel-app: value proposition and use case development [24].

Use case/service	Description
Travel planning	Travel with peace of mind: Plan your trip and view the most important entry test requirements for your destination
Corona regulations	Get informed about latest corona related behavior rules and news which could impact your trip
Find corona test	Find a lab and get tested: Quickly find medical providers in your surrounding area who offer required tests
Booking test	Book an appointment, conduct a test and receive the results directly via app
Share test results	View and share test results: Access your test results in your health record, share them with airlines and immigrations to receive permission to enter your destination

6 Conclusion and Outlook

Management of actor engagement is seen as a key dynamic capability for companies to cope with the challenges of digital transformation. This research aims to expand the knowledge base and theoretical foundations of SDA (Service Dominant Architecture) [11, 12]. SDA offers practitioners a framework and design environment to design and operate service systems and system of engagement [18]. Furthermore, it stimulates collaborative theorizing processes by involving decision makers, managers and practitioners in general as active participants in the research process [10: 418-19]. Our focus is on evolving and applying our framework and IT artifact SDA [11, 12]. SDA allows to build systems of engagement in terms of implementable and operable solution designs and IT artifacts.

References

1. Storbacka, K.: Actor engagement, value creation and market innovation. Ind. Mark. Manag. **80**, 4–10 (2019)
2. Lusch, S.L., Vargo, R.F.: Service-Dominant Logic – Premises, Perspectives, Possibilities. Cambridge University Press, Cambridge (2014)
3. Weiss, P.: Microfoundations for building systems of engagement: enable actor engagement using service dominant architecture. In: Spohrer, J., Leitner, C. (eds.) AHFE 2020. AISC, vol. 1208, pp. 116–122. Springer, Cham (2020). https://doi.org/10.1007/978-3-030-51057-2_17

4. Weiß, P., Warg, M., Zolnowski, A.: Building Systems of Engagement to Overcome the Challenges of Digital Transformation. Naples Service Forum 2019, Ischia/Naples, Italy, 4–7 June 2019
5. Grotherr, C., Semmann, M., Böhmann, T.: Using microfoundations of value co-creation to guide service systems design. a multilevel design framework. In: Proceedings of Thirty ninth International Conference on Information Systems, San Francisco, USA (2018)
6. Yin, R.K.: Case Study Research: Design and Methods, 5th edn. SAGE Publications, Los Angeles (2014)
7. Storbacka, K., Brodie, R.J., Böhmann, T., Maglio, P.P., Nenonen, S.: Actor engagement as a microfoundation for value co-creation. J. Bus. Res. **69**, 3008–3017 (2016)
8. Rittweger, R., Kronibus, A., Weiß, P.: Value co-creation as the core of service innovation: impacts of a case study of a fully digitized health insurance company. In: Spoh-rer, J., Leitner, C. (eds.) Advances in the Human Side of Service Engineering, pp. 86–92. Springer, Cham (2020)
9. Hevner, A.R., March, S.T., Park, J., Ram, S.: Design science in information systems research. MIS Q. **28**(1), 75–105 (2004)
10. Brodie, R.J., Peters, L.D.: New directions for service research: refreshing the process of theorizing to increase contribution. JSM **34**(3), 415–428 (2020)
11. Warg, M., Weiß, P., Zolnowski, A., Engel, R.: Service dominant architecture based on S-D logic for mastering digital transformation: the case of an insurance company. In: RESER Conference Proceedings, Naples, Italy, September 2016
12. Weiß, P., Zolnowski, A., Warg, M., Schuster, T.: Service dominant architecture: conceptualizing the foundation for execution of digital strategies based on S-D logic. In: Proceedings of the 51st Hawaii International Conference on System Sciences, Hawaii, January 2018, pp. 1630–1639 (2018)
13. Nunamaker, J.F., Chen, M., Purdin, T.D.M.: Systems development in information systems research. J. Manag. Inf. Syst. **7**(3), 89–106 (1990)
14. Normann, R.: Reframing Business: When the Map Changes the Landscape. Wiley, Chichester (2001)
15. Hedström, P., Swedberg, R. (eds.): Social Mechanisms. An Analytical Approach to Social Theory. Transferred to digital printing. Cambridge University Press (Studies in Rationality and Social Change), Cambridge (2006)
16. Meynhardt, T., Chandler, J.D., Strathoff, P.: Systemic principles of value co-creation: synergetics of value and service ecosystems. J. Bus. Res. **69**(8), 2981–2989 (2016)
17. Peters, L.D.: Heteropathic versus homopathic resource integration and value co-creation in service ecosystems. J. Bus. Res. **69**(8), 2999–3007 (2016)
18. Giddens, A.: The Constitution of Society. Polity Press, Cambridge (1984)
19. Chandler, J.D., Lusch, R.F.: Service systems: a broadened framework and research agenda on value propositions, engagement, and service experience. J. Serv. Res. **18**(1), 6–22 (2015)
20. Brodie, R.J., Fehrer, J.A., Jaakkola, E., Conduit, J.: Actor engagement in networks: defining the conceptual domain. J. Serv. Res. **22**(2), 173–188 (2019)
21. Chandler, J.D.: The future of service systems: from synergetics to multi-sided platforms. In: Maglio, P.P., Kieliszewski, C.A., Spohrer, J.C., Lyons, K., Patrício, L., Sawatani, Y. (eds.) Handbook of Service Science, Volume II. SSRISE, pp. 237–247. Springer, Cham (2019). https://doi.org/10.1007/978-3-319-98512-1_11
22. Lusch, R.F., Nambisan, S.: Service innovation: a service-dominant logic perspective. MIS Q. **39**(1), 155–175 (2015)
23. ottonova healthcare services. http://www.ottonova.de. Accessed 27 Jan 2021
24. ottonova's new Corona Travel App. https://ion-ah.com. Accessed 27 Jan 2021

HR-Management: Impacts from Service (Eco) Systems

Markus Frosch[1](✉), Markus Warg[2], and Maren Lange[3]

[1] Projekt 3T, Frankfurt, Germany
markus.frosch@projekt-3t.com
[2] Institut Für Service Design, University of Applied Sciences, Wedel, Hamburg, Germany
markus.warg@ifsd.hamburg
[3] Fresenius–University of Applied Sciences, Hamburg, Germany
mail@dr-maren-lange.de

Abstract. This paper demonstrates that human resource management (HRM) literature is moving towards a service science perspective in which value creation unfolds over time and is embedded in broader social systems [1, 2]. To contribute to this grounding, the authors draw on service science and service-dominant logic to promote a view of service (eco) systems and provide further theoretical foundation for hr and HRM [3, 4]. The purpose of this paper is to derive implications for the foundation, understanding and management of hr and HRM in terms of service exchange and value co-creation.

Keywords: HR management · Value cocreation · Foundation of HR concepts · Service exchange · Service (eco) systems

1 Introduction

In HR literature the term 'human resources' is controversial argued because it 'reduces people to the same category of value as materials and technology'. Many texts and dictionaries define 'human resources' as the people employed in an organization, a notion that obscures the meaning of human resources and that critics see as impugning the dignity and denying the independence of human agents. Therefore in recent HR literature HR is more accurately understood as the resources that are intrinsic to human beings, which they can apply to the various tasks of life [1, 5–7]. A comparably controversial discussion has already been held in marketing, service-dominant logic and service science. As a consequence, the traditional division between goods and services is long outdated. Customers do not buy goods or services; they buy offerings which render services which create value. The shift in focus to services is a shift from the means and the producer perspective to the utilization and the customer perspective [3, 4, 8, 9]. By adapting the perspectives of service science, service-dominant logic and institutional economics and introducing the systemic functioning of value cocreation, the existing core concepts of human resources are re-examined and impulses for further development are derived.

C. Leitner et al. (Eds.): AHFE 2021, LNNS 266, pp. 280–291, 2021.
https://doi.org/10.1007/978-3-030-80840-2_33

Therefore, the research question of our work is: What impacts results for the core concepts of human resources management from the adaptation of the perspectives of service science and service-dominant logic?

2 Research Methodology

The research methodology of this paper refers to the goal to derive implications for the foundation, understanding and management of key concepts of HR by linking with the theories and concepts of service science, service-dominant logic and institutional economics. Therefore we apply the research methodology of a conceptual paper for bridging existing theories and broaden the scope of thinking [10–12]. Starting point of this paper is the problemizing of the definition of HR and the resulting relation between employee and employer in current discussions of HR literature. By questioning the definition of HR also key concepts within HR and HR management are put into question. Examples are HR strategies, HR development or HR management.

As a consequence of our research objective to amend the foundation, understanding and management of key concepts of HR by using other theories, we focus on the research design of "theory adaptation". By introducing perspectives of service science, service-dominant logic and institutional economics a contribution for revising the existent knowledge within given key concepts of HR is intended. For this purpose, analogies in the systemic functioning of HR on the one hand and service science, service-dominant logic and institutional economics on the other hand will be developed.

3 Problemizing the Definition and Concepts of HR

"Many texts and dictionaries mistakenly define 'human resources" (HR) as the people employed in an organization, a notion that obscures the meaning of human resources and that critics see as impugning the dignity and denying the independence of human agents" [5]. This definition of HR is controversial argued because it 'reduces people to the same category of value as materials and technology. It can be regarded as a philosophy about the ways in which people are managed at work [5].

Furthermore, this understanding of HR is then transferred to the core concepts of HR and the facets of the relationship between employer and employee. The concepts transfer the understanding that the employee is unilaterally managed by the employer: "HR management is concerned with choices – choices that organizations make from a wide variety of possible policies, practices and structures for managing employees" [7, 13]. Another example is HR development, like for example: "a career is not something that should be left to each employee: instead it should be managed by the organization to ensure the efficient allocation of human and capital resources" [14]. These concepts, which are based on the traditional manufacturing understanding that value is embedded in goods and wealth is created by the acquisition of the good are reflected in HR management: the employee are people acquired, employed, managed and developed in an organization [4, 6, 15]. Even though approaches such as human capital management recognized the value of employees intellectual abilities (skills, knowledge, qualifications) the basic understanding of managing, controlling and allocating the resource 'humans' remained unchanged [16]. And thus contributed to the fundamental questioning of the usefulness of HR units and HR management expressed by "serious and widespread doubts about HR's contribution to organizational performance" [17] and statements like: "HR was not seen as adding strategic value" [18], "HR's activities appear to be - and often are - disconnected from the real work of an organization" [18], or "if we really took human capital seriously, we'd run companies in a very different way" [16].

In the recent HR literature, on the other hand, a different understanding of HR has prevailed „human resources are more accurately understood as the resources that are intrinsic to human beings, which they can apply to the various tasks of life, both in workplaces and in activities such as relationship building, child raising and community service. "Most obviously, human resources include the knowledge, skills, networks and energies that people may deploy in their various roles" [1, 15]. According to this understanding HR and HR management derive their social legitimacy from powering relationships, institutional factors and value cocreation of employee, employer and other actors and by thus "from its ability to serve as an effective steward of a social contract in employment relationships capable of balancing and integrating the interests and needs of employers, employees and the society in which these relationships are embedded" [19, 20].

4 Expanding the Application Domain

The controversial discussion about the definition of HR reflects a fundamental shift in the HR worldview and also challenges the existing HR concepts like for example HR management, HR staffing or HR development. In marketing the corresponding change in understanding was already introduced in the 90s, as Gummesson, for example, states: "Customers do not buy goods or services: They buy offerings which render services which create value. The traditional division between goods and services is long outdated. It is not a matter of redefining services and seeing them from a customer perspective; activities render services, things render services. "The shift in focus to services is a shift from the means and the producer perspective to the utilization and the customer perspective" [8].

Vargo and Lusch argue [4], that the fragmented and controversial understanding of service is due to the logic inherited from the industrial era where wealth and value were based upon tangible units of exchange. In this goods dominant logic goods [4, 21] were characterized by their exchange value, such that the transfer of a good transferred wealth what Adam Smith (1776) "called the relative or exchangeable value of goods" [22].

In goods dominant logic the producers - the firms - responsibility for creating value is considered to have ended when the production unit is exchanged and the ownership transferred to the consumer [9].

In the last decades market changes have put this goods-dominant logic under strain as the borderlines between goods and services are becoming increasingly fluid. Business models are undergoing a transformation from pushing products to actor-to-actor networks coming together to meet specific individual needs. In this emerging service (eco) systems value is created in use rather than in exchange. Value in use, enabled by technology, is now being co-created in a process of mutual value creation between multiple entities [9, 23].

By developing the idea of service-dominant logic (s-d l), Vargo and Lusch reconceptualised service. Service defined "as the application of specialized competencies (skills, knowledge), through deeds, processes, and performances for the benefit of another entity or the entity itself (self-service) [9]. In s-d l service is the value that is co-created by actors through an integration of resources accessible to them.

For this s-d logic serves as a meta-theoretical framework for explaining the process of value creation through service exchange among multiple resource-integrating actors forming institutionally coordinated service ecosystems [24, 25].

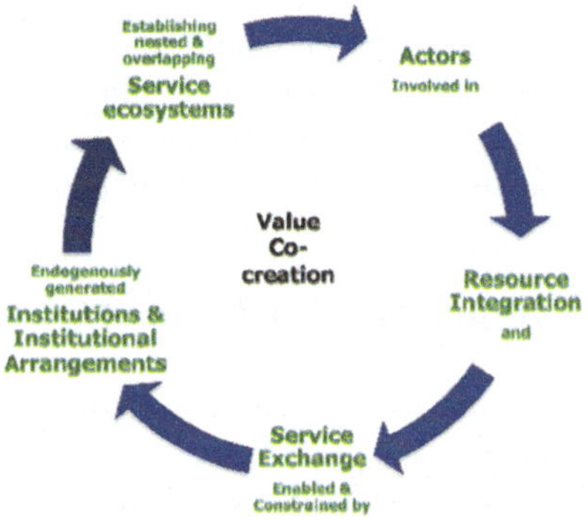

Fig. 1. The narrative and process of s-d logic [24]

For this new worldview of s-d logic, service science grounds the nature, scientific understanding, management principles and engineering discipline needed to understand and improve service and service innovation [26, 27]. With the service system a new unit of analysis is introduced by service science. Referring to the interactive character of service that involves at least two entities - one applying competence and another integrating the applied competences - these interacting entities are called service systems. More precisely, service systems are defined as dynamic value co-creation configurations of resources, including people, organizations, shared information (language, laws, measures, methods), and technology, all connected internally and externally to other service systems by value propositions [26]. From a service science perspective service (eco) systems can be described as a structure of interconnected service system entities. Together s-d logic and service science establish a general theory of service [28, 29]. Within this concepts, logics and theories five distinctive characteristics of service (eco) systems are defined [25]:

- service (eco) systems are relatively self-contained and have fuzzy boundaries,
- actors are relatively self-adjusting, as they show adaptive behaviour,
- actors are resource integrators that overcome the traditional view of enterprises too much focused on their internal resources and their mono-directional approach 'to' market,
- actors are connected by shared institutional logics,
- service exchange in service ecosystems results in mutual value creation.

Complementary holistic service (eco) system approaches lead from an analysis of the parts to the comprehension of the "whole" [30]. Holistic service systems provide "whole service" understood as capacity to operate for a period of time independently of all external service system interactions [31]. As actors within a service (eco) system are distant from each other, shared institutional arrangements are necessary in order to coordinate their otherwise unrelated behaviour [25]. For holistic service (eco) systems the existence of all necessary institutions as rules and norms to be independent and self-organizing is necessary. Institutions are the rules of the game in a society or, more formally, are the humanly devised constraints that shape interaction of human and technological actors. In consequence they structure incentives in actor exchange, whether political, social, economic or technological. Institutional arrangements and institutional change shape the way societies and service (eco) systems evolve through time [24, 32–35]. In the following, we will examine whether these characteristics of service (eco) systems facilitate new perspectives for HR and HR concepts.

5 Identifying New Perspectives for the Core Concepts of HR

By introducing perspectives of service science, service-dominant logic and institutional economics new perspectives for revising the existent knowledge and discussions within given key concepts of HR are demonstrated in the following.

5.1 Perspective: Towards a New Perception of the Employee

As stated, there are controversial argumentations in HR literature concerning the relationship between the employee and the organization. The adaptation of service science and s-d logic coincides with the newer HR approaches with regard to the role of the employee and the understanding of the interaction between employee (as permanent employed person or as free agent) and employer [1, 5, 15]. "Human resources are understood as the resources that are intrinsic to human beings and which they can apply to various tasks. "Most obviously, human resources include the knowledge, skills, networks and energies that people may deploy in their various roles" [1, 15]. This shows a change in understanding that is comparable to the change from goods-dominant to service-dominant logic.

Table 1. Towards a new perception of the employee

Goods-dominant logic	Service-dominant logic	HR traditional concept	HR new concept
Unit of exchange	Unit of exchange	Unit of exchange	Unit of exchange

(continued)

Table 1. (*continued*)

Goods-dominant logic	Service-dominant logic	HR traditional concept	HR new concept
Good as tangible unit of exchange [22]	Service is the fundamental basis of exchange [4] and defined as The application of resources (incl. competencies, skills and knowledge) to make changes that have value for another [4, 3, 36] -goods (tangible) and services render service [8]	HR as (tangible) people employed in an organisation [5]	HR as the resources that are intrinsic (intangible) to human beings, which they can apply to the various tasks of life, both in multiple workplaces and in private activities [1, 5, 15]

5.2 Perspective: HR as Steward Shaping New Forms of Cooperation

A service (eco) system is understood as "a relatively self-contained, self-adjusting system of resource-integrating actors connected by shared institutional arrangements and mutual value creation through service exchange" [24–26]. According to the recent comprehension HR and HR management derive their social legitimacy from an understanding that is very close to the understanding of s-d logic and service science, namely "from its ability to serve as an effective steward of a social contract in employment relationships capable of balancing and integrating the interests and needs of employers, employees and the society in which these relationships are embedded" [19, 20]. The value of HR management unfolds not only from attracting human workers to the company and managing them in the interest of the company (transactional view); rather, it is the starting point for building and developing mutually beneficial relationships and institutional arrangements for value cocreation. The shaping of these relationships, specifically as mutually agreed arrangements regarding the application of the intrinsic resources of employees

Table 2. HR as steward shaping new forms of cooperation

Goods-dominant logic	Service-dominant logic	HR traditional concept	HR new concept
Value creation	Value creation	Value creation	Value creation
Exchange value [22] Attributed to its current owner [37]	Value in use co-created through integration of resources accessible [4]	(exchange) transfer of value from employee to employer [2, 38]	developing and maintaining mutually-beneficial relationships in actor-to-actor networks [2]

in the context of the organization and society will become even more important in the future. As a result of technological possibilities and the opening up of organizations to new partnerships, the variety of actors to be involved will increase significantly (e.g. permanent employees, freelancers, out- and insourcing). And thus the expectations and requirements for HR management as steward.

5.3 Perspective: HR Management as Result and Process of Institutionalising New Forms of Work

Within the concepts, logics and theories of service science and s-d logic the core mechanisms of the systemic functioning of value cocreation are affected by institutions. Institutional logics are highly relevant: First for coordinating and connecting actors by shared rules and second as incentives and constraints of resource integration [24, 25, 32, 33, 39]. Institutions as framework and the embedding in institutional agreements enables the actors to act self-determined and self-organizing in larger social systems. Incentivated by institutions and value propositions actors engage in mutual value cocreation through service exchange and resource integration. Actors like employee and organizations as resource integrator have to overcome the traditional view of enterprises and being too much focused on exclusively internal resources and mono-directional approaches 'to' market.

Table 3. HR management as result and process of institutionalising new forms of work

Goods-dominant logic	Service-dominant logic	HR traditional concept	HR new concept
Value creation period			
Value creation ends when production unit is exchanged and ownership transferred to consumer [9]	Value creation unfolds over time and is embedded in broader social systems [2]	One-directional management of people and (parts) of their skills in mostly static and strict defined jobs. Main focus on permanent fulltime employees. Team collaboration is steered and coordinated by executives instead of self-determined [5][40]	Agile, self-contained with task-orientation within actor-to-actor networks; connected by shared institutional logics and service exchange [24, 40]

6 Three Implications for Practical HR Management

According to these new perspectives HR and HR management derives its value from shaping actor-relationships and institutional arrangements for enabling value cocreation of employee, employer and other actors. In order to comply to this role, three concrete implications can be derived. First HR management needs to open up for new and extended work models embedded in broader social, political and technological systems; second HR must come out of its box and position itself as an active driver of the change process and organizational development and third HR as enabler of the value cocreation process between multi-actors and thus as creator of new workforce design approaches.

6.1 Implication: HR Management as Multi-level and Interdisciplinary Approach

The new perspectives for HR and its core concepts require an understanding for being embedded in larger social contexts and consequently a multi-level redesign of the relationship between organizations and employees. It concerns the relationship to the individual employee (micro-level), to teams and groups (meso-level) and also to society (macro-level). Institutional arrangements for new work concepts which increasingly go beyond the boundaries of the organization must be negotiated and implemented with the employees and other human workers: e.g. new employer value propositions and new 'benefit components' which in many variants will supplement or in some cases replace 'money' in the sense of wages as the sole incentive system [41, 5].

For HR management, this also means reacting with much greater speed than before to social, political and technological developments and translating these developments into new instititutionalized work models and concepts (e.g. leading with context [42]) for their companies [43]. This results in a shift of activities away from the focus on HR administration and people management to the stewardship of new forms and concepts of work, which take into account the interests of the organization, of the people and of the society. Those new activities are associated with the cocreative and interdisciplinary approach and require new competencies and skills in HR management. The concept of the T-shaped professional, where existing deep business expertise is complemented by skills required for mutual value creation - e.g. collaboration, communication - could be a pattern for HRM [44].

6.2 Implication: HR as Facilitator of Change and Organizational Development

The shift from a traditional to a new perspective on the employee, that is seeing the HR in terms of the intrinsic (intangible) resources and capabilities of people rather than (tangible) people employed in an organization, demands a derivative within change and organizational development concepts. Since it is no longer sufficient to reduce people to their functionality, similar to the category of material and technology, it is essential to approach them from a holistic and mutually beneficial perspective like the holistic service (eco) system approach [30, 31] [45].

HR itself should therefore be capable of developing and implementing appropriate holistic change strategies considering the organization embedded in broader social systems "as an interdependent set of elements including tasks, individuals, formal organizational arrangements, and the informal organization" [46]. Taking this into account it is a major challenge for change management to align the incentives of the actors by shaping mutually agreed arrangements regarding the application of the intrinsic human resources of employees and within the context of the organization and the society. In particular, this means recognizing the values and ambitions of the actor, developing a common purpose and institutionalizing mutually agreed institutions. In this way HR succeeds by transforming formal change into a personal concern in order to empower value cocreation and to unleash the full potential of employee and organization.

6.3 Implication: HR as Creator of a New Workforce Design

Over the past four decades the service sector shaped advanced forms of interaction between human and technology actors with a wide range of actor combinations like human-to-human, human-to-technology, or technology-to-technology. The understanding of HR, HRM and workforce, on the other hand, has remained largely unchanged in recent decades. Workforce, workforce planning, workforce management, workforce diversity and so on, focus almost exclusively on the human actor. The new perspectives for HR and its core concepts do not limit workforce to the human being but stands for all actor combinations that contribute to an organization and its ability to apply resources [47, 48].

Especially human-technological collaboration and the demonstrated requirements for a redesign of the institutional arrangements of jobs and work operations will be relevant and challenging fields of HR management as steward of this transition. Examples are e.g. augmented working or new job designs [40].

7 Implications for Theoretical Research

Human resource and workforce management as part of broader institutionalization processes are nested in polycentric networked systems. As holistic service system approach of human resource – and workforce management the analysis of the parts leads to the comprehension of the "whole". Holistic service systems provide "whole service" to the actors within them and allow organizations to "run" for a period of time independently of all external service system interactions. The latter is also characterized by the existence of sufficient rules and norms for the holistic service system to be independent and self-organizing. The interplay between openness to value cocreation in broader networks and the need for stability and autonomy offers manifold opportunities for future research in the context of holistic service systems [30, 31].

8 Findings, Conclusion and Outlook

The concepts of service science, service-dominant logic and institutional economics essentially lead to three new perspectives for HR and HR management. First to a new perception of the employee and HR. HR understood as intrinsic resources which humans can apply for themselves or apply by mutual agreement in an organization. The implication for HR is to redesign the relationship between the organization and the human worker by approaching cocreative working models and the necessary institutional arrangements. The second new perspective is the understanding of HR as steward shaping and institutionalizing new forms of cooperation. The resulting implication is the relevance of HR as facilitator of change and thus as developer and implementer of mutually agreed common purpose within holistic change strategies. The third perspective is the relevance of institutionalizing new forms of work within actor-to-actor networks of value cocreation, e.g. agile, self-contained and self-organized instead of command and obedience. For the implementation this means the necessity for a fundamental redesign of jobs and work operations. In summary it can be stated that if these new perspectives will be established in the organization and its institutional arrangements, HR will take on the role of steward, unifying the interests of employees, the organisation and society.

References

1. Boxall, P., Purcell, J.: Strategy and Human Resource Management, 4th edn. Macmillan Education, London (2016)
2. Hartmann, N.N., Wieland, H., Vargo, S.L.: Converging on a new theoretical foundation for selling. J. Mark. **82**, 1–18 (2018)
3. Spohrer, J., Maglio, P., Bailey, J., Gruhl, D.: Steps toward a science of service systems. Computer **40**, 71–77 (2007)
4. Vargo, S.L., Lusch, R.F.: Evolving to a New dominant logic for marketing. J. Mark. **68**, 1–17 (2004)
5. Armstrong, M., Taylor, S.: Armstrong's Handbook of Human Resource Management Practice. Kogan Page Publishers (2020)
6. Inkson, K.: Are humans resources? Career Development International (2008)
7. Cascio, W.F.: The costs--and benefits--of human resources (2007)
8. Gummesson, E.: Relationship marketing: its role in the service economy. Underst. Serv. Manag. **244**, 68 (1995)
9. Ng, I., Vargo, S., Smith, L.: Reconceptualising Service through a Service-Dominant (2012)
10. Gilson, L.L., Goldberg, C.B.: Editors' Comment: So, What Is a Conceptual Paper? SAGE Publications Sage, Los Angeles (2015)
11. Flick, U.: An Introduction to Qualitative Research. Sage Publications Limited (2018)
12. Jaakkola, E.: Designing conceptual articles: four approaches. AMS Rev. **10**(1–2), 18–26 (2020). https://doi.org/10.1007/s13162-020-00161-0
13. Cascio, W.: Managing human resources: productivity, quality of work life, profits, 7th Edn. Tata McGraw-Hill. Abnorm. Soc. Psychol. **62**, 401–407 (2006)
14. Cascio, W.: Managing Human Resources: Productivity, Quality of Work Life, Profits. Mc Graw-Hill Book, Co-Singapore (1995)
15. Boxall, P.: Mutuality in the management of human resources: assessing the quality of alignment in employment relationships. Hum. Res. Manag. J. **23**, 3–17 (2013)
16. Lawler, E.E.: Make human capital a source of competitive advantage. In: Marshall School of Business Working Paper No. MOR 16–09 (2008)
17. Ulrich, D.: A new mandate for human resources. Harv. Bus. Rev. **76**, 124–135 (1998)
18. Cascio, W.F.: From business partner to driving business success: the next step in the evolution of HR management. Hum. Resour. Manag. **44**, 159-163 (2005)
19. Boxall, P.F., Purcell, J., Wright, P.M.: The Oxford Handbook of Human Resource Management. Oxford University Press on Demand (2007)
20. Kochan, T.A.: Social legitimacy of the HRM profession: A US perspective. In: Oxford Handbook of Human Resource Management, The 599 (2007)
21. Vargo, S.L., Lusch, R.F.: Why "service"? J. Acad. Mark. Sci. **36**, 25–38 (2008)
22. Smith, A.: The Wealth of Nations & The Theory of Moral Sentiments. e-artnow (2017)

23. Spohrer, J.C., Maglio, P.P. (eds.): Toward a Science of Service Systems. Springer, New York (2010)
24. Vargo, S.L., Lusch, R.F.: Institutions and axioms: an extension and update of service-dominant logic. J. Acad. Mark. Sci. **44**(1), 5–23 (2015). https://doi.org/10.1007/s11747-015-0456-3
25. Vargo, S.L., Lusch, R.F.: The SAGE Handbook of Service-Dominant Logic. SAGE Publications Limited (2018)
26. Spohrer, J.C., Vargo, S.L., Maglio, P.P.: The service system is the basic abstraction of service science. In: Proceedings 41st Hawaii International Conference on System Science, Big Island, pp. 104–114 (2008)
27. Maglio, P.P., Spohrer, J.: Fundamentals of service science. J. Acad. Mark. Sci. **36**, 18–20 (2008)
28. Spohrer, J., Fodell, D., Murphy, W.: Ten reasons service science matters to universities. EDUCAUSE Rev. **47**, 52 (2012)
29. Vargo, S.L., Akaka, M.A., Vaughan, C.M.: Conceptualizing value: a service-ecosystem view. J. Creating Value **3**, 117–124 (2017)
30. Golinelli, G.M., Barile, S., Spohrer, J., Bassano, C.: The evolving dynamics of service co-creation in a viable systems perspective. In: The 13th Toulon-Verona Conference proceedings of the International Conference in Coimbra, Portugal, pp. 2–4 (2010)
31. Spohrer, J., Piciocchi, P., Bassano, C.: Three frameworks for service research: exploring multilevel governance in nested, networked systems. Serv. Sci. **4**, 147–160 (2012)
32. North, D.C.: Institutions. Institutional Change and Economic, Cambridge (1991)
33. Ostrom, E.: Governing the Commons: The Evolution of Institutions for Collective Action. Cambridge University Press (1990)
34. Scott, W.R.: Institutions and Organizations: Ideas, Interests, and Identities. Sage publications (2013)
35. Edvardsson, B., Kleinaltenkamp, M., Tronvoll, B., McHugh, P., Windahl, C.: Institutional logics matter when coordinating resource integration. Mark. Theory **14**, 291–309 (2014)
36. Maglio, P.P., Vargo, S.L., Caswell, N., Spohrer, J.: The service system is the basic abstraction of service science. IseB **7**, 395–406 (2009)
37. Hill, P.: Tangibles, intangibles and services: a new taxonomy for the classification of output. Can. J. Econ. Revue canadienne d'Economique **32**, 426–446 (1999)
38. Hartmann, N.N., Wieland, H., Vargo, S.L., Ahearne, M.: Advancing Sales Theory Through a Holistic View: How Social Structures Frame Selling. Taylor & Francis (2020)
39. Scott, W.R.: Institutions and Organizations. Ideas, Interests and Identities. M@ n@ gement **17**, 136-140 (2014)
40. Oldham, G.R., Hackman, J.R.: Not what it was and not what it will be: the future of job design research. J. Organ. Behav. **31**, 463–479 (2010)
41. Kochan, T.A.: The Mutual Gains Enterprise: Forging a Winning Partnership Among Labor, Management, and Government. Harvard Business Press (1994)
42. Hastings, R., Meyer, E.: No Rules Rules: Netflix and the Culture of Reinvention. Penguin, New York (2020)
43. Ulrich, D., Allen, J., Brockbank, W., Younger, J., Nyman, M.: HR Transformation: Building Human Resources from the Outside in New York (2009)
44. Spohrer, J., Kwan, S.K.: Service science, management, engineering, and design (SSMED): an emerging discipline - outline and references. Int. J. Inf. Syst. Serv. Sect. **1** (2009)
45. Längle, A., Bürgi, D.: Existentielles Coaching: Theoretische Orientierung, Grundlagen und Praxis für Coaching. Facultas Verlags-und Buchhandels AG, Organisationsberatung und Supervision (2020)
46. Nadler, D.A.: Managing organizational change: an integrative perspective. J. Appl. Behav. Sci. **17**, 191–211 (1981)

47. Frosch, M.W., Markus: workforce Design - a resourcebased view. Whitepaper - Institut für Service Design April, 2019, 7 (2019)
48. Frosch, W., Markus, W.: A conceptual framework for workforce management: impacts from service science and S-D logic. In: AHFE, The Human Side of Service Engineering, p. 7. Springer, Sn Diego (2020)

Network Effects to Foster Service Innovation: An Empirical Analysis of Actual Practices

Benjamin Gregersen and Andreas Zolnowski(✉)

SIGNAL IDUNA Group, Neue Rabenstraße 15-19, 20354 Hamburg, Germany
{Benjamin.Gregersen,Andreas.Zolnowski}@signal-iduna.de

Abstract. The strength of associated network effects is not only an expression of the platform's initial success, but also a driving force for further growth and service innovation. However, many platform organizations struggle to create strong network effects to attract further platform participants. Therefore, organizations need to understand how to exploit the self-reinforcing character of different types of network effects and thus, to achieve a unique value proposition and deliver innovative solutions to their customers. The purpose of this work is to analyze how organizations can actively create and promote different types of network effects and thus, foster service innovation and enhance value propositions. To achieve the objectives set, we apply a qualitative approach with semi-structured interviews involving seven platform organizations. Conducting this research, we identified four main categories that play a key role regarding the creation of network effects: (1) user attraction, (2) user retention, (3) monetization and (4) governance.

Keywords: Service innovation · Service platform · Network effects · Mutual value co-creation

1 Introduction

Platform business models play a central role in the key industries due to their facilitating role in value creation and value exchange [1, 2]. They do not only provide the necessary infrastructure for valuable transaction between actors within the ecosystem, but also act as the foundation for the development of innovative and complementary services or products [3]. The mechanisms behind the growing relevance of platforms are network effects. Network effects describe the effect in which an increasing number of participants leads to an increase in the benefits of a good or a service [4, 5]. Therefore, the attraction of more and more users become a crucial task for platform businesses to create a self-reinforcing dynamic of growth [6].

Since network effects are based on personal interaction, the associated strength of network effects depends on several different factors. Platform companies need to understand the interrelationships both on and outside the platform and need to align their strategy accordingly [7]. Thus, the research questions that guided our study emerged as follows: What can platform businesses do to create strong network effects and thus, foster service innovation?

C. Leitner et al. (Eds.): AHFE 2021, LNNS 266, pp. 292–299, 2021.
https://doi.org/10.1007/978-3-030-80840-2_34

To answer this research question, we apply a qualitative approach conducting semi-structured interviews with seven platform business experts. The results were then grouped into four categories derived from the literature review that are crucial for the development of network effects: (1) user attraction, (2) user retention, (3) monetization and (4) governance. Based on the categories identified, the individual approaches of the platform businesses were assessed and compared with each other to identify similarities and difference between the companies considered.

This paper is structured as follows. The first chapter provides a literature review about the concept of platform businesses and the mechanisms of network effects. Second, the methodology approach will be explained followed by overview about actual practices of platform businesses to create strong network effects. In the fourth and last part the answers to the research question will be summed up within the conclusion.

2 Literature Review

2.1 The Concept of Platform Business Models

With the emergence of modern information and communication technologies the need for maintaining an own physical infrastructure or the ownership of assets has dramatically decreased and subsequently amplified the potential of platforms [8]. This has led to a development of innovative service business models and new forms of value creation [9] that can reshape existing customer interaction and value propositions [10].

Platforms come in many varieties. Therefore, the definitions alternate depending on the context. Basically, platforms effectively connect two or more sides of a market [11] that are connected by different types of network effects [12]. They are characterized by demand-side economies of scale and positive feedback loops [13, 14]. Based on a modular structure, platforms provide the necessary infrastructure that serve as the basis for the development of complementary services or products [15]. In the context of service innovation, platforms compromise and orchestrate different types of tangible and intangible resources, facilitate value co-creation, and thus enable value propositions [2, 16]. A more abstract definition describes platforms as systems, that "uses technology to connect people, organizations and resources [17].

Although platforms can take many different forms, they all have the same basic structure, which includes four types of roles. The platform owner holds the intellectual property and is responsible for the governance of the platform. The user interfaces are made available by a platform provider, usually a software company. Suppliers can offer their own products and services that are used by consumers. Platform participants may swap roles or perform multiple roles at the same time [18].

The key difference between a platform and an ecosystem is the existence of certain actors and network effects. Ecosystems are composed of "mostly loosely coupled social and economic (resource-integrating) actors connected by [...] mutual value creation through service exchange" [2]. To exploit the innovation potential of such ecosystems, platforms need to create strong network effects and enable valuable transactions between the various actors.

2.2 Network Effects as Driving Force Behind Platform Growth

The participants of a platform are connected to each other through different types of network effects as they influence each other's demand [19]. Network effects occur when the value of a good or a service increases in accordance with the number of the respective users [4, 12]. Basically, the more users adopt to a platform, the more attractive it gets for potential users. A basic distinction is made between direct and indirect network effects. Direct or cross-group network effects occur when an increase in participants on one side of the market attracts additional users of the same side e.g., users at social networks [20]. On the other hand, indirect network effects occur when an increase in the number of participants on one side of the market causes demand on the other side to rise [6].

Moreover, network effects are demand-side economies of scale that lead to positive feedback loops [13]. Cennamo and Santalo [21] observed, that platforms with a larger user base than their competitors potentially offer a higher average value per transaction. The more users interact via platform, the better the supply and demand can be matched. In addition, more user data is generated, which can be used for the further development of algorithms to promote both new and better interactions. This allows platforms to offer a better user experience and thus a higher average value per transaction [14, 19].

Fueled by direct or indirect network effects or even both, this dynamic can trigger a self-reinforcing growth spiral that encourages more users to adopt the platform as well as more complements to join the ecosystem [20]. Ultimately, this can lead to winner-take-all dynamics creating powerful monopolies and high entry barriers [22]. In this context platform become a "magnet for complementary innovators" [6], who contribute significantly to the value creation of the platform by developing and offering their products or services. Hence, strong network effects are the ultimate source to achieve service innovation and thus realize a superior value proposition.

3 Methodology

To answer the research question, we applied a qualitative approach conducting semi-structured interviews with platform business experts. A multi-case study scenario involving seven different platform organizations seemed the most appropriate for our exploratory study. This approach allows us both comparing actual practices of platform organizations and recognizing patterns between the cases.

The first step in preparing the questionnaire was to review relevant literature on network effects, platforms, ecosystems, and service innovation. Based on this, we identified the following four main impact factors, that act as the foundation of this study. (1) User attraction (2) User retention (3) Monetization and (4) Governance.

The interviews were organized with experts from business development or management to get a deeper understanding about the company's strategic decisions. All interviews were conducted via telephone or different video-conference platforms. The 45–60 min in depth interview was based on a semi-structured questionnaire which is represented in Table 1.

For a systematic analysis of both the qualitatively and the quantitatively collected data, the interviews were transcribed. Afterwards, the text segments were compiled in an Excel spreadsheet. A total of 221 text segments were summarized according to the

Table 1. Aggregated topics of the interview guideline

Type of question	Topic
Closed-ended questions	Firm age Core business/sector Number of employees Interviewee's role in the organization Target customers (B2B/B2C)
Open-ended questions	Platform Participant Chicken-egg problem Customer acquisition Core interaction / initial benefit Regular interaction Lock-in effects Revenue model including its evolution Degree of openness Incentive models Quality assurance Success factors and challenges

previously derived categories. For a greater validity of the study, a data triangulation was performed by direct observations during the interview and the gathering of additional information from secondary sources, such as websites, articles, and magazines. To gain broad insights about actual practices and to avoid bias in the sample, we aimed to analyze a diverse range of platform organization and contacted potential participants accordingly. The sample includes both transactional and innovation platforms, B2C and B2B as target customers, and varying levels of maturity from start-ups to market leaders with revenues of over €500 million.

Due to the limited sample size of seven platform organizations, our research provides an overview about actual practices of platform organizations to create strong network effects. Hence, findings derived from the study can only be indications of possible measures, for which the validity can only be confirmed by a more detailed examination.

4 Results and Discussion

A critical mass of users must be reached for network effects to occur at all. To achieve this, first platform owners must solve the chicken-egg problem. Within our sample different strategies can be observed. Especially in opaque markets actors of the ecosystem may find it difficult to find the right interaction partner. Platform owners that can find the right interaction partners, even with high manual efforts, can build trust by solving a real problem of potential platform participants. Building trust can convince consumers and suppliers to interact via the platform in the future. Another approach is to initially serve the supply side oneself. In the case of a comparison portal, users were able to compare insurance products based on publicly available information. The growing number of users persuaded insurance companies to provide further information and to enable direct sales of their products. As the number of consumers increases, the likelihood of connecting suppliers to the platform and extending the core interaction e.g., by direct sales of the compared products, grows. Especially in the early stages platforms owners need to provide value regardless of the number of users. The benefit must be perceived by both active and potential users in order to create enough attraction. This can be achieved, for example, through various marketing measures such as content marketing, search engine automation and optimization. Since connecting potential platform participants in the B2B market is generally more complex than in the B2C market, some of the companies observed have also significantly expanded their sales capacities. To efficiently match supply and demand, it is beneficial to have a balanced proportion of consumer and supplier. Nevertheless, an imbalance can also be used as an argument to attract users from the other side due to existing network effects. The common goal of maximizing ecosystem value enables a division of tasks between the actors. On the one hand, platform owners attract new users and promote interactions by connecting different platform participants. By this, interactions such as value co-creation activities with consumers or new product compositions with other suppliers are easier to realize. On the other hand, suppliers can access an existing customer base and gain feedback from interactions without having to build up significant resources. For this reason, they can pilot new products with low risk and thus focus more on the development of their products or services.

Network effects are based not only on the total number of users, but primarily on user activity and the number of interactions. Basically, there are two principle ways to retain users on the platform. On the one hand, platform participants can be retained through a unique service offering and a superior value proposition. On the other hand, users can be discouraged from leaving the platform due to additional lock-in effects and high switching costs. Often, however, these two different approaches cannot be treated separately. To achieve this, the observed companies focus on a user-friendly user experience, exclusive access to information or profound integration into the processes of the platform participants. Moreover, proactive interaction suggestions, newsletters, content marketing, and push notifications can increase the number of interactions if used effectively. Community features allow platform participates to interact with each other. This not only builds trust among the platform participants, but also has a positive effect on customer loyalty enhancing further interactions. By extending the core interaction with complementary offers or services, platform owners try to develop themselves as a "one-stop store" for satisfying multiple needs of their user groups at the same time. The modular structure and standards set by the platform owner determine the extent to which supplieres and various components can be linked to the platform. While platform-specific standards create lock-in effects, they also promote the development of complementary products and services and ultimately allow new and innovative forms of value creation. Nevertheless, interaction via the platform should always be more advantageous than direct contact between the interaction partners.

As user fees represent a significant barrier to use the platform, the monetization of customer relationships has a major influence on the emergence of network effects. The observed companies apply different revenue models regarding their user groups. Most companies started with transaction-based fees because they best reflect perceived value and scale as the platform grows. However, this approach led to uncertainty among some users due to low planning certainty. Furthermore, the transaction fee creates an additional incentive for partners to carry out direct transactions outside the platform. To counter this, two of the considered platforms have switched to a fixed user fee causing platform participants to perform more interactions in order to achieve the greatest cost-benefit. Furthermore, it may make sense to grant free access to the platform at the beginning to monetize the customer relationship at a later point in time. In this case, the less price-sensitive respectively the market side that profits less from participating should be subsidized, since the other market side need to come up for the costs. Meanwhile, most of the companies rely on a mix of fixed fees for the general use of the platform and transaction-based fees for the use of additional services with a high utility such as direct payment processing. In general, the stronger the network effects, the more likely it is that transaction-based revenue models can be implemented. However, monetization should not take place at the expense of network effects because it contradicts the basic idea of platforms. Usage fees should therefore always be based on the benefits offered by a platform. Therefore, platform participants usually benefit from relatively low costs compared to the benefits offered by the platform. Particularly at the beginning, this can be advantageous for both consumers and suppliers. However, this can change quickly as the platform has achieved a monopoly-like position.

Even though the number of users and the associated interactions have a significant impact on the growth of a platform, unregulated and unrestricted access can lead to negative customer experiences and thus to negative network effects. For this reason, platform owners must define within their governance who and how users can interact with each other. Potential platform participants can be pre-qualified and verified in advance as part of the onboarding process. For example, required information, verification documents or recommendations from market participants can be applied. Once on the platform, the platform owner can ensure appropriate quality e.g. by supporting suppliers with user statistics or demanding mandatory information for the presentation of certain offers. High-quality offers not only improve the customer experience, but also have a positive effect of network effects as they usually lead to more transactions. To influence the behavior of platform participants, platform owners can display customer ratings as well as the participant's performance in terms of service levels, such as response or payment duration or delivery times. Monitoring and displaying this information creates mutual trust among the users and enables the identification and, if necessary, sanctioning of misbehavior. In the case of serious misconduct, the measures taken by the companies considered ranged over various escalation levels from bilateral meetings for clarification, penalties, or even exclusion from the platform. In contrast, another task of the platform owner is to offer attractive incentives for the use of the platform. Most incentives such as discount campaigns, vouchers, bonus programs, etc. are of monetary nature. However, access to exclusive data, analyses, and customer information is also becoming increasingly important for innovative, customer-centric solutions and can already replace monetary incentives to some extent. Besides that, transparency both within the market and on the platform ensures increased competition between suppliers. For this reason, suppliers must differentiate themselves more strongly through innovative products and services as well as a great customer experience to remain competitive.

5 Conclusion

Since platform owners generally provide only a fraction of the products and services themselves, network effects are important for the platform to remain competitive, to attract suppliers and consumers and thus, foster service innovation on the platform.

This study has highlighted the role of network effects and provided actual practices that platform organizations can apply to (1) attract potential platform participants, (2) retain them on the platform, (3) use monetarization strategies without building up entry barriers, (4) ensure a great customer experience and build trust among the users and thus, create strong network effects.

The findings from the qualitative approach conducted empirically confirm some strategies and best practices that have been already highlighted by the existing literature. Among them, solving the chicken-and-egg problem, building lock-in effects, and establishing trust through transparency about performance on the platform.

Through the application of semi-structured interviews, we identified similarities and differences regarding the platform strategies of the investigated organizations. On the one hand, the strategies showed many similarities according to governance (e.g. sanctions for misbehavior, verification of actual users) or the retention of customers (e.g. user-friendly user interface, offering complementary products or services). On the other hand, the strategy varies as the platforms serve different user groups. Monetarization strategies for instance, highly depend on the price sensitivity of the user groups. To create network effects and attract the right users, understanding the needs and the relationships among the actors of the ecosystem become essential. Therefore, platforms must monitor the interactions and adapt their strategy accordingly. In the context of creating service innovation, platforms (1) allow suppliers to focus on value creation, (2) enable new forms of value creation, (3) increase productivity, as well as (4) innovation pressure. As the driving force behind a platform network effects are the ultimate source to achieve service innovation and thus realize a superior value proposition.

However, the results face some limitations. First, the limited sample size cannot reflect the whole variety of platform business models. As factors such as the competitive environment, the maturity of a platform or the customer base highly influence the platform strategy, this study does not attempt to draw conclusions based on statistical generalization. However, the purpose of this study is to provide an overview about actual strategies to create and promote the development of positive network effects. For further studies the results need to be reviewed as part of a more detailed analysis which focuses on individual aspects in order to confirm or disconfirm the different approaches as best-practices.

References

1. Evans, D.: The antitrust economics of multi-sided platform markets. Yale J. Regulat. **20**, 325–382 (2003)
2. Lusch, R.F., Nambisan, S.: Service innovation: a service-dominant (s-d) logic perspective. Manage. Inf. Syst. Quart. **39**, 155–175 (2015)
3. Gawer, A., Cusumano, M.A.: Industry Platforms and Ecosystem Innovation. J. Prod. Innov. Manag. **31**, 417–433 (2014)

4. Katz, M.L., Shapiro, C.: Systems competition and network effects. J. Econ. Perspect. **8**, 93–115 (1994)
5. Cusumano, M.A.: Platform wars come to social media. Commun. ACM **54**, 3–34 (2011)
6. Evans, P.C., Gawer, A.: The Rise of the Platform Enterprise. A global survey. The Emerging Platform Economy Series 1 (2016). https://www.thecge.net/app/uploads/2016/01/PDF-WEB-Platform-Survey_01_12.pdf
7. Evans, D.S., Schmalensee, R.: Matchmakers: The New Economics of Multisided Platforms. Harvard Business Review Press, Boston (2016)
8. Parker, G., van Alstyne, M.W., Choudary, S.P.: Platform revolution. How networked markets are transforming the economy - and how to make them work for you. W.W. Norton & Company, New York, London (2016)
9. Schallmo, D., Rusnjak, A.: Roadmap zur Digitalen Transformation von Geschäftsmodellen. In: Schallmo, D., Rusnjak, A., Anzengruber, J., Werani, T., Jünger, M. (eds.) Digitale Transformation von Geschäftsmodellen. SBMI, pp. 1–31. Springer, Wiesbaden (2017). https://doi.org/10.1007/978-3-658-12388-8_1
10. Weiß, P., Zolnowski, A., Warg, M., Frosch, M.: Becoming a platform organization: how incumbent companies stay competitive. Cutter Bus. Technol. J. **31**, 38–45 (2018)
11. Rochet, J.-C., Tirole, J.: Two-sided markets: a progress report. RAND J. Econ. **37**, 645–667 (2006)
12. Caillaud, B., Jullien, B.: Chicken and egg: competition among intermediation service providers. RAND J. Econ. **34**, 309–328 (2003)
13. Shapiro, C., Varian, H.R.: Information Rules: A Strategic Guide to the Network Economy. Harvard Business Press, Boston (1998)
14. van Alstyne, M.W., Parker, G.G., Choudary, S.P.: Pipelines, platforms, and the new rules of strategy. Harvard Bus. Rev. **62**, 54–60 (2016)
15. Gawer, A., Cusumano, M.A.: How companies become platform leaders. MIT Sloan Manage. Rev. **49**, 70–77 (2008)
16. Warg, M., Zolnowski, A.: Definition of Service Platforms based on a Service Systems Understanding. Whitepaper (2018)
17. Eastwood, B.: The return of platforms (and how to not fail at building one). MIT Sloan School of Management (2016). https://mitsloan.mit.edu/ideas-made-to-matter/return-platforms-and-how-to-not-fail-building-one
18. van Alstyne, M.W., Parker, G.G.: Platform business: from resources to relationships. GfK Mark. Intell. Rev. **9**, 24–28 (2017)
19. Weil, H.B., Lee, V.: Achieving platform dominance. MIT Sloan School of Management (2018). https://papers.ssrn.com/sol3/papers.cfm?abstract_id=3264998
20. Cusumano, M.A.: The evolution of platform thinking. How platform adoption can be an important determinant of product and technological success. In Commun. ACM 53, 32–34 (2010)
21. Cennamo, C., Santaló, J.: Platform competition: strategic trade-offs in platform markets. Strateg. Manag. J. **34**, 1331–1350 (2013)
22. Choudary, S.P.: The dark side of platform monopoly. When large platforms turn evil (2020). https://platformed.info/the-dark-side-of-platform-monopoly

Augmenting Service Capabilities in the Post-pandemic Era

Differences in Effect of Endorsement of Professional vs. Non-professional YouTuber Through Credibility and Parasocial Relationship

Hisashi Masuda[1](✉), Spring H. Han[1], and Jungwoo Lee[2]

[1] Graduate School of Management, Kyoto University, Kyoto, Japan
{masuda.hisashi.4c,han.hyunjeong.8r}@kyoto-u.ac.jp
[2] Center for Work Science, Institute of Convergence Science, Yonsei University, Seoul, Korea
jlee@yonsei.ac.kr

Abstract. In analyzing the effect of product/service recommendation advertising by social media influencers, the difference in the effect considering name recognition has not been sufficiently examined. This study examines the differences in the effects on product/service recommendation intention between professional and unidentified endorsers based on the trustworthiness and expertise of the information provider and the degree of parasocial relationship (PSR) for consumers who viewed advertisements by influencers on YouTube (YouTuber) and made a purchase. As a result, the recommendation intention for professional YouTuber was positively influenced by expertise and trustworthiness, whereas for recommenders of unknown type, it was positively influenced by PSR. The findings will provide service companies with formulating advertising strategies that consider the name recognition of the endorsers when conducting endorsement advertising using social media influencers.

Keywords: Endorsement · Credibility · Trustworthiness · Expertise · Parasocial relationship · Social media influencer marketing · YouTube

1 Introduction

It has long been studied that endorsement advertisements utilizing celebrities, such as entertainers and athletes, have a positive impact on marketing [1, 2]. In particular, how the characteristics of celebrity affect repurchase intentions has been analyzed using concepts such as the credibility of the information provider and the pseudo-human relationship (parasocial relationship, PSR) between the information provider and the listener through the media. The credibility of the information provider is mainly composed of the trustworthiness and expertise of the information provider [2], and has been reported that trustworthiness and expertise have a positive impact on marketing effectiveness, such as repurchase intention.

On the other hand, studies on the effects of endorsement by celebrities and famous people (social media influencers) on social media such as SNS, which are currently used by many consumers, have begun to appear [3–12]. Social media influencers have

C. Leitner et al. (Eds.): AHFE 2021, LNNS 266, pp. 303–310, 2021.
https://doi.org/10.1007/978-3-030-80840-2_35

built their careers by accumulating expertise in specific domains, and their information dissemination based on their own experiences may be more trusted by consumers than traditional celebrity endorsements.

However, the difference in endorsement advertising effectiveness considering social media influencers' name recognition has not been sufficiently examined in the academic literature. Many previous studies have focused their analysis on high-profile social media influencers. On the other hand, due to the diversity of their career development, social media influencers vary greatly in the types of consumers they attract attention from and the degree of their name recognition. Thus, it is expected that there will be differences in the effects on the endorsement advertising effects.

This study analyzes the impact of such differences in social media influencers' name recognition on their endorsement advertising effectiveness. Specifically, for consumers who purchased after watching an endorsement advertising video by an influencer (YouTuber) on video-sharing social media (YouTube), we compare the case where the endorser was recognized as a professional YouTuber to one who was as an unknown YouTuber, and then examine the differences in the effects on product/service recommendation intentions. The theories used to measure marketing effectiveness are the credibility of the endorsers constructed from the trustworthiness and expertise, and the parasocial relationship (PSR), which is the formation of a pseudo-human relationship between the information sender and the receiver through the media. This makes explicit the tendency of product/service recommendation intentions based on differences in influencer's name recognition in endorsement advertisements linked to purchase activities. As a result, insights that support more effective decision-making when formulating endorsement advertisements based on the name recognition of social media influencers can be obtained.

2 Previous Researches and Hypotheses Development

In communication research, studies have been conducted on the effect of the credibility of the informant. Credibility is a concept that relates to the degree to which information disseminated by an information provider is plausible. In particular, credibility is subdivided into dimensions based on trustworthiness and expertise [2]. On the other hand, in media research, parasocial relationships (PSR), which are pseudo-human relationships formed between information senders and receivers on media, have been studied. PSR is defined as the degree to which some kind of human relationship is considered to have been established between the sender of information on the media and its audience, without direct interaction [1].

In this study, we construct hypotheses about the effect of social media influencers on endorsement advertising and the difference in influencer's name recognition based on the perspectives of previous studies on trustworthiness, expertise, and PSR. As antecedents of the above constructs, we use homophily/similarity between the endorser and the audience and endorser's physical attractiveness and social attractiveness. The outcome of the marketing effect is the willingness to recommend the purchased product/service to friends and colleagues.

Proposed hypotheses based on previous literature can be summarized in the research model shown in Fig. 1 based on the following perspective. When the attitude homophily/similarity between the sender and receiver of information is high, they are more likely to interact [13, 14]. It has been verified that physical attractiveness, such as the information sender's appearance, affects the credibility of them [15–17]. The impact of the information sender's social attractiveness on the perceived credibility by the recipient of the information has been analyzed [16, 17]. It has been demonstrated that the trustworthiness and expertise of the information provider in endorsement advertisements affect the audience's intention to recommend the product to friends and colleagues [10]. Also, it has been demonstrated that the audience's PSRs affect their intention to recommend to their friends and colleagues [4]. Previous studies of endorsement advertising have been validated by giving focus to celebrities and famous social media influencers. In this context, it is reported that social media influencers are more effective in marketing than traditional celebrity endorsement advertising [9]. However, social media influencers' characteristics, such as their various career paths and accumulation of expertise in niche domains, are considered to be increasingly fragmented cases as well that do not have mass recognition.

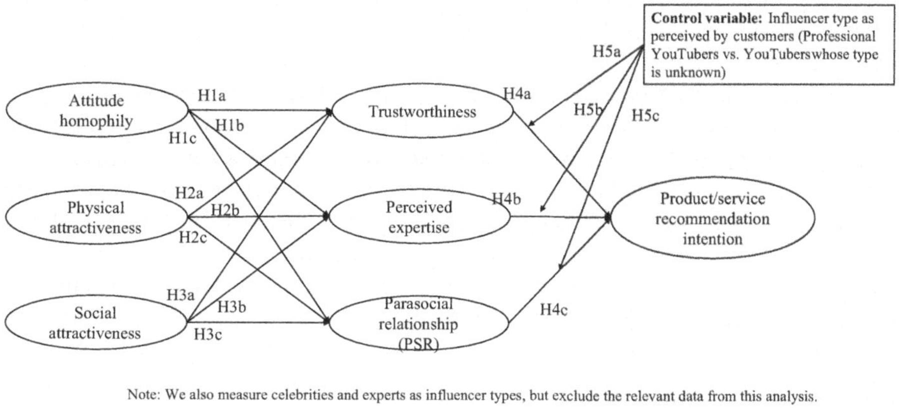

Fig. 1. Research model.

3 Methodology

This study examines how trustworthiness, expertise, and PSR affect product/service recommendation intention in the context of social media influencers on a video sharing platform (YouTube) based on the proposed research model. YouTube is a social media platform that allows users to upload their own videos and share them with a large audience [18]. Social media influencers create video content focused on a specific area and post it on YouTube, which is tied to their daily lives [7]. Because of these characteristics, a wider range of influencers besides celebrities and experts are active on this type of social media.

Here, partial least squares-based structural equation modeling (PLS-SEM) is used to test the hypotheses. There are two types of SEM (structural equation modeling): covariance-based (CB-) and PLS-SEM [19]. CB-SEM is mainly used to test theories based on how well a theoretical model can estimate sample data's covariance structure. The maximum likelihood method for CB-SEM estimation allows robust estimation even for data without normality, but it requires a large sample size. On the other hand, PLS-SEM does not require any assumptions on the distribution of the measured variables because the iterative algorithm performed is based on a series of least squares methods, and it also allows for robust estimation even when the sample size is small. In this study, we use PLS-SEM, which is more flexible for the distribution of data and provides robust estimation even for complex models.

In this study, the measurement items for each construct of the research model were adopted from previous studies and modified to fit this study's context. All items were measured using a 5-point Likert scale tied to "strongly disagree" and "strongly agree." The four dimensions of attitude homophily were measured using [11, 12]. Physical attractiveness was measured with four items extracted from [11, 12]. Social attractiveness was measured by four items extracted from [12]. Trustworthiness was measured by three items extracted from [11, 12]. Expertise was measured with four items extracted from [11, 12]. PSR was measured with 10 items extracted from [5, 11, 12]. As an outcome of marketing effectiveness, the degree of intention to recommend to friends was measured with one item.

A web-based questionnaire survey was constructed based on the proposed research model. The data was collected through a marketing research company in South Korea, where digital marketing is relatively widespread. All 313 respondents were from South Korea. All of them answered the questionnaire based on their experience of purchasing products or services after watching YouTuber's videos uploaded on YouTube. Of the 313 respondents, 112 were female (35.78%). The mean age of the participants was 46.35 years (SD = 11.83). The types of YouTubers identified by the respondents were "professional YouTubers" (57.00%), "celebrities" (10.24%), "experts" (11.95%), and "unknown type" (20.82%).

4 Result

The first step was the basic analysis, such as central tendency, variance, and correlation analysis. Next, the PLS-SEM was performed with smartPLS v3.3.2. In the first part of the PLS-SEM, we evaluated the measurement model for the constructs. In the second half of the analysis, we tested the hypotheses according to the proposed research model and then compared the differences based on the control variable.

4.1 Basic Analysis: Central Tendency and Variance, Correlation

The constructs were analyzed for central tendency and variance, and the score for each construct was calculated using the mean of the relevant items. The average score was highest for Expertise (3.61), followed by Trustworthiness (3.55) and Product/service recommendation intention. Attitude homophily had the lowest mean score (3.18), followed by Physical attractiveness (3.19) and Social attractiveness (3.33).

Correlation analysis of the target constructs was conducted using the arithmetic mean of the items associated with each construct. As a result, the correlation between each construct was confirmed at the 0.01 level of significance (2-tailed). The strongest correlation was between PSR and trustworthiness (0.765), followed by PSR and social attractiveness (0.736), and PSR and attitude homophily (0.686). Weak correlations were found between physical attractiveness and trustworthiness (0.483) and physical attractiveness and expertise (0.485).

4.2 Analysis of a Measurement Model

Then, PLS-SEM was applied to analyze the measurement model. The measurement models in this model are all reflective constructs (constructs affect the measurement scale as latent variables).

To evaluate the results of the reflective measurement model with all data (n = 313), outer loading, composite reliability, Cronbach's alpha, convergent validity with average variance extraction (AVE), and discriminant validity were employed [19]. All external loadings of the components are above the threshold value of 0.70. The composite reliability value is higher than 0.70. Cronbach's alpha is also above 0.70, so there is no problem in terms of reliability. The AVE exceeds the required threshold of 0.50 and has a high level of convergent validity. For discriminant validity, all values of the HTMT criteria were less than 0.9, indicating that discriminant validity was established.

4.3 Analysis of a Structural Model

To evaluate the proposed research model, we assessed VIF, the significance of path coefficients, level of R^2 values, f^2 effect size, and predictive relevance Q^2. First, all VIF values associated with the constructs are clearly below the threshold, and there is no collinearity problem. Second, we checked the R^2 values of the endogenous latent variables and found that the R^2 values for trustworthiness (0.504), expertise (0.422), PSR (0.740), and product/service recommendation intention (0.354) were moderate. Third, the f^2 values were all above 0.02 except for the relationships between attitude homophily and trustworthiness (0.003), attitude homophily and expertise (0.001), and trustworthiness and product/service recommendation intention (0.019). The effect of social attractiveness on PSR had the strongest effect size (0.277), and the effect of physical attractiveness on PSR had the second strongest effect size (0.234). Fourth, when the coefficients' significance was assessed, all paths were significant at the 1% level except for attitude homophily to expertise ($p = 0.515$) and to trustworthiness ($p = 0.434$). Fifth, the total effect of the main constructs on the intention to recommend was checked, with "social attractiveness" being the strongest (0.296), followed by "physical attractiveness" (0.190) and "attitude homophily" (0.063). Sixth, the path model's predictive validity was assessed using a blind-folding procedure, and the Q^2 values associated with each construct were all greater than zero.

The path coefficients (H1 to H4) for the hypothesis testing are shown in Table 1.

Table 1. Results of hypotheses testing (H1–H4).

	Hypotheses	β	t-value	p-value	Result
H1a	Attitude Homophily → Trustworthiness	0.063	0.863	0.389	Rejected
H1b	Attitude Homophily → Expertise	0.047	0.666	0.506	Rejected
H1c	Attitude Homophily → PSR	0.144	2.632	0.009	Accepted
H2a	Physical Attractiveness → Trustworthiness	0.172	2.712	0.007	Accepted
H2b	Physical Attractiveness → Expertise	0.286	4.487	0.000	Accepted
H2c	Physical Attractiveness → PSR	0.362	7.599	0.000	Accepted
H3a	Social Attractiveness → Trustworthiness	0.528	7.148	0.000	Accepted
H3b	Social Attractiveness → Expertise	0.378	5.055	0.000	Accepted
H3c	Social Attractiveness → PSR	0.445	7.185	0.000	Accepted
H4a	Trustworthiness → Product/service recommendation intention	0.183	2.009	0.045	Accepted
H4b	Expertise → Product/service recommendation intention	0.180	0.028	0.028	Accepted
H4c	PSR → Product/service recommendation intention	0.296	3.207	0.001	Accepted

Note: Two Tailed Test; Significance Level = 5%

The hypotheses were tested by validating the research model for the case of professional YouTuber (n = 167) and YouTuber of unknown type (n = 61). The results are shown in Table 2.

Table 2. Results of hypotheses tests (H1 to H5) for each group.

		Professional Youtuber (n=167)		YouTuber of unknown type (n=61)		H5a,b,c
	Hypotheses	β	Result	β	Result	Result
H1a	Attitude Homophily → Trustworthiness	0.132	Rejected	0.289*	Accepted	
H1b	Attitude Homophily → Expertise	0.020	Rejected	0.459*	Accepted	
H1c	Attitude Homophily → PSR	0.201**	Accepted	0.282**	Accepted	
H2a	Physical Attractiveness → Trustworthiness	0.085	Rejected	0.303	Rejected	
H2b	Physical Attractiveness → Expertise	0.258**	Accepted	0.290	Rejected	
H2c	Physical Attractiveness → PSR	0.313***	Accepted	0.509***	Accepted	
H3a	Social Attractiveness → Trustworthiness	0.563***	Accepted	0.061	Rejected	
H3b	Social Attractiveness → Expertise	0.423***	Accepted	-0.057	Rejected	
H3c	Social Attractiveness → PSR	0.438***	Accepted	0.201	Rejected	
H4a	Trustworthiness → Product/service recommendation intention	0.260*	Accepted	0.222	Rejected	Accepted
H4b	Expertise → Product/service recommendation intention	0.185	Rejected	0.040	Rejected	Rejected
H4c	PSR → Product/service recommendation intention	0.188	Rejected	0.447*	Accepted	Accepted

Note: Two Tailed Test; Significance Level = 5%

5 Discussion

In this study, we compared professional YouTuber and YouTuber of unknown type in influencer types and analyzed the difference in influence on product/service recommendation intentions. For YouTubers who were perceived as professional, trustworthiness affected product/service recommendation intention. Besides, the social attractiveness of the influencer has an impact on trustworthiness. On the other hand, for YouTuber of unknown type, PSR formed by customers had an effect on product/service recommendation intention. In addition, attitude homophily and physical attractiveness had an effect on the PSR.

This study's academic contribution lies in the fact that it considers the differences in the name recognition of social media influencers for the measurement of their marketing effectiveness. In the case of influencer marketing among YouTubers who are not particularly well known, PSR has an effect on the intention to recommend a product/service, but in the case of YouTubers recognized as professionals, trustworthiness has an effect on the intention. This makes it important to explore and demonstrate a more detailed model of influencer marketing effectiveness that considers the name recognition of social media influencers.

As a practical implication, when using social media influencers on video platforms such as YouTube in the promotion of service companies, it may not be possible to use well-known influencers from the beginning due to cost considerations. On the other hand, social media influencers with low cost can be said to have low name recognition. In such a case, it is important to increase PSR, which is a pseudo-human relationship formed between influencers and viewers, in order to increase marketing effects such as recommendation intention by social media influencers with low name recognition. In order to increase the PSR, the attitude homophily between the influencer and the audience as well as the physical attractiveness of the influencer are involved; therefore, it is effective to use influencers close to the audience and to devise their proper appearance according to the target customers. When the influencer is well known and recognized as a professional, the trustworthiness of the influencer will affect the marketing effect such as recommendation intention. Since social attractiveness is involved in enhancing trustworthiness, it becomes important to have content that can match social needs of the targeted customers.

The limitation of this study is that it does not sufficiently analyze the subdivision of influencer types as perceived by customers and the structure of these categories. Research that can extract the characteristics of influencer types that are recognized by customers and the rules for recognizing them is needed in the future. Another point is that the discussion is based only on the comparison between the extreme cases of unknown type and professional. There might be a change in marketing effectiveness associated with the increased recognition of influencers. A more dynamic model of social media influencer marketing can be developed by considering the more segmented stages of awareness and the temporal context of relationship building between the customers and the influencer.

Acknowledgments. This work has been supported by the National Research Foundation of Korea (NRF-2019K2A9A2A08000175) and the Society for Serviceology.

References

1. Horton, D., Richard Wohl, R.: Mass communication and para-social interaction: observations on intimacy at a distance. Psychiatry **19**(3), 215–229 (1956)
2. Sternthal, B., Phillips, L.W., Dholakia, R.: The persuasive effect of scarce credibility: a situational analysis. Public Opin. Q. **42**(3), 285–314 (1978)

3. Kim, J., Song, H.: Celebrity's self-disclosure on Twitter and parasocial relationships: a mediating role of social presence. Comput. Hum. Behav. **62**, 570–577 (2016)
4. Hwang, K., Zhang, Q.: Influence of parasocial relationship between digital celebrities and their followers on followers' purchase and electronic word-of-mouth intentions, and persuasion knowledge. Comput. Hum. Behav. **87**, 155–173 (2018)
5. de Berail, P., Guillon, M., Bungener, C.: The relations between YouTube addiction, social anxiety and parasocial relationships with YouTubers: a moderated-mediation model based on a cognitive-behavioral framework. Comput. Hum. Behav. **99**, 190–204 (2019)
6. Munnukka, J., Maity, D., Reinikainen, H., Luoma-aho, V.: "Thanks for watching" the effectiveness of YouTube vlog endorsements. Comput. Hum. Behav. **93**, 226–234 (2019)
7. Chapple, C., Cownie, F.: An investigation into viewers' trust in and response towards disclosed paid-for-endorsements by YouTube lifestyle vloggers. J. Promot. Commun. **5**(2), 110–136 (2017)
8. Dhanesh, G.S., Duthler, G.: Relationship management through social media influencers: effects of followers' awareness of paid endorsement. Public Relat. Rev. **45**(3), 101765 (2019)
9. Djafarova, E., Rushworth, C.: Exploring the credibility of online celebrities' Instagram profiles in influencing the purchase decisions of young female users. Comput. Hum. Behav. **68**, 1–7 (2017)
10. Ki, C.-W.C., Kim, Y.-K.: The mechanism by which social media influencers persuade consumers: the role of consumers' desire to mimic. Psychol. Mark. **36**(10), 905–922 (2019). https://doi.org/10.1002/mar.21244
11. Lou, C., Kim, H.K.: Fancying the new rich and famous? Explicating the roles of influencer content, credibility, and parental mediation in adolescents' parasocial relationship, materialism, and purchase intentions. Front. Psychol. **10**, 2567 (2019)
12. Sokolova, K., Kefi, H.: Instagram and YouTube bloggers promote it, why should I buy? How credibility and parasocial interaction influence purchase intentions. J. Retail. Consum. Serv. **53**, 101742 (2020). https://doi.org/10.1016/j.jretconser.2019.01.011
13. Rogers, E.M., Bhowmik, D.K.: Homophily-heterophily: relational concepts for communication research. Public Opin. Q. **34**(4), 523–538 (1970)
14. McPherson, M., Smith-Lovin, L., Cook, J.M.: Birds of a feather: homophily in social networks. Ann. Rev. Sociol. **27**(1), 415–444 (2001)
15. Reis, H.T., Nezlek, J., Wheeler, L.: Physical attractiveness in social interaction. J. Pers. Soc. Psychol. **38**(4), 604–617 (1980)
16. Rubin, A.M., Perse, E.M.: Audience activity and soap opera involvement a uses and effects investigation. Hum. Commun. Res. **14**(2), 246–268 (1987)
17. Rubin, A.M., Step, M.M.: Impact of motivation, attraction, and parasocial interaction on talk radio listening. J. Broadcast. Electron. Media **44**(4), 635–654 (2000). https://heinonline.org/HOL/P?h=hein.journals/jbem44&i=645
18. Freeman, B., Chapman, S.: Is "YouTube" telling or selling you something? Tobacco content on the YouTube video-sharing website. Tob. Control **16**(3), 207–210 (2007)
19. Hair, J.F., Hult, G.T.M., Ringle, C., Sarstedt, M.: A Primer on Partial Least Squares Structural Equation Modeling (PLS-SEM). Sage publications, Thousand Oaks (2016)

Restaurant Reservation System: Allocating Customers with Space Management Under the Impact of COVID-19

Bingxin Du(✉), Nariaki Nishino, Koji Kimita, and Kohei Sasaki

Department of Technology Management for Innovation, School of Engineering, The University of Tokyo, Tokyo, Japan

Abstract. The service industry takes an essential and necessary position in business and provides various types of entertainment to consumers. Restaurant is recognized as one of the most influenced service industries under the COVID-19 impact, consumers who goes to the restaurant are strictly decreased since the preventing from COVID-19. This paper strives to focus on designing a new reservation system help the restaurant manage the space efficiently and maximize the social value. Considering that customers may have incentives to misreport their values, the VCG mechanism is implemented to ensure that each customer takes rational actions. As a result, we expect that such a reservation system with the VCG mechanism applied can allocate customers efficiently and raise the revenue of the restaurant on condition that the proper prevention of COVID-19 contagion.

Keywords: COVID-19 · Restaurant service · VCG mechanism

1 Introduction

The service industry takes an essential and necessary position in business and provides various types of entertainments to consumers. In the economy, service industry contributes more than half of the global GDP since 1995, especially in Japan, services valued almost 70% of Japan's GDP in recent years. COVID-19, however, brings a disastrous change in the world economy, a 5.2% decrease in the world GDP is expected by the World Bank. Research related to COVID-19 has also been done academically, an SIR model using to estimate the evolution of COVID-19 is introduced by Atkeson [1], and it shows that social distance measurements are strictly required for almost one and a half year, resulting in huge economic cost of a cumulative loss of working and consuming. Especially, keeping social distance is formally avoided in the service industry, since communication is a significant tool for service provider to deliver the service. Nevertheless, not all business is hit by COVID-19 badly, Fernandes [2] states that the effect of COVID-19 is not distributed equally throughout the economy, some businesses actually benefit from the breakout, hospitality related service sectors, however, are particularly crippled, sectors such as transport, entertainment, retail, hotels and restaurants are influenced even badly. Fernandes [2] also points that the financial loss is unrecoverable in service sectors. For example, because of the breakout, you can delay the time

C. Leitner et al. (Eds.): AHFE 2021, LNNS 266, pp. 311–319, 2021.
https://doi.org/10.1007/978-3-030-80840-2_36

of buying a TV, but it is unlikely that you will go to restaurants every day after the pandemic disappears to complement the lost consumption during COVID-19. Because of this, service activities extremely decrease globally, and in Japan, more than 95% of the service business has been influenced by COVID-19.

Based on this situation, how to recover the service business rapidly and effectively under a controllable infection rate of COVID-19 becomes a popular issue. For example, Jordi et al. [3] suggest that because of fears of contagion in closed indoor space, services industries and restaurants might need to consider developing outdoor space in order to recover their business activities and ensure the prevention of COVID-19 infection. Such comment can be classified as the method which focuses on creating new space for serving customers under COVID-19 circumstances. Nevertheless, many restaurants, such as restaurants locating in the building, failed to create new space to serve customers. Moreover, Lu et al. [4] analyze an outbreak case of COVID-19 in an air-conditioned restaurant in Guangzhou, China, and conclude that the restaurant need to increase the distance between tables and improve restaurant ventilation. This can be an effective way to reduce the risk of infection, however, improving ventilation may be expensive, and restaurant holders may hesitate to do so. Therefore, this paper focuses on managing the indoor space capacity of restaurants as significant methods to avoid infections and to recover the service. Mentioning about the restaurant that contains a large proposition of the service industry, consumer behavior has changed by COVID-19, most customers prefer to use delivery services or take away rather than eat-in, but the truth is that the number of customers who go to or use restaurants is still strictly decreasing, and many restaurants are suffering and have to close during this tough period.

Following the fact and considerations above, we strive to explore a new reservation system using mechanism design, trying to help restaurants to allocate consumers efficiently with their limited indoor capacity when there is congestion. Furthermore, the reservation system designed is also expected to increase restaurant revenue simultaneously, since it gives restaurants the right to select their customers in order to maximize the value of their products and services. To achieve this objective, we implement the VCG (Vickrey–Clarke–Groves) mechanism to the new reservation system. The reason why we wish to adopt auction system in the restaurant reservation system is that the value of the space is ambiguous, in other words, the space is not priced in most services of current society. Especially, in COVID-19 situation, the value of the space is gradually recognized, and consumers are willing to pay an extra money to keep social distance with each other. Although the number of customers decreases and it would be easier to reserve a table than usual times, for customers who prefer to go to the restaurant and also care about COVID-19, this reservation system provides an option for them to control the number of customers in the restaurant indirectly. If such customers exist, and they are willing to pay for the space or extra tables in the restaurant, then the number of customers who are dining together can be decreased, and the social value also can be maximized under the prevention of COVID-19. Since the auction system is decided to be adopted in order to label the space with the specific price, we find that the VCG mechanism can efficiently achieve our goal and give us a desirable result.

2 Literature Review

The VCG mechanisms is well known in auctioning and public good problems. Fundamentally, the VCG mechanism belongs to the Groves mechanism [5] since the Clarke mechanism [6] is a special case of Groves mechanism and the Vickrey mechanism [7] is a special case of the Clarke mechanism [8]. The VCG mechanism is useful because it promises the maximization of the social value and it also has other desirable properties. Generally, there are N agents in the game, and agent $i \in N$. The VCG mechanism can be defined as a directive mechanism $\mathcal{D} = \left((\Theta_i)_{i \in N}, f(\cdot)\right)$, where Θ_i is the set of each possible profile of agent's type, and $f(\cdot) = (x(\cdot), t_i(\cdot), \ldots, t_n(\cdot))$ is a social choice function, and $x(\cdot)$ is an allocated outcome and $t(\cdot)$ is a monetary transfer function, which are respectively satisfying

$$x(\theta) \in arg \max_{x \in K} \sum_{i=1}^{n} v_i(x, \theta_i). \tag{1}$$

$$t_i(\theta) = \sum_{j \neq i} v_j\left(x^*(\theta), \theta_j\right) - \sum_{j \neq i} v_j\left(x^*_{-i}(\theta_{-i}), \theta_j\right). \tag{2}$$

Equation (1) shows the allocative efficiency of the VCG mechanism, ensuring that for any given θ, the sum of the valuations of agents is maximized over an outcome x. Equation (2), known as the *Vickrey Payment*, guarantees the budget balance, and it is defined by the difference between the sum of valuation of agents expect agent i when i is participating in the game and the sum of valuation of agents expect agent i when i is not participating in the game.

Normally, budget balance is defined as $\sum_{i=1}^{n} t_i(\theta) = 0$. In the VCG mechanism, however, the Eq. (2) results in $\sum_{i=1}^{n} t_i(\theta) < 0$, which is called weak budget balance by Garg et al. [8]. Although the budget balance is weak, under such payment rule and allocatively efficient $f(\cdot)$, the VCG mechanism is still capable to give us a desirable property called dominant strategy incentive compatibility. This property ensures that truth-telling by each agent will reach a dominant strategy equilibrium by applying the mechanism $\mathcal{D}$, in other words, agents are willing to report their true value over an outcome set K.

Because of the desired properties of the VCG mechanism, it has been applied widely throughout social and economic issues. Parking system which is a popular field with VCG mechanisms applied has been studied by many researchers, Chen et al. [9] use the VCG mechanism in parking system to manage the car park position efficiently, resulting in minimized total social value. Specifically, the VCG mechanism guarantees the truth telling for each driver when they are asked to report their destinations, and based on the destination and valuation provided, an additional fee which is calculated by the VCG mechanism monetary transfer is expected to be paid to regulate drivers' reporting. Under this constitution, the parking system is capable to allocate each drive to different parking stations, in ordering to achieve the minimized total social cost. A multiple stage VCG mechanism has also been studied in the parking system, Shao et al. [10] provide a two-stage auction with VCG mechanisms applied. They divide drivers into three types: neutrality, loss aversion, and gain seeking, and calculate monetary transfers towards these three different type drivers. The conclusion they find is that the utility of neutral drivers reaches the highest, and the payment they receive stills the lowest, comparing to the loss aversion drivers and gain seeking drivers. The VCG mechanism is also applied to the online system for advanced ad slot reservations, this is a trading between advertiser and publisher. Constantin et al. [11] also add cancellations to such online system, and they find that an implementable online mechanism with the VCG mechanism applied can derive prices and bump payments efficiently and also prove incentives and revenue. As above studies show, the VCG mechanism is widely used in the parking system and gives the desirable performance. When applying in restaurant reservation system, however, such sound studies are rare. Then, we construct the following model to explore the value of implementation of the VCG mechanism.

3 Modeling

3.1 Assumptions

The premise of the model assumes that the actual number of consumers are decreased under the COVID-19 circumstance because of the anxiety of customers, however, the number of consumers who are willing to go to the restaurant are not decreased. If the social distance can be made sure when the customers are dining, they prefer to go to the restaurant as usual. Moreover, customers who are willing to pay for the extra tables to ensure the social distance are existed. Therefore, the model is established for the space management and social value maximization. In the model, there exist a service provider and consumers. The service provider runs a restaurant business where course dinners and tables are provided. Each consumer thinks of making a reservation in the restaurant based on their preferences of courses and dinning environment, in other words, consumers need to submit a reservation with course and number of tables combined, and the choice of consumer is called outcome x.

Supposing the measurement of preventing contagion of COVID-19 is implemented, implying that social distance is required. Specifically, it is realized by giving the right to consumers to choose the number of tables, and a payment that is calculated by the VCG mechanism is received by the consumer based on the information provided. Such a model with the following booking system is introduced to the restaurant business.

- Considering reality, the number of tables (M) that can be chosen will always be greater or equal to one, and it is assumed that a table comes along with two seats. For instance, if there is only one consumer who is trying to make a reservation, the minimum number of tables that he or she can choose is one, and the maximum is the total number of tables in the restaurant. On the other hand, if the number of customers is three or four, the minimum number of tables that they can choose is two, since the number of customers goes beyond the capacity of one table.
- Various types of courses (C) are provided by the restaurant, denoted by c, where $c \in C$. The cost of courses is considered to be different, and the restaurant has its own valuation for courses.

- A start-price of the booking fee (b) that can be thought of as an entry fee is assumed to be paid, and it is denoted as the function $\bar{b}(m)$, implying the minimum price of booking fee is different between the customer who reserves one table and one who reserves two tables. In the reservation system, such a start-price ensures the show-up probability of customers whose reservation is confirmed by the restaurant. An additional booking fee is also allowed to be added, and it is considered to be a crucial factor when the restaurant is comparing the value among customers who choose the same course.
- The price of menu courses (p_i) is expected to be set by customers based on the valuation of the restaurant to each course. Like auction in the case of the auctioneer has the valuation of the good, we assume that the restaurant has positive values for courses, that is, the cost of courses is included, $cost_c > 0$. The restaurant announces a reserve price $\bar{p}_c$, which may not equal to $cost_c$. Customers see the reserve price as the valuation of the course for the restaurant, and they bid based on the reserve price $\bar{p}_c$.
- It is assumed that customers can only make a reservation for the next day. To avoid conflict of reservations, customers are asked to select possible time periods when making a reservation, and the restaurant will confirm the exact reservation time to the customer at the end of the day. One hour is assumed to be the basic unit to measure the time period. For example, customers i and j are making reservations for tomorrow's dinner, customer i chooses a time period from 6 p.m. to 8 p.m., meaning that customer i can have dinner at any time between 6 p.m. to 8 p.m., and customer j chooses a time period from 7 p.m. to 8 p.m. If there is only one available table at 7 p.m., then the restaurant will arrange the dinner for customer j at 7 p.m. and customer i at 6 p.m. If all the information provided by the customers is the same, then the allocation would be decided by the ordering sequence.

3.2 VCG Mechanism Implementation

The reservation system is considered in terms of the auction, in that sense, the restaurant is not expected to reply to the customer immediately once the customer finalizes their reservations. As assumed above, customers are only allowed to make a reservation for the next day, and the restaurant will confirm their reservation results by the end of the day, depending on the information provided by the customer. Customers are asked to make a collective choice from a bundle of numbers of tables (m), menu courses (c), a reservation fee willing to pay ($p + b$), and available time periods when they are making a reservation. Therefore, the utility function of customer i is defined as

$$u_i = v_i(x) - (p_i + b_i). \tag{3}$$

The profit function of the restaurant is considered as

$$\Pi = \sum_{i=1}^{N} (p_i + b_i) - cost_c \times \textit{the number of courses sold}. \tag{4}$$

With the VCG mechanism applied, a *Vickrey Payment* is actually paid by the customer, denoted by t_i for customer i. Therefore, the utility of customer i and the profit function of the service provider are changed to the following:

$$u_i = v_i(x) - t_i. \tag{5}$$

$$\Pi = \sum_{i=1}^{N} t_i - cost_c \times the\ number\ of\ courses\ sold. \tag{6}$$

3.3 Basic Model

To start, a simple model is established. There exist three customers $i, j,$ and k who are going to make a reservation for tomorrow's dinner at the same restaurant for the same time period. The restaurant provides two courses to the customer, course A, and course B, which are denoted by C_A and C_B, respectively. In addition, for the restaurant, C_A is more valuable than C_B. Only three tables are offered to the customer, and each customer can choose an amount of tables having dinner from one to three, and the total amount of tables allocated to the customer is no larger than three. Each customer is asked to enter a reservation fee that he or she is willing to pay for the course and the table, defined as course price p, and booking fee b. To apply the VCG mechanism, we need to list the valuation of customers $i, j,$ and k over different social outcomes. For instance, the following table shows the valuation of each customer over different outcomes x, and the number in the valuation function shows the number of tables.

For customer i,

$v_i(C_A, 1) = 15$	$v_i(C_B, 1) = 10$
$v_i(C_A, 2) = 15$	$v_i(C_B, 2) = 10$
$v_i(C_A, 3) = 15$	$v_i(C_B, 3) = 10$

For customer j,

$v_j(C_A, 1) = 10$	$v_j(C_B, 1) = 8$
$v_j(C_A, 2) = 12$	$v_j(C_B, 2) = 10$
$v_j(C_A, 3) = 14$	$v_j(C_B, 3) = 12$

For customer k,

$v_k(C_A, 1) = 8$	$v_k(C_B, 1) = 8$
$v_k(C_A, 2) = 16$	$v_k(C_B, 2) = 16$
$v_k(C_A, 3) = 20$	$v_k(C_B, 3) = 20$

From the valuation functions, each customer's preference over different outcomes can be ascertained. Customer i prefers menu courses to tables (or space), since its valuation does not change when the number of tables changes. Conversely, customer k prefers tables to menu courses because its valuation stays the same when the course changes. On the other hand, the preference of customer *j* is in the balance of preferences of customer i and k. Considering this valuation distribution, the social outcomes are shown in the Table 1.

Table 1. Possible outcomes with three customers, two courses, and three tables

Outcome	Customer *i*	Customer *j*	Customer *k*	Social value
x_1	C_A, 1	C_A, 1	C_A, 1	33
x_2	C_A, 1	C_A, 2	*Nothing*	27
x_3	C_A, 1	*Nothing*	C_A, 2	31
x_4	C_A, 2	C_A, 1	*Nothing*	25
x_5	C_A, 2	*Nothing*	C_A, 1	23
x_6	C_A, 3	*Nothing*	*Nothing*	15
x_7	*Nothing*	C_A, 1	C_A, 2	26
x_8	*Nothing*	C_A, 2	C_A, 1	20
x_9	*Nothing*	C_A, 3	*Nothing*	14
x_{10}	*Nothing*	*Nothing*	C_A, 3	20

*Since every customer has a positive value against the number of tables, and the restaurant wants to maximize the total value, only outcomes including full capacity (three tables are all occupied) are considered.
**Only course A is included in the list since the value of course A is no less than the value of course B for each customer.

Obviously, x_1 is the optimal allocation and gives us the largest social value which is equal to 33. Note that the restaurant actually will provide course B to customer k, since the value of course A and course B is indifferent to customer k, and the cost of course B is less than the cost of course A for the restaurant. Applying the VCG mechanism to this case, customer *i* receives a monetary transfer of $18 - 26 = -8$, customer *j* receives a monetary transfer of $23 - 31 = -8$, and customer *k* receives a monetary transfer of $25 - 27 = -2$. Meaning that the actual amount that these three customers need to pay for the course and the table are 8, 8, and 2, respectively. Recall that the utility function of the customer under the VCG mechanism is defined as $u_i = v_i(x) - t_i$. Therefore, the utility for customer *i*, *j*, and *k* are 7, 2, and 6, respectively. Moreover, the VCG mechanism also proves that truth-telling is a dominant strategy. For example, if customer *i* reports his or her value of the outcome $(C_A, 2)$ to be 30, then allocating $(C_A, 2)$ to customer *i*, and $(C_A, 1)$ to customer *j* gives us a total social value of 40. Since 40 is greater than 33, which is the prior optimal allocation, the restaurant will serve customers based on the new optimal allocation. Customer *i*, however, faces a monetary transfer of $10 - 26 = -16$, which is larger than his true value of the outcome $(C_A, 2)$, 15, resulting in a negative value of his or her utility. Since the customer is rational and intelligent, customer *i* is not expected to lie about the value, otherwise, his or her utility would decrease comparing to telling the truth. Therefore, in such reservation systems with the VCG mechanism applying, there are no incentives for customers to misreport their valuation towards different outcomes.

In the basic model, it shows how the VCG mechanism can be applied to the reservation system. The optimal social choice, however, allocates each customer a table under outcome x_1. Note that in COVID-19 circumstances, the congestion is avoided, then the first outcome x_1 is eliminated. Hence, the new optional optimal outcome would be x_3 which gives the second largest total social value of 31, and in this case, a combination of course A and one table $(C_A, 1)$ is allocated to customer i, and a combination of course A and two tables $(C_A, 2)$ is allocated to customer k. The monetary transfer for each customer i, j, k would be 10, 0, and 12, respectively. Moreover, if customer i misreports the value of outcome $(C_A, 2)$ to 30, then he or she will receive a monetary transfer of $10 - 26 = -16$, which is larger than the true value, 15, resulting in a negative value of its utility. Therefore, applying the VCG mechanism to the second optimal allocation also proves the dominant strategy incentive compatibility under this assumed valuation distribution.

4 Conclusion

In this paper, we have introduced a new restaurant reservation system with the VCG mechanism applied, aiming to help restaurants manage the space efficiently and maximize the total social value under the impact of COVID-19. More importantly, a full implementation of this reservation system is expected to increase the restaurant revenue and the activity index, further stimulates the economy. The process of auction is imported to the reservation system, bidding on the combinations of courses and number of tables. With the VCG mechanism applied, the truth-telling is guaranteed, and the outcome with the highest total social value can be achieved. Considering the COVID-19 circumstance, the full capacity of the restaurant is prohibited, meaning that the restaurant cannot serve customers with the same number of tables. This action reduces the risk of infection compared to that of usual times, and alleviates the anxiety of customer, further proves the number of customers who are still willing to go to the restaurant under the COVID-19 circumstance. In the further research, we also plan to consider the real case when implementing such a new reservation system to the restaurant. The general VCG mechanism requires agent information of valuations towards all combinations of courses and number of tables, considering the real situation of every restaurant, however, the procedure of reservation may seem cumbersome for customer. Therefore, we strive to find the optimal number of report combinations when consumer making a reservation for the restaurant in the next stage.

References

1. Atkeson, A.: What will be the economic impact of covid-19 in the us? Rough estimates of disease scenarios (tech. rep.). National Bureau of Economic Research (2020)
2. Fernandes, N.: Economic effects of coronavirus outbreak (covid-19) on the world economy. Available at SSRN 3557504 (2020)

3. Jordi, H.-R., et al.: The impact of covid-19 on public space: a review of the emerging questions (2020)
4. Lu, J., et al.: Covid-19 outbreak associated with air conditioning in restaurant, Guangzhou, china, 2020. Emerg. Infect. Dis. **26**(7), 1628 (2020)
5. Groves, T.: Incentives in teams. Econom.: J. Econom. Soc. **41**, 617–631 (1973)
6. Clarke, E.H.: Multipart pricing of public goods. Public Choice **11**(1), 17–33 (1971)
7. Vickrey, W.: Counterspeculation, auctions, and competitive sealed tenders. J. Financ. **16**(1), 8–37 (1961)
8. Garg, D., Narahari, Y., Gujar, S.: Foundations of mechanism design: a tutorial part 1-key concepts and classical results. Sadhana **33**(2), 83 (2008)
9. Chen, Z., Yin, Y., He, F., Lin, J.L.: Parking reservation for managing downtown curb-side parking. Transp. Res. Rec. **2498**(1), 12–18 (2015)
10. Shao, S., Xu, S.X., Yang, H., Huang, G.Q.: Parking reservation disturbances. Transp. Res. Part B: Methodol. **135**, 83–97 (2020)
11. Constantin, F., Feldman, J., Muthukrishnan, S., Pal, M.: An online mechanism for ad slot reservations with cancellations. In: Proceedings of the Twentieth Annual ACM-SIAM Symposium on Discrete Algorithms, pp. 1265–1274 (2009)

Empathy-Based CE Strategy to Tackle Complex Challenges

Hiromi Iida[1(✉)], Kensaku Ishibashi[1], Akino Inoue[1], and Yuriko Sawatani[2]

[1] Panasonic, Panasonic Laboratory Tokyo, 8-21-1 Ginza, Chuo-ku, Tokyo 1040061, Japan
{iida.hiromi,ishibashi.k}@jp.panasonic.com, ainoue@bizdone.co.jp
[2] Business School, Nagoya University of Commerce and Business, Nagoya, Japan
yuriko_sawatani@nucba.ac.jp

Abstract. Circular Economy (CE) is a concept that transforms the consumption of resources in corporate activities into a new form of symbiosis for sustainable economic growth. Efforts towards CE require multi-departmental collaboration and transformation of existing systems, so it is difficult to proceed unless clear goals are set. In an organization that has aimed to expand its business as a manufacturing industry leader in the linear economy, it is necessary to make a major change in thinking in order to consider the value of the entire CE ecosystem. Also, the digital transformation of the business in a platform society where value co-creation with ecosystem partners is increasingly important, created additional challenges and opportunities for alignment with the CE strategy.

We try to develop a strategy from the bottom up based on the effectuation of entrepreneurial theory and human-centered design. In this paper, we discuss the background, approach, and practical results.

Keywords: Circular economy · Radical innovation strategy · Human centered design · Design thinking · Systems thinking · Effectuation

1 Introduction

Globalization is advancing as economic activities expand, and global environmental problems such as climate change cannot be ignored. Many governments have come up with "Green Growth Strategy" aimed at creating a virtuous cycle between the economy and the environment, and have called on society, companies, and people to make changes. Manufactures face complex challenges that cannot be solved by conventional business growth strategies. There are many points of contact with society and people, such as resource procurement, production, post-sales maintenance, and recycling. As a result, new business models to address these issues tend to be complex.

This paper takes up "Circular Economy" as an example of such social issues, and shows the strategic approaches based on empathy in a technology-oriented company. Circular economy is a concept that transforms the consumption of resources in corporate activities into a new form of symbiosis for sustainable economic growth. Based on these changes in the external environment, we aim to establish a bottom-up, human-centered strategy for the future.

C. Leitner et al. (Eds.): AHFE 2021, LNNS 266, pp. 320–327, 2021.
https://doi.org/10.1007/978-3-030-80840-2_37

2 Background

In December 2015, the European Commission announced “Circular Economy Package” that will form the core of its growth strategy for 2030. Unlike the conventional one-way economy that consumes and disposes of resources (“Linear Economy”), “Circular Economy (CE)” is a new economic model that realizes economic growth independent of resource consumption (decoupling) by recovering consumed resources and continuing to reuse and recycle them. In addition, as people’s lifestyles change, new values that make use of service rather than physical products are spreading globally. Governments outside Europe are also focusing on CE as part of their sustainable growth strategy.

In order to realize a sustainable society, Japanese companies are taking advanced measures focusing on 3R (= reduction, reuse, recycling) and individual elemental technologies and products. For example, it has a system based on advanced recycling technology, and maintains a high recycling rate internationally as well as the national character of Japan’s “Mottainai = Mind to treasure things”.

CE emphasizes an approach that focuses on society and people’s lives. For example, there are “Products as a Service” schemes that emphasize the use of shared platforms and business models such as maintenance and refurbish in which physical products can be utilized on shared platform over the long term. Furthermore, it is necessary to design the entire production cycle from the procurement of resources to product life, sales, distribution, use, and recovery, and to build a supply chain that achieves zero waste in society. These challenges are characterized by open, complex, networked dynamic systems. As a result, various new business models for addressing these issues tend to be complex. In addition, it is difficult to promote CE from the top down in large companies, unlike addressing environmental regulation issues or customer needs. Because it is difficult to set the future vision of CE and the concrete goal to be realized. Indeed, the future vision of CE is being debated among governments, academia, and companies.

The reason why it is difficult for large companies to tackle new businesses, including changing their existing business models are described as follows:

First, it is not easy for large companies, which emphasize technology and are optimized for product development and sales, to tackle social issues. It is difficult to solve social problems with a single technology or product, and discussions for this purpose often involve friction between organizations. Top-down strategies are not feasible plan to be implemented because organizational friction prevents substantive discussion.

Second, unlike technical problem setting, knowledge and information within a large vertical organization cannot analytically advance the problem setting of social issue origin. The procedure of social observation and qualitative investigation on the problem setting of the social issue cause has not been established.

Third, even if a problem can be set, it is difficult to solve the problem and implement it for future business. For complex problems, experimenting with small ideas is effective, but it is difficult in a large vertical organization.

As described above, it is difficult for existing large companies to promote long-term themes involving social innovation for the future. However, they have the potential to take the lead in drawing up visions for the future, proposing necessary mechanisms and ways to conduct business ahead, and forming and implementing an ecosystem with their business partners. Therefore, a new approach suitable for large enterprises is necessary, and we tried to establish it.

3 Hypothesis and Approach

In the following, the hypothesis and approach of the strategy planning for the new business of the social issue origin in the large enterprise are described taking the strategy decision on Circular Economy as a theme.

As we have seen in Chapter 2, it is difficult to derive a top-down strategy for the radical innovation of the social issue origin in large enterprises. Therefore, the approach in the bottom up is investigated, but there was no method which is a basis in the bottom up until now. And, in order to carry out the problem setting with the aim of the radical innovation focusing on social issue, the analytical approach based on the technology roadmap alone is insufficient, and the method for social observation and qualitative investigation is required. In addition, it is necessary to decide the strategy roadmap for the problem solution and to experiment with small ideas. We made the following hypotheses and approaches to tackle the challenges.

3.1 Hypotheses and Approaches to Organizational Challenges

In the bottom-up approach in large enterprise, we assume that the principle of Effectuation which is the theory of the entrepreneurship [1, 2] which produces the radical innovation in the world could be applied in order to promote the activity without depending on the organization system.

As a concrete approach, members who consider CE by themselves are gathered and building a small community starting with sharing their thoughts among them. We assumed activities corresponding to each principle of "Effectuation" as shown in Table 1.

Table 1. Expected activities corresponding to each principle of "Effectuation".

Effectuation principle	Meaning	Expected activity
Birds in hand	Take advantage of the resources	Engage members who are thinking about CE by themselves
Affordable loss	Limit acceptable losses	Leverage the expertise of members to promote autonomous action
Pilot in the plane	Concentrate on the controllable range	Focus on "Creating Value Proposals for Future Users"
Crazy quilt	Expand resources through partnerships	Expand the scope of efforts through a network of members
Lemonade	Take advantage of unexpected events	

3.2 Hypotheses and Approaches to Problem Setting Challenge

For the second challenge for the problem setting of the social issue origin, we made a hypothesis that it could be effective to create future picture by the human centered process with focusing on the person who has already anticipated the future world of circular economy.

As a concrete approach, we make the interview to the extreme user who takes a lead in the concept of Circular Economy and obtain a future picture by design thinking for future customers [3].

3.3 Hypotheses and Approaches to Problem Solving Challenge

For the third challenge that the problem solution does not advance, if the future picture was obtained by some means, how to promote experiments with small ideas in each department was examined. Then, we thought that it would be effective to grasp the whole management problem from a bird's-eye view, to obtain the clue of the innovation, and to bring back the content which projected the strategy hypothesis to the responsible region of each department on the basis of the understanding of the whole picture. Thus, we constructed the two-stage process based on the hypothesis.

As a concrete approach, the 1st stage applies the service design which utilized design thinking and system thinking, and it clarifies the gap between future picture ("To-be") and present mechanism of organization ("As-is"). Then, management issues are highlighted and an initial strategy is formulated as a scenario of how "As-is" can be transformed to "To-be". As the 2nd step, the future picture and the strategy for the realization are incorporated into the social trend and the technology roadmap which each technology division possesses to carry out the forced connections. This is equivalent to projecting a future picture onto each division's territory. By this, the technology roadmap of each division and the future picture can be associated. Each member brings back to own division the association between the concept of the future picture obtained through the activities and the technology roadmap of assigned area.

An overview of these approaches is shown in Fig. 1.

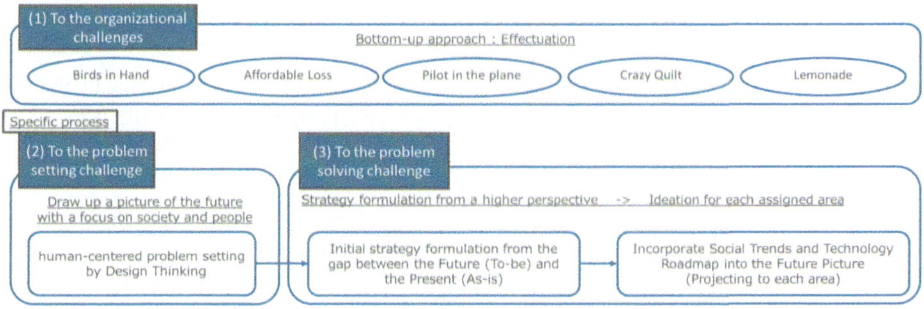

Fig. 1. Overview of approaches to the challenges.

4 Practical Results and Considerations

The result of the practice based on the above approach and the consideration are described.

4.1 Results of Efforts for Organizational Issues

As a result of tackling the bottom-up activity which we aim on the basis of the hypothesis that "Effectuation" which is the entrepreneur theory could be effective, the result as aimed by the hypothesis was almost obtained. The members who sympathized with the aim of the better world through solving the social issue, and tried to positively advance the preparation in the standpoint of themselves by regarding CE as an opportunity of the business innovation gathered, and the base of the community was formed. By sharing and confirming the meaning of CE for the company among members, convincement feeling for the aim direction and confidence to the team were obtained. The practical results corresponding to each principle of "Effectuation" are shown in Table 2.

Table 2. Practical results corresponding to each Effectuation principle.

Effectuation principle	Our expectation	Practical result
Birds in hand	Engage members who are thinking about CE by themselves	A good community was formed. They gained a sense of satisfaction with the direction they were aiming for and a sense of trust in the team
Affordable loss	Leverage the expertise of members to promote as autonomous action	There was little organizational friction, and there were no concerns about the cancellation of initiative
Pilot in the plane	Focus on "Creating Value Proposals for Future Users"	A tentative vision for future users was created
Crazy quilt	Expand the scope of efforts through a network of members	The human network of each key person was effectively utilized, and sufficient input was obtained
Lemonade	N.A.	Though it was affected by the pandemic, the activity was accelerated by the on-line communication and the diversity of the member advanced

4.2 Results of Efforts for Problem Setting Issues

For the second challenge that the problem setting of the social issue origin cannot be achieved by the analytical technique based on the knowledge in the organization alone, we created the future picture of the human-center for the future user by the setting of persona which symbolizes the world-view of Circular Economy, and the hypothesis was confirmed.

Specific results are as follows. The user interview was carried out as "empathy process" in the design thinking in order to collect information outside the organization. We conducted depth interviews with extreme users who are now a bit strange but will be common in the circular society of the future, and came up with ideas while understanding their values and desires in a circular society.

After obtaining sympathy points and insight from the interview, the authors examined various ideas including what kind of problem solution should be made for the future, and the trend in which the sign has already appeared. For example, for the scenario of "Creation of an ecosystem to meet the expectations of users who do not dispose of products immediately after failure and want to continue to use them after repair", a hypothesis was obtained to reinforce the original scenario such as "With the aging of the population and the increase in the number of working elderly people, the number of repair workers will increase.".

Through these verifications, multiple future scenarios and human-centered future picture (to-be) were set. By this activity, the value hypothesis which the core member can sympathize with and the future problem were derived. That is to say, we could draw pictures what CE aims at and identify user value and problem which will occur there.

4.3 Results of Efforts for Problem Solving Issues

As a result of the problem setting, the process of obtaining the clue of the innovation from the bird's-eye view in the 1st stage and projecting the future picture and the strategic hypothesis to the assigned area of each department in the 2nd stage was tackled in respect of the third challenges. On the first stage process, the strategy roadmap for realizing the future picture was obtained as a convincing initial hypothesis, and we can say that it functioned effectively. On the process of the second stage, as a result of the intended activity, the satisfaction of the participation members was obtained, and the effectiveness was confirmed. However, it is necessary to verify how much it led to the next action in the future.

Specific efforts are as follows.

In the 1st stage, system thinking was applied to the human-centered future picture (to-be), and it was analyzed as a system dynamic. The system diagram was drawn by hearing and analysis in the office on what the present mechanism (as-is) of organization. By this, each member understood the figure which became a deadlock from constraints such as sense of value, key performance index and role sharing in the existing business. For example, in the as-is structure, business evaluation is only an economic indicator, and it does not lead to business evaluation even if customer satisfaction is raised by sufficiently responding to users who wish to repair. It became clear that this was one of the management problems for promoting CE. On the other hand, the whole road map on how to approach to "to-be" was considered, where to start and what steps to realize it were discussed together with time axis. By this process, the members realized the feasibility of the created future picture, and the will to promote this theme by themselves heightened.

As a process of the second stage, social trend and technology trend which each division possesses were incorporated into the future picture and strategy. As a result of the forced connections, many small hypotheses which detailed the future picture were created. By this process, it was possible for participants to find the relation between their development activity and the future picture. As a result of the activity, they brought back to their division the collection of ideas which projected the future picture and strategy in their assigned area. From this, it is expected to be incorporated into the strategy of each department, and further connected to the development activity of bottom up. The overall result of the above practice is shown in Fig. 2.

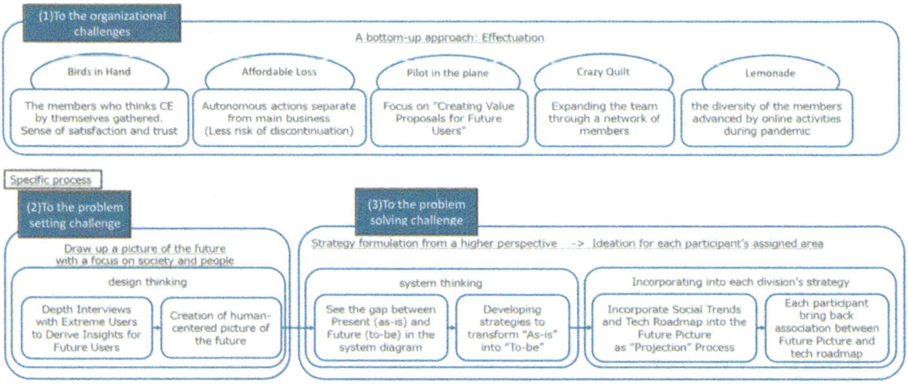

Fig. 2. Overview of the results of implementation

5 Conclusions and Future Directions

In this paper, under the theme of developing strategies for Circular Economy, we implemented strategies for new businesses towards solving social issues at large enterprises in a human-centered, bottom-up approach with a future perspective.

Regarding "Organizational challenges", members who were highly aware of social issues and positively prepared for themselves became core members and promoted initiatives. The diverse expertise of the core members also helped them to develop a strategy that included the diverse perspectives of the participants.

As for "Problem setting issues", by practicing design thinking based on interviews with future-oriented extreme users, we were able to create a concrete "future picture" that the participants could sympathize with.

For the "problem-solving challenge", we were able to develop a strategy by analyzing the as-is and to-be gaps, and to create a group of ideas for concepts and solutions that can be mapped to the technical areas of each department to which the study members belong.

The process thus far has created a bottom-up activity in which participating members can act autonomously. However, there are still a limited number of members willing to work on Circular Economy, so it is not yet possible for divisions to prioritize the

Circular Economy theme and incorporate it into their technology strategy. Therefore, it is necessary to associate with a top-down strategy in order to promote it throughout the company. Therefore, we would like to reach out top management to develop the CE strategy further.

References

1. Sarasvathy, S.D.: Causation and effectuation: toward a theoretical shift from economic inevitability to entrepreneurial contingency. Acad. Manag. Rev. **26**(2), 243–263 (2001). https://doi.org/10.5465/AMR.2001.4378020
2. Dew, N., Sarasvathy, S.D., Read, S., Wiltbank, R.: Immortal firms in mortal markets?: an entrepreneurial perspective on the "innovator's dilemma." Eur. J. Innov. Manag. **11**(3), 313–329 (2008). https://doi.org/10.1108/14601060810888982
3. Sawatani, Y.: Innovation of service system by human-centered design and effectual evolution: hypothesis development and case verification. Naples Forum on Service (2019)

Education, Training, and Technology Facilitating a Future Workforce

Exploring Disciplinary Technologies for Increased Accessibility in the Civil Engineering and Construction Industry: Starting the Conversation

Cassandra McCall[1(✉)] and Denise R. Simmons[2]

[1] Engineering Education Department, Utah State University, 4160 Old Main Hill Rd, Logan, UT 84322, USA
cassandra.mccall@usu.edu
[2] Department of Civil and Coastal Engineering, University of Florida, 365 Weil Hall, Gainesville, FL 32611, USA
denise.r.simmons@essie.ufl.edu

Abstract. Due to an aging workforce and national recruitment initiatives, the number of individuals with disabilities in the civil engineering and construction industries is expected to increase. However, little work has examined existing access barriers that discourage disabled and aging individuals from entering into or remaining in the civil engineering and construction fields, with the majority of extant work focusing on individual accommodation. In this paper, we explore the ways existing and emerging workplace technologies that are reshaping disciplinary norms also improve the overall workplace accessibility in civil engineering and construction. Specifically, we focus our discussion on three technologies impacting civil engineering and construction: drones, augmented reality, and prefabricated and modular construction. In this discussion, we introduce points of alignment observed across these technologies and disability studies perspectives and conclude with identifying areas for future work and considerations for industry for creating universally accessible environments without perpetuating colonizing or ableist ideals of workplace participation.

Keywords: Civil Engineering · Workplace accessibility · Universal design · Assistive technologies · Broadening participation

1 Introduction

Currently available statistics show that approximately 10% of employed engineers in the US report having one or more disabilities [1]. This number is expected to rise due to increases in disability reporting as the current workforce continues to age. The average age of civil engineering and construction professionals is approximately 42 years old [2], with workers aged 55 years and older experiencing increases in workforce representation from 17% in 2011 to 22% in 2018 [3]. While we do not equate age with disability, we acknowledge that reported disabilities tend to increase with age [1].

C. Leitner et al. (Eds.): AHFE 2021, LNNS 266, pp. 331–337, 2021.
https://doi.org/10.1007/978-3-030-80840-2_38

At the same time, national initiatives to diversify the engineering workforce have consistently underscored the need for disabled individuals in these fields and positioning their perspectives as necessary for advancing innovation and broadening the societal impacts of engineering research and practice [4, 5]. While they are still severely underrepresented at the disciplinary level and across institutions, individuals with disabilities are slowly matriculating into higher education and science and engineering fields. Statistics from agencies such as the National Science Foundation (NSF) and the National Center for College Students with Disabilities (NCCSD) have reported an almost 10% increase in university enrollment for individuals with disabilities from approximately 11% in 2008 to 20% in 2018 [1, 6, 7]. Similar trends are observed in the field of science and engineering, where college students with disabilities are about as likely to major in these fields than in non-science and engineering fields [6].

Recent research is just beginning to understand the experiences of undergraduate science, technology, engineering, and math (STEM) students during college and their implications for higher education culture and practice [8, 9]. From this work, we know that underlying disciplinary cultures typically associated with engineering (e.g., perceiving accommodations as 'extra' help by limiting academic independence; positioning engineering education as a meritocracy of difficulty) contribute to students' stigmatizing experiences as they learn to navigate academic environments [8]. At the same time, studies examining sociocultural factors that influence the participation of individuals with disabilities in industry have found that employers often harbor sincere yet pessimistic views when comparing individuals' performance abilities with the costs associated with hiring them [10–12]. To improve industry recruitment and retention, researchers have examined accommodation practices to improve the accessibility of workplace environments. However, a majority of this work has been dominated by conversations of individual accommodation rather than on shifting industry cultures and workplaces to meet the needs of disabled professionals [13].

In this paper, we move beyond concepts of individual accommodation by exploring the ways workplace technologies can promote universal accessibility. Informed by the social and social relational models of disability, we will use this exploration to critically examine existing and emergent technologies that are reshaping disciplinary norms and overall workplace accessibility in civil engineering and construction. From this work, we introduce points of alignment observed across these technologies and disability studies perspectives, identify areas for future work, and provide industry with considerations for creating universally accessible environments without perpetuating colonizing, ageist, or ableist ideals of workplace participation.

2 Models of Disability

We first frame our discussion of promoting universally accessible workplace environments in civil engineering and construction by drawing from three existing constructs within the field of Disability Studies: 1) the medical model of disability, 2) the social model of disability, and 3) the social-relational model of disability.

Medical models of disability position physical, cognitive, and developmental difference as a "sickness" or "condition" that must be "treated" [14]. From this perspective, disability is perceived as an impairment that must be accommodated so that individuals can obtain a dominantly accepted sense of normality. Within broader workplace cultures, literature that focuses on individual accommodation typically follows this model in that an individual must adapt to their surroundings.

Social models of disability position disability as a dynamic and fluid identity that consists of a variety of physical, cognitive, or developmental differences [14]. Dissenting from assumptions of normality and the focus on individual bodily conditions (hallmarks of the medical model), the social model focuses on the political and social structures that inherently create or construct disability in our daily lives [14]. A focus on the universal design of materials and tools accessible to all employees within a given workplace environment exhibits an awareness of disability as a social construct, thus shifting the responsibility of adaption from the individual to the environment.

The social-relational model positions disability as generated through social relationships that are controlled and constrained by social structures and actions [15]. For example, in an attempt to reduce the barriers experienced by individuals with disabilities, disability advocates helped pass Section 504 of the Rehabilitation Act of 1973. Section 504 was the first disability civil rights law that prohibited discrimination of people with disabilities in programs that receive federal funding, which later led to the Americans with Disabilities Act (ADA) of 1990. From this perspective, the social-relational model combines aspects of the medical model (i.e., disability as condition) with those of the social model (i.e., generated by surroundings) to capture the lived experiences of individuals with disabilities as they interact with the world around them.

Here, we draw from the social and social-relational models of disability to shift conversations of workplace accessibility from individual accommodation to universal accessibility by positioning technologies not as tools for individual accommodation, but for changing industry culture.

3 Enhancing Inclusivity Through Industry Technologies

A number of emerging technologies are shaping the ways civil engineering and construction work is conceptualized and conducted. Some of these technologies include connected systems, artificial intelligence, and machine learning [16–18]. In this paper, we focus our discussion on three existing and emergent technologies in the field: 1) drones, 2) augmented reality, and 3) prefabricated and modular construction. We focus on these three technologies due to their increasing and prominent use in the field as well as their potential impacts on overall civil engineering and construction workplace culture, particularly in relation to promoting accessible workplace environments. In this section, we provide a brief description of each technology and its anticipated implications for disabled professionals in the field.

Drones. Drones are slowly emerging as a useful tool in the construction and civil engineering industries [19, 20]. In the construction and civil engineering context, drones have been predominantly used for inspection and surveying applications. They allow companies to conduct expansive and detailed inspections within a matter of hours for large projects that would take humans days or even weeks to conduct on foot. Moreover, drones can provide inspection viewpoints not accessible or achievable by traditional means. Overall, this can dramatically cut labor costs and ensure construction and building quality while improving employee safety.

Drones are not typically classified as an inclusive technology; however, they can significantly broaden the participation of individuals with disabilities in construction and civil engineering for individuals with a variety of physical disabilities including mobility impairments, hearing impairments, and specific types of visual impairment as well as developmental impairments such as autism. A significant benefit to drone-based inspections for these groups includes reduced environmental stimuli that are often experiences on job sites and can be overwhelming for these individuals. For example, in the case of Stargardt's Disease, a form of early onset macular degeneration that reduces central vision and heightens light sensitivity, drones would allow inspectors to capture close-up views of project sites, limit exposure to bright, sunny environments, and eliminate exposure to a variety of job site hazards. In this context, applying a social and social-relational perspective of disability would prompt companies to offer all employees options for completing such inspections using this technology and allow employees to conduct their work in environments that are most advantageous to them.

Augmented Reality. Augmented reality (AR) has become an innovative teaching tool in civil and construction engineering education by supplementing traditional education techniques with avenues for viewing and engaging in recordings of active construction sites [21, 22]. In these instances, augmented reality provides learning opportunities for students that could not be achieved using traditional practices alone. Similarly, AR has been used in industry to enhance the capabilities of standard computer-aided-drawing (CAD) and other visualization tools [23], to conduct as-built and as-planned comparisons of in-progress projects [24], and to improve remote design collaboration across geographically-separated teams [25]. While this is not an exhaustive list of the ways AR has been applied in the design, planning, building, and review phases of construction (see [26] for an in-depth overview of these applications), it demonstrates the potential for AR to significantly shape the sociocultural norms that govern the ways civil engineering and construction professionals interact with one another and conduct their work.

The use of AR in university and industry settings establishes a strong foundation for application of this technology to improve the inclusion of people with disabilities in the civil engineering and construction field. Combined with current technologies and infrastructure improvements necessitated by the COVID-19 pandemic, these technologies can be used to disrupt traditional office or field-based work to allow employees to work from a variety of locations while maintaining the ability stay actively engaged and involved in a project. This technology can be particularly useful for individuals with intermittent physical disabilities (i.e., physical disabilities with which individuals do not continuously experience symptoms such as Lyme disease or postural orthostatic tachycardia syndrome [POTS]) because they allow employees to flexibly shift from in-person to virtual working environments and vice versa as needed.

Prefabricated and Modular Construction. Over the past few decades, prefabrication and modular construction have emerged as viable options for construction practices [27]. Prefabrication, or prefab, is a manufacturing process in which materials are assembled to create a component of an installation [27, 28]. Modular construction, a process similar to prefab but occurs on a larger scale, refers to a process in which items are assembled off-site to create fully furnished modules that are then transported to the construction site. These processes have been predicted to benefit the construction industry by improving productivity and quality, reducing waste, and resolving issues related to lower profit margins (see [27] for an in-depth discussion). A significant benefit of these processes is the environments in which they occur. Because they involve manufacturing components for transfer to other sites, they can be carried out in controlled factory environments that protect workers from the elements such as excessive heat and high winds.

For individuals with disabilities, allowing opportunities for employees to conduct their work in controlled environments is significant and can reduce access barriers for not only individuals with limited mobility but also individuals who experience various forms of neurological and physiological conditions that cannot be easily managed in outdoor settings. For example, individuals with multiple sclerosis are often quite sensitive to heat, which can cause dizziness, numbness, and reduced strength. Individuals with Lyme disease can experience significant joint pain that is exacerbated by cold temperatures or experience increased headaches due to heat. Approaching environmental barriers from the social and social-relational models of disability could include creating university-industry partnerships that connect disabled graduates to companies and positions that conduct their work in these controlled environments. Similarly, companies may provide individuals with options for conducting components of their work within controlled environments or ensure employee access to spaces with controlled environments.

In this paper, we only highlight just a few benefits and potentials of these technologies for professionals with disabilities in the workplace. While the implications highlighted here are fairly intuitive, we use these examples to shift disciplinary discourse toward an intentional inclusion of accessibility as a necessary component of workplace cultures and environments. We examine these technologies as a way to oppose views that creating accessible work environments is out of our reach and position them as an existing yet underutilized pathway from inclusion and the de-stigmatization of individuals with disabilities in these fields.

4 Conclusions and Future Work

In this paper, we examine three technologies (i.e., drones, augmented reality, and pre-fabrication and modular construction) that are significantly influencing the ways civil engineering and construction professionals interact and conduct their work. We focus on these technologies to initiate conversations that explicitly link technological advancements to advancements in universal accessibility in civil engineering and construction. From this perspective, we position these technologies as indicators of cultural change that disrupts normative conceptions of who civil engineering and construction professionals are and the type of work they can conduct.

While we speculate on the advantages of utilizing these technologies for increased accessibility in civil engineering and construction work environments, more research is necessary to understand how workplace culture, technology use and development, and employee/employer training inform the universal accessibility of the workplace. In order to truly achieve an accessible workplace for those with physical, cognitive, and developmental disabilities, technology must operate in harmony with the nature of the work to be conducted, employee abilities, and task outcomes. Future work includes advancing the design of intelligent work technologies to remove barriers that discourage disabled persons from entering and remaining in work as civil engineering and construction professionals. Intelligent Tutoring Systems (ITS), which are similar to computer-based learning environments, have the potential to answer this call. ITS exhibit a high degree of interactivity with a worker, construct a worker's psychological state, and adapt tutoring functions according to responses from the worker. The key to this technology is its ability to provide individualized learning opportunities. Moreover, our future work aligns with prior literature that calls for training programs to educate employers on techniques for implementing such technologies as well as approaches to creating inclusive workplace cultures and practices [29].

Overall, we contend that by adjusting the lenses by which technologies are conceptualized, developed, and applied in the civil engineering and construction industries, we can leverage them to not only advance the technical prowess of the field, but also to promote the inclusion of all individuals. Repositioning these technologies will further promote inclusive workplace cultures, contribute to efforts that seek to broaden participation in the field, and further augment the impacts of engineering research and practice on society.

Acknowledgments. This material is based upon work supported by the National Science Foundation under Award No. 1733636. Any opinions, findings, and conclusions or recommendations expressed in this material are those of the author(s) and do not necessarily reflect the views of the National Science Foundation.

References

1. National Science Foundation: Women, Minorities, and Persons with Disabilities in Science and Engineering (2019). https://ncses.nsf.gov/pubs/nsf19304/digest/enrollment
2. Civil Engineers, DataUSA. https://datausa.io/profile/soc/civil-engineers
3. Vaquera, K.H.: An aging construction workforce: recognition and response. Natl. Law Rev. **19** (2019)
4. Groen, C., McNair, L.D., Paretti, M.C., Simmons, D.R., Shew, A.: Exploring professional identity development in undergraduate civil engineering students who experience disabilities. Paper presentation, American Society for Engineering Education Annual Conference and Exposition (2018)
5. National Science Foundation, Dear Colleague Letter: Research to Improve STEM Teaching and Learning, and Workforce Development for Persons with Disabilities. https://www.nsf.gov/pubs/2019/nsf19033/nsf19033.jsp
6. Women, Minorities, and Persons with Disabilities in Science and Engineering: 2011, National Science Foundation (2011)
7. NCCSD Quick Facts: National Center for College Students with Disabilities. http://www.nccsdonline.org/uploads/7/6/7/7/7677280/quick_facts.pdf
8. McCall, C., Shew, A., Simmons, D.R., Paretti, M.C., McNair, L.D.: Exploring student disability and professional identity: navigating sociocultural expectations in US undergraduate civil engineering programs. Australas. J. Eng. Educ. **25**, 79–89 (2020)
9. Pearson Weatherton, Y., Mayes, R.D., Villanueva-Perez, C.: Barriers to persistence for engineering students with disabilities. Paper presentation. American Society for Engineering Education Annual Conference and Exposition (2017)

10. Bonaccio, S., Connelly, C.E., Gellatly, I.R., Jetha, A., Martin Ginis, K.A.: The participation of people with disabilities in the workplace across the employment cycle: employer concerns and research evidence. J. Bus. Psychol. **35**(2), 135–158 (2019). https://doi.org/10.1007/s10869-018-9602-5
11. Gold, P.B., Oire, S.N., Fabian, E.S., Wewiorski, N.J.: Negotating reasonable workplace accommodations: perspectives of employers, employees with disabilities, and rehabilitation service providers. J. Voc. Rehab. **37**, 25–37 (2012)
12. Hernandez, B., McDonald, K., Divilbiss, M., Horin, E., Velcoff, J., Donoso, O.: Reflections from employers on the disabled workforce: focus groups with healthcare, hospitality and retail administration. Empl. Responsib. Rights J. **20**, 157–164 (2008)
13. Butterfield, T.M., Ramseur, J.H.: Research and case study findings in the area of workplace accommodations including provisions to assistive technology: a literature review. Technol. Disabil. **62**, 657–675 (2018)
14. Adams, R., Reiss, B., Serlin, D.: Keywords for Disability Studies. New York University Press, New York (2015)
15. Thomas, C.: Rescuing a social relational understanding of disability. Scand. J. Disabil. Res. **6**, 22–36 (2004)
16. Qi, C., Fourie, A., Ma, G., Tang, X., Du, X.: comparative study of hybrid artificial intelligence approaches for predicting hangingwall stability. J. Comput. Civ. Eng. **32** (2018)
17. Salehi, H., Burgueño, R.: Emerging artificial intelligence methods in structural engineering. Eng. Struct. **171**, 170–189 (2018)
18. Luckey, D., Fritz, H., Legatiuk, D., Dragos, K., Smarsly, K.: Artificial intelligence techniques for smart city applications. In: International Conference on Computing in Civil and Building Engineering, pp. 3—15. Springer, Cham (2021)
19. Li, Y., Liu, C.: Applications of multirotor drone technologies in construction management. Int. J. Const. Manage. **19**, 401–412 (2019)
20. McCabe, B.Y., Hamledari, H., Shahi, A., Zangeneh, P.: Roles, benefits, and challenges of using UAVs for indoor smart construction applications. In: ASCE International Workshop on Computing in Civil Engineering (2017)
21. Behzadan, A.H., Kamat, V.R.: A framework for utilizing context-aware augmented reality visualization in engineering education. In: 12th International Conference on Construction Application of Virtual Reality, pp. 292–299. National Taiwan University Press, Taipei (2012)
22. Shirazi, A., Behzadan, A.H.: Design and assessment of mobile augmented reality-based information delivery tool for construction and civil engineering curriculum. J. Prof. Issues Eng. Pract. **141** (2015)
23. Wang, X., Dunston, P.S.: Potential of augmented reality as an assistant viewer for computer-aided-drawing. J. Comput. Civ. Eng. **20**, 437–441 (2006)
24. Shin, D.H., Dunston, P.S.: Identification of application areas for augmented reality in industrial construction based on technology suitability. Autom. Constr. **17**, 882–894 (2008)
25. Wang, X., Dunston, P.S.: Tangible mixed reality for remote design review: a study understanding user perception and acceptance. Vis. Eng. **1** (2013)
26. Behzadan, A.H., Dong, S., Kamat, V.R.: Augmented reality visualization: a review of civil infrastructure system applications. Adv. Eng. Inform. **29**, 252–267 (2015)
27. Hanna, A.S., Mikhail, G., Iskandar, K.A.: State of prefab practice in the electrical construction industry: qualitative assessment. J. Constr. Eng. Manage. **143** (2017)
28. Winch, G.: Models of manufacturing and the construction process: the genesis of re-engineering construction. Build. Res. Inform. **31**, 107–118 (2003)
29. Baker, P.M., Linden, M.A., LaForce, S.S., Rutledge, J., Goughnour, K.P.: Barriers to employment participation of individuals with disabilities: addressing the impact of employer (mis)perception and policy. Am. Behav. Sci. **62**, 657–675 (2018)

Human-Technology Frontier: Measuring Student Performance-Related Responses to Authentic Engineering Education Activities via Physiological Sensing

Idalis Villanueva Alarcón[1(✉)], Edwin Marte Zorrilla[1], Jenefer Husman[2], and Matthew Graham[2]

[1] Engineering Education Department, University of Florida, Gainesville, Fl 32611, USA
i.villanueva@ufl.edu

[2] Education Studies, University of Oregon, Eugene, OR, USA

Abstract. Multi-modal approaches have increasingly shown promise in exploring the human side of engineering via the assessment of authentic responses to learning or working environments. This study explores the utility of non-invasive physiological wrist sensors in measuring the reactive and regulatory responses of a group of 161 engineering students taking an authentic engineering practice exam. The practice exam was categorized into Conceptual problems (e.g., rote memorization) and Analytical problems (e.g., requiring application of learned concepts through equations and free-body diagrams). Responses were measured through electrodermal activity and indicators of performance. Findings identified that the type of practice exam problem, even if designed to be within the moderate range of difficulty, influenced how students reacted to and regulated their performance to the problem (as seen by stronger positive correlations in the Analytical problems) and that these may occur via multi-componential processes.

Keywords: Human-technology frontier · Human side of engineering · Engineering education · Physiological sensing · Student performance

1 Introduction

In science, technology, engineering, and math (STEM), workforce development initiatives necessitate that trainees learn about and perform as intended by their leads in their immediate learning environment. In turn, the actions and decisions of STEM trainees will determine how these individuals will manage future societal and complex problems post-graduation. In the classroom environment, learning and performance are traditionally studied through various techniques such as self-report instruments, artifact retrieval and analysis, sampling of experiences, interviews, or observational procedures [1–3]. These techniques, while valuable, are constrained to participant or observer subjectivity and restricts its understanding of individuals' authentic responses to a task in real-time.

C. Leitner et al. (Eds.): AHFE 2021, LNNS 266, pp. 338–345, 2021.
https://doi.org/10.1007/978-3-030-80840-2_39

The latter limitation poses a challenge for many scholars seeking to objectively measure such events in real-time.

The situational contexts individuals may find themselves in, for example in the classroom, are important to consider when examining students' performance. Educational psychology scholars posit that in the context of the classroom, learners experience test-related, class-related, or learning-related multi-component input processes such as affect, physiological responses, motivational changes, and cognitive fluctuations [2, 3]. Subsequently, these alter outcome processes such as decision-making and performance [2, 3]. Yet, there is a scarcity of empirical investigations exploring the relationships of these input and outcomes processes in real-time.

Recent and nascent literature has shown the promise of multi-modal approaches to explore the authentic responses of students to their learning environments and tasks [2–9]. For example, Pekrun and his colleagues examined stress responses during a final oral exam [2] using salivary cortisol; we have used this same method during engineering exams [5–7]. Azevedo and others have examined students' metacognitive judgments using physiological sensors when individuals interact with a meta-tutor environment [8]. Villanueva and colleagues have begun to tie engineering content (i.e., engineering statics and design) to physiological responses during in-classroom activities (active learning; [9]) and out-of-classroom exam-taking experiences (statics practice exams; [7]). Recently, Strohmeier and colleagues [10] utilized electrodermal activity sensors with self-reports to explore state (*momentary*) anxiety to mathematical tests. Together, these studies are pointing to an increasing desire from scholars to develop studies and methods that are more authentic to the situations and contexts that different learners are tasked to perform in.

For this work, the utility of electrodermal activity as a non-invasive electrodermal activity sensing tool to explore students' input physiological processes (i.e., reactive or regulatory emotions, motivations, etc.) related to its output processes (i.e., effort, performance, etc.) was explored. In addition, this study also tabulated the instructors' expected difficulty levels for the exam questions and students' actual performances to better contextualize the practice exam used for this study.

Electrodermal activity (EDA) is an umbrella term used to define changes in the electrical properties of the skin. The most widely studied electrical property of the skin is known as *skin conductance*. Skin conductance is quantified by applying an electrical potential between two contact points in the skin and measuring its electrical current flow. The resultant flow, stemming from sympathetic neuronal activity, is captured as both a background tonic (skin conductance level or SCL) and fast-varying phasic (skin conductance response or SCR) signal [11–13]. EDA is useful in identifying changes in sympathetic arousal, either through emotional or cognitive states, and it is the only autonomic psychophysiological variable that is not connected to parasympathetic activity [11–13]. In other words, when a person becomes nervous or anxious about a task, their body stimulates metabolic outputs to deal with external challenges (e.g., their palms become sweaty). Since emotional regulation and cognitive processes, among other brain functions, can influence the control of sweating, a fluctuation in the electrical conductivity of the skin can be measured. Also, these measurements can be collected rather cheaply and non-invasively [14, 15] (e.g., EDA wrist sensors) making physiological EDA sensing tools attractive for scholars.

2 Methods

Research Design. This study is composed of a subset of data derived from a larger study exploring the motivational and performance experiences of engineering students performing on an authentic practice exam in the topic of statics [7]. The study was quasi-experimental and integrated EDA sensing with salivary biomarkers of stress along with self-reports in the practice engineering exam for triangulation. Per Institutional Review Board (IRB) guidelines and in alignment with human subjects' research, a practice exam equivalent in content to the actual exam was provided to prevent unnecessary harm to students' grades in the course.

The total number of participants reported in this study was 161 engineering students. These students took an engineering statics course for the first time in a Mountain West institution in the United States between Fall 2018, Spring 2019, and Fall 2019. The instructor has been teaching engineering statics for several years and the exam used for this study was designed by this faculty member. The 15-question exam (6 conceptual and 9 analytical) used was the same throughout the semesters in this study. For this study, students were given ample time to complete the questions, as approved by the instructor.

Data Collection and Measures. The setup of the experiment and its considerations is described elsewhere [7] but in summary, involved a time-stamped research design that measured the time students took to perform per practice exam question, their performance to each item, the raw EDA data from each practice exam question and event (before and after the exam), salivary data (not reported in this work) and self-reports (not reported in this work). For this study, we wanted to explore how reactive input processes, in this case, sympathetic emotional arousal (measured through skin conductance responses), are related to output processes (measured through difficulty index scores of performance) in authentic engineering examination settings.

There were several measurements collected in our study. The first measurement was the *event-specific skin conductance response* (es.SCR), which allowed us to attribute a specific eliciting stimulus [14] to its fast-varying reactive responses. The second measurement was the *difficulty index* (DI), which is the measure of the number of questions that students answered correctly on the practice exam divided by the total number of responses. DI is also an indication of whether a problem created by the instructor was difficult ($DI < 0.3$), moderate ($0.3 < DI > 0.8$) or easy ($DI \geq 0.8$) [16]. The third measurement was the instructor's self-reported difficulty level for the exam, attained through a three-point Likert scale where '1' was reported as 'easy', '2' was reported as 'moderate' and '3' was reported as 'difficult'. The fourth measure was the average time spent in Conceptual and Analytical exam problems; this measured time is an indication of effort executed by the student. Finally, we tracked the percentage of students who received fully correct responses to Conceptual and Analytical exam problems to understand students' effectiveness in their performance.

Data Processing and Analysis. For the collected and continuous raw EDA data, we first visually identified and removed any data tied to potential manufacturer sources of outliers, such as sensor malfunctions, incomplete collection, or poor contact of the electrodes in the skin. These were evidenced by negative values or constant near-zero continual data segments [7]. Then, additional EDA outliers due to user-generated sources such as erratic movements (e.g., hand hitting desk or nervous tapping), survey or salivary biomarker collection periods, or large changes in body temperatures or blood volume pressure readings [7] were removed. This required that the temperature and blood volume pressure sensors found in the wrist sensor (E4, Empatica, Inc.) was used in parallel to the collected EDA data. Next, the authors used a custom-developed process to remove any potential contributions to the raw EDA signal via fine motor body movements from the participant (e.g., typing) [7]. This process required that the 3-axis accelerometer found in the EDA sensor was used in parallel to the collected galvanic skin response data.

To summarize, this processing method used an L2-norm accelerometer transformation of the 3-axis of movement (in the x, y, z directions) to calculate the Euclidean distance of movement (total movement). From the collected signal, the standard deviation and coefficient of variation (CV) of the accelerometer data were calculated, ranked-ordered to identify its upper and lower limits, and a threshold equivalent to 95% of the standard deviation of the accelerometer data was established. The corresponding raw EDA values to this transformed accelerometer data were further used for filtration. Filtration of this signal was conducted through continuous deconvolution methods developed in the open-source MATLAB software, Ledalab [17]. The resultant continuous data was then separated into baseline tonic (skin conductance level-SCL) and fast-varying phasic (skin conductance responses-SCR) EDA signals. For this study, we were interested in the deconvoluted continuous phasic EDA signal, although we acknowledge that this represents "a small proportion of the overall EDA complex" [17, p. 4]. The resultant processed and filtered EDA phasic signal labeled as a non-specific skin conductance response (ns.SCR_i) was aligned to the timestamps of a particular event (i.e., duration of time for a practice exam question) to extract the *event-specific SCRs* (es.SCR_i), which measures the specific eliciting stimulus [17] of a given event to its fast-varying reactive responses. The number of peaks present in these es.SCRs (es.SCR_{peak}) was used to measure the reactivity of the response while the mean of these es.SCRs (es.SCR_{mean}) was used to measure the regulatory potential of these responses.

For the processing the DI data, we calculated the number of exam questions answered correctly on the exam divided by the total number of responses. This DI value was divided by time for each event. For the instructor assessment, we calculated the average scores from the Likert Scale responses for the three semesters. The practice exam questions were categorized as Conceptual (those requiring rote memorization or interpretation of terminology) or Analytical (those that required application of the concepts learned in the form of equations, identification of parameters provided, and drawing of free body diagrams to attain an answer). The conceptual and analytical questions were timestamped and aligned to its corresponding es.SCR_i and DI data. Time and percentages of performance for the exam problems were tabulated and a t-test analysis was conducted.

For analysis of the es.SCR_i data, we followed recommended practices to transform and standardize EDA data [7, 11, 14, 17]. For a normalized transformation, the Log(es.SCR_i + 1) was calculated to remove any problems with skewness, kurtosis, and heterogeneity of variance in the EDA data [17]. For standardization of the EDA data, the $\text{Log}(\text{es.SCR}_i + 1)$ was divided by the time per practice exam question to make group comparisons by event. Descriptive statistics (a t-test for equality of means) was determined for the transformed and standardized mean and peak es.SCRs, DI, the instructor averaged estimated scores, time, and percentage of students who answered the practice exam questions correctly; data are presented as mean ± standard error of the mean. Additionally, a Pearson correlation analysis was conducted for the transformed and standardized mean and peak es.SCRs with DI scores per type of practice exam question (Conceptual and Analytical).

3 Results and Discussion

T-test analysis was conducted for the Conceptual and Analytical problems and significance was determined by a p-value under 0.05. The time spent on the exam problems was tabulated as a measure of effort from students for the Conceptual and Analytical questions. There was a 3.5x-fold increase in the Analytical problems compared to the Conceptual problems (372.98 ± 8.47 s versus 80.58 ± 1.22 s; $p < 0.001$). Also, 65% of the students answered the Conceptual questions correctly while 58% of the students answered the Analytical questions correctly; no statistical significance was found between these groups ($p = 0.69$). The DI scores were 0.64 ± 0.01 for the Conceptual questions and 0.68 ± 0.00 for the Analytical questions and these differences were significant ($p < 0.001$).

The instructor estimated the difficulty level of practice exam questions as being twice as higher for the Analytical problems (2.04 ± 0.01; moderate) compared to the Conceptual problems (1.21 ± 0.02; easy). These estimates were significantly different from each other ($p < 0.001$). A summary of these t-tests is provided in Table 1.

Table 1. Descriptive statistics for conceptual and analytical problems in an engineering statics practice exam; N = 161; data presented as Mean ± Standard Error of Mean; statistical significance as calculated through T-test is * $p < 0.05$, ** $p < 0.01$, and *** $p < 0.001$ when comparing analytical to conceptual problems.

Measurement	Exam problem type	Mean ± SEM
Time spent	Conceptual	80.58 s ± 1.22 s
	Analytical	372.98 s ± 8.47 s***
Students who answered correctly	Conceptual	65% ± 0.00%
	Analytical	58% ± 0.00%
Instructor guess	Conceptual	1.21 ± 0.02
	Analytical	2.04 ± 0.02**
Difficulty index	Conceptual	0.64 ± 0.01
	Analytical	0.68 ± 0.00***

Pearson correlation analysis (Table 2) was conducted for the transformed and standardized es.SCR_{mean} and es.SCR_{peak} data and DI scores (both divided by time) for the Conceptual and Analytical problems. For reactive responses (through es.SCR_{peak} values), both Conceptual and Analytical exam problems resulted in strongly positively correlated reactive responses when compared to DI/time ($r = 0.64$, $p < 0.01$ and $r = 0.85$, $p < 0.01$). For regulatory potential responses (through es.SCR_{mean} values), a strong positive correlation for the Conceptual problems ($r = 0.59$, $p < 0.01$) and a weak positive correlation for the Analytical problems ($r = 0.21$, $p < 0.01$) compared to DI/time was found.

Table 2. Pearson correlation analysis between log(es.SCR_{peak} + 1) or log(es.SCR_{mean} + 1) versus DI/time to measure students' reactive or regulatory responses, respectively; $N = 161$; data presented as Mean ± Standard Error of Mean; statistical significance is * $p < 0.05$, ** $p < 0.01$, and *** $p < 0.001$.

Response type	Exam problem type	Correlation
Reactive	Conceptual	0.59
	Analytical	0.21**
Regulatory	Conceptual	0.64
	Analytical	0.85**

Collectively, the data suggest that: (1) instructors' assessment of the difficulty level of the practice exam problems does not necessarily match students' performance to the same practice exam problems; (2) more effort (as seen by increases in time) by the student to practice exam problem types may slightly improve performance but not sufficiently to cause an improvement in the number of students excelling; (3) the type of practice exam problem, even if perceived to be within the moderate range of difficulty (DI scores of 0.64 and 0.68, respectively) influences how students react to and regulate their execution of the problem (stronger correlations in Analytical problems).

These findings suggest the need to better align exam difficulty levels with students' zone of proximal development [18, 19], a learning theory suggesting that students best perform and learn when instructors understand the zone by which students best learn and can be 'pushed' to gain further motivation to learn. The zone of proximal development states that when faculty pushes students through unattainable perceived boundaries, a students' motivation decreases, and their performance may be negatively impacted. Thus, finding a right balance between perceived difficulty and actual performance of students is important in examination contexts.

Furthermore, the findings in this study demonstrate that responses to an exam experience can be measured, quantified, and connected to its context (e.g., exam problem types, instructor creation of 'just-right' levels of difficulty for students). This suggests the presence of multi-componential processes tying motivational with cognitive constructs. This parallels the educational psychology framework of the Control-Value Theory of emotions [2, 20] that posits that emotions are inextricably linked to multi-componential processes such as motivation, affect, cognition, and psychophysiology. Thus, it is important to not disconnect individuals' output processes (i.e., performance) from these input processes (e.g., emotions) in their learning or working environments. Collectively, the data points to the need to guide students to cope with unexpected challenges and to perform in ways originally anticipated by the instructor.

4 Implications to the Human-Technology Frontier

Taleb suggests [21] that 21st century skills (e.g., complex problem solving) are both applicable and antifragile (involving individual growth when exposed to "volatility, randomness, disorder, and stressors… risk, and uncertainty." [21, p. 3]. Even the National Academy of Engineering has recognized that 21st century skills for engineering can't function without equipping its workforce to manage emotional experiences such as social distress, mental and emotional health, which influence decision-making and performance [22]. As suggested by these publications and the Accreditation Board for Engineering and Technology (criterion 7; [23]), there is and will be a need to holistically integrate learning experiences to include appropriate teaching/training strategies that help students gain an ability to multi-modally acquire and apply new knowledge effectively.

There seems to be a growing recognition of the importance of multi-componential processes in learning and performance, not just in the classroom but also in the way society views the future of engineering. Thomas Insights published an article delineating the top 20-in demand skills for employees in unexpected environments and scenarios such as those expected to be experienced in a post-COVID-19 world [24]. Some of these skills include critical thinking, complex problem solving, and stress tolerance. In a similar vein, Qadar and Al-Fuqaha [25] conducted an in-depth literature review of the seven steps needed to succeed in engineering education post-COVID-19. Step one of their model suggests the need for students to start with the end in mind by aligning efforts with product outcomes and motivations with effective performance.

The findings demonstrated in this work demonstrate the utility that physiological sensing can have in helping scholars/educators to cross the human-technology frontiers in workforce training and development. Applying multi-modal approaches such as the ones presented here provide a potential pathway to study more authentically the human side of learning and performance in engineering.

Acknowledgment. This material is based upon work supported in part by the National Science Foundation (NSF) No. EED-1661100/1661177. Any opinions, findings, and conclusions or recommendations expressed in this material do not necessarily reflect those of NSF.

Author contributions in this paper: Villanueva Alarcón (research design, data collection and analysis, writing, editing); Marte Zorrilla (data collection, analysis, writing, editing); Husman (research design, data collection, editing); Graham (writing and editing).

References

1. Jarrell, A., Harley, J.M., Lajoie, S., Naismith, L.: Success, failure, and emotions: examining the relationship between performance feedback and emotions in diagnostic reasoning. Educ. Tech. Res. Dev. **65**(5), 1263–1284 (2017)
2. Pekrun, R., Bühner, M.: Self-report measures of academic emotions. In: Pekrun, R., Linnenbrink-Garcia, L. (eds.) International Handbook of Emotions in Education, pp. 561–566. Routledge Press, London (2014)

3. Roos, A.L., et al.: Test anxiety and physiological arousal: a systematic review and meta-analysis. Educ. Psychol. Rev. **33**, 1–40 (2020)
4. Spangler, G., Pekrun, R., Kramer, K., Hofmann, H.: Students' emotions, physiological reactions, and coping in academic exams. Anxiety Stress Coping **15**(4), 413–432 (2002)
5. Husman, J., Cheng, K.: When performance really matters: future time perspective and career aspirations. In: European Association of Learning and Instruction, Limassol, Cyprus (2015)
6. Husman, J., Cheng, K.C., Barnes, M.E., Espino, K.: Managing emotions when it counts: stress, academic achievement emotions, and perceptions of instrumentality. In: American Educational Research Association, Chicago, IL (2016)
7. Villanueva, I., et al.: A cross-disciplinary and multi-modal experimental design for studying near-real-time authentic examination experiences. J. Vis. Exp. **151**(e60037), 1–10 (2019)
8. Azevedo, R., Taub, M., Mudrick, N., Farnsworth, J., Martin, S.A.: Interdisciplinary research methods used to investigate emotions with advanced learning technologies. In: Zembylas, M., Schutz, P.A. (eds.) Methodological Advances in Research on Emotion and Education, pp. 231–243. Springer (2016)
9. Villanueva, I., Campbell, B., Raikes, A., Jones, S., Putney, L.: A multi-modal exploration of engineering students' emotions and electrodermal activity in design activities. J. Eng. Educ. **107**(3), 414–441 (2018)
10. Strohmaier, A.R., Schiepe-Tiska, A., Reiss, K.M.: A comparison of self-reports and electrodermal activity as indicators of mathematics state anxiety. An application of the control-value theory. Frontline Learn. Res. **8**(1), 16–32 (2020)
11. Boucsein, W., Backs, R.W.: Engineering psychophysiology as a discipline: historical and theoretical aspects. In: Backs, R.W., Boucsein, W. (eds.) Engineering Psychophysiology. Issues and Applications, pp. 3–30. Lawrence Erlbaum, Mahwah (2000)
12. Boucsein, W., Backs, R.W.: The psychophysiology of emotion, arousal, and personality: methods and models. In: Duffy, V.G. (ed.) Handbook of Digital Human Modeling, pp. 35–38. CRC, Boca Raton (2009)
13. Sano, A., Picard, R.W., Stickgold, R.: Quantitative analysis of wrist electrodermal activity during sleep. Int. J. Psychophysiol. **94**(3), 382–389 (2014)
14. Braithwaite, J.J., Watson, D.G., Jones, R., Rowe, M.: A guide for analyzing electrodermal activity (EDA) & skin conductance responses (SCRs) for psychological experiments (Technical report, 2nd version). Behavioural Brain Sciences Centre, Birmingham (2015)
15. Empatica. Empatica E4 wristband (2016). https://www.empatica.com/e4-wristband
16. Johari, J., et al.: Difficulty index of examinations and their relation to the achievement of programme out comes. Procedia Soc. Behav. Sci. **18**, 71–80 (2011)
17. Benedek, M., Kaernbach, C.: A continuous measure of phasic electrodermal activity. J. Neurosci. Methods **190**(1), 80–91 (2010)
18. Vygotsky, L.S.: Mind in Society: The Development of Higher Psychological Processes. Harvard University Press, Cambridge (1978)
19. Vygotsky, L.S.: Thought and Language. MIT Press, Cambridge (1986)
20. Pekrun, R.: The control-value theory of achievement emotions: assumptions, corollaries, and implications for educational research and practice. Educ. Psychol. Rev. **18**(4), 315–341 (2006)
21. Taleb, N.N.: Antifragile: Things That Gain from Disorder. Random House, New York (2012)
22. National Academy of Engineering. https://www.nae.edu/230195/Call-for-Engineering-Response-to-the-COVID19-Crisis
23. ABET Engineering Accreditation Commission. Criteria for accrediting engineering programs effective for reviews during the 2020–2021 accreditation cycle (2019)
24. Thomas Insights. https://www.thomasnet.com/insights/top-10-in-demand-skills-in-a-post-covid-world-according-to-world-economic-forum/
25. Qadir, J., Al-Fuqaha, A.: A student primer on how to thrive in engineering education during and beyond COVID-19. Educ. Sci. **10**(9), 1–22 (2020)

Students Who Stand Out: Faculty Perspectives on the Competencies that Define Outstanding Students Entering the Workforce

Madeline Polmear(✉), Denise R. Simmons, Elizabeth Volpe, and Danielle Weisenfeld

Department of Civil and Coastal Engineering, University of Florida, Gainesville, USA
{mpolmear,denise.r.simmons,evolpe1,dweisenfeld}@ufl.edu

Abstract. The future of work will be defined by increased connectivity between technology and people. The engineering workforce is situated at this intersection and needs a broad and ever-evolving system of competencies to address socio-technical issues. Although engineering practice continues to adapt at the human-technology frontier, education has been slower to respond. As a result, there is a purported competency gap between academia and industry. Within engineering education, faculty members play a critical role in the academic preparation and professional formation of students. Thus, it is instructive to consider faculty members' perspectives on the competencies that define a student's ability to succeed in the workplace. This study employed a qualitative approach through interviews with 11 engineering faculty members to characterize these perspectives. The findings indicate that faculty members emphasize students' intra- and interpersonal competencies and their involvement in competency-building activities outside of the classroom. This paper concludes by leveraging the findings into a broader discussion of competencies needed for the future of work and equity and access related to competency-building activities.

Keywords: Competency · Human factors · Undergraduate education · Engineering

1 Introduction

Engineering is a dynamic profession that is constantly evolving in response to technological, environmental, and societal changes. To adapt and thrive in the workforce, engineers need a range of competencies, which are defined as knowledge, skills, and attributes. The undergraduate curriculum plays a key role in competency development through academic training and professional socialization. The importance of technical and nontechnical competencies in engineering education is reflected in the accreditation criteria [1] and aspirations for the Engineer of 2020 [2]. The combination of these competencies enables individual engineers to be leaders and the engineering profession to take a leadership role in addressing the most pressing societal challenges. Leadership education in engineering is motivated both by the need to be competitive in the international job market and to responsibly serve communities [3]. Although there is

C. Leitner et al. (Eds.): AHFE 2021, LNNS 266, pp. 346–353, 2021.
https://doi.org/10.1007/978-3-030-80840-2_40

agreement on the importance of leadership, there is less consensus on how to define and develop it in the context of engineering [4]. Within the engineering leadership literature, there are different competency-based definitions of leadership [3]. As an example, leaders are conceptualized as big thinkers, risk takers, decision makers, team builders, and strong communicators [5]. In addition to these qualities, attributes such as courage, honesty, and adaptability have been associated with leadership in engineering [6]. Taken together, these studies suggest the importance of competencies in defining leadership and developing leadership capacity in future engineers.

Technical skills have historically been the bedrock of engineering education while professional skills have been increasingly integrated into curricula in the last 20–30 years via the Washington Accord on a global scale [7] and accreditation agencies on a national scale [1, 8, 9]. However, there are concerns from industry that academia has been slow to adapt to the ever-evolving competencies needed for engineers of the present and future [10]. It has been suggested that this lag contributes to a gap between the competencies of engineering graduates and the expectations of employers [11, 12]. Given the crucial role that engineering faculty members play in preparing students for their careers, it is insightful to understand the competencies that they value in their students and that they believe define the students who are ready for the workforce.

2 Methods

This research was designed to address the following question: what competencies do engineering faculty members identify in students who are prepared to enter the workforce? The framework guiding this study is a set of leadership-coupled professional competencies that engineering graduates need for entering the workforce as identified by industry leaders [4]. The framework includes the interpersonal, intrapersonal, professional, and technical competencies that provide a definition of leadership contextualized in engineering practice.

2.1 Project Context

This study is part of a larger project exploring leadership development to prepare engineering students for the 21st century workforce. The project draws on qualitative and quantitative data from industry professionals, faculty members, and undergraduate students to provide a comprehensive understanding of workforce and leadership development and to explore areas of alignment and misalignment between these three groups with the aim of more effectively bridging academia and practice. The present study focuses on the faculty perspective through a qualitative approach.

2.2 Data Collection

An online survey was sent to STEM faculty members at five large U.S. institutions and asked respondents to indicate the importance of various competencies to their students' future career and to share their perspectives on leadership education. The survey concluded with an opportunity for respondents to provide their contact information if they were interested in participating in a follow-up interview. Semi-structured interviews were completed with 11 engineering faculty members, each lasting approximately one hour. Interviews were conducted over Zoom, recorded, and transcribed verbatim. As a part of the interview process, participants were asked about the outstanding seniors in their program and what separated them from their peers. The bulk of the data for this study was collected from this specific prompt within the context of the interview and larger project.

2.3 Data Analysis

The 11 transcribed interviews were uploaded into Dedoose, a qualitative coding software, for analysis. The analysis team, composed of the first, third, and fourth authors, established familiarity with the data through the process of memoing and open conversation as recommended by [13]. A combination of deductive and inductive coding [14] was implemented. The codebook was developed based on the leadership-coupled professional competencies that engineering industry leaders identified as essential for engineers entering the workforce [4]. Emergent codes of competencies that participants identified and were not included in the framework were added to the codebook throughout the process. Two members of the analysis team coded full transcripts individually then met weekly to discuss their findings and any discrepancies. The third member of the team was consulted if any discrepancies arose. The researchers took turns leading these conversations and sharing their screen and coding on Zoom to reduce bias from just one researcher leading the analysis. The research team then met to discuss the major themes that could be deduced from the data and reach alignment in the key findings.

3 Findings

Findings from this study emerged through two major themes that are detailed in the following sections.

3.1 Faculty Reflect on Interpersonal and Intrapersonal Competencies More Than Technical Competencies When Describing Attributes of Their Outstanding Students

Participant responses varied in terms of the competencies that they reflected upon in their outstanding students and no single competency was mentioned by half or more of the participants. Despite this variability, every participant mentioned an interpersonal or intrapersonal competency as characteristic of their "super star" students. These competencies are more closely associated with personality traits, habits, and ways of being and are not necessarily related directly to the technical requirements and tasks of a job [15]. By applying the theoretical framework to the data, the following competencies were identified: Ambition/Drive, Big Picture Thinking, Critical Thinking/Problem Solving, Learning/Curiosity, Adaptability, People Focus, Assertiveness, Teamwork/Collaboration/Networking, and Professionalism.

Figure 1 depicts the relative prevalence of the competencies; the larger size indicates the greater number of faculty members who discussed the competency as a characteristic of outstanding students. There was one emergent competency, maturity, that was mentioned by two participants. Ambition/Drive was discussed by six faculty members and Adaptability was mentioned by one.

Fig. 1. The word cloud based on the number of faculty members who mentioned each competency

Ambition/Drive occurred the most out of all of the codes in the faculty transcripts. When asked, "What separates those [rock star] students from their peers?" Dr. Natalie responded:

> The fire in the belly and then the right side of opportunity... So as long as there is this, then everything else, at least the students that we see at [institution], anybody can be a star.

Here, Dr. Natalie emphasized that along with opportunities, the "fire in the belly" or ambition/drive of students is really what allows them to excel. She even goes as far as to say that anyone with that sort of drive "can be a star".

Big Picture Thinking was discussed in 5 of the 11 faculty transcripts. Dr. Drew reflected on his outstanding students stating:

> They think very much in terms of 'This task is a piece of a bigger project, which is a piece of a bigger program, which is a piece of a bigger impact on something much bigger than us.'

Dr. Drew emphasized the value of a student's ability to draw connections from smaller tasks to the larger issues at hand as a vital competency he sees in his outstanding students. Dr. Drew also goes on to emphasize the importance of communication and people focus stating:

> I think the next thing that we really see from those stellar students is an ability to relate to people. They are very able to communicate with a wide range of folks, and they're very good listeners.

Dr. Drew focused on the interpersonal competencies that he attributed to his stellar students. A number of other participants focused specifically on the inter- and intrapersonal competencies that differentiate outstanding students and none of the participants described their star students as solely technically competent engineers. Technical competencies in the framework, including Computer Skills, Legal Knowledge, Economic Principles and Trends, and Safety and Risk Management, were not mentioned by any of the participants.

3.2 Faculty Emphasize Experiences Outside of the Classroom for Building Competencies Over Students' Classroom Performance and GPA

Six of the eleven faculty discussed life experiences and involvement in activities outside of the classroom, such as internships and clubs, as characteristic of their "outstanding" students. Several faculty members also provided anecdotes of students' individual interactions with them that had made them stand out, and almost no faculty brought up GPA or classroom performance as attributes of their "outstanding" students. Their outstanding students were doing more than just getting good grades. Dr. Drew stated:

> A student from our program graduating with a 3.8 GPA who has not been involved in student activities, has not done internships, has not been in leadership positions is probably, even though they're coming out of [institution] with a high GPA in a well regarded program, they're probably going to have tough time getting a job.

Dr. Drew continued to emphasize that a student with a lower GPA and a number of extra-curricular competency building experiences is much more likely to be prepared for the workforce and get a job. When asked to describe what separates his standout students from their peers, Dr. Norm stated:

> ...they typically were ones that had co-op experience [...] an international co-op thing, study abroad, everything, and those students, that experience is just amazing for them. I would have to say that the students that, in general, do really well, take advantage of something extra, whatever that is.

Dr. Norm noted that students' participation in activities outside of the classroom enabled them to excel. Dr. Norm went on to say:

> So the students who participate in clubs [...] they get an experience that is ... I kind of think that the curriculum is maybe not the bare minimum, but it's on that track, so whatever these enrichment things are, that's where the students just take off.

Dr. Norm stressed the importance of student involvement. He noted that completing the curriculum is "on that track" of being near the bare minimum. Therefore, the students that he views as outstanding are involved outside the classroom and do not focus solely on classroom achievement.

4 Discussion

The findings presented above provide starting points for a broader discussion on the competencies students need for the future of work and their opportunities to pursue activities outside of the classroom to develop them. First, the findings indicate faculty members reflect upon interpersonal and intrapersonal competencies more than technical competencies when describing attributes of their outstanding students. As technology continues to reshape the landscape in which engineers work, "companies expect a significant shift on the frontier between humans and machines when it comes to existing work tasks between 2018 and 2022" [16]. As tasks are increasingly completed by machines and algorithms, it is estimated that 75 million jobs worldwide will be displaced by 2022 while 133 million new roles will be created [16]. As machines replace some jobs, the new roles will increasingly rely on "human" skills, such as creativity, critical thinking, resilience, flexibility, emotional intelligence, and leadership [16]. These competencies align with those identified by the faculty participants as characteristic of outstanding students.

The competencies identified by faculty members in the study and those reported by industry stakeholders suggest the need to align student expectations with the realities of the profession. As part of the larger project in which this study is embedded, it was found the civil engineering students considered communication, humility, teamwork, adaptability, people focus, and management as the most important competencies for their future success [17]. The faculty findings indicate some overlap with the interpersonal and intrapersonal competencies that are noticeable attributes of outstanding students. The nature of engineering curricula signals to students the topics and skills they should value; consequently, "the technical content of the curriculum is a massive black hole whose gravitational pull inexorably absorbs the students' attention, time, and fidelity" [18]. This prioritization can create dissonance with the interpersonal, intrapersonal, and professional competencies that are valued for workforce preparation but are given less space in the curriculum. As a result, it is incumbent on faculty members to communicate the importance of nontechnical competencies and provide opportunities for students to develop them.

Second, faculty members emphasized experience outside the classroom to build these key competencies more than classroom achievement. The importance that faculty members place on out-of-class involvement aligns with previous research suggesting these settings as high-impact practices for competency development [19, 20]. However, it is important to contextualize this finding in a broader consideration of equity and access. Opportunities to participate in out-of-class activities are not uniformly distributed across all engineering students. For example, engineering competition teams help develop technical and professional skills, but are associated with a White, male culture that excludes underrepresented students from gaining access to these advantages [21]. Within civil engineering, Simmons et al. (2018) found differences in out-of-class involvement based on race/ethnicity and socio-economic status with African American students and students from lower income families having lower participation. As a result, relegating competency development to outside the formal curriculum or evaluating students' workforce preparedness based on their out-of-class involvement can serve to further marginalize students. Instead, educators and administrators need to consider how to expand access to out-of-class activities and future work can explore how to reduce systematic barriers to participation.

A related consideration is expanding the understanding of what counts as a competency-building activity and how such engagement can be rewarded. For example, Dr. Annabelle described her "rock star" students as those who have experience outside of academia and emphasized students' nontraditional roles in developing competencies. She stated:

> Often they're the ones that are non-traditional. They've worked, they've continued to work. Sometimes they're vets [...] a soccer player [...] Yeah, so the rock stars are so often nontraditional. They have wives, they have kids.

Dr. Annabelle provided a perspective that was unique in the data: competency building activities are not only in the form of clubs or internships. Nontraditional roles such as caretaking or military service are impactful in students' development and workforce preparation. There are a number of ways in which students can build their competencies outside of the classroom. This leads to the important question of how these nontraditional roles might be measured or valued in an academic setting to acknowledge students developing competencies in more unique ways.

5 Conclusion

The future of engineering work will look markedly different. By 2030, 85% of the jobs held by today's learners will not have been invented yet [22]. To prepare the future workforce, it is important to understand the competencies needed for engineers to be adaptive leaders on the human-technology frontier. It is thus the role of engineering programs and educators to understand the competencies needed and provide opportunities to develop them that are accessible to all students.

Acknowledgments. The authors gratefully acknowledge the National Science Foundation for supporting this work under grant number [1463802 and 1931371]. Any opinions, findings, conclusions, or recommendations expressed here are those of the authors and do not necessarily reflect the views of the National Science Foundation.

References

1. Criteria for Accrediting Engineering Programs, 2019–2020 | ABET. https://www.abet.org/accreditation/accreditation-criteria/criteria-for-accrediting-engineering-programs-2019-2020/. Accessed 29 Jan 2021
2. The Engineer of 2020: Visions of Engineering in the New Century. https://nae.edu/25876/The-Engineer-of-2020-Visions-of-Engineering-in-the-New-Century. Accessed 29 Jan 2021
3. Rottmann, C., Reeve, D.W., Sacks, R., Klassen, M.: An intersubjective analysis of engineering leadership across organizational locations: implications for higher education. Can. J. High.Educ. **46**, 146–173 (2016)

4. Simmons, D.R., McCall, C., Clegorne, N.A.: Leadership competencies for construction professionals as identified by construction industry executives. J. Constr. Eng. Manag. **146**, 04020109 (2020). https://doi.org/10.1061/(ASCE)CO.1943-7862.0001903
5. Farr, J., Brazil, D.: Leadership skills development for engineers. IEEE Eng. Manag. Rev. **38**, 110–118 (2010). https://doi.org/10.1109/EMR.2010.5645763
6. Goodale, M.J.: The right stuff: traits and skills of effective leaders. Leadersh. Manag. Eng. **5**(1), 7–8 (2005)
7. 25 Years of the Washington Accord, IEA. https://www.ieagreements.org/assets/Uploads/Documents/History/25YearsWashingtonAccord-A5booklet-FINAL.pdf
8. Accreditation Criteria Procedures, Engineers Canada. https://engineerscanada.ca/sites/default/files/accreditation-criteria-procedures-2016-final.pdf
9. Accreditation of Higher Education Programmes third edition (1).pdf. https://www.engc.org.uk/engcdocuments/internet/Website/Accreditation%20of%20Higher%20Education%20Programmes%20third%20edition%20(1).pdf.
10. An Industry Perspective on STEM Education for the Future: ISSIP-NSF Workshop. The International Society of Service Innovation Professionals (2019)
11. Male, S.A., Bush, M.B., Chapman, E.S.: Perceptions of competency deficiencies in engineering graduates. Australas. J. Eng. Educ. **16**, 55–68 (2010). https://doi.org/10.1080/22054952.2010.11464039
12. Nair, C.S., Patil, A., Mertova, P.: Re-engineering graduate skills – a case study. Eur. J. Eng. Educ. **34**, 131–139 (2009). https://doi.org/10.1080/03043790902829281
13. Birks, M., Chapman, Y., Francis, K.: Memoing in qualitative research: probing data and processes. J. Res. Nurs. **13**, 68–75 (2008). https://doi.org/10.1177/1744987107081254
14. Creswell, J.W., Miller, D.L.: Determining validity in qualitative inquiry. Theory Pract. **39**, 124–130 (2000). https://doi.org/10.1207/s15430421tip3903_2
15. Bancino, R., Zevalkink, C.: Soft skills: the new curriculum for hard-core technical professionals. Tech. Connect. Educ. Careers **J1**(82), 20–22 (2007)
16. WEF_Future_of_Jobs_2018.pdf. http://www3.weforum.org/docs/WEF_Future_of_Jobs_2018.pdf
17. Polmear, M., Simmons, D.R., Clegorne, N.A.: Undergraduate civil engineering students' perspectives on skills for future success. In: 2020 IEEE Frontiers in Education Conference (FIE), pp. 1–9 (2020). https://doi.org/10.1109/FIE44824.2020.9274269
18. Newberry, B.: The dilemma of ethics in engineering education. Sci. Eng. Ethics **10**, 343–351 (2004). https://doi.org/10.1007/s11948-004-0030-8
19. Burt, B., et al.: Outcomes of engaging engineering undergraduates in co-curricular experiences. In: 2011 ASEE Annual Conference & Exposition Proceedings, pp. 22.1132.1–22.1132.11. ASEE Conferences, Vancouver, BC (2011). https://doi.org/10.18260/1-2--18498
20. Polmear, M., Chau, A.D., Simmons, D.R.: Intersection of diversity, out-of-class engagement, and engineer of 2020 outcomes for civil engineering students. J. Eng. Manag. **37**, 04021019 (2021)
21. Pan, R., Shehab, R.L., Trytten, D.A., Foor, C.E., Walden, S.E.: Barriers to broadening participation in engineering competition teams. Presented at the 2016 ASEE Annual Conference & Exposition, June 26 (2016)
22. IFTF: The Next Era of Human | Machine Partnerships. https://www.iftf.org/humanmachinepartnerships/. Accessed 29 Jan 2021

Service Design of Ward Nurse Station Based on Kansei Engineering

Siwen Xie[1], Huifang Shang[1](✉), and Chuanshun Wang[2]

[1] School of Design and Media, East China University of Science and Technology, Shanghai 200237, China
hfshang@ecust.edu.cn

[2] Shanghai Xian Dai Architectural Design (Group) Co., Ltd., Shanghai 200237, China

Abstract. The nurse station is the hub of various activities in the hospital ward space. The service design of nurse station plays an important role in patients' hospitalization experience and nurses' work experience. In this paper, Kano model is introduced into the Kansei Engineering method to study the service of ward nurse station. A questionnaire survey is conducted among nurses and patients in three hospitals. This study compared nurse and patient evaluations to reveal the relationship between ward nurse station service attributes and user perception, and to determine the priority of nurse station service attributes that need to be improved. This study can help hospital managers improve the service quality of ward nurse station. When hospital managers want to improve the experience of patients and nurses, they can improve the service quality by improving the service attributes that affect the experience.

Keywords: Hospital ward · Nurse station · Service design · Kansei engineering

1 Introduction

The nurse station is located in the core of the hospital ward and plays an important role in the hospital ward. It is the physical station where nurses work and interact with other nurses, doctors, patients, family members and other hospital staff [1]. The design of the nurse station affects various factors such as the work efficiency of nurses, the collaboration between nurses, the nursing mode of the ward, the response speed of nurses to patients, and the hospitalization experience of patients [2–4]. In general hospitals in China, the form of ward nurse stations is mainly open nurse stations. Past studies have shown that open nurse stations can improve the ability of nurses to help patients by improving the availability and accessibility of patients, so as to improve the experience and satisfaction of patients [5]. In recent years, the research on nurse stations has mostly focused on the spatial structure and layout [1, 6, 7], and there is a lack of relevant research discussions on the comprehensive service design of nurse stations.

In recent years, medical services have received great attention. Continuously improving the quality of medical services and better serving patients have become the top priority

S. Xie and H. Shang—Contributed equally to this work.

C. Leitner et al. (Eds.): AHFE 2021, LNNS 266, pp. 354–362, 2021.
https://doi.org/10.1007/978-3-030-80840-2_41

of medical institutions and relevant departments [8]. According to the analysis report of China's fifth national health service survey, with the continuous strengthening of the construction of China's medical and health service system, people's demand for medical and health services has increased rapidly [9], and the level of medical services also needs to be improved to meet people's increasing needs for medical services. As a part of medical service, the service of ward nurse station greatly affects the satisfaction of inpatients to medical service. Using the existing service design framework to carry out a comprehensive service innovation of the hospital ward nurse station, we need to bring the patients, medical staff, hospital managers and other stakeholders into the service design, emphasize the people-centered concept, not only pay attention to the service experience perceived by patients in the nurse station, but also pay attention to the work experience of nurses in the nurse station. Because the daily work of nurses is extremely complicated, the layout of nurse station should be reasonable to create a healthy working environment for nurses. When the job satisfaction of nurses is improved, the burnout of nurses can be alleviated, so as to better serve patients and improve the medical experience of patients [4, 6, 10].

The purpose of this study is to apply Kano model and Kansei engineering method to explore the priority of service attributes improvement and to explore the relationship between service attributes and user feeling. Based on the service dominant logic, this study takes the patients, medical staff, hospital managers and other stakeholders into consideration under the existing service design framework, emphasizes the human-centered design concept, and promotes the service innovation of the nurse station in the ward through the method of Kansei engineering.

2 Methods

2.1 Research Methodology Framework

In this paper, the integration framework of Kansei Engineering and Kano model proposed in the previous discussion was used to study the service of nurse station in hospital ward [11–14]. According to the framework of the method, after the service domain is determined, the service semantic space and service attribute space need to be constructed respectively, and then the questionnaire survey is conducted. Through data analysis and modeling, the service semantic space and service attribute space are integrated. Finally, the relational model is established to determine the service attributes that need to be improved first (Fig. 1).

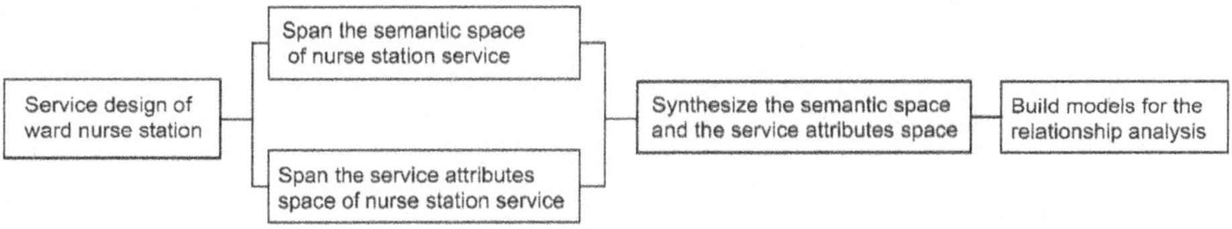

Fig. 1. Research method framework.

2.2 Span the Semantic Space and the Service Attributes Space

This research conducted in-depth interviews with some nurses and patients, and initially collected service elements and some Kansei words. In the stage of constructing the service semantic space, a total of 78 Kansei words were collected through interviews and literature review, and then the KJ method and expert group discussions were used to screen the collected Kansei words, and finally semantic space containing 6 groups of Kansei word pairs was constructed. They are amiable-alienated (K1), efficient-inefficient (K2), reliable-unreliable (K3), professional-nonprofessional (K4), functional- deficient (K5), thoughtful-inattentive (K6). In the stage of building service attribute space, firstly, according to the research on the service dimension of medical furniture by Kao jihchang et al. (2020), combined with the actual situation, the service dimension was determined, and four service dimensions including supportability, usability, technicality and convenience were summarized [15]. The supportability dimension is used to evaluate the medical service experience and support ability from the perspective of patients; the usability dimension is used to evaluate the satisfaction of medical staff and patients with the layout of medical space and the medical furniture products used; the technicality dimension is used to reflect the support intensity of medical space for new technologies and whether the medical institutions are ready to face the new medical model; the convenience dimension measure the degree of convenience for users to move, store and communicate in the medical space. According to the above four service dimensions, 18 service attributes are summarized, and the questionnaire is made according to these 18 service attributes (Table 1).

Table 1. Service dimension and service attribute classification.

Service dimension	Service attribute
Supportability	A1. When patients have questions to ask the nurse station, they can get the reply from the nurse station
	A2. When negative emotions, they go to the nurse station to communicate with nurses, nurses can give comfort and response with empathy
	A3. When the patient presses the call bell of the nurse station in the sickbed, the response of the nurse station can be obtained
	A4. Patients can feel warm and positive service when they go to the nurse station
	A5. The nurse station respects the patient's privacy and does not disclose the patient's information to others
	A6. The nurse station can agree with the special service requirements of patients

(*continued*)

Table 1. (*continued*)

Service dimension	Service attribute
Usability	A7. A shared Pantry is set up beside the nurse station for the rest of medical staff, patients and their families to increase the communication among doctors, nurses and patients
	A8. The nurse station set up a special working point for patients' consultation
	A9. The nurse station bar with high and low countertops ensures that the general nurses in charge of computer work such as medical orders have a quiet working space
	A10. Nurse station furniture round and safe, will not hurt people
Technicality	A11. The wall of the nurse station is equipped with a large intelligent display, which can view patient information, nursing plan arrangement and other information in real time
	A12. The nurse station is equipped with an intelligent display screen, which can play rolling videos that are helpful to understand the disease for patients to view and learn A13. Each nurse is equipped with an iPad, nurses can view and input patient information anytime and anywhere, as well as timely check the doctor's orders A14. The nurse station has an intelligent guidance system, which can help patients guide the way A15. The nurse station has a storage space and the location of items marked
Convenience	A16. The ward is close to the nurse's station, which can be reached quickly
	A17. A logistics pipeline is set between the nurse station and other medical rooms (such as pharmacy, examination room, etc.). The nurse does not need to go to various departments to fetch things, but can receive supplies at the nurse station
	A18. When patients or their family members go to the nurse station to deal with affairs (such as exit and admission procedures), the process is simple and the affairs can be quickly processed

2.3 Questionnaire

This research questionnaire consists of three parts. The first part is the survey of personal information, including personal identity, hospital name, gender, age, education information. The second part is the Kano questionnaire based on 18 service attributes. The third part is the 7-point Semantic-Difference questionnaire. This part of the questionnaire designs 6 different service scenarios according to 18 service attributes, and respondents need to score Kansei words for each scene.

2.4 Data Collection

In this study, a total of 46 nurses and 35 patients were recruited as respondents to fill in the questionnaire. The respondents were from the wards of Fudan University Shanghai Cancer center, the Second Affiliated Hospital of Zhengzhou University and Yuyao People's Hospital. The reliability analysis of the questionnaire shows that the Cronbach α of the Kano model questionnaire is 0.928, and the Cronbach α of the Semantic-Difference questionnaire is 0.980. The Crobach values in the second part and the third part of the questionnaire are both greater than 0.8, indicating that the reliability of the overall research data of the questionnaire is very high and suitable for the next step of data analysis [16]. This study use Excel 2016 and SPSS 26 to analyze the questionnaire data.

3 Results and Discussion

3.1 Evaluation Results of Kano Model

Data analysis was conducted on the evaluation results of Kano model of nurses and patients, and the results were obtained according to the calculation method of Kano model better worse coefficient [18]. With Better as the vertical axis and Worse as the horizontal axis, the Better-Worse analysis charts of patients and nurses were drawn respectively. Among the must-be qualities, patients and nurses pay more attention to the service attribute of supportability dimension. When nurses evaluate the service of nurse station from the perspective of patients, the number of nurses' demand for the service attribute of supportive dimension is higher than that of patients. Therefore, compared with patients, nurses have higher requirements for the service of supportability dimension of ward nurse station. In addition, in the technical dimension, only one service attribute entered the must-be qualities, and this attribute was the result of nurse evaluation. It can be found that the demand of nurses for hospital intelligence is higher than that of patients, and the demand of nurses for portable intelligent devices is higher than that of fixed large-scale smart display.

According to the service attributes of the three quadrants of must-be qualities, one-dimensional qualities and attractive qualities, patients' demand for service is concentrated in the supportability dimension and convenience dimension. Based on this, when the hospital needs to improve the service experience of patients in the ward nurse station, it needs to focus on improving the service attributes of supportability and convenience dimensions, in which the consulting service of the nurse station is the first improvement item, followed by the response speed of the nurse station to the needs of patients. Improving the service efficiency of the nurse station helps to improve the satisfaction of patients. In addition, the improvement of patient satisfaction is inseparable from nurses. Nurses are service providers and emotional contact points with patients. Providing nurses with appropriate medical furniture and intelligent devices can help improve nurses' work efficiency and provide better services for patients (Figs. 2 and 3).

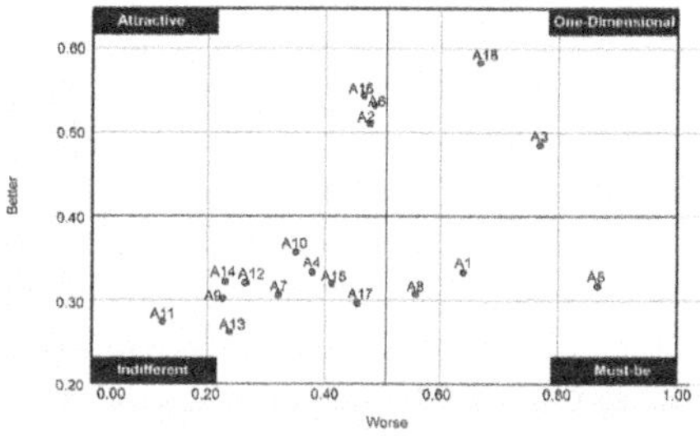

Fig. 2. Better-Worse analysis of patients.

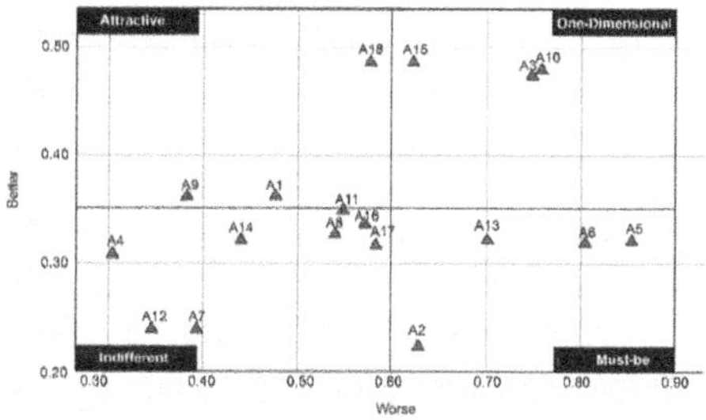

Fig. 3. Better-Worse analysis of nurses.

3.2 Stepwise Linear Regression Model: Relationship Between Service Attributes and Kansei Words of Nurse Station

Independent sample t-test was conducted for nurses and patients on the data of the third part of the questionnaire. The results showed that there was no significant difference in Kansei words between patients and nurses. Therefore, this study carries out a stepwise linear regression model for the evaluation data of the overall service attributes and Kansei words, and the statistical results are shown in Table 2.

Table 2. Significant regression model of Kansei words.

Kansei words	R^2	Significant model	Description of service attributes affected
Amiable (K1)	0.57	k1 = 7.277 + A7*0.862-A8*0.772 + A18*0.737-A1*0.596-A13*0.482	A1, A7, A8, A13, A18
Efficient (K2)	0.572	K2 = 7.475-A8*1.002 + A7*0.695 + A3*0.459 + A18*0.492	A3, A7, A8, A18
Reliable (K3)	0.566	K3 = 7.387-A8*1.138 + A7*0.717 + A18*0.618-A3*0.450	A3, A7, A8, A18
Professional (K4)	0.664	K4 = 7.298-A8*0.803 + A18*0.926-A15*0.394 + A7*0.546-A1*0.466 + A2*0.300-A10*0.329	A1, A2, A7, A8, A10, A15, A18
Functional (K5)	0.544	K5 = 7.404-A8*1.038 + A7*0.693 + A18*0.588-A3*0.490	A3, A7, A8, A18
Thoughtful (K6)	0.521	K6 = 7.532-A8*0.997 + A7*0.462 + A9*0.292-A1*0.319 + A6*0.266	A1, A6, A7, A8, A9

According to the stepwise linear regression model, the service attributes that affect each Kansei words are not single, but multi-dimensional. The most significant model is the Kansei word "professional". Seven service attributes have a statistical impact on "professional". These seven service attributes come from the supportability dimension, usability dimension and convenience dimension respectively. From the perspective of multiple service dimensions, in order to improve the "professional" emotional response of patients to the service of nurse station, we should start with improving the quality of consultation service of nurse station, improving the work efficiency of nurses, and improving the space planning and furniture of nurse station. In addition, Kansei words "efficient", "reliable" and "functional" are affected by service attributes A3, A7, A8 and A18 in different degrees. Therefore, improving these four service attributes will help to improve the experience of patients and nurses from multiple feeling levels.

3.3 Continuous Improvement Strategy of Nursing Station Service

Integrating the results of stepwise linear regression model and Kano model, the service attributes are connected with the affected Kansei words, and the service attributes are arranged from left to right in the way of Kano model priority increasing step by step, and Fig. 4 is obtained. Nurses and patients have different priorities for the service attributes. Nurses' priority service attributes include the related service needs that can improve work efficiency, while patients focus more on the related service attributes that affect their medical experience. In terms of feeling, the emotional experience of patients and nurses is related to the improvement of service attributes, and the feeling linked by the service attributes with priority improvement will also be improved.

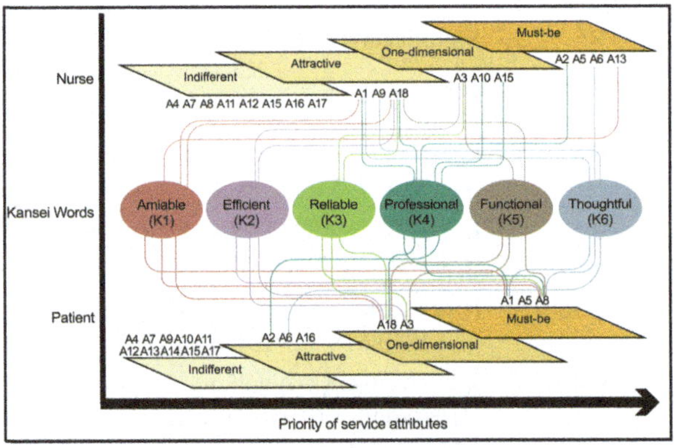

Fig. 4. The order of service attribute improvement and its relationship with Kansei words.

In order to improve the service quality of nursing station, we should consider the multi-dimensional needs of patients and nurses. First of all, hospital managers should give priority to meeting the must-be qualities needs of patients and nurses. Improving the consulting service quality of ward nurse station helps to create a more cordial atmosphere, and providing nurses with portable smart working equipment helps to increase nurses' job satisfaction. Secondly, the hospital ward should improve the service efficiency of the nurse station, which helps to improve the multiple emotional experience of patients, and meeting the one-dimensional qualities needs of nurses can help nurses improve their work efficiency, so as to provide better services for patients. Finally, when the hospital needs to improve the emotional experience of patients and nurses in specific aspects, it can gradually improve the emotional experience of users according to the priority of service needs.

4 Conclusion

In order to improve the service quality of nurse station in hospital ward, this paper revealed the priority of service improvement and the influence of service attributes on patients' and nurses' feelings through Kano model and Kansei engineering method. To improve the service of ward nurse station, we need to consider many factors. Starting from the demand level of patients and nurses for the service of nurse station, combined with the impact of service attributes on the perceived response of patients and nurses, we can improve the comprehensive service quality of nurse station according to the actual situation and improvement goals of the hospital.

This study has brought some enlightenment to the service design of the ward nurse station, but there are still some limitations. First, the study was conducted in specific hospitals, and the results did not apply to hospitals in other parts of the world. Secondly, the number of respondents included in this study is small, which has a certain impact on the results, so a wider range of respondents should be included in the future research. Finally, this study is carried out for specific service dimensions, and the service attributes are not comprehensive. In the future research, we need to consider the service dimensions and service attributes of ward nurse station more comprehensively.

Acknowledgments. Siwen Xie and Huifang Shang contributed equally to this work.

References

1. Fay, L., Hui, C., Real, K.: A systematic literature review of empirical studies on decentralized nursing stations. Herd Health Environ. Res. Des. J. **12**(1), 44–68 (2018)
2. Gurascio-Howard, L., Malloch, K.: Centralized and decentralized nurse station design: an examination of caregiver communication, work activities, and technology. HERD **1**, 44–57 (2007)
3. Bayramzadeh, S., Alkazemi, M.F.: Centralized vs. decentralized nursing stations: an evaluation of the implications of communication technologies in healthcare. HERD **7**, 62–80 (2014)
4. Zborowsky, T., Bunker-Hellmich, L., Morelli, C., O'Neill, M.: Centralized vs. decentralized nursing stations: effects on nurses' functional use of space and work environment. HERD **3**, 19–42 (2010)
5. Shattell, M., Bartlett, R., Beres, K., et al.: How patients and nurses experience an open versus an enclosed nursing station on an inpatient psychiatric unit. J. Am. Psychiatr. Nurses Assoc. **21**, 398–405 (2015)
6. Chen, X.X., Xuan, X.D., Li, Z.F.: The effect of nursing unit spatial structure on nurse behavior: a comparative study based on the theory of evidence-based design. Urbanism Archit. **301**(32), 51–55 (2018)
7. Chen, X.X., Xuan, X.D., Li, Z.F.: The influencing factors of nursing staff satisfaction on nursing unit. Huazhong Archit. **10**, 45–50 (2019)

8. Tian, C.-J., Tian, Y., Zhang, L.: An evaluation scale of medical services quality based on "patients' experience". J. Huazhong Univ. Sci. Technol. [Med. Sci.] **34**(2), 289–297 (2014). https://doi.org/10.1007/s11596-014-1273-5
9. Center for Health Statistics and Information, NHFPC: An Analysis Report of National Health Services Survey in China (2013)
10. Ge, L., Luo, X.: Nursing unit design research based on evidence-based design. Urbanism Architect. **22**, 25–27 (2014)
11. Cai, M., Wang, Q.Q.: Research on service design method based on Kansei engineering. Des. Res. **9**(06), 60–67 (2019)
12. Chen, M.C., Hsu, C.L., Chang, K.C., Chou, M.C.: Applying Kansei engineering to design logistics services–a case of home delivery service. Int. J. Ind. Ergon. **48**, 46–59 (2015)
13. Hsiao, Y.H., Chen, M.C., Liao, W.C.: Logistics service design for cross-border E-commerce using Kansei engineering with text-mining-based online content analysis. Telematics Inform. **34**(4), 284–302 (2017)
14. Hartono, M., Chuan, T.K.: How the Kano model contributes to Kansei engineering in services. Ergonomics **54**(11), 987–1004 (2011)
15. Kao, J.C., Xin, X.Y., Xie, M.H.: Study on the service design of medical furniture based on the theory of open innovation. Des. Res. **10**(05), 55–61 (2020)
16. Eisinga, R., Grotenhuis, M.T., Pelzer, B.: The reliability of a two-item scale: Pearson, Cronbach, or Spearman-Brown? Int. J. Public Health **58**(4), 637–642 (2013)

A Proposed Roadmap to Close the Gap Between Undergraduate Education and STEM Employment Across Industry Sectors

Yassi Moghaddam[1(✉)], Stephen Kwan[2], Louis Freund[3], and Martha G. Russell[4]

[1] International Society of Service Innovation Professionals, Santa Clara, USA
yassi@issip.org
[2] Lucas Graduate School of Business, San José State University (Emeritus), San Jose, USA
stephen.kwan@sjsu.edu
[3] Industrial and Systems Engineering, San José State University (Emeritus), San Jose, USA
louis.freund@sjsu.edu
[4] Graduate School of Education, Stanford University, Stanford, USA
martha.russell@stanford.edu

Abstract. Automation, powered by advances in machine learning and data science, is super-charging innovation and disrupting industries and jobs. Work in America looks very different from a couple of decades ago, and it will look even more different going into the future decades. For new jobs to offset the losses, particularly in the fields of Science, Technology, Engineering, and Mathematics (STEM), America will require an adaptable workforce [1] with a new set of skills [2]. To gain an industry perspective on how STEM undergraduate educational institutions might best meet the expected demands for new skills, the International Society of Service Innovation Professionals (ISSIP) was awarded a contract by the National Science Foundation (NSF). The goal of this project was to seek the perspectives of industry leaders using a survey and a workshop focusing on workforce re-skilling and educational trends that industry views as important and necessary for a thriving future workforce. These perspectives were summarized and provided to leaders in higher education. In this paper, the authors present the outcomes of the workshop, the stakeholder feedback, a roadmap for the future of STEM education, and a set of recommendations to the leaders of higher education institutions.

Keywords: STEM jobs · STEM education · STEM skills · Future of skills · Future of undergraduate education · Future of work · Work practice skills · Soft skills · Industry-university relations · Engineering education

1 Introduction

Innovation drives economic growth, and high-skill talent in science, technology, engineering, and math (STEM) fields is critical to the nation's innovation leadership. In many

C. Leitner et al. (Eds.): AHFE 2021, LNNS 266, pp. 363–373, 2021.
https://doi.org/10.1007/978-3-030-80840-2_42

industries the most in-demand occupations, specialties, and skills of today did not exist ten or even five years ago, and the pace of change driven by technological advancements [3] is set to further accelerate. This rapid pace of change will have a tremendous impact on how the workforce of the future acquires and applies new skills [4].

To gain an industry perspective on how undergraduate educational institutions might best meet these expected new demands, the International Society of Service Innovation Professionals (ISSIP) conducted a project for the National Science Foundation's program, "Accelerating Discovery: Educating the Future STEM Workforce." The project set out to understand the views of industry leaders on both the preparedness of current graduates from STEM educational programs to enter the workforce and the workforce readiness requirements for jobs in a thriving 21st-century America.

Through insights gained from experts, surveys and a workshop, the project team examined future work environments and the types of expertise needed, together with preferred formats for skills acquisition and work rules. It also examined industry-driven certification systems (including standards, benchmarks, and badging) and industry practices for skills validation. Furthermore, the project explored the challenges faced today by recent 4-year STEM graduates entering the workforce, with regards to meeting current industry demands as well as successfully navigating future career pathways, and whether such persons are likely to face competency issues, shortfalls or omissions in their skills. A final goal of the project was to examine whether a gap will exist between current undergraduate STEM educational program content and the job-readiness of future STEM graduates. If so, the project was to consider how this gap might be closed by modifying STEM educational content and processes.

The overall motivation for this discovery project was to help to assure that American higher education in the future will be able to supply talent for a thriving US workforce. This project was intended to provide input to help higher education leaders develop programs that will increase America's innovation capacity for the high-paying jobs of today and tomorrow.

This paper summarizes the methods, findings, and recommendations of this project.

2 Method

This discovery project was implemented through a series of expert interviews, surveys and a workshop. ISSIP carried out this project in 2 phases outlined below. Phase 1 was conducted to prepare for and produce the workshop. Phase 2 used the workshop outcomes to further refine the project findings.

In Phase 1, the project team developed a survey consisting of ten items which were based on prior industry research [5, 6]. The Phase 1 survey items had been constructed and presented to NSF in the project proposal. When the project began, the team refined these items and then created a Qualtrics survey instrument that could be easily distributed and which would tabulate some response options automatically.

As the survey refinement process was underway, a panel of experts ("Expert Panel") was assembled by developing a list of 50 industry leaders from the ISSIP membership roster who were known to the team to have responsibility or visibility into the corporate college hiring process, either as HR leaders, as direct hiring managers, or as business strategists. The main screening criteria for participants was having direct or significant visibility into their organization's future workforce development.

The size of the Expert Panel was aimed to be at a minimum of 25 and a maximum of 35 persons (based on the need to keep the size of the planned Phase 1 workshop between 30–40 participants). Of the list of 50 industry leaders, 35 passed the screening criteria to join the Expert Panel. They represented different industry sectors including technology (Cisco, IBM, Facebook, LinkedIn, ServiceNow, Equinix, Accenture, AT&T); banking (CapitalOne); education (Axelos, BeBraven, and others); transportation & infrastructure (Contra Costa Transportation, HNTB, Kane Environmental Inc); and manufacturing (Flex).

Eight of the Expert Panel members were selected and asked to pilot the initial survey. Based on their feedback, the team modified the survey and finalized the "Survey Phase 1" list of items, which was then distributed to all members of the Expert Panel for Phase 1 data gathering.

Also, as a part of the Phase 1 protocol, the team conducted dialogues with a subset of the Expert Panel to identify crucial workshop topics. The subset was selected based their survey responses. In Phase 1, 35 surveys were distributed and 23 responses were submitted prior to the workshop. The results of Survey Phase 1 were summarized for presentation and discussion at the workshop, which was conducted on Dec 10, 2019.

As planned, the workshop brought together 36 people, including members of the selected Expert Panel from industry, academics, and representatives of government foundations. Specifically, of the 35 Expert Panel members who were invited to the workshop, 24 made themselves available to participate. Other workshop participants included the project PI, Co-PI and two senior personnel, three faculty members, San Jose State University (2), Arizona State University (1), all with rich industry perspectives, 2 representatives from NSF, and 2 SJSU graduate students (to help with the workshop operation). Additionally, one guest, an ISSIP Board member, also a Fellow at Japan Science of Technology Agency, attended the workshop as an observer. The workshop agenda consisted of two speaker presentation sessions, three Q&A panel sessions, and one breakout session. The workshop completed Phase 1 of the project. Content and comments in the various workshop sessions is summarized below in the "Findings – Workshop" section.

In Phase 2, the project team made minor modifications to the survey based on feedback gathered in the workshop, and fielded Survey Phase 2 to increase the number of stakeholder responses. Survey Phase 2 was sent to 362 experts – the remaining 12 Expert Panel members who had not responded in Phase 1, plus 350 ISSIP member Experts. This final survey distribution yielded a cumulative Phase 1 and Phase 2 total of 75 survey responses across eighteen industry sectors. More than 50% of respondents were from Information Technology related industries; others identified with healthcare, education, banking, government, consulting, manufacturing, media, energy, and transportation. Partial responses were accepted. The team then analyzed the consolidated results from Survey Phase 1 and Survey Phase 2 and the workshop to develop and disseminate the final report.

No members of the Expert Panel, survey respondents, or workshop participants (with exception of the project team, and the 2 students) were given any compensation or incentives for completing the survey or participation in the workshop.

3 Key Findings

Across the depth of experience and the breadth of perspectives collected from the Expert Panel, surveys and the workshop, a framework emerged (Fig. 1). This framework[1] provides a roadmap for understanding the requirements for STEM education for a thriving 21st century workforce. Following presentation of this framework, the findings of each contributing component of this discovery process are detailed below.

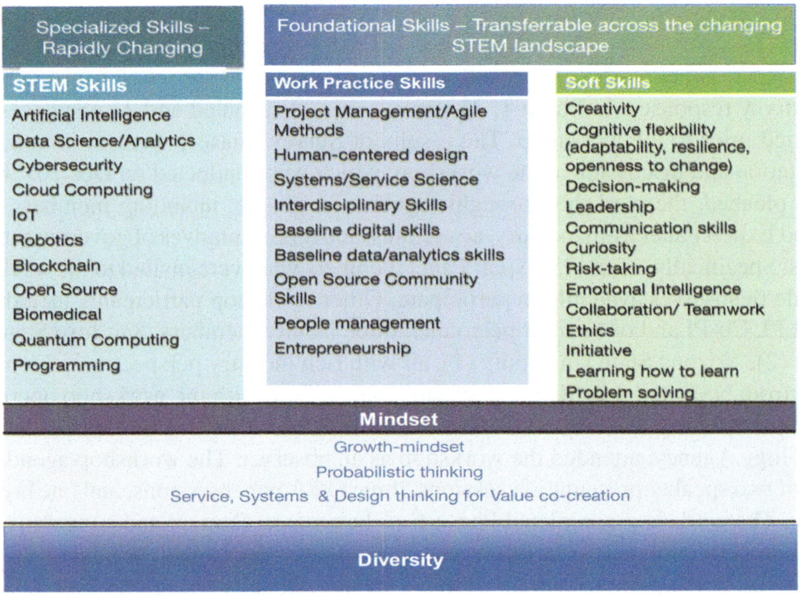

Fig. 1. An industry perspective on skills requirements for the future

Key Survey Results Takeaways. One important component of the project was the survey. The results of a sampling of survey results (same questions in Phase 1&2) identifying key skills from industry perspective, and key takeaways from the survey results are presented here.

> ***Question 2.1*** *"Considering the new hires in your organization this year, how prepared are they in each of the following skills? Also, please identify which are the top 5 most important skills in the list."*

[1] The framework dimensions are similar to those of the "T-shape" paradigm [10], in which the "I" of the "T" includes Specialized Skills and the bar of the "T" Foundational Skills.

Responses were selected from a predefined list [6]. Of the thirteen skill categories shown in Fig. 2A, "Analytical Skills" and "Judgment and Decision Making" were chosen as top skills by the highest number of respondents. These were followed by "Collaboration," "Creativity" and "Cognitive Flexibility." "Forecasting" and "Negotiation Skills" were selected least frequently as one of the top five skills desired for new hires. "Customer Service Skills," "Leadership," "Computation," "Data Skills," "Design," and "People Management" were also selected as shown.

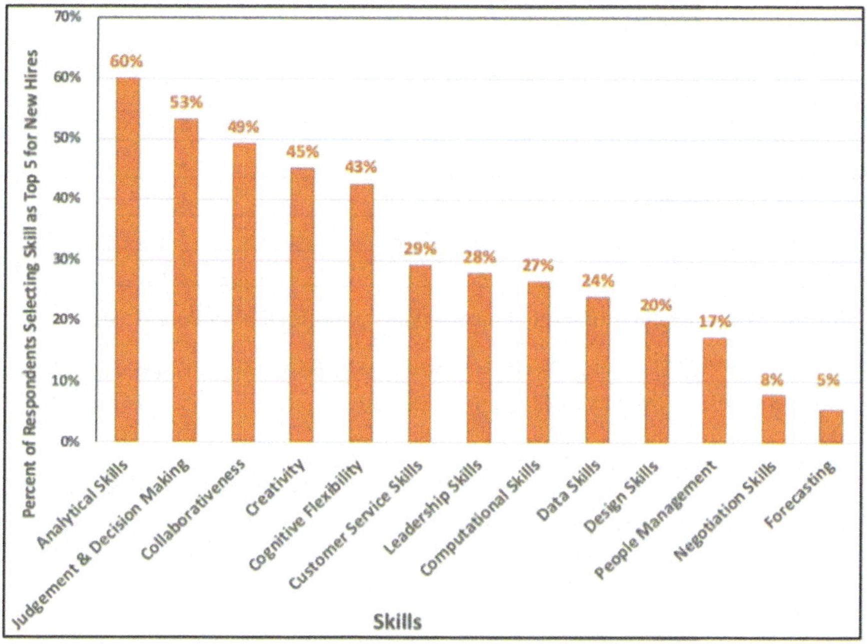

Fig. 2A. Percent of respondents selecting skill as one of top 5 for new hires. (N = 75)

It is important to note that of the five skills considered most important, four are Foundational Skills (Fig. 2A), and only one, "Analytical Skills," is a Specialized Skill. These top priority skills are the same that the workshop participants identified as Foundational Skills, meaning that they provide the foundation upon which technical and specialized skills are built.

Skills	*N*	*Average Ratings of Level of Preparedness*
Computational Skills	67	5.5
Collaborativeness	71	5.1
Analytical Skills	70	5.0
Data Skills	68	5.0
Creativity	68	4.7
Cognitive Flexibility	69	4.5
Judgement & Decision Making	71	4.5
Customer Service Skills	70	4.3
Design Skills	69	4.0
Leadership Skills	67	3.9
People Management	67	3.8
Forecasting	65	3.8
Negotiation Skills	67	3.4

Fig. 2B. Average ratings of level of preparedness of skills (1: Very Poor, 7: Very Well)

Question 2.2 *"What other critical skills not listed above are considered important and the new hires' preparedness in these skills?".*

More than 50 other "critical skills" were listed by respondents to this question. The top two of those listed were "Communications" and "Adaptability". Other responses included "Emotional intelligence", "Curiosity", "Innovative", and "Social skills".

Key takeaways from the full set of surveys are presented in Table 1 below.

Table 1. Key survey takeaways, an industry perspective on STEM education for the future

An industry perspective on STEM education for the future
Rapid technology advances are changing the nature of jobs, resulting in jobs that are driving demand for new skills
Universities are lagging behind in teaching undergraduates some of the skills needed for industry entry-level jobs
High demand skills span a broad range that can be categorized as "Specialized Skill" (technical in nature), and "Foundational Skills" (general in nature.) Industry is dependent on both categories of skills
Undergraduate hires come to work well-prepared with Specialized Skills but not well-prepared with Foundational Skills
Industry demand is increasing for candidates that have practical experience through internships and co-ops
US universities are not producing enough American STEM graduates to satisfy industry needs, and therefore industry has turned to candidates with H1B visas
Badging and verifiable skills are considered by industry as augmenting, not replacing, 4-year degree programs
"Life-long learning" and "pervasive learning" are viewed as very important
A basic level of computing skills is required, even for non-technical jobs, in STEM work environments

3.1 Workshop

Another main component of this project was a one-day Workshop, "An Industry Perspective on STEM Education for the Future," held on Dec 10, 2019.

Workshop Key Takeaways. Skills can be viewed as belonging to two major categories as shown in Fig. 1: 1) Specialized Skills designating those technical and scientific skills for which requirements keep changing as technologies advance, and 2) Foundational Skills which include those skills that are long lasting and transferable across technologies, jobs, companies, and even industries. Two types of Foundational Skills were identified: a) "Work Practice Skills," and b) "Soft Skills. In this view, Work Practice Skills are those skills mostly pertinent to work processes, such as human-centered design, agile, service thinking, and interdisciplinary skills, while Soft Skills are those skills such as interpersonal & communications skills, risk taking, resilience, collaboration, etc. [6, 7]. In addition to both types of skills, according to the industry participants in this study, new Mindsets are required to constantly reskill and upskill across these skills dimensions. These Mindsets, according to the workshop panelists include the Growth Mindset [8]. Probabilistic Thinking, Service, Systems and Design Thinking. More details about the results of the workshop is outlined below.

Workshop Sessions Summaries: In the **first session,** speakers representing high tech, media, telecom, financial and professional services acknowledged that STEM job opportunities today, and for the foreseeable future, center on many areas of STEM – including artificial intelligence, cybersecurity, cloud computing, data science, IoT (Internet-of-Things), open-source, and blockchain. Additionally, jobs in emerging fields such as quantum computing, biomedical, control systems, robotics, telerobotics, augmented reality, and virtual reality are already present today. The specialized skills required for these current and emerging jobs are constantly changing as technologies continue to advance. They, therefore, according to industry stakeholders, lend themselves better to micro-learning modes of education than to established curricula, which can rapidly respond to evolving reskilling needs.

While differences in pedagogical approaches and curricula across institutions were acknowledged by the speakers in the first session, they concurred that, in general, higher education needs to re-focus the 4-year STEM degree programs to provide the broad "Foundational Skills" (Fig. 2A and 2B) that are essential over the lifespan of careers and actually prepare learners for life-long micro-learning. The panel also made a distinction between STEM jobs and STEM degrees. According to these industry representatives, it is anticipated that while the majority of good paying entry-level jobs will be in STEM, not everyone who enters that job market needs to be a graduate of a STEM academic degree program. That being the case, these speakers advised that higher education leaders consider how to prepare both STEM and Non-STEM students for the new STEM jobs.

Accordingly, the panelists stated that it would be a mistake to focus on STEM education at the exclusion of the liberal arts. Many of the Foundational Skills such as abstract thinking, conceptual problem solving, human-centered design, and creativity are fostered by a liberal arts education. Plus, the panelists concurred that liberal arts graduates with a baseline level of "digital skills" will be in greater demand in the coming years, particularly as job-supporting AI becomes more pervasive in the workplace [9].

Finally, the panelists characterized STEM graduates as coming to work for their first post-graduation job fairly well-prepared with technical Specialized Skills, but often lacking the Foundational Skills necessary to perform well in those jobs and beyond.

In the **second session,** speakers representing healthcare, infrastructure, state government, and insurance acknowledged that their sectors are going through major transformations resulting in increased demand for STEM graduates. They indicated that, based on their observations, today's STEM graduates enter the workforce fairly well-prepared with technical skills. However, they are often lacking in the Foundational Skills that are necessary to apply those skills effectively to solve problems in the business context. These panelists emphasized the opportunity for higher education to partner with industry to foster experiential job-relevant learning through structured apprenticeship and internship programs that can be integrated into university curricula.

The speakers in the **third session** represented views from several innovative educational organizations that are leveraging successful partnerships aimed at rapidly filling the skills gap between education and employment. This panel seconded the idea that higher education can better partner with industry. These partnerships can help to ensure the relevance of academic content to industry and can also foster job-relevant experiential learning through well-structured apprenticeship and internship programs. Pioneering industry-education collaboration models come in different forms. Some universities and community colleges are currently partnering with novel learning organizations and local businesses to bring experiential and career-enhancing learning to students. Other universities are partnering with educational startups to provide career pathways to STEM jobs for underserved student populations. Also, innovative service organizations are offering the content of their high-demand industry certifications to universities for integration into academic curricula in order to develop rapid micro-learning capabilities.

Finally, a dire need for greater representation of women and members of minority groups in upskilling efforts was repeatedly mentioned by multiple panelists in all sessions (as was the case in the survey responses); it is also evident in industry trends [11].

Workshop Breakout Sessions: To explore these ideas this in more depth, the workshop participants were asked to form four breakout groups. Each group was asked to perform its own SWOT analysis (Strengths, Weaknesses, Opportunities, Threats) of today's US higher education systems and institutions and to provide comment on this claim: US higher education is considered "the best in the world" [12]. A combined view of the report-outs of the breakout sessions is presented below.

Strengths. The strengths of US higher education lie in the high-quality standard of higher education and in the diversity and flexibility of degree programs for learning and research. This opinion extends to both STEM and Liberal Arts, as well as those higher education environments that allow learners to build social and cultural networks beyond the classroom walls.

Weaknesses. The weaknesses noted by the workshop participants include the high cost of education that is making access to higher education (a characteristic that historically has been a strength) increasing more challenging for many Americans. The weaknesses also include an imbalance of American vs. international graduates, as well as a disconnect from industry and from skills requirements for the workforce. Additionally, an under-emphasis on liberal arts and humanities education in STEM curricula is resulting in lack of Foundational Skills preparedness in new college hires. Finally, academic institutions are typically seen as risk-averse cultures with siloed curriculum structures resulting in slow response to industry changes.

Opportunities. The opportunities for higher education lie in building more bridges to industry, including internships, co-ops and apprenticeships. In addition, forming partnerships with other learning organizations to pave the way for offering more industry focused project-based/problem-based and life-long learning opportunities. Furthermore, higher education leaders are advised to explore opportunities for providing lifelong learning beyond the degree, so current students can become "customers" for life, not just for 4-years.

Threats. The democratization of information raises the bar for the traditional academics, as knowledge from top scholars is available free or at costs far lower than many higher education degree programs. In addition, the perceived lack of relevance in traditional higher education disciplines and structures has contributed to an eroded confidence in the 4-year STEM degree as the pathway to high paying jobs and has widened the perceived gap between expectations for 4-year STEM degrees and confidence in their ability to fulfill those demands.

4 Concluding Remarks and Future Directions

Specialized STEM Skills are required as jobs constantly change and technologies advance. Therefore, a micro-learning mode for rapidly developing fields can better facilitate upskilling for graduates' job readiness. Because of this, industry leaders encourage educators to consider re-focusing 4-year STEM degree programs more concretely on providing the broad Foundational Skills that transfer across jobs and prepare the learner for micro-learning throughout a lifetime. Many of the Foundational Skills such as abstract thinking, conceptual problem solving, human-centered design, and creativity can be fostered by a well-integrated liberal arts education into STEM undergraduate programs. Companies are offering high-demand industry certifications to universities for integration into existing curricula; some skills verifications are offered in partnership with higher education STEM programs.

Industry representatives acknowledge the difficulty of change for academic institutions and indicated strong receptiveness to partnering to co-create solutions. Their recommendations fall into three categories:

1. Incentivize

   - Reward faculty for cross-disciplinary and higher education/industry engagements.
   - Incentivize building bridges between higher education and industry.
   - Catalyze and offer students job-relevant experiential learning opportunities.

2. Strengthen

- Leverage strengths inherent in the diversity and flexibility of STEM degree programs.
- Emphasize the importance of liberal arts/humanities and their integration into STEM.
- Expand STEM curricula to track and align with industry-relevant applications.
- Increase applied learning through apprenticeship programs, Co-Ops, and paid internships.
- Foster lifelong learning.

3. Create

- Integrate Foundational Skills into student learning opportunities.
- Develop and offer micro-learning programs for Specialized Skills development.
- Offer project-based learning opportunities in every year of the undergraduate program.
- Involve industry in curriculum reviews (e.g., industry advisory boards).

This project began an important discovery process needed to inform responsive change to the requirements needed for a strong pipeline of talented STEM employees. Future research questions include: How are industry recruiters and hiring managers screening STEM graduates for Specialized, Foundational Skills? And how do those criteria and processes map into academic curricula? What are the short and long-term impacts of COVID-19 on the findings of this pre-pandemic report? Further research could provide valuable guidance for educational institutions and organizations empowered to create incentives for change.

Further insights from industry are needed. Additional insights from educational leaders are needed. Dialogue between industry employers and academic educators will be essential. Both are responsive to incentives, and incentives to drive needed changes remain a national opportunity.

References

1. Herman, A.: America's High-Tech STEM Crisis. Forbes, September 2018
2. McKinsey Global Institute: The Future of Work in America. McKinsey and Company (2019)
3. Kelly, K.: The Inevitable: Understanding the 12 Technological Forces That Will Shape Our Future. Viking (2016)
4. Dell Technologies: Realizing 2030: A Divided Vision of the Future (2018). https://www.delltechnologies.com/en-us/perspectives/series/a-divided-vision-of-the-future/. Accessed 1 May 2020
5. Accenture. It's Learning. Just Not as we Know it (2018). https://accntu.re/37PGj7O. Accessed 29 July 2019
6. World Economic Forum Report 2016: The Future of Jobs, Employment, Skills and Workforce Strategy for the Fourth Industrial Revolution. https://bit.ly/37RUWaI. Accessed 29 July 2019
7. Azhar, A., Droog, C.: Resilience is a skill that's just as important as tech know-how. Strategy+Business, PWC, 11 February 2020

8. Dweck, C.: Mindset: The New Psychology of Success. Penguin Random House LLC (2006)
9. Hartley, S.: The Fuzzy and Techie, Why Liberal Arts Will Rule the Digital World. HMH Books & Media Publishing (2017)
10. Moghaddam, Y., Demirkan, H., Spohrer, J.: T-shaped professionals: Adaptive Innovators. Business Expert Press (2018)
11. Simonite, T.: AI Is the Future—But Where Are the Women? Wired Magazine, August 2018
12. The Economist Special Report: Secrets of success, America's system of higher education is the best in the world. The Economist, 14 August 2018

Artificial Intelligence Based Co-creation

Conceptualizing Artificial Intelligence-Based Service Ecosystems

Alfred Zimmermann[1(✉)], Rainer Schmidt[2], Kurt Sandkuhl[3], Dierk Jugel[1], Christian Schweda[1], Michael Möhring[2], and Barbara Keller[2]

[1] Herman Hollerith Center, Reutlingen University, Boeblingen, Germany
{alfred.zimmermann,dierk.jugel,
christian.schweda}@reutlingen-university.de
[2] Munich University of Applies Sciences, Munich, Germany
{rainer.schmidt,michael.moehring,barbara.keller}@hm.edu
[3] University of Rostock, Rostock, Germany
kurt.sandkuhl@uni-rostock.de

Abstract. Platforms and their surrounding ecosystems are becoming increasingly important components of many companies' strategies. Artificial Intelligence, in particular, has created new opportunities to create and develop ecosystems around the platform. However, there is not yet a methodology to systematically develop these new opportunities for enterprise development strategy. Therefore, this paper aims to lay a foundation for the conceptualization of Artificial Intelligence-based service ecosystems exploiting a Service-Dominant Logic. The basis for conceptualization is the study of value creation and particularly effective network effects. This research investigates the fundamental idea of extending specific digital concepts considering the influence of Artificial Intelligence on the design of intelligent services, along with their architecture of digital platforms and ecosystems, to enable a smooth evolutionary path and adaptability for human-centric collaborative systems and services. The paper explores an extended digital enterprise conceptual model through a combined, iterative, and permanent task of co-creating value between humans and intelligent systems as part of a new idea of cognitively adapted intelligent services.

Keywords: Digitalization · Value-co-creation · Artificial Intelligence · Service ecosystems · Digital architecture reference cube

1 Introduction

Platforms are considered one of the most important business models of the 21st century [1]. They play an increasingly important role in the development of corporate strategies. In 2019, seven of the ten most valuable companies in the world provided digital platforms (e.g., Microsoft, Facebook, Alibaba, Amazon) [2]. Platforms generate value by facilitating actions between external consumers and producers of products or/and services, platforms [3]. They are spreading rapidly in many different sectors such as sports, transportation, and accommodation; trading both physical goods and services

C. Leitner et al. (Eds.): AHFE 2021, LNNS 266, pp. 377–384, 2021.
https://doi.org/10.1007/978-3-030-80840-2_43

[4]. Moreover, platforms have become the main basis for new business ideas through crowdsourcing [5], enabling the outsourcing of labor, resources, and money from the crowd. Thus, they establishing the gig economy. For example, 60% of all startups in 2019 are based on platform strategies [6]. Platforms are an important source of innovation and play an essential role in many economies in the digital age [7].

Artificial Intelligence (AI) is the most important general-purpose technology of our time [8], catalyzing complementary innovations. Artificial Intelligence is receiving a high level of attention due to recent advances in various areas such as image recognition, translation, and decision support [4]. AI is also a key to the creation of platforms and automation opportunities for intelligent service ecosystems.

The paper raises an important research question: What are AI-based approaches and core concepts for digital platforms and ecosystems to effectively determine and actively adapt or readjust value in the use of smart services ecosystems? Typically, improving digital services and products and their architectures is a manual design task that is extremely complex and requires specialized, infrequent experts. But can this architectural design process be significantly shortened and improved in content by Artificial Intelligence algorithms, especially machine learning with deep learning and semantic support?

In this paper, we present a new conceptual framework for the design of Artificial Intelligence-based ecosystems and their digital platforms, which helps to analyze and actively enable faster and deeper usage value discovery, as well as semi-automated intelligent adaptation and agile engineering of smart service ecosystems.

2 Artificial Intelligence

Fundamental Artificial Intelligence (AI) capabilities include automatically generated solutions from previous application patterns as well as from solution elements derived via causal knowledge structures such as rules and ontologies, results from learned solutions based on data analytics with machine learning as well as deep learning with neural networks. The changing role of technology from a supporting tool to an actor in value creation is highlighted by [9]. Intelligent service ecosystems can be pulled into platforms and curated [4]. AI enables innovative applications such as predictive maintenance [10], logistics optimization, and improved customer service management [11]. Artificial Intelligence is critical for business competitiveness, although choosing the right strategy to implement is a challenging task [9]. AI applications also identify user requirements and adapt them in order to provide the best possible product or service.

Artificial Intelligence development has long been characterized by the desire to shape human intelligence according to achievements. In particular, symbolic Artificial Intelligence, which is based on formalized knowledge and rules, was strongly influenced by the imitation of demanding intellectual tasks. Examples include solving mathematical problems and taking on decision-making tasks similar to those of human experts in the form of expert systems. In this symbolic orientation of Artificial Intelligence, the desire to clearly replicate the human expert in all his competence becomes quite clear.

Machine learning [12] applies an inductive approach, using recurrent data patterns to guide decisions. Machine learning efforts had less ambitious goals. The goal was not to automate all the expertise of human experts, but to automate decision making in narrowly defined domains, i.e., to replace the human decision maker with a machine decision maker for a given decision. This classical machine learning is characterized by the fact that the learning process itself is still very much supported by human experts who identify the relevant features for the decision in the right direction.

The basic mechanism of neural networks is to adjust the weights representing the strength of the connection between neurons until the transformation of input signals into output signals shows the desired behavior [13]. This adjustment is called "training" and is performed by using training data [13]. In this way, approaches based on neural networks also automate the selection of features relevant to the decision. All that remains for the human involved is the identification of the objective function and the basic configuration of the neural networks. In the meantime, however, there are also approaches that attempt to automatically select the architectures used and their configuration.

3 Platforms and Ecosystems

Platforms enable direct interactions between multiple otherwise unconnected groups of actors [14]. Their operators act as intermediaries that enable transactions between two or more groups of actors. Platform strategies are becoming a critical component for many business models [15]. The creation of network effects and ecosystems through platforms provide a new source of competitive advantage [5]. A shift in strategic focus to building communities and engaging resources of platform members is identified as a new axiom in [16].

Significant foundations for the success of a platform are not laid within the platform by its technical design and by the use of technologies. They are created in the so-called ecosystem of the platform. The ecosystem of the platform is based on the totality of users who collaborate via the platform. These users can be distinguished into several groups. In order to form a platform, it is necessary to achieve sufficient growth in several groups. Network effects play a central role in this process. They can occur within but also between user groups. Fostering these network effects is a critical task in the development of a platform. An ecosystem is characterized by a set of interacting resources, such as organizations, individuals, and autonomous actors and systems, that jointly develop their capabilities and roles according to the Service-Dominant logic (S-D logic) [17]. Based on the S-D logic [18], a service ecosystem is a self-contained and self-adaptive system of loosely coupled resource-integrating actors connected by shared institutional logic and mutual value creation through service exchange. From a technical perspective, digital services are slices of code that perform a specific functionality to enable business offerings [1] of digital products and services, which are composed of business services, data services, and infrastructure services.

Many companies fear ignoring the impact of platform formation and the emergence of new ecosystems, as passivity allows external actors to disrupt incumbent markets [19]. However, companies face a dilemma because creating a platform is a risky task: a large percentage of platforms fail because they do not attract enough users [1]. Therefore, it is necessary to develop methods and procedures that allow companies to develop platforms in a reliable way and minimize the risk of failure. Previous studies present recommendations for building platforms [19]. User engagement, active users, and incentives to participate are cited as more important to platform success than pushing goods to customers [5]. However, these extensive and comprehensive studies focus on the operational processes of platforms: specifically, how to attract participants, match supply and demand, and facilitate transactions [5].

Few existing research studies address the inner workings of platforms, particularly how value is created via platforms in a sustainable manner. The primary model for studying value creation on platforms is based on network effects [20]. A more detailed view of value creation mechanisms on platforms is presented in [21]. However, other views are increasingly emerging. Building on the SD-Logic [17], liquefaction and resource density [21] are identified as important factors to increase value creation on platforms. Resource liquefaction is the decoupling of information from its physical form or device. Resource density is the mobilization of information in the most effective and efficient way on platforms.

4 Value Co-creation

Research on value creation on platforms began with a focus on low marginal costs and network effects [5]. This combination differentiates platforms from other business models: Marginal cost is the cost of producing an additional unit of service. Low marginal costs allow platforms to expand their business with minimal costs [16]. Recently, research on value creation on platforms has focused on network effects [20]. Network effects increase the value of a product when the number of customers of the product increases [20]. Network effects can be divided into direct and indirect network effects [20]. Direct network effects increase the value for a group of actors by their increasing number [22]. Indirect network effects refer to the value created for one group of actors in the platform by the increasing number of another group of actors.

An extensive literature review in [23] argues "that the commonly used research models and frameworks for observing the value creation mechanisms of Web 2.0 and social media applications are partially incomplete and fallacious". The complex structure of value creation in digital ecosystems was identified by Suseno et al. [24]. Therefore, we chose a new perspective for our investigation. A rather fresh view of value creation on platforms focuses on value co-creation. We draw on the definition of value creation in the Service-Dominant Logic [17], which breaks with the traditional goods-dominant logic and focuses on services as a means of economic exchange. The S-D logic was introduced by Vargo and Lusch [25] and further explored in [26]. Value co-creation is defined as a set of actors integrating resources (operant and operand) that are integrated, accepted, and exchanged in an ecosystem of resources regulated by institutional arrangements [21]. Therefore, a network-centric perspective [20] is the basis of value co-creation. This view contradicts the classical perspective that there is a strict separation of producer and consumer that either creates or destroys value [21]. S-D logic focuses on the processes of interaction rather than the outcome in the form of a product.

Among the S-D logic-based platform models focusing on service innovation the model of Lusch and Nambisan [21] provides an extended view of service innovation: platforms support the co-creation of value by allowing participants to present value propositions and exchange services. Platforms "liquefy" resources and increase resource density [21] by facilitating access to them. Resource integration is a key capability for service systems. Actors and resources interact through platforms and jointly create value for individual actors and platform providers. The value creation process consists of the following steps shown in Fig. 1: value proposition information exchange (1), value proposition filtering (2), service and currency exchange (3), resource liquefaction (4), and resource density (5).

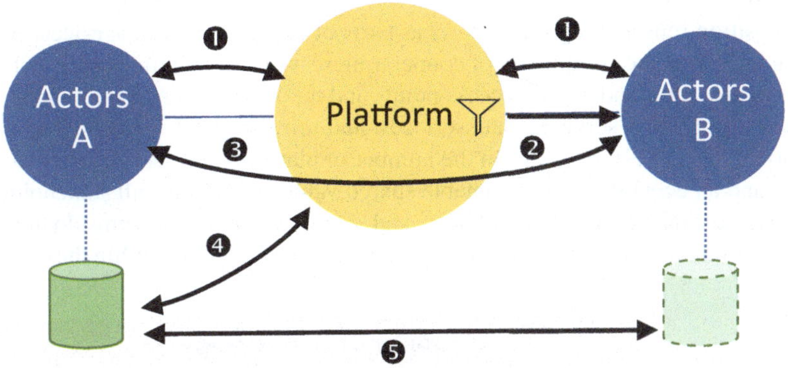

Fig. 1. Platforms value co-creation mechanisms

Based on these theoretical considerations, platforms can be interpreted as value co-creation mechanisms based on S-D logic [21]. On platforms, value co-creation takes place through the exchange of outputs between actors who have integrated resources [27]. Institutions and institutional arrangements govern the exchange of services in service ecosystems [27]. Platforms are an emergent Actor to Actor (A2A) structure that provides an organizational logic for actors to exchange services and co-create value [20]. A first example of analyzing platforms in this way can be found in [28].

5 Framework for AI-Based Service Ecosystems

A successful service platform [4] should support a network of actors and host a set of loosely coupled open services and software products as part of a rapidly growing digital ecosystem [29]. The DEA Cube (**D**igital **E**nterprise **A**rchitecture Reference **Cube**) in Fig. 2 extends our holistic architecture reference and classification framework from [30] to drive bottom-up integration of dynamically composed micro-granular architecture services and their models. The DEA Cube provides a comprehensive architectural reference model to compose micro-granular architecture service models for supporting intelligent digital services and products. We have extended our service-oriented enterprise architecture reference model for the evolving digital transformation context by micro-granular structures, like the Internet of Things, and Microservices.

DEA structures a set of close related architectures like: platform and ecosystem architecture, digital strategy and governance, digital transformation management, business and information architecture, context and human-centric architecture, AI and cognitive architecture, data and analytics architecture, application services architecture, cybersecurity architecture, technology architecture, and operations architecture.

The DEA Cube abstracts from a particular business scenario or technology because it can be applied to different architectural instantiations to support intelligent ecosystems independently of different domains. The DEA cube covers the top of the Platform and Ecosystem Architecture [30]. A digital platform is in our understanding a repository of business, data, and infrastructure services used to rapidly configure digital offerings from digital services. Digital services and components are slices of code that perform a specific task.

We position reusable digital services as parts of an ecosystem of services. A digital platform linearizes the complexity of cooperating services. It integrates core technology services to provide standardized access points and repositories for an intelligent service ecosystem of business services, data services, and infrastructure services. The value of a platform to users [29] results from the number of platform and service users. A digital platform and an ecosystem should enable shared value creation for all stakeholders and facilitate the exchange of goods, services, and social currency. Platforms do not own or control their resources and are therefore well suited for scalability within the ecosystem.

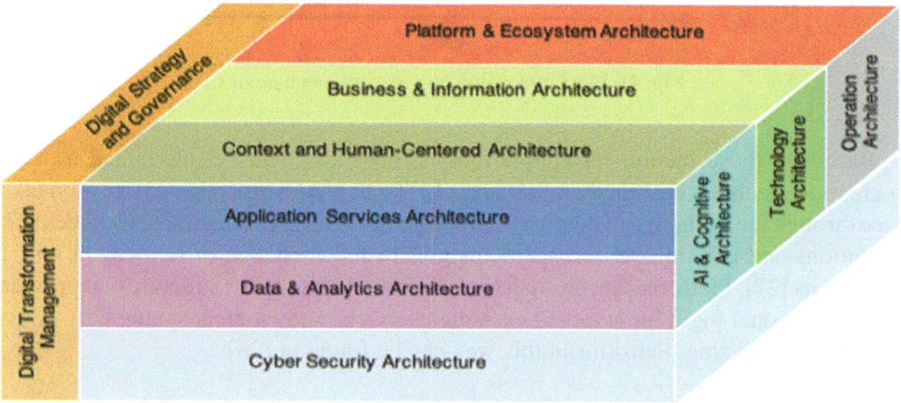

Fig. 2. Digital Enterprise Architecture Reference Cube (DEA Cube) [30]

Metamodels and their architectural data [29, 30] are the core part of the digitization architecture. Architecture metamodels should support analytics-based architectural decision management and the strategic as well as IT/business alignment. Three quality perspectives are important for an adequate IT/business alignment and are differentiated as: (I) IT system qualities: performance, interoperability, availability, usability, accuracy, maintainability, and suitability; (II) business qualities: flexibility, efficiency, effectiveness, integration and coordination, decision support, control and follow up, and organizational culture; and finally (III) governance qualities: plan and organize, acquire and implement deliver and support, monitor and evaluate.

6 Conclusion and Future Research

The novel introduced architectural concepts presented to support intelligent digital service address an integral set of aligned domains of the DEA Digital Architecture Reference

Cube. Our new solution outlined above has important practical implications for innovative smart services supported by artificial intelligence. This solution promises a tightly coordinated and adaptive integration of essential automatisms and human tasks through cognitive value creation for smart service ecosystems. Furthermore, we extended previous research (e.g., [30]) and show how the concepts can be integrated and extended to build a new framework upon it. We complement our results by an Intelligent Control Compliance Cockpit as an ecosystem of intelligent services, which is currently being explored with the background of an international insurance company.

Limitations can be found in a deeper empirical evaluation of our framework as well as definitions of a may industry-specific use. As part of an on-going research project, future research should cover these limitations and empirically evaluate the framework (e.g., through case studies and expert surveys) as well as work on extensions. Furthermore, influencing factors like value co-destruction should be investigated in future research.

References

1. Yoffie, D.B., Gawer, A., Cusumano, M.A.: A study of more than 250 platforms reveals why most fail. Harvard Business Review, May 2019
2. Alt, R., Zimmermann, H.-D.: Electronic markets on platform competition. Electron. Markets **29**, 143–149 (2019). https://doi.org/10.1007/s12525-019-00353-y
3. Subramaniam, M., Piskorski, M.J.: How Legacy Businesses Can Compete in the Sharing Economy. https://sloanreview.mit.edu/article/how-legacy-businesses-can-compete-in-the-sharing-economy/. Accessed 14 June 2020
4. McAfee, A., Brynjolfsson, E.: Machine, Platform Crowd. Harnessing Our Digital Future. . W. W. Norton & Company, New York (2017)
5. Parker, G., van Alstyne, M.W., Choudary, S.P.: Platform Revolution: How Networked Markets are Transforming the Economy–and How to Make Them Work for You. Norton & Company, New York (2016)
6. Cusumano, M.A., Gawer, A., Yoffie, D.B.: The future of platforms. MIT Sloan Manag. Rev. Spec. Issue Disruption **61**, 46–54 (2020)
7. Evans, G.P., Van Alstyne, M., Peter: How Platform Strategies Continue to Create Value. https://sloanreview.mit.edu/article/how-platform-strategies-continue-to-create-value/. Accessed 01 Apr 2019
8. Brynjolfsson, E., McAfee, A.: The business of artificial intelligence. Harvard Bus. Rev., 1–20 (2017)
9. Davenport, T.H., Ronanki, R.: Artificial intelligence for the real world. Harv. Bus. Rev. **96**, 108–116 (2018)
10. Möhring, M., Schmidt, R., Keller, B., Sandkuhl, K., Zimmermann, A.: Predictive maintenance information systems: the underlying con-ditions and technological aspects. Int. J. Enterp. Inf. Syst. (IJEIS) **16**(2), 22–37 (2020)
11. Chui, M., et al.: Sizing the potential value of AI and advanced analytics—McKinsey & Company. https://www.mckinsey.com/featured-insights/artificial-intelligence/notes-from-the-ai-frontier-applications-and-value-of-deep-learning. Accessed 11 May 2018
12. Provost, F., Fawcett, T.: Data Science for Business: What You Need to Know about Data Mining and Data-analytic Thinking. O'Reilly Media, Sebastopol (2013)
13. Chollet, F.: Deep Learning with Python. Manning Publications Co., Greenwich (2017)
14. Hagiu, A.: Strategic decisions for multisided platforms. MIT Sloan Manag. Rev. **55**, 71–80 (2014)

15. Wan, X., Cenamor, J., Parker, G., Van Alstyne, M.: Unraveling platform strategies: a review from an organizational ambidexterity perspective. Sustainability. **9**, 734–752 (2017)
16. Alstyne, M.W.V., Parker, G.G., Choudary, S.P.: Pipelines, platforms, and the new rules of strategy. Harv. Bus. Rev. **94**, 54–62 (2016)
17. Vargo, S., Lusch, R.: Why "service"? J. Acad. Mark. Sci. **36**, 25–38 (2008). https://doi.org/10.1007/s11747-007-0068-7
18. Subramaniam, M., Iyer, B., Venkatraman, V.: Competing in digital ecosystems. Bus. Horiz. **62**, 83–94 (2019)
19. Giessmann, A., Legner, C.: Designing business models for platform as a service: towards a design theory. In: ICIS 2013 Proceedings, November 2013
20. Shapiro, C., Carl, S., Varian, H.R.: Information Rules: A Strategic Guide to the Network Economy. Harvard Business Press (1998)
21. Lusch, R.F., Nambisan, S.: Service innovation: a service-dominant logic perspective. MISQ **39**, 155–175 (2015). https://doi.org/10.25300/MISQ/2015/39.1.07
22. Eisenmann, T., Parker, G., Van Alstyne, M.W.: Strategies for two-sided markets. Harv. Bus. Rev. **84**, 92–101 (2006)
23. Ketonen-Oksi, S., Jussila, J.J., Kärkkäinen, H.: Social media-based value creation and business models. Ind. Manag. Data Syst. **116**, 1820–1838 (2016)
24. Suseno, Y., Laurell, C., Sick, N.: Assessing value creation in digital innovation ecosystems: a Social Media Analytics approach. J. Strateg. Inf. Syst. **27**, 335–349 (2018)
25. Vargo, S.L., Lusch, R.F.: Institutions and axioms: an extension and update of service-dominant logic. J. Acad. Mark. Sci. **44**, 5–23 (2016)
26. Haki, K., Blaschke, M., Aier, S., Winter, R.: A Value co-creation perspective on information systems analysis and design. Bus. Inf. Syst. Eng. **61**, 487–502 (2019)
27. Vargo, S., Maglio, P., Akaka, M.: On value and value co-creation: a service systems and service logic perspective. Eur. Manag. J. **26**, 145–152 (2008)
28. Camilleri, J., Neuhofer, B.: Value co-creation and co-destruction in the Airbnb sharing economy. Int. J. Contemp. Hosp. Manag. (2017)
29. Ross, J.W., Beath, C.M., Mocker, M.: Designed for Digital. How to Architect Your Business for Sustained Success. The MIT Press, Cambridge (2019)
30. Zimmermann, A., Schmidt, R., Sandkuhl, K., Masuda, Y.: Architecting intelligent digital systems and services. In: Zimmermann, A., Howlett, R. J., Jain, L. C. (eds.) Human Centred Intelligent Systems. SIST, vol. 189, pp. 127–137. Springer, Singapore (2021). https://doi.org/10.1007/978-981-15-5784-2_11

5G Assisted Smart Manufacturing and Industrial Automation

Abdellah Chehri[1(✉)] and Alfred Zimmermann[2]

[1] Department of Applied Sciences, University of Quebec in Chicoutimi, Saguenay, QC, Canada
achehri@uqac.ca

[2] Faculty of Informatics, Reutlingen University, Reutlingen, Germany
alfred.zimmermann@reutlingen-university.de

Abstract. The digitization of factories will be a significant issue for the 2020s. New scenarios are emerging to increase the efficiency of production lines inside the factory, based on a new generation of robots' collaborative functions. Manufacturers are moving towards data-driven ecosystems by leveraging product lifecycle data from connected goods. Energy-efficient communication schemes, as well as scalable data analytics, will support these various data collection scenarios. With augmented reality, new remote services are emerging that facilitate the efficient sharing of knowledge in the factory. Future communication solutions should generally ensure connectivity between the various production sites spread worldwide and new players in the value chain (e.g., suppliers, logistics) transparent, real-time, and secure. Industry 4.0 brings more intelligence and flexibility to production. Resulting in more lightweight equipment and, thus, offering better ergonomics. 5G will guarantee real-time transmissions with latencies of less than 1 ms. This will provide manufacturers with new possibilities to collect data and trigger actions automatically.

Keywords: Industry 4.0 · Smart manufacturing · 5G networks · Industrial internet of things · Systems engineering · Human factor

1 Introduction

The term Industry 4.0 encompasses the promise of a new industrial revolution [1] that aims to create the fourth industrial revolution's challenges. With a focus on the smart factory concept, new consists such as big data and analytics, CPS (Cyber-physical systems), cloud, fog, edge computing, IIoT (Industrial Internet of things), augmented reality, intelligent robotics, additive manufacturing, simulation, and cybersecurity are being developed. Incorporating the technologies together and bringing a cohesive technological environment is a challenge [2].

The Internet of Things (IoT) consist of all of the things which are able to connect to the Internet without considering size and type, such as transport devices, medical devices, humans, animals, plants, houses, power, water, etc. IoT objects with recognizing themselves and others can exchange information about themselves by making decisions in the relevant field using intelligent space. And access to data gathered by other things can

C. Leitner et al. (Eds.): AHFE 2021, LNNS 266, pp. 385–393, 2021.
https://doi.org/10.1007/978-3-030-80840-2_44

be part of a complex service [3]. Due to the ability of this technology to connect a variety of heterogeneous devices, there is a significant focus on IoT technology and innovation in this field as well as it is one of the infrastructures in information and communication technology and economic developments and one of the strategies facing the world of science. This technology has taken a big step in performing complex operations, processing, and exchanging information through programs and management systems located in data centers or network clouds, without human intervention and intelligent designs and architecture and selecting appropriate protocols [4, 5].

Among other goals, the new Industry 4.0 production paradigm seeks to support or enhance human capabilities and skills and assist in training workers. The human factor will remain essential in production systems, and design must be centered on these needs. The terms human factors and ergonomics describe the interactions between workers, the tasks to be performed, and the work itself. These terms are interchangeable. We also define human factors as studying the elements that make it easier to get a job correctly. Another possible definition is exploring the relationships between people, the tools and equipment they use in the workplace, and their environment. The manufacturing sector faces challenges like tailor-made mass production, productivity, competition, labor shortage, trade attractiveness, and an aging worker population. All production systems are covered by Industry 4.0.

In this surge of modernism, tremendous driving forces motivate scientists and engineers to move their ideas beyond conceptual wireless generations towards a long-term evolution (LTE) of mobile communications, namely 5G (of fifth generation). The ITU-R working group released a document in 2015 naming the IUT-R M. 2083 recommendation, which marks three years of work on 5G standardization globally. This document defines the necessary bases for the research and development of 5G [6]. 5G will push traditional mobile broadband to the extreme in terms of data rates, capacity, and availability. 5G will also enable new services, including industrial Internet of Things (IoT) connectivity and critical communication. The goals of 5G are set very high with data rates of up to 20 Gbps, and capacity increases up to 1000 times with flexible platforms for device connectivity, ultra-low latency, and high reliability. 5G is expected to have a fundamental impact on all society by improving efficiency, productivity, and safety.

2 5G Network Objectives

For the first time in the history of wireless telecommunications, 5G promises a speed and level of reliability comparable to wired. In the years to come, this technology will offer speeds ten times faster than 4G. It will serve a much larger number of subscribers and IoT objects. Besides, it will guarantee real-time transmissions with latencies of less than 1 ms. This will offer manufacturers new possibilities to collect data and trigger actions automatically.

In the medium term and beyond, the future development of IMT will undoubtedly evolve according to market requirements. By studying the different use cases and user needs, three main categories of use have been grouped [7]:

- Massive Machine Type Communications (mMTC): Industrial process automation with massive machine type communications are examples of use cases that will be possible in future 5G systems. This usage scenario requires good quality of service in terms of reduced energy consumption, good connectivity, and high reliability.

- Enhanced Mobile Broadband (eMBB): in large urban centers with high user density, there is an increasing demand for access to multimedia content, services, and data. This demand will lead to needs in terms of bandwidth, which will lead to advanced mobile broadband. This use also encompasses requirements in terms of coverage and constant very high mobile speed whatever the reception context.
- Ultra-Reliable, Ultra-Reliable, and Low Latency (uRLLC) communications: This use scenario will be critical in future applications requiring very high responsiveness along with extremely reliable transmission. These needs are mainly found in autonomous transport, in the digitalization of industry, remote surgery, and smart electricity grids.

Figure 1 summarizes the three use cases for 5G in the different application areas.

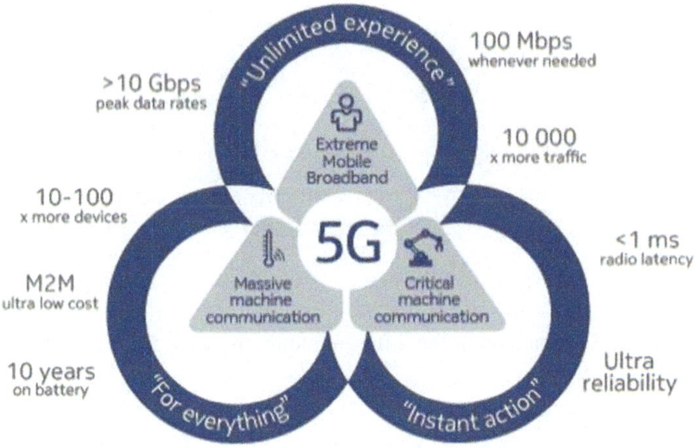

Fig. 1. 5G Diversity of service, use case and requirements [8].

3 Different Requirements of 5G

ITU members, including major players in the sector, industry forums, national and regional standards organizations, regulators, network operators, equipment manufacturers as well as representatives from universities and research institutes and the Member States, have met within the framework of the activities of the working group and have completed a series of studies relating to the main requirements concerning the performance of 5G technologies and fields of application IMT systems according to the three categories of uses mentioned in the previous paragraph.

It is necessary to collect and accept the requirements to get a clear idea of the characteristics of systems, users, and operations. The main requirements for future IMT systems are described in Recommendation ITU-R M.2083-0 [9].

Figure 2 summarizes a comparative study of these 5G requirements versus those of 4G. We note that the change in some of these requirements will not be too significant because the current network has already provided some solutions. On the other hand, requirements such as traffic capacity per area, energy efficiency, and peak data rates will need to see giant leaps forward in IMT. This shows that the arrival of the next generation of mobile networks will not be just about speed to support growing demand like previous generations did.

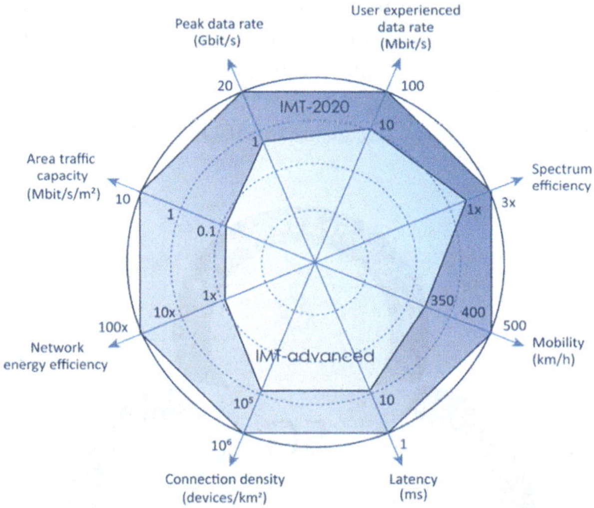

Fig. 2. Comparison of the different requirements of 5G versus 4G.

4 5G Assisted Smart Manufacturing

5G will facilitate the implementation of advantageous features. For example, network slicing allows manufacturers to define multiple virtual networks on shared physical infrastructure and optimize each for a specific group of applications. This division is essential to guarantee the quality of service and the end-to-end security characteristics of the network. 5G will also allow computers and devices to be seamlessly integrated into the network. Manufacturers will thus be able to use edge computing to process data at the edge of the network, close to the source, at the factories' very heart. Although it is possible to work in edge computing without 5G, this new generation of mobile networks will bring this technology more speed, efficiency, reliability, and responsiveness.

a) *Very high data rates:* This is a requirement that is defined primarily for evaluation purposes in the eMBB use categories. It is defined as the maximum theoretically achievable data rate assigned to a single mobile station, assuming error-free conditions when all available radio resources are used for the corresponding link. This rate excludes the resources used for physical layer synchronization, reference signals or pilots, guard bands, and guarantees).

b) *Energy efficiency:* is defined as the number of bits transmitted by Joule of energy, and it has two aspects:

- On the network side, energy efficiency corresponds to the number of information bits received or transmitted by users, per unit of energy consumption of the radio access network (RAN) (in bit/joule);
- On the device side, the energy efficiency corresponds to the number of information bits per unit of the communication module (in bit/joule). Energy efficiency should be significantly improved compared to 4G to increase the battery life of connected objects (batteries that can be up to 100 times less energy-consuming).

A compromise between the energy efficiency of the network and the energy consumption of connected objects can relate to two scenarios:

- Good energy efficiency in the case of data transmission
- Low energy consumption in the absence of data.

c) *Spectral efficiency:* the future environment of telecommunications systems is likely to be very dense and, above all, heterogeneous. Still, it should accommodate the three categories of uses mentioned above and have efficient transmission to allow access to many services everywhere. And at any time. The transmission of data and multimedia applications requires high spectral efficiency. The latter is defined as the amount of binary data (bits) transferred during one second over a bandwidth of 1 Hz (measured in bit/s/Hz/cell).
The improvements in spectrum efficiency should apply to small and large cells, in the low and high-frequency bands, in high and low mobility scenarios.
d) *Connection density:* Connection density can support the successful delivery of a message of a specific size in a period, even in very dense places like a train station, a football stadium, etc. Unlike 4G, which only offered thousands of connections per square kilometer, 5G will multiply that number by a factor of ten to reach one million connections per square kilometer.
e) *Latency:* is defined as the time required for a data packet to travel from source to destination through a network. This new standard's arrival should allow ultra-short latency, equivalent to less than a tenth of current communication systems' latency time. Very low latency will be driven by the need to support new applications. Some envisioned 5G use cases, such as connected and autonomous cars, emergency services, industrial internet of things, or Industry 4.0, connected drones, remote medical operations, etc., may require much lower latency than is possible with today's mobile communication systems.

With higher speed, increased real-time capacity, and improved flexibility, several industrial applications become much easier to consider:

- *The interaction between machines* is an essential concept to increase the distribution of intelligence in factories. Equipment capable of self-monitoring can communicate with other parts of the process to optimize overall performance.

- *Control loops* that require reliable real-time control and a short cycle time can operate within their network slice without experiencing the disruptive effect that another component, such as a camera, might have on network stability.
- *Wireless sensors* allow better control of operating temperature, vibrations, and noise emissions on production machines, triggering an alert if any of the parameters exceed set thresholds within a given period.
- *Autonomous Vehicles (AGVs)* are easier to operate with 5G, using cloud-based intelligence to deliver materials just-in-time to the plant or even synchronizing to distribute large loads.
- *Augmented reality (AR)* will allow operators to visualize in real-time and directly on the machine all the critical information on its operating status. And if they can't solve a problem, what they see can be transmitted over 5G to an expert who, remotely, can annotate the images or point to a component.

5 Adoption of the 5G Geographic Architecture

The infrastructure inherent in the deployment of autonomous 5G (standalone, SA) will impose a global step in integrating 5G for various geographies. High-tech regions, such as North America, Asia, and Europe, have already started limited deployment, followed closely by other countries worldwide. In total, 55 commercial networks should have been put into service by the end of 2019. The proximity of neighboring countries and the strong multiplication of carriers will make these deployments particularly difficult in Europe. To address these challenges, the European Commission created the 5G Action Plan for Europe to spur rapid progress and provide a roadmap planning the rollout of 5G in all European countries by the end of 2020 [10].

Currently, there are two types of 5G network standardized by the global mobile communications standards organization 3GPP: non-standalone access (NSA) and standalone access (SA) for 5G New Radio (NR). NSA 5G networks are built on top of existing 4G structures to facilitate the early adoptions to 5G due to its easy and fast deployment. A device compatible with an NSA 5G network will most likely have 4G capabilities and be able to switch to the NSA 5G when convenient to do so, which is determined by the 4G base station. However, SA 5G networks work independently, relying on a dedicated 5G structure with a dedicated 5G base station. Devices can be designed to work on any combination of 4G, NSA 5G, and SA 5G technologies, which increases complexity when designing circuits and antennas capable of switching between the different modes [11].

When it was standardized in 2017, SA 5G was conceived as a natural follower to the NSA option, providing significant speed improvements and reduced latency in the network. The advantages of SA 5G come from the fact that a dedicated 5G core is built to enhance the antennas' performance using beamforming techniques and allow significantly more data being transferred at the same time.

Industrialized nations such as China, Japan, and India are investing heavily in the practical and financial implications of converting to 5G. New antennas, software technologies, and infrastructure equipment are a boon to electronics and software design and production markets worldwide. Therefore, the emphasis is on rapid deployment. One of

India's largest telecommunications providers has already upgraded its entire network to be 5G compatible, and China Mobile plans to build 10,000 5G base stations.

6 5G Security Architecture

The implementation of 5G will generate incredible benefits in terms of performance and diversity of applications thanks to the widespread use of cloud-based resources, virtualization, network slicing, and network slicing. 'other emerging technologies. However, with these changes come new security risks and add additional "attack surfaces" to the 5G security architecture.

5G is built on security practices from previous generations of mobile technologies. Yet, the model of trust has broadened considerably, and more stakeholders are now involved in the process of providing services. The Internet of Things (IoT or IoT) and the proliferation of users create an exponentially higher number of devices, and many of these traffic entries are no longer supervised by humans [12, 13].

Some of the enhanced 5G security features detailed in 3GPP standards include unified authentication to decouple access point authentication, extensible authentication protocols to support secure transactions, flexible security policies for responding to a more significant number of uses and Subscriber Permanent Identifiers (SUPI) to ensure confidentiality on the network.

As the 5G rollout continues and the most critical performance nodes become increasingly virtualized, operators will need to monitor and evaluate their security performance continuously. Adherence to best practice requires end-to-end network security control over the entire system architecture, equipment, and applications [14].

There is no doubt that 5G can deliver the exponential improvement in throughput that users have become accustomed to with each new generation of mobile networks. Still, the increased speed is just the beginning [15]. The expected changes will affect various markets (ranging from personal transport to industrial production and agriculture). They will be so significant that many are already seeing 5G as a new industrial revolution. At the heart of this conceptual revolution is a multi-faceted 5G architecture, with MEC, NFV, Massive MIMO (Massive MIMO), and service-based, cloud-aligned core architecture working together to deliver a new generation of services. The 5G test solutions designed to respond to these fundamental architectural changes will be the real stepping stones in the transition to 5G.

7 Conclusion

Information and Communication Technologies(ICTs) are now an integral part of our society. ICTs have a profound socioeconomic impact and enrich our daily lives with a plethora of services ranging from media entertainment (e.g., video) to more sensitive and critical public applications. If analysts' prognoses are correct, just about every physical object we see (clothes, cars, trains, etc.) will also be connected to networks by the end of the decade (the Internet of Things).

Industry 4.0 is a rapidly developing industry, and we are only at its beginnings. It must face new demands from consumers, particularly with personalization, which takes more and more market share, yet it is challenging to produce this product in large quantities. The profitability of companies will, therefore, be impacted if they fail to reduce their production costs. To cope with this, artificial intelligence in the manufacturing processes and control becomes essential not to explode the prices. However, the advances currently being made in artificial intelligence and 5 G, for industry, suggest that we are no longer very far from this democratization, mainly thanks to large private companies like Amazon who want to achieve this quickly.

References

1. Lasi, H., Fettke, P., Kemper, H.-G., Feld, T., Hoffmann, M.: Industry 4.0. Bus. Inf. Syst. Eng. **6**(4), 239–242 (2014). https://doi.org/10.1007/s12599-014-0334-4
2. Derhamy, H.: Architectural design principles for industrial Internet of Things, Luleå University of Technology (2018)
3. Patel, K.K., Patel, S.M.: Internet of things-IOT: definition, characteristics, architecture, enabling technologies, application & future challenges. Int. J. Eng. Sci. Comput. **6**(5) (2016)
4. Abdul-Qawy, A.S., Pramod, P., Magesh, E., Srinivasulu, T.: The Internet of Things (IoT): an overview. Int. J. Eng. Res. Appl. **1**(5), 71–82 (2015)
5. Mueller, E., Chen, X.-L., Riedel, R.: Challenges and requirements for the application of industry 4.0: a special insight with the usage of cyber-physical system. Chin. J. Mech. Eng. **30**(5), 1050–1057 (2017)
6. Ghosh, A.: 5G new radio (NR): physical layer overview and performance. In: IEEE Communication Theory Workshop. http://ctw2018.ieee-ctw.org/files/2018/05/5G-NR-CTW-final.pdf. Accessed 20 Aug 2019
7. Popovski, P., Trillingsgaard, K.F., Simeone, O., Durisi, G.: 5G wireless network slicing for eMBB, URLLC, and mMTC: a communication-theoretic view. IEEE Access **6**, 55765–55779 (2018)
8. Mallinson, K.: 3GPP-The Path to 5G: As much Evolution as Revolution. The Mobile Broadband Standard. http://www.3gpp.org/news-events/3gpp-news/1774-5g_wiseharbour. Accessed 10 Dec 2020
9. Iwabuchi, M., et al.: 5G field experimental trials on URLLC using new frame structure. In: Proceedings of the 2017 IEEE Globecom Workshops, Singapore, 4–8 December 2017
10. Lema, M.A., et al.: Business case and technology analysis for 5G low latency applications. IEEE Access **5**, 5917–5935 (2017)
11. Ericsson. NR URLLC Rel-16 Use Cases and Requirements. https://list.etsi.org. Accessed 10 Dec 2020
12. Slalmi, A., Kharraz, H., Saadane, R., Hasna, C., Chehri, A., Jeon, G.: Energy efficiency proposal for IoT call admission control in 5G network. In: 2019 15th International Conference on Signal-Image Technology & Internet-Based Systems (SITIS), Sorrento, Italy, pp. 396–403 (2019). https://doi.org/10.1109/SITIS.2019.00070
13. Slalmi, A., Saadane, R., Chehri, A., Kharraz, H.: How will 5G transform industrial IoT: latency and reliability analysis. In: Zimmermann, A., Howlett, R.J., Jain, L.C. (eds.) Human Centred Intelligent Systems. SIST, vol. 189, pp. 335–345. Springer, Singapore (2021). https://doi.org/10.1007/978-981-15-5784-2_27

14. Slalmi, A., Chaibi, H., Saadane, R., Chehri, A., Jeon, G.: 5G NB-IoT: efficient network call admission control in cellular networks. Concurr. Comput.: Pract. Exp., e6047 (2020). https://doi.org/10.1002/cpe.6047
15. Slalmi, A., Chaibi, H., Chehri, A., Saadane, R., Jeon, G., Hakem, N.: On the ultra-reliable and low-latency communications for tactile internet in 5G era. In: 24th International Conference on Knowledge-Based and Intelligent Information & Engineering Systems, Verona, Italy, 16–18 September 2020

Qualitative Reasoning for Service Interactions

Marco Aiello[1(✉)], Alidéou Prosper Ayégnon[2], and Oliver Jackson[3]

[1] Service Computing Department, IAAS, University of Stuttgart, Stuttgart, Germany
marco.aiello@uni-stuttgart.de
[2] Guest Service Computing Department, IAAS, University of Stuttgart, Stuttgart, Germany
[3] Bachelor Student Informatik, University of Stuttgart, Stuttgart, Germany

Abstract. The realization of human-centered intelligent services requires a conceptualization of space and time that is intuitive for the human and employed in a concrete way by the services. In this position paper, we argue that qualitative temporal and spatial representation and reasoning must play a role in service interactions and service compositions in particular. We describe an initial prototype dealing with qualitative temporal relations and illustrate the added value of these kind of approaches in the tourism domain, as an exemplification.

Keywords: Qualitative reasoning · Service-oriented computing · Service composition

1 Introduction

The availability of on-line, always available, interoperable services has changed the way in which we build complex software and also how users can satisfy their information and functional needs. If before the advent of the Web and Web services, developing a system meant writing code to cover all necessary functionalities, today one can take advantage of online services and focus on the logic of the system, rather than the detailed functionality development [1]. It is a shift of going from programming in the small to large scale system development.

An important technique that has come with the 'servicification' of information ecosystems is that of *service composition*. The idea is that of putting together several service invocations following some business logic in order to obtain added value services. The way in which the logic and sequence of invocations is put together can range from prefixed compositions all the way to run-time generated compositions [2, 3]. The research on service composition has flourished in the past two decades and great progress has been made [4], though no one solution has emerged a general-purpose, widely adopted one.

Several tools and systems have been successful in supporting the software engineering of service compositions. Similarly, a great effort has gone in supporting the translation of user requests directly into service compositions. Our research efforts have gone in the latter direction. In 2002, we have proposed an approach based on Artificial Intelligence Planning for service composition [5]. Since then we have developed

C. Leitner et al. (Eds.): AHFE 2021, LNNS 266, pp. 394–399, 2021.
https://doi.org/10.1007/978-3-030-80840-2_45

many frameworks and prototypes that translate formalized user requests into planning problems that in turn dynamically interact with service ecosystems. The strength of our approach lies in the ability to compose at run-time and to handle uncertainties tied to the loosely-coupled nature of service ecosystems. In other terms, if to satisfy a user we need a service that offers tickets for a train, but when we go to execute the service returns tickets that are too expensive, our system can generate alternative compositions by invoking new services, such as for instance a bus service [6].

This research line has been motivated by the hypothesis that Artificial Intelligence (AI) techniques can support users in satisfying their needs by creating intelligent sequences of service compositions on behalf of the users. The compositions are aware and react to the run-time conditions. Up to now we have focused on the planning and scheduling aspects. Here, we argue for a further necessary development, that is, the temporal and spatial interpretation of the user needs and their taking into account when creating service compositions. Take for example the following user request: "I want to go to Strasbourg next weekend; stay two nights in a four start hotel; and lodge in the city center, very close to the Notre-Dame cathedral, but also to the train station." A satisfactory reply to this request would be a Web service where the temporal and the spatial constraints of the query are used for the invocation. For the sake of the example, assume that the invocation returned no results, but in fact, there is a GT hotel available on Friday but not Saturday and there is a BW hotel available on Saturday and Sunday. What is the correct answer to the original query? Most likely the user would still like to go to Strasbourg and accept to have to change hotel in between days. So the correct answer to the query is the composition of two bookings that satisfy the spatial constraint of being in Strasbourg and that temporally cover the full period of the stay.

To realize human-centered intelligent service compositions to support the user, we therefore need a formal notion of time and space that is adequate for the user and the domain [7, 8]. In the case of traveling, we need a combination of metric and qualitative reasoning techniques to translate user requests with spatio-temporal aspects into concrete spatial and temporal domains and in turn into service invocations. The system has to be able to understand that a time range of three days is the same as three individual days one after each other. The system has to understand what it means to be "very close" to the Notre-Dame cathedral and translate it into a geographic region. We remark here that qualitative, human, spatial notions do not have a direct metric translation. So saying "near" to something is, from a cognitive point of view, not the same as centering a sphere around the target point and considering all interior points as near and all others as not near, see, e.g., [8, 9].

The present article takes the form a position paper arguing that formal spatio-temporal reasoning, seen as a branch of AI, is crucial for achieving user satisfying service compositions. It provides some initial results and examples. The organization of the paper follows the mentioned dissemination goal: Sect. 2 illustrates an initial approach and related prototype for temporal service composition; an example of how a temporally aware composition would work in the domain of travel organization is presented in Sect. 3; while Sect. 4 reports final considerations on service compositions using AI techniques.

2 A First Approach to Service Composition with Concrete Domains

In 1981, James Allen proposed a qualitative representation of time based on temporal intervals and their thirteen relations [10]. The representation—independently proposed by van Benthem [6]—has been very influential in the AI field of qualitative temporal reasoning. The core idea is to move from a metric representation of time to extended intervals of time (e.g. modeled by their starting and ending instants) and to consider the relationships among these independently of their exact duration (e.g. if the time intervals are disjoint, they overlap, or one is contained in the other). We resort to these ideas in our initial attempt of brining qualitative temporal reasoning to service composition.

We realized a prototype that can interact with services in the domain of travel and is able to reason about temporal equivalence [11]. We consider user requests and service results to have temporal extensions in the following way. Let D be a demand with defined start-time D_s and end-time D_e as well as a set O which contains offers o that have a start- and end-time o_c, respectively. The goal is to find a match between the given demand time window $[D_s, D_e]$ and a subset of $O = \{o_1, \ldots, o_m\}$ such that the offers are temporally ordered (in Allen's terms, $\forall$ i, j: $I < j \rightarrow o_i(\text{before} \vee \text{meets})o_j$) and have delays between them of at most D_{md}. Whenever there is a trade-off between multiple attributes such as offer cost o_c, rating o_r or waiting time between two offers $O[x]_s - O[y]_e$, the best solution is defined through user preferences. This requires all attributes to have an assigned weight w_a with $0 \leq w_a \leq 1$ and $\sum_W = 1$ which signifies their relative importance.

The temporal reasoning now becomes a weighted scheduling problem. A valid solution is an ordered list L of length n such that no wait time exceeds the maximum allowed delay D_{md}:

$$L[1]_s - D_s \leq D_{md} \quad \text{demand start to first offer}$$

$$D_e - L[n]_e \leq D_{md} \quad \text{Last offer to demand end}$$

$$\forall_{1 \leq i < n}\ L[i+1]_e - L[i]_s \leq D_{md} \quad \text{Between any two adjacent offers}$$

Algorithms to find such matches are presented in [11]. In the same reference, we provide algorithms that take into account user preferences expressed as weights on the

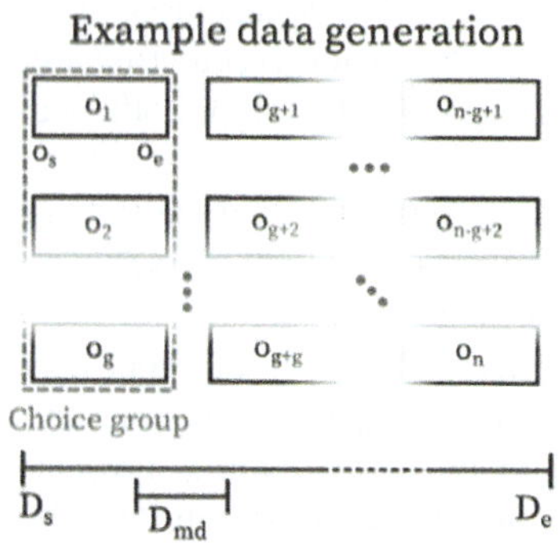

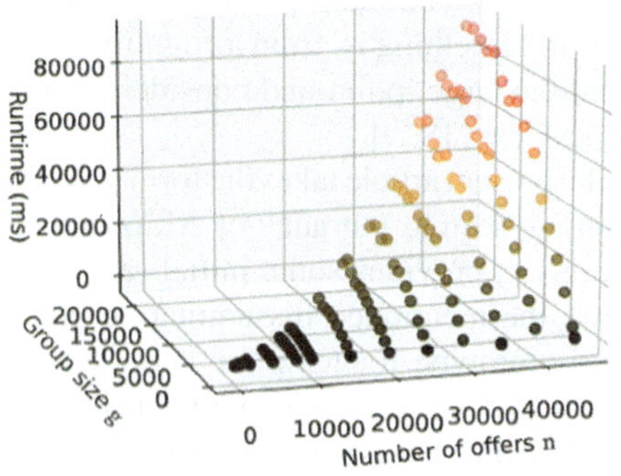

Fig. 1. Matching algorithm (a) data generation and (b) performance [11]

offers' attributes. The computational complexity of such algorithms falls in the $O(n^2)$ class. This means that for very large sets of offers, the algorithm execution will be long and one might need to simplify the search space and accept sub-optimal solutions. Nevertheless, for medium and even large sets the approach is feasible, as we showed with a small performance evaluation run on a standard computer (Intel Core i5-6600K CPU clocked at 3.50 GHz running Node.js v12.16.1 on Windows 10 and 8 GB RAM). Figure 1(a) shows how the data is generated. We start with a set number n of unique offers and arrange their start- and end-times equally until we have choice groups of size g. D_{md} is set such that one offer from each choice group has to be included to form a valid solution. Figure 1(b) shows the execution time of the algorithm for growing size of group and number of offers.

Algorithm 1 covSearch(T imeInterval D, Set < T imeInterval > oSeq)

```
BEGIN
 referenceIntervalItem := biggestFirstOverInterval(D; oSeq)
 coveringSeq.add(referenceIntervalItem.D)
 if (covers(referenceIntervalItem:D; D))
  return planningOf(coveringSeq)
 do
maxIntervalItem :=
maxOverlapOrMeet(referenceIntervalItem.D,referenceIntervalItem.inde
x,oSeq)
  coveringSeq.add(maxIntervalItem.D)
  if (covers(coveringSeq; D))
   return planningOf(coveringSeq)
 else
  referenceIntervalItem := maxIntervalItem
 endif
 while (maxIntervalItem6= NULL)
 return planningOf(coveringSeq)
END
```

To exemplify the approach, let us consider a slightly modified algorithm which we name "Coverage Search Algorithm (covSearch)," presented in pseudocode as Algorithm 1. Given hotel availabilities, the algorithm provides a coverage of the specific booking period D, if it exists, with available offers O. Firstly, we filter all the offers and only keep those offers in $o_i \in O$ such that $o_i(overlaps \vee starts \vee during \vee finishes \vee equals)D$. Then, we order the remaining offers based on their starting time and obtain the sequence $oSeq$.

In a greedy fashion, we select first the offer with best fitting interval o_b that overlaps or starts D. If the temporal relation is: contains or equals, we return as result the offer. In the latter case, one booking will cover the whole period. If the offer does not satisfy the booking by itself, then we iteratively search for more offers to complete the coverage of the period. The intervals that either overlap or meet with o_b are found. From all these intervals, the one that has the biggest non overlapping section with o_b is selected; we call it o_{b2}. If the combination of the two now covers the period D, we return the two offers, otherwise we search for a third offer. This proceeds until we have found a minimal coverage or no coverage exists. In the next section, we illustrate the running of the algorithm with an example.

3 An Example Run for Hotel Booking

Say that Alice wants to stay in Strasbourg from the 4th of June 2021, 13:00AM up to the 15th of June 2021, 11:00AM. After querying a booking system, she obtained the following availabilities (to make thinks easier, we consider that rooms are in the same area of Strasbourg):

(room$_4$, 08/06/2021 12:00 AM, 11/06/2021 12:00 AM), (room$_2$, 02/06/2021 12:00 AM, 08/06/2021 12:00 AM), (room$_1$, 01/06/2021 12:00AM, 06/06/2021 12:00 AM), (EQ room\s\do5(3), 07/06/2021 12:00 AM, 10/06/2021 12:00 AM) and (room$_5$, 09/06/2021 12:00 AM, 16/06/2021 12:00 AM). The stay duration and all availabilities are provided as input to Algorithm 1 covSearch.

First the algorithm orders the time intervals of the offers:

(room$_1$, 01/06/2021 12:00AM, 06/06/2021 12:00AM), (room$_2$, 02/06/2021 12:00AM, 08/06/2021 12:00AM), (room$_3$, 07/06/2021 12:00AM, 10/06/2021 12:00AM), (room$_4$, 08/06/2021 12:00AM, 11/06/2021 12:00AM) and (room$_5$, 09/06/2021 12:00AM, 16/06/2021 12:00AM). The time interval with maximum overlapping time with the stay period is found and it is: o_b= *(room$_2$, 02/06/2021 12:00 AM, 08/06/2021 12:00 AM).* This time interval does not contain and is not equal to the whole stay period, so we move to the next step.

The time interval which is met by o_b and which has the biggest non overlapping section with o_b is o_{b2}= *(room$_4$, 08/06/2021 12:00AM, 11/06/2021 12:00AM).* By combining o_b and o_{b2} we have the interval *(02/06/2021 12:00 AM, 11/06/2021 12:00 AM)* which does not cover the entire stay period, so the process continues.

The next time interval that is overlapping with o_{b2} and which has the biggest non overlapping section with it is o_{b3}= *(room$_5$, 09/06/2021 12:00AM, 16/06/2021 12:00AM).* By combining o_b, o_{b2}, and o_{b3}, we have the time interval *(02/06/2021 12:00AM, 16/06/2021 12:00AM0* which contains the stay period. By considering the stay period constraint, the final result is the booking:

P = [(room$_2$, 04/06/2021 13:00AM, 08/06/2021 12:00AM), (room$_4$, 08/06/2021 12:00AM, 11/06/2021 12:00AM),(room$_5$, 11/06/2021 12:00AM, 14/06/2021 11:00AM)].

4 Concluding Remarks

The current mainstream way for a human to interact with online services is based on remote function calls with syntactically prescribed parameters to be passed. To make things easier for the human, a (web-based) user interface can facilitate the task of expressing requests and receiving responses. In this paper, we argued for using explicit concrete domains for representing quantities that are temporal and spatial in nature. These domains should have a qualitative language that goes with it which is close to the way humans deal with time and space. We have shown an initial prototype to deal with time and illustrated, by means of an example, why such forms of qualitative representation can increase the expressive power of service composition frameworks.

References

1. Aiello, M.: The Web Was Done by Amateurs. Springer (2018)

2. Weerawarana, S., Curbera, F., Leymann, F., Storey, T., Ferguson, D.F.: Web services platform architecture: SOAP, WSDL, WS-policy, WS-addressing, WS-BPEL, WS-reliable messaging and more. Prentice Hall PTR (2005)
3. Kaldeli, E., Lazovik, A., Aiello, M.: Domain-independent planning for services in uncertain and dynamic environments. Artif. Intell. **236**(7), 30–64 (2016)
4. Lemos, A.L., Daniel, F., Benatallah, B.: Web service composition: a survey of techniques and tools. ACM Comput. Surv. (CSUR) **48**(3), 1–41 (2015)
5. Aiello, M., et al.: A request language for web-services based on planning and constraint satisfaction. In: Buchmann, A., Fiege, L., Casati, F., Hsu, M.-C., Shan, M.-C. (eds.) TES 2002. LNCS, vol. 2444, pp. 76–85. Springer, Heidelberg (2002). https://doi.org/10.1007/3-540-46121-3_10
6. van Benthem, J.: The Logic of Time: A Model-Theoretic Investigation into the Varieties of Temporal Ontology and Temporal Discourse, 2nd edn. Springer, Cham (2010)
7. Aiello, M., Pratt-Hartman, I., van Benthem, J.: Handbook of Spatial Logics. Springer, Dordrecht (2007). https://doi.org/10.1007/978-1-4020-5587-4
8. Freksa, C.: Qualitative spatial reasoning. In: Mark, D.M., Frank, A.U. (eds.) Cognitive and Linguistic Aspects of Geographic Space, pp. 361–372. Springer, Dordrecht (1991). https://doi.org/10.1007/978-94-011-2606-9_20
9. Stock, O.: Spatial and Temporal Reasoning. Springer, Dordrecht (1998). https://doi.org/10.1007/978-0-585-28322-7
10. Allen, J.F.: An interval-based representation of temporal knowledge. In: Hayes, P.J. (eds.) Proceedings of the 7th International Joint Conference on Artificial Intelligence, IJCAI '81, Vancouver, BC, Canada, 24–28 August 1981, pp. 221–226. William Kaufmann (1981)
11. Jackson, O.: Development of an API for temporal coordination of multiple services. Bachelor Thesis. University of Stuttgart (2020)

Transfer of Service Research into Companies

Thomas Meiren(✉), Michaela Friedrich, and Christian Schiller

Fraunhofer Institute for Industrial Engineering IAO, Nobelstr. 12, 70569 Stuttgart, Germany
thomas.meiren@iao.fraunhofer.de

Abstract. The systematic transfer of research results into practice is indispensable for solving a wide range of operational problems. It is also a key driver of innovation – an increasingly vital economic factor. This applies equally to the field of service research, which provides a key source of knowledge for service companies. A current study is investigating how this knowledge is obtained and used by German service providers. The study also provides a starting point to discuss implications for service research.

Keywords: Service research · Service transfer · Information channels · Empirical study

1 Introduction

For a long time, only a relatively small community was concerned with questions of service research [1]. In the past two decades, however, scientific engagement in this area has grown substantially, as evidenced by the rising number of specialist publications and journals dedicated to this topic [2]. The result has been a steady growth in knowledge in this field, accompanied by the emergence of important subdisciplines such as service marketing, service operations management and service innovation – to name but a few [3].

To a large extent, service research has always been an applied discipline. From the beginning, companies have been an important object of investigation. Similarly, many solutions are now developed with a clear focus on their practicability. Indeed, for a lot of publicly funded projects, a transfer into concrete practice is explicitly prescribed – for example, in the research projects covered by the major EU-funded programs [4]. By the same token, service research associations have always included companies as a key target group. Moreover, in recent times, more and more young academics from the field of service science are finding their way into management positions at service companies.

Yet certain questions remain. How do companies regard service research? What do companies undertake to facilitate the transfer of this research? What tools are best suited for facilitating this transfer? What support do companies require? These and similar questions are now being examined in an investigation of German companies. This study is based on an empirical survey, itself supplemented by expert interviews and workshops. The key findings are presented in greater detail below.

C. Leitner et al. (Eds.): AHFE 2021, LNNS 266, pp. 400–405, 2021.
https://doi.org/10.1007/978-3-030-80840-2_46

2 Results of the Empirical Study

A number of countries have a well-established tradition of service research. These include the United States of America, the United Kingdom and the countries of Scandinavia. In Germany, however, there were only occasional publications in this field until the mid-1990s, after which a more concerted examined of the issues surrounding the service economy first emerged [5]. One of the key drivers of this shift was the launch of the Service 2000 plus initiative in 1995. Here, 300 experts from research, industry and society joined forces to trace out the future path to a service economy [6]. Germany subsequently became one of the first countries in Europe to launch its own publicly funded programs of research in this field, which are still in existence today [7]. These programs have not only produced fundamental research but also given rise to applied solutions that have then found their way into practice. The funding portal of German federal government lists over 920 service research projects for the German Federal Ministry of Education and Research (BMBF) alone [8]. Today, service research rests upon a firm foundation and has generated extensive knowledge across a broad range of service-related fields.

A new study is now investigating how German companies obtain and use the knowledge generated by service research. The related survey is targeted at all services companies and runs from January 2021 to April 2021. The current paper is based on an interim evaluation conducted on March 15, 2021. As of this date, 107 companies had participated. Of these, 75 were from the services sector and 32 from manufacturing industry (providers of product-related services). In terms of company size, 65 of the respondents are to be classified as small and medium sized enterprises (SMEs), with less than 250 employees, and 42 as large enterprises.

A focus of the survey is how companies obtain and then use information. For example, companies are asked which information channels they use in order to find out about new service topics (Fig. 1). Given the restrictions resulting from the COVID-19 pandemic (e.g., with respect to events), companies are being asked to provide an assessment both for 2019 (before the COVID-19 pandemic) and for the current period (during the COVID-19 pandemic).

The results show that, in the period before the COVID-19-pandemic, survey respondents tended to favor direct channels in order to find out about new service topics. Most popular were face-to-face events (78%) such as conferences, workshops and seminars, with exhibitions and fairs (69%) also achieving a high score, as did business networks (65%). In addition, the Internet is an important source of information, reported by 71% of respondents.

The COVID-19 pandemic seems to have substantially altered, over a very short space of time, the way in which the companies surveyed access information about new service topics. As a result of travel and contact restrictions, face-to-face events (14%) and exhibitions and fairs (15%) have fallen to the very bottom of the table, whereas digital channels have risen to the top. The biggest leap is for online events (e.g., webinars), which polled a mere 29% before the pandemic but now receive one of the leading scores (91%).

Looking specifically at the SMEs participating in the survey, it is clear that their use of information channels, both before and during the COVID-19 pandemic, is broadly similar to that of large enterprises. However, given the small sample in the current

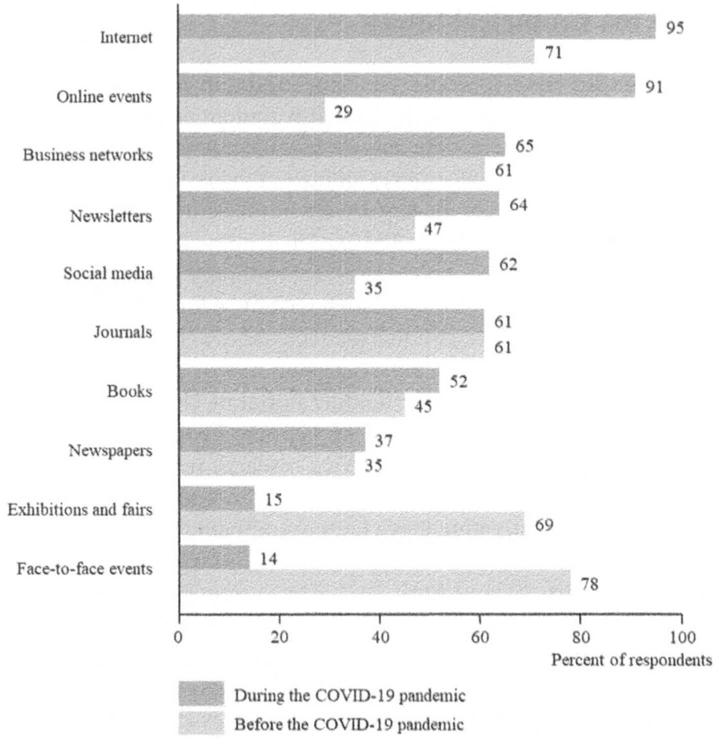

Fig. 1. Information channels to find out about new service topics

interim evaluation, this will require a more detailed investigation once the survey has been completed.

In a further question, companies were asked to assess how they would obtain information following the COVID-19 pandemic. This reveals that respondents would like to resume face-to-face events but also intend to continue making frequent use of digital channels. From the perspective of service research, this means that face-to-face and digital channels are likely to play an equally important role in the transfer of research into practice.

The study also investigates what kind of information companies would like to obtain from the research community (Fig. 2).

Top of the list here, by quite some margin, are practical examples (80%). This is followed by guidelines (64%) and industry benchmarks (55%). These are all forms of information that involve an additional level of processing and – unlike basic information or studies and trend information – can not always be directly derived from research findings.

There is only one significant area of deviation among SMEs in the interim sample. This is in respect of information about funding opportunities, which is desired by 57% of SMEs, as opposed to 29% of large enterprises.

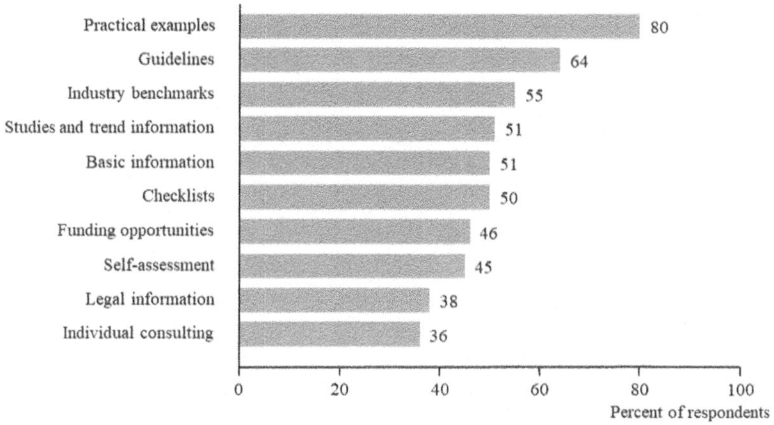

Fig. 2. Forms of information desired for new service topics

Companies were also asked about the difficulties they face in obtaining information about new service topics (Fig. 3).

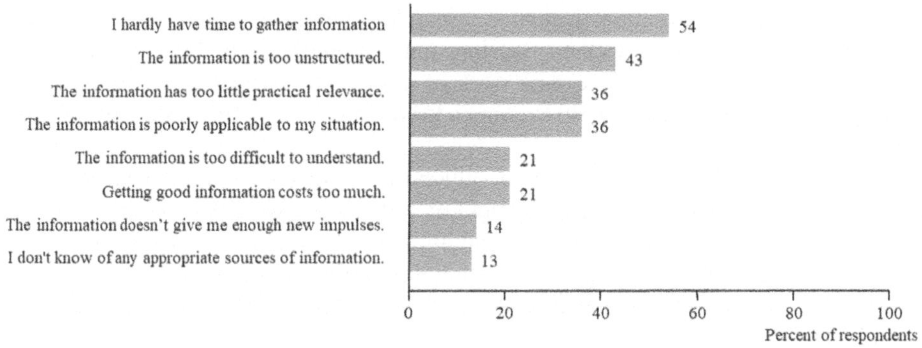

Fig. 3. Typical problems in obtaining information on new service topics

Just over half of respondents (54%) say they hardly have time to gather information. The reasons for this deficit lie with the companies themselves. By contrast, the remaining problems named by companies are more likely to have their origin in the field of service research. Companies frequently find that the information available is too unstructured (43%), has too little practical relevance (36%) or is poorly applicable to their situation (36%).

The survey also asks companies which institutions they consult for support on new service topics (Fig. 4).

Here, universities and research organizations are the most frequently named (67%). Industrial associations and regional networks are also popular sources of support (52% and 46% respectively).

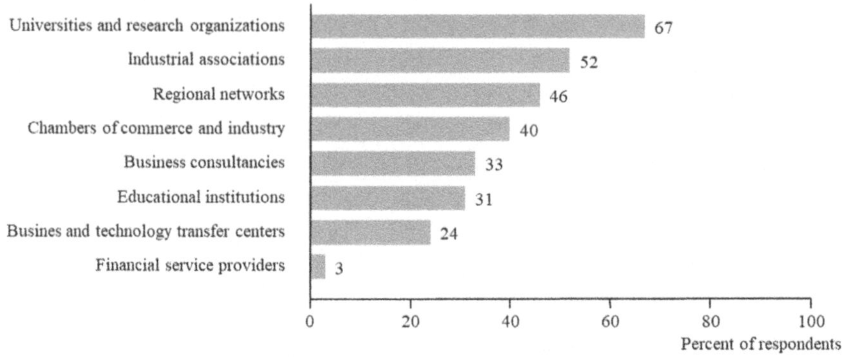

Fig. 4. Support for new service topics

The responses from SMEs were similarly weighted, although universities and research organizations received a significantly lower score (53%).

3 Implications for Service Research

Rather than being an end in itself, service research should always generate added value for society as a whole. As well as boosting innovation and efficiency in companies, it can also help improve the lives and work of individuals. With the shift to a service economy, service research is set to become a key area of knowledge production for the 21st century.

Service research should let itself be guided by practical requirements. Research programs and research activities in the field of services are worthless if they only lead to more paperwork rather than initiating and implementing necessary change processes. To ensure effective service research, it is therefore all the more important to ensure a transfer into practice from the very outset and to use a set of tools appropriate to the specific target group.

This interim evaluation of the study purposely adopts the perspective of those at the receiving end of research transfer. In so doing, it offers an initial insight into how service providers obtain and use the information generated by research on new service topics. It is evident that companies desire research that is as practical as possible and suits their specific needs, and that they obtain this research via a variety of information channels. At the same time, new requirements are now emerging with regard to the use of digital transfer tools – not least as a result of the COVID-19 pandemic.

In expert talks to accompany the study, the following recommendations are to be discussed with a view to improving the transfer of research into practice:

1. To meet the challenges of successful research transfer, universities and research establishments should develop new guiding principles and undertake innovative organizational changes. Successful transfer of service research presupposes the existence of adequate incentives and the requisite competencies on the part of those involved.

2. The generation of new knowledge should not merely be a one-way process undertaken by research on behalf of society. Instead, practical considerations should be incorporated, to an appropriate degree, both in research planning and in the actual research phase.
3. New tools to support the transfer process should be developed and implemented. Here, hands-on research can play a key role (e.g., laboratories and service demonstration centers), as can interactive events (e.g., innovation workshops, makeathons and experimental workshops) and new digital channels to facilitate transfer. In order to generate the greatest possible acceptance, it is also vital to ensure that results are delivered in a form that is appropriate to the specific target group.
4. New modes of cooperation along the value chain from basic research to applied research to business practice should also be tested. The involvement of multipliers (e.g., intermediary actors) will play a key role here.

The potential for innovation in the service economy is far from being exhausted. At present, there is still a lack of readiness on the part of companies to embrace systematic research and development in the service sector. This needs to change. On the other hand, research itself must cooperate more closely with the service sector and ensure that its findings are put into practice on lasting basis.

References

1. Fisk, R.P., Grove, S.J.: The evolution and future of service: building and broadening a multidisciplinary field. In: Maglio, P.P., Kieliszwski, C.A., Spohrer, J.C. (eds.) Handbook of Service Science, pp. 641–663. Springer, New York (2010)
2. SCOPUS. https://www.scopus.com
3. Maglio, P.P., Kieliszwski, C.A., Spohrer, J.C. (eds.): Handbook of Service Science. Springer, New York (2010)
4. European Commission: Horizon (2020). https://ec.europa.eu/programmes/horizon2020/en
5. Ganz, W., Meiren, T. (eds.): Service research today and tomorrow: spotlight on international activities. IRB, Stuttgart (2002)
6. Bullinger, H.-J. (ed.): Service 2000 plus. IRB, Stuttgart (1998)
7. Federal Ministry of Education and Research: Federal Report on Research and Innovation 2020. BMBF, Berlin (2020)
8. Federal Ministry of Education and Research: Foerderkatalog. https://foerderportal.bund.de/foekat/

Smart Tool Use in Knowledge Intensive Work Situations – An Information Technology Review

Katja Gutsche(✉) and Michael Weltin

Furtwangen University, Robert-Gerwig Platz 1, 78120 Furtwangen, Germany
{Katja.gutsche,michael.weltin}@hs-furtwangen.de

Abstract. Smart tools for assistive information providence promise positive effects on service delivery and quality. This paper describes results of a broad empirical study, where actual service technicians were subjects, analyzing the effect on service quality by information tools which are different in their degree of adaptiveness for customized use. The paper delivers guidelines on what tools best to develop and use dependent on the service technician's age and critically reflects on the potential of AR-technology within service operation.

Keywords: Assistance · Digitization · Information technology · Service quality

1 Motivation

Tools in general are objects to enhance the human ability to perform a certain task as i.e. repairing a bike or eating a steak. In business context they are in place to support employees in fulfilling their tasks more safely and efficient under less strain. As the success within industrial work situations becomes more and more dependent on the availability of knowledge and information, on-site as well as remote, the use of innovative information technologies becomes attractive. This development is fueled as information technology becomes more available and less pricey.

Industrial services define a typical work place where numerous and distributed information are needed for a safe and fast service operation. As technologies like augmented reality, virtual reality or basic remote communication with or without AI become more advanced and available, companies point out their intensions for a company-wide rollout of smart information tools. Some research has been done, especially pointing out the potential of Augmented Reality within service operation [1, 2]. However, a guideline for technology managers of what is best to install is missing. An empirical review of smart tools with respect to work design, human demands and task requirements as well as age within service operation has not been conducted.

This paper summarizes the results of an empirical study looking at (mis-)matches of smart tools and service quality measures. The emphasis of this study lays upon the technological opportunities as well as the expected differences in attitude towards digitalization within different age groups of service technicians (digital natives vs. digital immigrants).

C. Leitner et al. (Eds.): AHFE 2021, LNNS 266, pp. 406–413, 2021.
https://doi.org/10.1007/978-3-030-80840-2_47

2 Delivering High Service Quality

2.1 Service Operation as Knowledge Intensive Work

Service operation is understood as the delivery of services in accordance to customer requirements and business limitations as i.e. given resources. Managing service processes is one of the responsibilities within service operation [3]. Seen from the process perspective service operation starts with the customer need, i.e. car repair, and ends with a hopefully satisfied customer after successful service fulfillment.

In the Business-to-business (B2B) service sector meeting the customer demands is crucial for the economic success of the service customer. From the customer point of view, the economic impact of a successful repair of an industrial robot in a car production line, which is unavailable as long as the robot is not fixed and causing reduced output, is much higher than a delayed repair in a car workshop. B2B-service customer demands strongly correlate with the service technician's competence to fulfill the tasks correct, safe and in a short period of time.

Delivering competent B2B-services requires more knowledge. This is due to the following facts.

- Service tasks are more demanding (customer demands, time restrictions, hazardous work environment, etc.),
- service objects are more complex and technological diverse (i.e. nuclear power plant vs. consumer car),
- wrong service delivery mostly has a higher economical and safety impact therefore more risk (i.e. faulty plane maintenance vs. a faulty repair of a coffee machine)
- legal instructions are sometimes numerous and must be followed by the technician

2.2 Service Quality

From the customer perspective the service result can best be described by the understanding of service quality. Quality in general describes how well requirements towards a product or process are met [4]. Service quality is understood as the ability of a service provider to meet customer expectations and fulfil the customer requirements towards a primarily intangible service product under the involvement of the service customer [5]. The SERVQUAL instrument sums up different ways of how to characterize service quality and defines (1) tangibles, (2) reliability, (3) responsiveness, (4) assurance, (5) empathy as quality elements [6, 7, 8].

For achieving safe, economic and compliant B2B-service operation smart tools offering help along the service process chain have a high attraction. Not all of the service quality elements are directly influenced by the use of different information tools for service assistance (Table 1). It is expected that smart information tools increase service reliability, responsiveness and assurance.

2.3 Information Technology for Service Assistance

Service technicians often need special knowledge to perform the needed service tasks, especially in non-standardized, rare service scenarios. This need is fueled by the current developments within the industrial sector towards Industry 4.0, where unexpected,

Table 1. Effects of information tools on service quality.

	effected by information tool?
Tangibles	
Reliability	x
Responsiveness	x
Assurance	x
Empathy	

complex, rare service tasks for diverse, advanced assets become more common [9, 10, 11]. For retrieving service information, different kinds of instruction tools are available.

Up to now, the most common 'tools' [12] are (A) static instructions either as paper or as pdf on a digital device, both of them allowing no or very low kind of interaction by the user. Upcoming information solutions are the (B) instruction videos, (C) mobile instructive applications and (D) augmented reality (AR) applications. Whereas instruction videos become more and more seen in practical use, especially AR so far is mostly in prototypic use. Besides their use intensity, the four types of information technology also differ in their level of sensor integration allowing more or less adaptiveness and personalized use. The more sensors, the higher the capability of the technology to adapt automatically to changed work settings and modify instructions [13]. Therefore, the smarter the tool, the more individualized the service assistance.

3 Research

Information technologies used within service operation are in place to make a technician's workplace safer and his/her work more efficient leading to a higher quality. However, developers as well as users have difficulties to decide which information technology for service assistance best meets the technicians' needs and has best impacts on the service result.

3.1 Research Questions

The following research questions shall be studied to give advice on this topic.

R1: Does the use of smarter information tools result in higher service quality? To answer this question process time, number of failures and safety level shall be the studied quality measures.

R2: Which tool is to be preferred in either age group looking at mental demand?

R3: Does the same information tool best meets the user requirements for digital natives as well as digital immigrants?

3.2 Empirical Study

Method

This study took place in November and December 2020. The authors set up the testing scenario in four different German companies. In total, 40 subjects took part. All but one object were male, typical gender distribution within the industrial service world [14]. The average age of all subjects was 40 years. The study paid attention on differences towards the use of information tools depending on age [15]. The subjects were asked to use a certain information tool as TEXT, VIDEO, mobile APP or an AR-Head-Mounted-Display-Application. In contrast to previous studies, all state-of-the-art technologies are evaluated in one study, subjects are real service technicians.

In each testing group about half of the subjects were digital natives and the other half so-called digital immigrants (born before 1980) [16, 17, 18].

The study included demographic variables, quality measures (a) time, (b) safety level, (c) error rate and task load measures outlined by the NASA [19]. Measuring the NASA TLX looks at the hoped influences of assistive information tools on lowering the technician's strain level within service operation. As service operation creates knowledge intensive work situations, especially the effect of mental workload as the sum of all external parameters influencing the employee mentally [20] measured within the TLX is of interest. Mental workload has a direct input on the task output, here on the service quality [21, 22].

Setting

All subjects were asked to repair a chain saw whose starter pull cord had been broken (Fig. 1). They were asked to fulfill the task as quickly as possible. None of the subjects had ever done this repair before. For each of the four information tools 10 test persons were asked to fulfil the repair scenario under the help of the given service assistance tool.

Tested information tools:

- The *text instruction* was not the one provided by the manufacturer, but designed to be more equivalent to the other three information tools. Text and pictures were sorted in 18 sequences.
- The *video instruction* using picture and voice had a total length of 10 min, all information were given as one sequence, subjects could pause or rewind the video wherever needed.
- Within the *mobile application* 23 information sequences were given using video, picture and sound. The subjects are free to re-run and stop sequences as well as pushing the help button for receiving further information using touch functionality implemented on the given tablet device.
- The *Augmented Reality application* was installed on the first generation Microsoft Hololens®. Through the use of object recognition the technician gets detailed, accurately fitting information within 20 sequences helping to fulfil the work routine which the technician confirms and executes. Besides AR-information this solution had sound, picture and video information as well as help buttons integrated. The objects could

pause, re-play or stop the assistive tool by gesture or voice control. The subjects using the AR solution got a brief introduction in how to operate the Microsoft Hololens®.

To introduce a stress factor into the test scenario and making the service task more realistic [14] a timer using light and beep was installed (see Fig. 1). The timer started with the beginning of the test, showing green light and giving a beep every 9 s. The longer the service scenario went on the shorter the periods between the beep-sounds. At the beginning the shuttle was every 9 s, after 20 min a beep was set every 4 s. As a time of 10 min was set as target time, the light kept green for the first 10 min. After this period the light showed orange for three more minutes. After 13 min the timer light switched to red.

Fig. 1. Test setting (here: mobile APP as information tool shown)

Results

To answer the research question associated with the smart assistive tool use within service operation, basic descriptive and inferential statistical analyses were reviewed. The ANOVAs used an alpha level of 0,05.

Table 2 sums up the most interesting results achieved through the test measures. On the top of the table the number of subjects in each age group for the four information tools are indicated. In total 23 digital natives and 17 digital immigrants took part.

Processing Time. The results show differences in the average process time of digital natives vs. digital immigrants of about 2 min. However, ANOVAs comparing the effect of information tool on process time within each age group showed no statistical significance (digital natives: $F(3,19) = 0{,}973$; $p = 0{,}426$) (digital immigrants: $F(3,13) = 1{,}175$; $p = 0{,}357$).

Number of Failure. Looking at the number of failures as wrong tool use, faulty repair step etc. we see big differences in the average values. For digital natives the mobile APP shows best results whereas for digital immigrants the VIDEO tool fits best. The differences are statistically significant within the group of digital natives, $F(3,19) = 3{,}356$; $p = 0{,}041$.

Table 2. Effects of information tools on service quality

		TEXT	VIDEO	mobile APP	AR APP
digital natives		6	6	6	5
digital immigrants		4	4	4	5
		Processing Time			
digital natives	Mean	12:16	11:48	13:00	13:54
	SD	02:58	01:45	01:19	02:14
digital immigrants	Mean	12:51	12:50	15:30	16:17
	SD	03:01	01:26	04:53	03:30
		Number of faults			
digital natives	Mean	2,33	1,33	0,67	0,6
	SD	1,506	0,816	0,816	0,894
digital immigrants	Mean	2,5	0,75	2	1
	SD	0,577	0,5	2,708	1,414
		Safety level			
digital natives	Mean*	2	1,83	1	1,8
	SD	1,095	0,753	0	1,095
digital immigrants	Mean*	2,5	1,75	1	2,8
	SD	1	0,975	0	0,447
	*mean out of 3: high - medium - low				
		Mental demand			
digital natives	Mean	8,555	11,3333	2,2217	10,8
	SD	7,10832	6,25666	1,91687	9,98393
digital immigrants	Mean	12,8325	4,335	6,75	12,8
	SD	8,49307	2,65135	3,59398	7,7958
		Performance			
digital natives	Mean	16,5533	1,89	12,445	9,4
	SD	10,44948	0,69036	9,57681	4,00306
digital immigrants	Mean	7,6675	7,1675	3,25	8,2
	SD	5,03477	4,53272	1,96705	5,65587

Safety Level. Three different safety levels were defined (1 – high, 2 – medium, 3 – low) If all safety measures indicated throughout the repair process were met, the subject achieved a high safety level, were none met a low level was realized and if only some safety measures were followed, the subject reached a medium safety level. As all safety measures addressed about the same criticality, no further differentiation was needed.

The average safety level was higher within the digital native group. Within both groups the mobile APP as information tool achieved best results safety-wise. Within the group of digital immigrants the effect of information tool on safety level showed a statistical significance, $F(3,13) = 5{,}551$; $p = 0{,}011$.

Mental Demand. Looking at the mental demand measured by using the NASA TLX we see once again big differences within the age groups and looking at the different information tools. Even if the differences are not statistically significant due to a high deviation (Digital Natives: $F(3,19) = 2{,}252$, $p = 0{,}115$; Digital Immigrants: $F(3, 13) =$

1,966, p = 0,169) we see the lowest demand level by using mobile APP as information tool for digital natives. Digital immigrants show best results with VIDEO and mobile APP.

Performance. Interesting enough looking at the performance measure within the NASA TLX we see a significant difference for digital natives (F(3,19) = 4,079, p = 0,022) As the performance measure asks the subject to give a feedback on how successful he or she thinks the task was accomplished. This finding is of interest as previous research on the work design of service technicians identified that the ability to successfully accomplish a task is the most relevant job motivator [14]. Performance is best evaluated by the subjects using TEXT. This result is unexpected and a clear reason cannot be stated at that point of research.

4 Conclusion and Outlook

Coming back to the research question, the following conclusions can be made.

R1: No general statement can be made upon if the use of smarter information tools result in higher service quality? The service quality measured in process time, number of failures and safety level are better when not using the AR-APP which is the smartest tool tested. Looking at safety a mobile APP using animations and video showed best results in both age groups. Processing time was best using VIDEO in each age group. In average less faults were made by the digital natives using mobile APP or AR-APP, digital immigrants showed less faults when assisted by VIDEO.

R3: Mental demand for digital natives is lowest using the mobile APP, for digital immigrants using VIDEO as information tool.

R2: Information tools should be different within the two age groups. This study shows a clear preference towards VIDEO for digital immigrants and a mobile APP solution for digital natives.

For developers this basically should lead into an information tool which can easily be transferred in either one. In addition, this research underlies the inappropriateness of paper as information tool, independent of the technician's age.

The study also indicates, that expectations towards AR-technology cannot be met at the current status of this technology. But as the tests using AR with a HMD were done with the Microsoft Hololens® 1 and the next generation is already released shifts within the quality and load measures might be possible.

Future research should address service quality and ICT-use dependent on technicians experience instead of age group. The effect of subjective technology acceptance on service quality also has to be studied.

This study also has some limitations. The results shown were drawn from singular testings, long-term effects as i.e. on mental workload or work moral have not been studied. The testing scenario did not include customer expectations, as needed for a comprehensive service quality measure.

References

1. Gutsche, K., Droll, C.: Enabling or stressing? – smart information use within industrial service operation. In: Duffy, V. (ed.) Digital Human Modeling and Applications in Health, Safety, Ergonomics and Risk Management, pp. 119–129. Springer (2020)
2. Meluzov, N., Friedewald, A., Mundt, C., Lödding, H.: Produktivitätssteigerung in der instandhaltung durch digitale assistenzsysteme. In: Biedermann, H. (ed..) Digitalisierte Instandhaltung. Stand und Perspektiven, 33. Instandhaltungsforum (Hrsg. ÖVIA), Tagungsband, TÜV-Verlag, Köln , pp. 181–197 (2019)
3. Johnston, R., et al.: Service Operations Management: Improving Service Delivery, Pearson (2020)
4. ISO 9000: Quality management systems—Fundamentals and vocabulary (2015)
5. Bruhn, M.: Qualitätsmanagement für Dienstleistungen, Springer (2020)
6. Zeithaml, V.A., Parasuraman, A., Berry, L.L.: Qualitätsservice. Was Ihre Kunden erwarten – was Sie leisten müssen. Frankfurt a. M. (1992)
7. Parasuraman, A., Zeithaml, V.A., Berry, L.L.: SEVQUAL: a Multiple-Item scale for measuring consumer perceptions of service quality. J. Retail. **64**(1), 12–40 (1988)
8. Grönroos, C.: A service quality model and its marketing implications. Eur. J. Market. **18**(4), 36–44 (1984)
9. Drucker, F.: The Age of Discontinuity: Guidelines to Our Changing Society; Edison. Transaction Publ. (1992)
10. Hirsch-Kreinsen, H., ten Hompel, M.: Digitalisierung industrieller Arbeit - Entwicklungsperspektiven und Gestaltungsansätze. In: Vogel-Heuser, B., et al. (ed.) Handbuch Industrie 4.0.. Springer (2015)
11. Acatech (ed.): Smart Maintenance für Smart Factories - Mit intelligenter Instandhaltung die Industrie 4.0 vorantreiben. Acatech Position. München (2015)
12. Oehmig, P.: Digitalisierug von arbeitsprozessen in der technischen redaktion. In: Hennig, J., Tjarks-Sobhani, M. (eds.) Perspektiven der Technischen Kommunikation., pp. 23–36. Stuttgart (2020)
13. Zoughbi, S., Al-Nasrawi, S.: Regional development getting smarter with ICT. In: Encyclopedia of Information Science and Technology (Third edition). IGI Global (2015)
14. Gutsche, K, Griffith, J.: Automating motivation: a workplace analysis of service technicians and the motivational impact of automated assistance, In: Hara, Y., Karagiannis, D. (eds.) Serviceology for Services, pp. 101–108. Springer (2017)
15. Charness, N., Boot, W.R.: Aging and information technology use: potential and barriers. Curr. Dir. Psychol. Sci. **18**, 253–258 (2009)
16. Prensky, M.: Digital natives, digital immigrants, part 2. On Horizon **9**(5), 1–6, MCB UP Ltd. (2001)
17. Tapscott, D.: Growing Up Digital: The Rise of the Net Generation. McGraw-Hill, New York (1998)
18. Palfrey, J., Gasser, U.: Born Digital: Understanding the First Generation of Digital Natives. Basic Books, New York (2008)
19. Hart, S.G., Staveland, L.E.: Development of NASA-TLX (Task Load Index): result of empirical and theoretical research. In: Hancook, P.A., Meshkati, N. (eds.) Human Mental Workload. North Holland Press, Amsterdam (1988)
20. ISO 10075–3: Ergonomische Grundlagen bezüglich psychischer Arbeitsbelastung. Berlin: Beuth (2018)
21. Schlick, C., Bruder, R., Luczak, H.: Arbeitswissenschaft. Springer (2010)
22. Johannsen, G.: Mensch-Maschine-Systeme. Springer, Heidelberg (1993)

Creating Value in Teams and Organizations

Robotic Baby Doll with Minimal Design for Interactive Doll Therapy in Elderly Dementia Care

Nobuo Yamato[1(✉)], Hidenobu Sumioka[2], Masahiro Shiomi[2], Hiroshi Ishiguro[3], and Youji Kohda[1]

[1] Graduate School of Knowledge Science, Japan Advanced Institute of Science and Technology, 1-1 Asahidai, Nomi, Ishikawa 923-1292, Japan
{s2040016,kohda}@jaist.ac.jp

[2] Advanced Telecommunications Research Institute International, 2-2-2 Hikaridai, Seika-cho, Soraku-gun, Kyoto 619-0288, Japan
{sumioka,m-shiomi}@atr.jp

[3] Graduate School of Engineering Science, Osaka University, 1-3 Machikaneyama, Toyonaka, Osaka 560-8531, Japan
ishiguro@sys.es.osaka-u.ac.jp

Abstract. We designed HIRO, a robotic baby doll, to be used in an interactive, non-pharmacological intervention that combines doll therapy with robot technology for elderly people with dementia. We took a minimal design approach; only the most basic human-like features are represented on the robotic system to encourage users to use their imagination to fill in the missing details. The robot emits baby voice recordings as the user interacts with it, giving the robot more realistic mannerisms and enhancing the interaction between user and robot. In addition, the minimal design simplifies the system configuration of the robot, making it inexpensive and intuitive for users to handle. In this paper, we discuss the benefits of the developed robot for elderly dementia patients and their caregivers.

Keyword: Elderly care · Therapy robot · Human-robot interaction · Welfare care · Dementia

1 Introduction

Dementia is an ever-growing concern for health and social welfare in the 21st century. As of 2015, 47 million people around the world suffer from dementia [1], and the Japanese government estimates that by 2025, 20% of the elderly population in Japan will have dementia [2]. Elderly people with dementia often exhibit problem behaviors such as verbal abuse and violence, and they may also experience depression and delusions, i.e., behavioral and psychological symptoms of dementia (BPSD). These symptoms, in turn, require caregivers to monitor elderly people with dementia more closely [3]. However, the field of nursing care is on the brink of collapse as the demand for caregivers continues

C. Leitner et al. (Eds.): AHFE 2021, LNNS 266, pp. 417–422, 2021.
https://doi.org/10.1007/978-3-030-80840-2_48

to rise. Implementing interventions to reduce BPSD is important for elderly people with dementia, as well as their caregivers and elderly care facilities.

Although pharmacological interventions are often used to reduce BPSD, non-pharmacological interventions are generally used first to avoid side effects from drugs [4]. Doll therapy, in which a human baby doll is provided to an elderly person with dementia, is a non-pharmacological intervention that involves pseudo-social stimulation [5]. Studies have shown that actions such as hugging, talking, feeding, holding, and dressing a doll can increase interaction with others and reduce BPSD in elderly people with dementia [6]. However, while interacting with real babies is an ideal intervention for a broader range of older adults [7], doll therapy is primarily targeted at women with parenting experience. This indicates that a motionless doll may not provide enough stimulation for dementia patients to imagine interacting with a baby.

We used a minimal design approach [8] to design a baby-sized robot, HIRO, so that it can be used by a wide range of elderly people with dementia. The minimal robot contains the most basic elements of human features to encourage users to rely on their imagination when interacting with the robot. Facial features were eliminated so that dementia patients are free to imagine the robot's face. We added the recorded voices of real babies and gave the robot the capability to interactively vocalize in response to human interaction. This enables the robot to behave more realistically and enhances the feeling of interacting with a human baby.

Our previous experiments showed that dementia patients, including male patients, reacted positively to HIRO [9]. We also found that in addition to the therapeutic effects common to non-pharmacological interventions, HIRO provided the additional benefits of interacting with a baby. In this study, we report the benefits of the minimal design of HIRO for the elderly users as well as their caregivers and discuss the new value it brings to the nursing home. First, we describe the development of HIRO, and then we report on the effects of the interaction between HIRO and the elderly, for both the elderly and their caregivers. Finally, we discuss how the minimal design approach can play an important role in the future of therapy robots.

2 Minimal Baby Design

After developing the prototype HIRO (W210 × D165 × H300 mm; 610g), the commercialized HIRO (W230 × D170 × H320 mm; 460 g) was created, as shown in Fig. 1. The control module of the robot is covered with polyester fabric, contains a computer, a three-axis accelerometer, and a speaker.

HIRO[1] is roughly the size of a one-year-old baby and has the baseline appearance necessary for being recognizable as a human baby. It does not have facial features such as eyes or a mouth. Although eye contact and facial expressions are considered to contain important information in emotional interactions, we excluded them because they could be misinterpreted without proper control.

The human-like emotional expressions of the robot were supplemented by using the voices of actual infants. We recorded the voices of one-year-old infants and collected

[1] We refer to both prototype version and commercialized one as HIRO in this paper since basic functions are shared between them.

Fig. 1. HIRO (left) and interaction between HIRO and elderly woman (right)

around 100 voice patterns including laughing and crying. The voices were classified into four categories (negative, weak negative, weak positive, positive), and HIRO chose one of voices from the four categories based on its emotional state, which changes based on changes of three-axis accelerometer [9]. Weak positive speech includes babbling, such as "ma-ma-ma."

3 Benefits of User Interaction with HIRO

HIRO is an interactive robot that uses human-like emotions conveyed through minimal visual information and voice to foster positive interactions with the elderly. Since 2018 we have been using HIRO and its commercial version to investigate the benefits of a baby robot for elderly people with dementia and their caregivers in five nursing facilities. The results showed that the elderly was able to positively engage with HIRO.

The participants' responses to HIRO did not differ significantly from when interacting with a robot with a face. The elderly participants with dementia imagined HIRO's emotions when engaging with it though it had no face. We conducted an experiment in a nursing home in which we compared the reactions of elderly dementia patients to HIRO with their reactions to a conventional baby robot with a face [9]. We divided 21 elderly dementia patients (18 women, 3 men) into two groups at a nursing home for the elderly in Hyogo prefecture. One group was given HIRO (no-face group), and the other group was given a baby robot with a typical appearance (face group). Each participant was given five minutes to care for the robot, and their reactions to the robot were examined. The results showed that 60% of the face group and 81.8% of the no-face group held the robot for the entire five minutes and treated it as if it were a real infant. They caressed the robot, sang to it, lifted it, kissed it, hugged it, and rocked it.

Interestingly, the elderly people with dementia who could speak interacted with the robot by talking to it, even though the robot could not speak. The participants exhibited two main behaviors; one was to mention the robot's impressions and characteristics such as its name. For example, many participants talked to the robot, making comments such as "Cute" and asking "What's your name?" A few participants pointed out that there were "no eyes or mouth" on HIRO's face. However, they continued to hold HIRO and maintained a positive attitude. On the other hand, some participants appeared to be unnerved by the lack of hair during long-term observations, so further investigation and alterations to the design may be needed.

The second main behavior was to respond to the robot's voice. Participants responded to the robot's cries and laughter with "Don't cry" or "It's laughing." They also attempted to interpret the babbling of the robot. For example, one participant said, "Please don't say no" to the robot, while another said, "You said your name was Kentaro." In the case of elderly people with severe dementia who had difficulty communicating with the robot, they parroted the robot's voice instead.

Many of the participants seemed to recognize it was a robot though they actively took care of it. One participant said, "It's well made." Another participant asked, "What is the name of this doll?" On the other hand, one participant with severe dementia seemed to be confused about whether the robot was an actual human or not. For example, when HIRO started crying, one participant thought it was hungry and said, "I can't breastfeed." Another elderly participant tried to show his coloring session to HIRO. One participant repeatedly patted the robot's underside and tried to undress it when it began to cry.

Regardless of whether the participants were aware they were interacting with a robot, they seemed to engage in "motherese," the characteristic speech that parents use when speaking to their babies. This is supported by one staff's comment on the clarity of speech: "I think the participants tried to speak more clearly than usual because they thought they were talking to a baby" (Fig. 2).

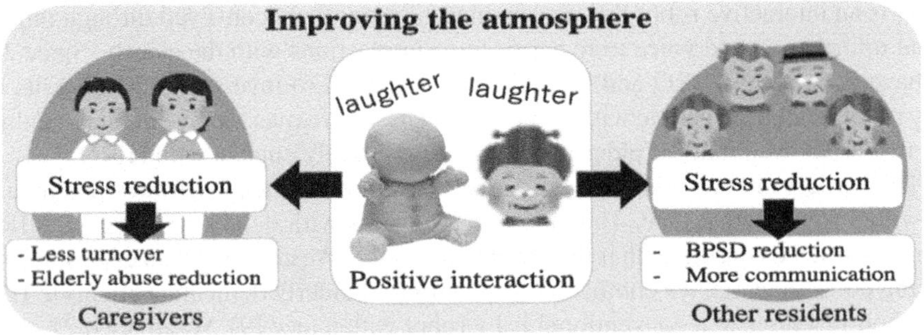

Fig. 2. Ripple effect of baby robots on the people around them

4 Benefits of HIRO for Caregivers

The experiments showed that unlike typical nursing therapy robots, HIRO could reduce the mental burden on the elderly users as well as their caregivers. Robots are typically heavy and expensive because they contain complex mechanical systems. As a result, caregivers need to pay attention when elderly users are interacting with a robot to prevent them from dropping the robot and damaging it, or to prevent the users from injuring themselves. In contrast, HIRO is a light, inexpensive, and durable robot, because of its minimal design. Caregivers do not need to spend as much time monitoring users interacting with HIRO, and as a result, they can take their eyes off the users and perform other tasks for a short period of time.

Additionally, listening to HIRO's baby-like voice and observing HIRO's interaction with the elderly may reduce the mental stress of caregivers. Staff members stated that they felt happy when they heard the robot's laughter while they were at work, and it was soothing them to see an elderly person playing with HIRO. It is known that hearing laughter can reduce stress [10], and HIRO's laughter may create new value regarding the introduction of robots into nursing home such as improving the atmosphere and reducing serious problems such as caregivers abusing the elderly (Fig. 2).

5 Individual Therapy Robots

Due to the high cost of conventional therapy robots used in nursing homes, one robot is typically shared by several users. However, this is not ideal given the circumstances of the COVID-19 pandemic as the robots would need to be thoroughly disinfected between uses to prevent the spread of the virus. Going forward, alternative ways of providing robot therapy need to be considered to avoid risking the health and safety of patients and caretakers. A minimal therapy robot such as HIRO can be more easily provided to all people who need it, such as elderly people, managers of elderly facilities, and families of dementia patients, because the robot is designed at a low cost and can be provided to each user. In addition, the use of robots can potentially reduce the mental stress of the elderly users as well as their caregivers. Therefore, a minimal design approach is important for future therapy robots as nursing homes adjust to new lifestyles in the midst of the COVID-19 pandemic (Fig. 3).

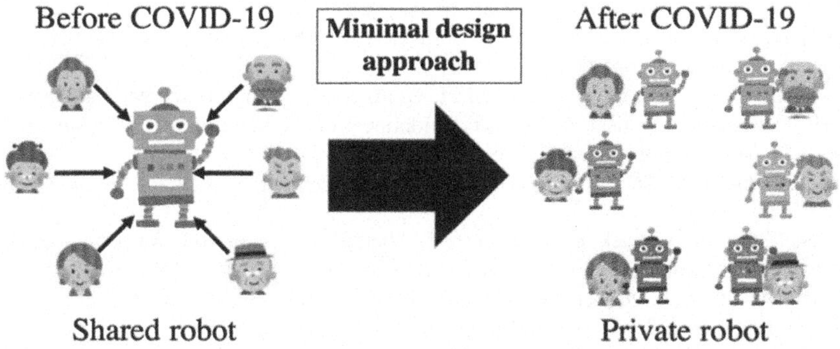

Fig. 3. Interactive doll therapy with individual robots before/after COVID-19

6 Summary

We introduced a minimal approach to design HIRO, a baby-sized robot for interactive doll therapy with elderly people suffering from dementia. HIRO was designed to have an abstract body representation and no facial features. Emotional interaction was achieved with a real baby's voice to activate the imagination of the elderly users with dementia. We evaluated the efficacy of HIRO with elderly people with dementia and their caregivers in a nursing home and found that the elderly participants were receptive to HIRO. Furthermore, we observed various secondary benefits. HIRO was durable and easy for users to handle. The caregivers' workload was reduced, and their mood was improved when they heard HIRO's laughter. We determined that elderly men who were not the target of the conventional doll therapy could use HIRO, though not all elderly people were equally receptive. The individual preferences may be due to individual age, gender, and the degree of dementia. Some of the participants did not seem to know how to care for HIRO, so some users may need encouragement to learn how to care for the doll by watching other people interact with it. This should be investigated in longitudinal experiments. Social isolation during the COVID-19 pandemic has had a serious impact on the mental health of elderly people with dementia and their caregivers. We hope that HIRO will not only help improve their mental health but also serve as a new cost-effective tool for robotic therapy.

Acknowledgments. This work was supported by JST CREST Grant Number JPMJCR18A1, Japan. We thank all staffs and residents in Ryusei Fukushikai Yume Paratice for helping our experiment.

References

1. Livingston, G., et al.: Dementia prevention, intervention, and care. Lancet **390**, 2673–2734 (2017)
2. Japanese Ministry of Health, Labor, and Welfare (MHLW): White Paper on Aging Society 2017, Office of Government Public Relations (2018)
3. Cerejeira, J., Lagarto, L., Mukaetova-Ladinska, E.B.: Behavioral and psychological symptoms of dementia. Front. Neurol. **3** (2012)
4. Azermai, M., Petrovic, M., Elseviers, M.M., Bourgeois, J., Van Bortel, L.M., Vander Stichele, R.H.: Systematic appraisal of dementia guidelines for the management of behavioural and psychological symptoms. Ageing Res. Rev. **11**(1), 78–86 (2012)
5. Mitchell, G.: Doll Therapy in Dementia Care: Evidence and Practice. Jessica Kingsley Publishers (2016)
6. Mitchell, G., McCormack, B., McCance, T.: Therapeutic use of dolls for people living with dementia: a critical review of the literature. Dementia **15**, 976–1001 (2016)
7. Cohen-Mansfield, J., Marx, M.S., Dakheel-Ali, M., Regier, N.G., Thein, K.: Can persons with dementia be engaged with stimuli? Am. J. Geriatric Psychiatry **18**, 351–362 (2010)
8. Sumioka, H., Nishio, S., Minato, T., Yamazaki, R., Ishiguro, H.: Minimal human design approach for sonzai-kan media: investigation of a feeling of human presence. Cogn. Comput. **532**(6), 760–774 (2014)
9. Sumioka, H., Shiomi, M., Yamato, N., Ishiguro, H.: Acceptance of a minimal design of a human infant for facilitating affective interaction with older adults: a case study toward interactive doll therapy. In: 29th IEEE International Conference on Robot and Human Interactive Communication, pp. 775–780 (2019)
10. Fujiwara, Y., Okamura, H.: Hearing laughter improves the recovery process of the autonomic nervous system after a stress-loading task: a randomized controlled trial. BioPsychoSocial Med. **12** (2018). Article number: 22

Value Creation in Aikido from a Service Perspective

Makoto Saito(✉), Youji Kohda, and Tsutomu Fujinami

School of Knowledge Science, Japan Advanced Institute of Science and Technoogy, 1-1 Asahidai, Nomi, Ishikawa 923-1292, Japan
{umitaro,kohda,fuji}@jaist.ac.jp

Abstract. Aikido is a modern Japanese martial art of self-defense style. Though there are no tournaments, practitioners of all ages and genders around the world train their technique and discipline their body and mind. People train through embodied knowledge; after observing an Aikido master's demonstration, each practitioner pair up with a training partner and practice mimicking the observed technique. Despite the lack of personalized coaching, there are many enthusiastic, long-term practitioners of Aikido around the world. To clarify their motivations, this study focused on value creation in Aikido from the perspective of a service. We conducted a qualitative analysis by interviewing several advanced practitioners and found that rich emotional value was created both physically and mentally rather than functional value, and the created value has resulted in many enthusiastic practitioners.

Keywords: Aikido · Embodied knowledge · Service perspective · Value creation · Emotional value

1 Introduction

Aikido is a Japanese martial art created during the 1920s by Morihei Ueshiba (1883–1969), an expert martial artist who reached the highest levels of mastery in the classical Japanese martial arts. Presently, people of all ages and genders train their technique and discipline their mind and body at Aikido dojos in 140 countries around the world [1].

Aikido consists of self-defense techniques such as throwing or pinning an attacker (*uke*) with one's bare hands in retaliation to the attacker grabbing or striking the defender (*tori*). Specifically, instead of directly striking back against an attacker's frontal attack, the defender (*tori*) tries to "enter" a favorable position behind or to the side of the attacker and "blend" with the attacker's position, movement, and force. The defender then subtly uses his/her own force to shift the direction of the attacker's force and "lead" it to break the attacker's stance in such a way that is as unnoticeable as possible to the attacker. Then the defender throws or pins the attacker to the ground. Unlike most other sports, Aikido is a purely demonstrative sport with no matches, tournaments, or contests [2, 3].

Aikido has a relatively simple training method. Practitioners acquire embodied knowledge by taking turns to mimic the techniques with a training partner after observing an Aikido master's demonstration. This is based on the philosophy of Aikido, as the

C. Leitner et al. (Eds.): AHFE 2021, LNNS 266, pp. 423–428, 2021.
https://doi.org/10.1007/978-3-030-80840-2_49

goal is not to compete and win but to learn through two-way training in which both sides practice the techniques on each other. This cultivates a spirit of recognition and respect for the other person, teaches how to respond to others, and promotes an awareness of distance that can be applied to daily social life [4]. In Aikido, there is no personalized coaching or extensive support often seen in other sports, which generally results in high customer satisfaction. Non-practitioners may find it difficult to understand the appeal of Aikido. Nevertheless, there are many enthusiastic long-term practitioners around the world who practice more than 100 days a year. Why is Aikido so enjoyable and how do people maintain their long-term enthusiasm for it?

In this study, we examined the value creation that takes place in Aikido from the service perspective. We interviewed five practitioners who have been practicing Aikido diligently for many years (6–30 years) and asked them to reflect on the period from when they started to the present. Then, through qualitative analysis, we extracted features of value found in Aikido.

2 Aikido from the Service Perspective

Figure 1 shows the general flow for one cycle of the Aikido training method. After warm-up exercises, (1) the Aikido master (*Sensei*) demonstrates a technique in front of the practitioners in a dojo. The practitioners focus on observing and memorizing the physical movements of the master. (2) Then they select a partner to practice with regardless of gender, skill level, or age. (3) After pairing up, they practice the techniques repeatedly by mimicking the master's movements observed and memorized. Each person alternates between the roles of the attacker (*uke*) and defender (*tori*) four times. One cycle from (1) to (3) takes about approximately 10 min. The cycle is repeated several times, with the practitioners switching their partners each time, until the end of the class (1–1.5 h). A practitioner can practice approximately 6–10 different techniques with different partners in the span of one class.

From the perspective of a service based on S-D logic [5], Aikido can be considered an "Aikido teaching service" that takes place in a dojo. The Aikido master is a "service provider" and the practitioners are "customers." All of the stakeholders make up an "Aikido teaching service ecosystem" in which each actor acquires value accordingly through the resource integration between actors.

During one Aikido class, the combinations of practitioners (actors) paired up change roughly 6–10 times corresponding to the techniques practiced. While a pair is training one technique, they alternate between the roles of the attacker and defender after practicing a technique four times, i.e., approximately every minute. This leads to a variety of actor-to-actor resource integration and collaborations between individual actors, resulting in a high density of resources.

In general, other sports divide athletes by skill level, gender, age, body weight, or physique due to the nature of the competition. In Aikido, however, practitioners are paired up regardless of skill level, gender, age, body weight, physique, etc. Thus, even if they are practicing the same technique, they can receive completely different responses depending on their partners, or conversely, they need to be flexible enough to match their partner at the time. It may seem like mere repetition of the same technique, but in fact, there is a momentary aspect of a once-in-a-lifetime experience in each training.

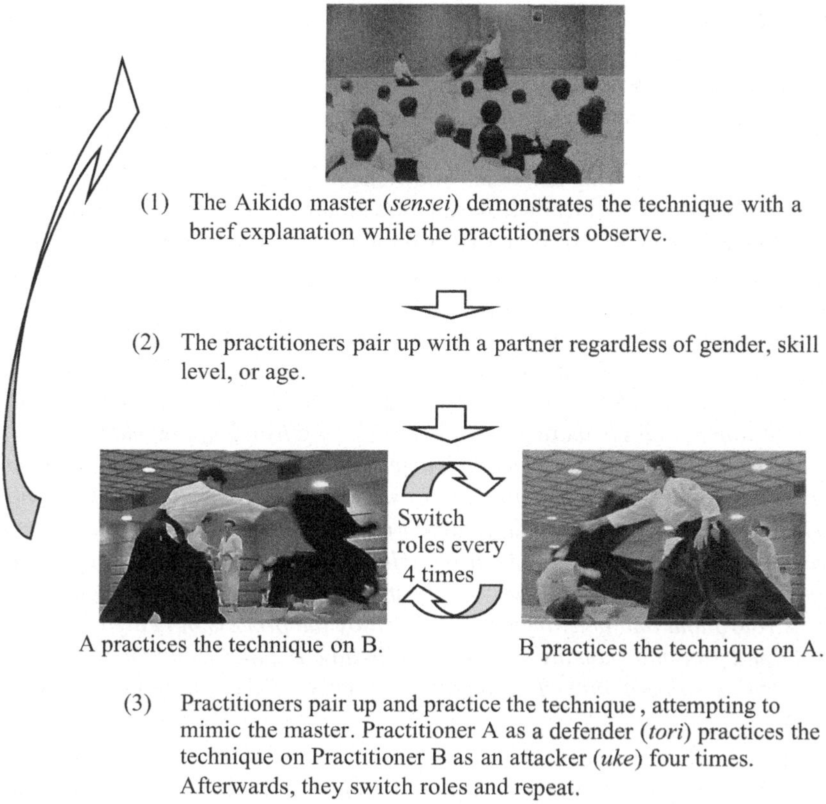

Fig. 1. Aikido training method (1 cycle)

3 Methods

This study was conducted using a qualitative analysis in the form of semi-structured interviews. The five interviewees (2 male and 3 female) were selected from Aikido practitioners who have trained for 6–30 years (black belt holders 2nd–5th dan). They were asked about their motivation and reasons for continuing to practice Aikido, their thoughts on the training methods, and what they have gained through long-term training. The responses were coded, fragmented, and recontextualized to derive the results.

The interview transcripts were analyzed using the Steps for Coding and Theorization (SCAT) method, a qualitative data analysis technique for generative coding and theorization [6]. SCAT consists of four steps of coding in which a researcher edits segmented text and fills in the following four columns: (1) focused words from the segmented text; (2) words outside of the text that are replaceable with the words in (1); (3) words which explain the words in (1) and (2); (4) themes and constructs. Finally, the themes and constructs coded in (4) are woven together as a storyline and theories are extracted from the storyline. SCAT is suitable for analyses that include chronological elements of a story. Thus, we used SCAT because the interviews in this study included chronological elements such as changes that occur to the body and mind as a result of training over time.

4 Results

The theories below were extracted from the storyline derived by SCAT and their representative interviewee's texts *(italics)* are as follows.

(1) The process of acquiring embodied knowledge is an effective and enjoyable way to learn a skill. Embodied Knowledge enables practitioners to learn hidden insights of Aikido techniques on a trial-and-error basis with guidance from the master's brief verbal explanations.

"I started to think about how to use my body and other things that I hadn't thought about in my daily life. It's interesting to discover things such as the physics behind effective movements, how to use body weight efficiently in Aikido techniques, loosening certain muscles to increase your overall strength and power, and how to control movement based on one's mental image. It's an enjoyable process of receiving hints from masters and other practitioners, gradually connecting them, forming a hypothesis, and acquiring knowledge. I enjoy the sense of accomplishment of reaching an understanding in both my mind and body. It's not enough to only understand with your head. However, it cannot be said that you truly understand unless your mind and body become one. I enjoy the process of being taught by a master and gradually coming to an understanding. It's like solving a riddle. Techniques can also be improved through this process." (Male, training for 8 years, 2nd dan)

(2) Collaborative learning through Aikido training can foster the ability to anticipate the thoughts of others based on one's perception and the ability to act with the other person's perspective in mind. Even without talking, practitioners often feel connected to each other mentally, making them feel at ease in practice.

The perception of the relationship with the opponent is very important in Aikido practice. Learning from each other's physical reactions. Physical interaction and communication based on the actions and reactions of the body. Thinking about the other person's feelings, both the uke and tori, and trying to facilitate the other person's progress as well as your own. The feeling of being connected to each other mentally is enjoyable. (Female, training for 6 years, 2nd dan)

(3) Training in silence and concentrating on one's training partner can help teach mindfulness and can be useful for improving one's ability to cope with interpersonal stress in society [7].

My master said to me, "You don't have time to look away. You have to concentrate on what to do about the person in front of you who is attacking you. You don't have any time. Concentrate!" During training, I try to concentrate on the opponent in front of me. Now I am able to concentrate better. (Female, training for 20 years, 4th dan)

(4) The enjoyment and value in Aikido stems from continued practice over time rather than winning or losing on the spot. Practitioners feel that they can improve their techniques continuously, even at an old age, thus fostering sustainable learning.

I chose Aikido because I thought it was a martial art that I could continue for a long time. I chose Aikido because I saw a lot of old people practicing energetically. I was attracted to the endless training, the effort, and the fact that it is something that can be practiced every day with no clear end. (Male, training for 10 years, 2nd dan)

It's bothersome to constantly think about winning and losing. [The people who gather for tournaments are all high energy, and it makes me cringe. I don't want to compete anymore. Always striving to win is tiring. (Female, training for 10 years, 3rd dan)

(5) Experiential learning is the gradual discovery of issues, awareness, answers, understanding, and enlightenment through many experiences over a long period of time. The appeal lies in the fact that it is up to the individual to decide how to experience and grow.

If I don't start the learning process and continue to gain a lot of experience, I can't become who I want to be. That's the interesting part. There is a long way to go. I'm enjoying the process of gradually finding the answer by myself. There is likely value in enjoying the process itself rather than the result. Those who seek immediate results may quit Aikido. (Male, training for 30 years, 5th dan)

To further clarify the mechanism of value creation in Aikido from the theories above, we attempted to extract the topics implied in each theoretical description. The extracted topics are as follows: "acquiring embodied knowledge", "collaborative learning", "learning mindfulness", "sustainable learning", and "experiential learning". From the perspective of services, the Aikido practitioner is considered an actor who learns through resource integration and performs value creation through these integrated resources.

In the next step, the topics were grouped into physical and mental aspects based on the functional value (e.g., skill improvement) of the topics and the characteristic of Aikido [8]. Then we evaluated the groups on the basis of emotional value (e.g., comfort and pleasure) and emotional strength of the description of the theories above. The results are shown in Fig. 2.

It is noteworthy that the four mental topics have low functional value but high emotional value. At first glance, these topics seem to be meaningless because they have no functional value. However, they make sense in the long run because they strengthen one's inner motivation, which motivates practitioners to train long term even though Aikido does not have competitions.

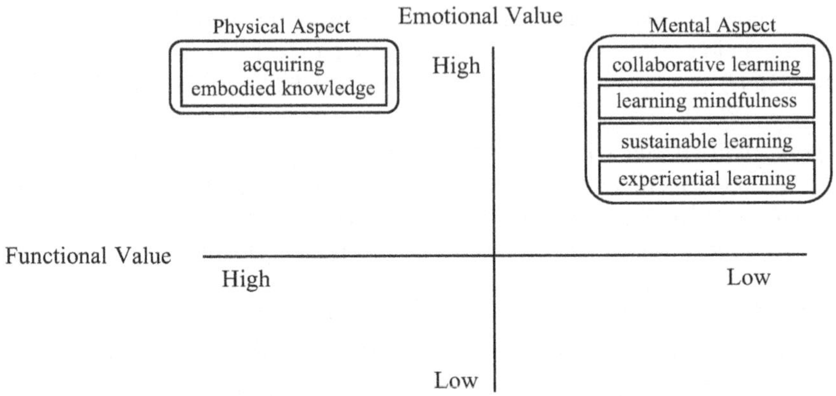

Fig. 2. Four-quadrant graph showing positioning of each topic. The topics are grouped by functional value (horizontal axis) and emotional value (vertical axis)

5 Summary

We investigated the reasons people continue to practice Aikido without losing interest, even though it is a sport that does not involve winning or competing. The emotional value of Aikido lies in its mental aspects rather than its physical aspects, stemming from the training style in which practitioners cultivate mutual understanding, assistance, and enhancement with other practitioners. The rich emotional value generates strong inner motivation that encourages practitioners to continue training long term.

References

1. Aikikai Foundation. http://www.aikikai.or.jp/eng/aikido/about.html
2. The International Aikido Federation (IAF). https://www.aikido-international.org/
3. Global Association of International Sports Federations (GAISF), https://gaisf.sport/
4. Ueshiba, M.: Aikido, the Contemporary Martial Art of Harmony; Training Methods and Spiritual Teachings. Naigai Publishing (2019)
5. Lusch, R.F., Vargo, S.L.: Service-Dominant Logic: Premises, Perspectives, Possibilities. Cambridge University Press, Cambridge (2014)
6. Otani, T.: "SCAT" A qualitative analysis method by four-step coding: easy startable and small-scale data applicable process of theorization. Bulletin of the graduate school of education and human development (educational sciences), vol. 54(2), pp. 27–44. Nagoya University (2007)
7. Kabat-Zinn, J.: Full Catastrophe Living: Using the Wisdom of Your Body and Mind to Face Stress, Pain, and Illness. Delta (1991)
8. Lee, J.-S., Lee, C.-K., Choi, Y.: Examining the role of emotional and functional values in festival evaluation. J. Travel Res. **50**(6), 685–696 (2011)

Industry-Academia Research Collaborations in the Post-corona Era: A Case Study of Remote Operations in a Japanese State-of-the-Art Research Facility

Takashi Onoda[1(✉)] and Yasunobu Ito[2]

[1] School of Life Science and Technology, Tokyo Institute of Technology, Nagatsuda 4259 Mail box J3-02, Midori-ku, Yokohama, Kanagawa, Japan
onoda.t.ac@m.titech.ac.jp

[2] Japan Advanced Institute of Science and Technology, Asahidai 1-1, Nomi, Ishikawa, Japan
itoyasu@jaist.ac.jp

Abstract. Recently, research facilities in academia take on a crucial role in the innovation ecosystem where facilities are promoting research collaborations through shared use with other research organizations, academia, and firms. We focus on the remote operation of measurement methods and speed-up technologies as a countermeasure against COVID-19 setback, whereby face-to-face contact and interactions are restricted, thus making it impractical for users to visit and use research facilities. We discuss how these new technologies and methodologies promote and obstruct industry-academia research collaborations during the post-corona period. Specifically, the focus is on NMR (Nuclear Magnetic Resonance Device), a state-of-the-art research facility that promotes shared use to external users in Japanese academic research institutes.

Keywords: Industry-academia relation · Research collaboration · Academic research facilities · Epistemic cultures

1 Introduction

"National Innovation System" recognizes science and technology as a whole system that establishes relationships with industries, aligns with government's policies and supports research [1–4]. Under the system, academia plays crucial roles. So far, the academia has continued basic research where fruits of its research are used in applied sciences that naturally lead to the creation of new technologies and industries in order to secure economic developments; this type of model is called the linear model [5–8]. The idea has shifted to forming strong relations with firms where academia's basic research is being applied in business; this situation is called Mode 2 in knowledge production [9, 10].

Under the concept of "Innovation Ecosystem", which defines innovation as a whole system that breaks away from the idea of relying on the inspiration and achievements

C. Leitner et al. (Eds.): AHFE 2021, LNNS 266, pp. 429–435, 2021.
https://doi.org/10.1007/978-3-030-80840-2_50

of a handful of geniuses but rather constantly creates innovation in a system [11, 12], further emphasis is on academia to promote innovations as ever before. In the innovation ecosystem, public research establishments which have large-scale research facilities should take on a vital role in promoting research collaborations not only in academia but also with firms [13–15].

The Japanese government and public research establishments decided to open their research facilities to encourage research collaborations, though most of them have not been shared to external users, i.e., researchers from other research organizations [16, 17]. Research facilities in academia focus as hubs of research collaborations between academia and firms. However, in Japan, several shared activities are handled by scientists in facilities and there is concern that promotion will not proceed because many facilities only offer services (e.g. measurement services) which do not lead to research activities. We demonstrate beyond the dualistic distinctions between service and research activities in Japanese research facilities by ethnographic analysis that can perform deep and narrow qualitative analysis under solid relationship with informants.

Furthermore, it has been difficult to visit research facilities directly because of the COVID-19 pandemic since the beginning of 2020, whereby face-to-face contact and interactions are restricted. This paper focuses on the remote operation of measurement methods and speed-up technologies that are appealing to external users. This paper also shows how these new technologies and methodologies promote and obstruct industry-academia research collaborations during the post-corona period.

2 Literature Review

The literature review on service and research collaborations in academia has three streams. The first is the service-oriented approach which is shown in the discussions of managing research facility called core-facility, and big research projects in genome science and high-energy physics. The second is the research-oriented approach which is exhibited in the discussions of research collaborations. The last is the user-oriented approach which is indicated in the discussions of LSRIs for sharing and co-creation of values in management of technology (MOT) studies.

The first approach is seen in the discussions on R&D management and the management of research facilities. Research facilities and analytical instruments, such as next-generation DNA sequencers and NMR spectroscopies, which are much expensive and challenging for beginner use, are managed mainly through efforts in improving customer services [18–20]. The second one is the research-oriented approach which is discussed in R&D management. Several discussions analyze that collaborative-based papers have more citations, high-impact than has ever been published by a single author using bibliometric approaches [21, 22]. The last one is a user-oriented approach which emphasizes the existence of "others" and partners in research collaborations. This approach is not uncommon in discussions of MOT studies [23–25].

The first approach is the service-oriented approach shown in management of research facility and big research projects, such as in genome science and high-energy physics. The approach emphasizes service elements and does not pinpoint on research collaborations because it considers service activities are engaged by technicians who are not involved in research activities. The second is the research-oriented approach which is exhibited in the discussions of research collaborations. The approach highlights research elements which enhance researchers' own research capacities; it lightly touches upon service elements because it does not consider having partners in research collaborations. The last is the user-oriented approach which is indicated in the discussions of LSRIs for sharing and co-creation of value in MOT studies. The approach highlights both service and research collaborations; it does not refer to frictions between service and research collaborations.

3 Method and Material

In this article, we demonstrate the relationships beyond dualistic distinctions between service and research collaborations in ethnographic analysis that can perform deep and narrow qualitative analysis under solid relationship with informants. These analyses

Fig. 1. An NMR spectrometer in the facility. Source: RIKEN NMR Facility 2020

would reveal under the iceberg the hidden conflicts and benefits of external sharing that is not expressed in bibliometric approaches. This article tries to analyze epistemic cultures of shared activities. Epistemic culture is a concept that has influenced how to recognize and analyze in each scientific discipline [26–29].

We conducted ethnographic and interview surveys of four years since 2015 at the NMR facility in RIKEN, with a facility director, scientists, and technicians in the facility, and examined profits and tasks of remote operations in research institutes.

We focus on an NMR (Nuclear Magnetic Resonance) facility, a state-of-the-art research facility, which promotes shared use to external users in Japanese academia, research institutes. NMR spectroscopy can measure compound materials and biological macromolecules at the atomic level. The methodology, used in a wide range of fields in chemistry, materials, food, environment, and life sciences has expanded its usage to include drug discovery, drug metabolite kinetics, and analysis of new generation batteries in the recent years. This paper focuses on a Japanese NMR facility in the public research institute, RIKEN (the Institute of Physical and Chemical Research). RIKEN is one of the Japanese representative research public institutes on physical and chemical sciences. The NMR facility can measure compound materials and biological macromolecules at the atomic level (Fig. 1).

4 Results and Discussions

Due to the COVID-19 incidents since April 2020, it was impossible not only for external users but also for staff to utilize the NMR facility; with about thirty requests postponed until after July 2020. Now that it is no longer feasible to visit the facility, there is a demand for installing new samples by remote operations. Normally, it takes about 30 min per sample to set, load and adjust new samples together with the person in charge at the facility.

Remote operation offers the solution to the dilemma by allowing users to measure samples without visiting the facility and operators can operate the device even when working from home. It is possible to save more than five hours per day, combining a sample changer and automatic parameter adjustment function in the remote operation system. In case the device was not used a day before, samples can be set manually in advance, even at night or on weekends so it can automatically perform measurement from several hours up to two days at the maximum. Remote operation and automatic measurement can improve convenience for external users as well as facility staff in saving machining time and the burden of setting up samples. The automation technology spawns new devices that enable measurement in a facility where users cannot visit directly in the post corona era (Fig. 2).

On the other hand, current technology on remote operations often are mainly routine measurements for screening. Facility staff considers these measurements as routine and measurements do not lead to research collaborations. These measurements tend to be one-directional service for them, though the technology of remote operations is attractive in the post-corona era. In the future, it would require additional process in remote operations such as setting up a virtual screen where users and facility staff can see the measurement at the same time. These technologies would foster reciprocal communication between

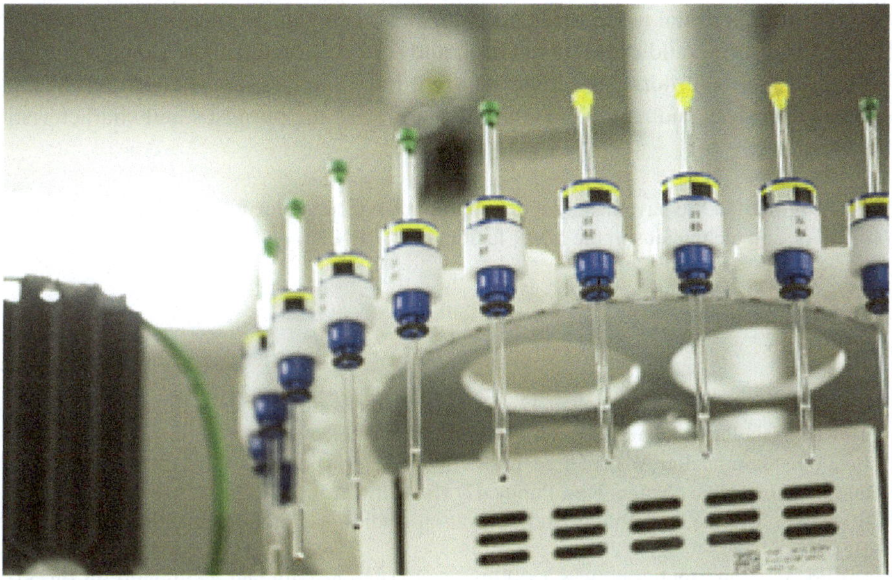

Fig. 2. The sample changer in the facility. Source: RIKEN NMR Facility 2020

facility staff and users to create more opportunities that lead to research collaborations. On the other hand, remote operations entail an issue of security policies regarding the data calculated at the facility. It is cumbersome to deal with the firms' security policies because every firm's policy is different; the use of video-communication tools is often restricted by firm's network policy and it is difficult for facility staff to support these network policies accordingly for the sake of cost-efficiency. This is one reason why working with firms do not promote shared use; in addition, their modest use of the facility has deferred solution on this issue. Therefore, in overcoming this issue, facilities should cooperate with other facilities and institutes to build a virtual environment on servers which enables sharing or exchanging data with users, separate from the on-premises servers which is managed by each facility.

5 Conclusion

Remote operations in state-of-the-art research facilities has been recognized as one of vital technologies which promotes user-efficiency as well as shorten machining time for facilities even in the era of COVID-19. This paper indicates the case study of an NMR facility in Japanese public institute. As stated in the introduction, in the innovation ecosystem, research facilities in academia encourage research collaborations. It is required to expand shared use to external users in various research fields including firms not accustomed to using research institutes, as well as to academic users who are using the facility. Thus, research collaborations between users and facility staff are the key to achieve innovation ecosystem. However, there is concern that current shared use of measurements do not lead to research collaborations for facility staff. Thus, research collaboration is mandatory in order to promote a win-win solution between facility staff and users. It is also essential to set the evaluation axis to encourage research collaborations where facility staff could evaluate research and service from a comprehensive perspective. The evaluation criteria should be set according to the wishes of each staff because the ratio of services and research is unique along with the purposes and characteristics of each facility.

Acknowledgments. We have had a lot of support and encouragement from directors, scientists, coordinators, and technicians in the NMR facility at RIKEN.

References

1. Liu, T.-H.: The philosophical views of national innovation system: the LED industry in Taiwan. Asia Pac. Manag. Rev. **24**, 291–297 (2018)
2. Schot, J., Steinmueller, W.E.: Three frames for innovation policy: R&D, systems of innovation and transformative change. Res. Policy **47**, 1554–1567 (2018)
3. Soete, L.: Science, technology and innovation studies at a crossroad: SPRU as case study. Res. Policy **48**, 849–857 (2019)
4. Cirillo, V., et al.: Only one way to skin a cat? Heterogeneity and equifinality in European national innovation systems. Res. Policy **48**, 905–922 (2019)
5. Ravetz, J.: The No-Nonsense Guide to Science. New Internationalist, Oxford (2006)
6. Stokes, D.E.: Pasteurs Quadrant: Basic Science and Technological Innovation. Brookings Institution Press, Washington, DC (1997)
7. Rosenbloom, R.S., Spencer, W.J.: Engines of Innovation: U.S. Industrial Research at the End of an Era. Harvard Business School Press, Cambridge (1996)
8. Ziman, J.M. (ed.): Prometheus Bound: Science in a Dynamic "Steady State." Cambridge University Press, Cambridge (1994)
9. Gibbons, M., et al.: The New Production of Knowledge: The Dynamics of Science and Research in Contemporary Societies. SAGE, London (1994)
10. Etzkowitz, H.: The Triple Helix: University-Industry-Government Innovation in Action. Routledge, London (2008)
11. Riesener, M., et al.: Innovation ecosystems for industrial sustainability. Procedia CIRP **80**, 27–32 (2019)
12. Sun, S.L., et al.: Enriching innovation ecosystems: the role of government in a university science park. Glob. Transit. **1**, 104–119 (2019)
13. Arundel, A., et al.: Advancing innovation in the public sector: aligning innovation measurement with policy goals. Res. Policy **48**, 78–798 (2019)
14. OECD/Eurostat: Oslo Manual Guidelines for Collecting and Interpreting Innovation Data, Paris (2005)
15. Autio, E., et al.: A framework of industrial knowledge spillovers in big-science centers. Res. Policy **33**, 107–126 (2004)
16. Japanese Council for Science and Technology: 'Leading in the New System of Research Equipment and Machines for Shared and core Combined by the Management in Research Organization' (2015)
17. Cabinet Office Government of Japan: Comprehensive Strategy on STI 2014 (executive summary), Science and Technology Policy Council for Science, Technology and Innovation (2014)
18. Haley, R.: A framework for managing core facilities within the research enterprise. J. Biomol. Tech. **20**, 226–230 (2009)

19. Farber, G.K., Weiss, L.: Core facilities: maximizing the return on investment. Sci. Transl. Med. **3**, 1–4 (2011)
20. Hockberger, P., Meyn, S., Nicklin, C., et al.: Best practices for core facilities: handling external customers. J. Biomol. Tech. **24**, 87–97 (2013)
21. Lozano, S., Rodríguez, X.-P.: Arenas A, Atapuerca: evolution of scientific collaboration in an emergent large-scale research infrastructure. Scientometrics **98**, 1505–1520 (2014)
22. Wuchty, S., Jones, B.F., Uzzi, B.: The increasing dominance of teams in production of knowledge. Science **316**, 1036–1039 (2007)
23. von Hippel, E.: The dominant role of users in the scientific instrument innovation process. Res. Policy **5**, 212–239 (1976)
24. Prahalad, C.K., Ramaswamy, V.: The Future of Competition: Co-Creating Unique Value with Customers. Harvard Business Review Press, Cambridge (2004)
25. Vargo, S.L., Lusch, R.F.: Service-dominant logic: continuing the evolution. J. Acad. Mark. Sci. **36**, 1–10 (2008)
26. Knorr-Cetina, K.D.: Epistemic Cultures: How the Sciences Make Knowledge. Harvard University Press, Cambridge (1999)
27. Brosnan, C.: Epistemic cultures in complementary medicine: knowledge-making in university departments of osteopathy and Chinese medicine. Health Sociol. Rev. **25**, 171–186 (2016)
28. Mørk, B.E., Aanestad, M., Hanseth, O., Grisot, M.: Conflicting epistemic cultures and obstacles for learning across communities of practice. Knowl. Process Manag. **15**, 12–23 (2008)
29. Kasperowski, D., Hillman, T.: The epistemic culture in an online citizen science project: programs, antiprograms and epistemic subjects. Soc. Stud. Sci. **48**, 564–588 (2018)

Practices and Issues of Corporate Ethnographers: A Case Study of a Corporate Laboratory of a Japanese Telecommunications Service Provider

Tomoko Oto(✉) and Yasunobu Ito

Japan Advanced Institute of Science and Technology, Asahidai 1-1, Nomi, Ishikawa, Japan
{o-tomoko,itoyasu}@jaist.ac.jp

Abstract. Japanese companies started to use the cultural anthropological method, namely ethnography, from the latter half of 2000. However, there are only some literature on how ethnography is introduced into companies in Japan, how it is operated, and what the challenges are. The purpose of this study is to identify the practices and challenges of ethnographic research in Japanese companies from the perspectives of in-house ethnographers. The main investigation field is a corporate laboratory of a telecommunications service provider. The data of this paper were collected from the participant observations and interviews in the company. We compared the ethnography practiced in companies to academic/traditional anthropological ethnography, which often involves anthropologists living in an isolated community for an extended period of time to understand culture. The comparison revealed both the ingenuity and challenges of ethnography as it was practiced in companies.

Keywords: Business ethnography · In-house ethnographer · Corporate laboratory

1 Introduction

In recent years, ethnography, the methodology grounded in cultural anthropology, has been adopted and spread in various academic and business fields. In particular, ethnography have been used for the improvement and optimization of business operations and for the derivation of concepts for product and service development in the form of "business ethnography", which retains the essence of conventional academic ethnography but has been reconstructed for business purposes [1]. One of the reasons why ethnography usage has started in business is that UX (User eXperience) has become more important in the provision of services as well as in the pursuit of product functionality. In user research, the method of ethnography has been chosen and introduced in business in order to collect the latent needs of users and other information that could not be captured by the conventional interviews and questionnaire in the past. Ito pointed out that this sort of approach was first seen in the early 1990s in Europe and North America (especially in the U.S.)

C. Leitner et al. (Eds.): AHFE 2021, LNNS 266, pp. 436–443, 2021.
https://doi.org/10.1007/978-3-030-80840-2_51

and from there, a style of ethnography similar to that currently in use in industry spread to Japan since around the late 2000s [2, 3].

With an increasing number of corporates now implementing ethnography, the advantages of adopting ethnographic techniques in the field of business are noted and the problems encountered when applying ethnographic concepts in business have been widely debated, and there have been discussions on the various differences in methodology between academic and business applications in the field. For example, it points out that one of the challenges of introducing ethnography into the field of business is that in many cases, various restrictions is imposed such as having to prepare the protocol to be executed by the team in advance, along with the duration, the budget, and the composition of the team [4]. In addition, corporate management has pointed out that research tends to take time and costs too much. Therefore, it is difficult to reconcile such perceptions between the corporate management side and ethnographers in the short term, and there is a suggestion that a long-term plan to change the corporate culture is necessary in order for the corporate management side and ethnographers to build a good working relationship [5].

In order to achieve the "human understanding" that conventional academic ethnography demands, it is necessary to introduce some time-cost considerations into the research. Conventional academic ethnographic research usually entails a long-term study involving several years in a single field. On the other hand, corporate ethnographic research is conducted in a short time period; ranging from a few weeks to a few months (often around three months) per study. In addition, a company may conduct one research or several completely different field studies at the same time or in a short time span. This style of corporate research is related to the Japanese normal business practice that requires the reporting of results according to managerial deadlines (*hanki* - half year or *shihanki* - quarter). Therefore, hired ethnographers are required above all to conduct research in a short period of time and report the results to the company.

Another reason of challenge may have to do with the hiring of cultural anthropologists as a skilled ethnographer. Within Japan, the employment of anthropologists, as trained and experienced experts of ethnography have not progressed a great deal. In the U.S., the employment of such experienced expert anthropologists is becoming increasingly common in government, business corporates, non-profit sectors, etc., [5]. As Ito pointed out, in many cases in Japan, fresh university graduates are hired en masse and receive education tailored to their respective companies [1]. This style of Japanese employment is normally called "membership-based employment", whereas western firms' employment based on specific duties is called "job-based employment" [6]. Therefore, regarding ethnographers, companies hire non-specialists rather than specialists of ethnography. Although firms do not employ any anthropologists or skilled ethnographers, a few companies, such as Fujitsu collaborate with anthropologists and other experts to provide their own ethnographic education and train ethnographers within their organizations [7, 8]. In some cases, employees learn how to do ethnography on their own.

In this context, it is meaningful to clarify how each company introduces and implements ethnography and to highlight what its characteristics are. This will enable us to compare ethnographic methods among companies and, eventually, to make international comparisons. However, thus far, there have been practically no research done on how Japanese companies have adopted ethnography, how they are implementing it, and what its characteristics are.

The purpose of this study is to identify the practices and challenges of ethnographic research in Japanese companies from the perspectives of in-house ethnographers.

2 Method

The main investigation field is a corporate laboratory of a telecommunications service provider, KDDI Research. As of April 1, 2020, the company has 308 employees and conducts a wide range of research activities from basic to applied research on information and communications. One of the authors (Oto) was employed by the telecommunications firm's laboratory as an ethnographer from 2015 to 2020, and was involved in various ethnographic studies. For example, the author was involved in the research on mobile phone use among seniors and on friendship network among child-rearing mothers.

In this paper, we will focus on two research teams (Team A and Team B) conducting qualitative research at KDDI Research (hereafter referred to as the laboratory). It is important to note that the laboratory does not have a dedicated team for qualitative research. When qualitative research becomes necessary, researchers and research teams will participate in qualitative research as needed.

The research period for the A team was from August 2015 - December 2018, and for the B team from December 2017 - March 2018. Regarding the A team, I was involved in ethnography research projects as a team member. The A team conducted research with a total of five: (two members from the laboratory and three members from the research company) and the B team consisted of a total of eight: (five members from the laboratory and three members from the research company). All the members' disciplines were different and included cognitive psychology, anthropology, and sociology. Regarding my involvement with the B team, when permitted, I took part in field observations and attended various related events such as examinations, meetings, debriefing sessions, and analytical workshops.

3 Results

The analysis revealed several characteristics of the ethnographic research conducted in KDDI Research.

3.1 Collaboration with an External Research Company

As mentioned above, the laboratory does not have its own teams or divisions that specialize in qualitative research, so when qualitative research project is initiated, each researcher has to manage the entire research project so that everything is successfully completed within the deadline. However, the laboratory involves many researchers with background in engineering, and not everyone possesses specialist knowledge on qualitative research. For this reason, corporates search externally for "knowledge of qualitative research". This means teaming up with an external research company. In the laboratory, researchers are always conscious of the cost. Therefore, when they compare the time expended and the costs incurred for nurturing ethnography experts internally with the cost of importing ethnography knowledge externally, the option of importing ethnography knowledge may seem preferable. Rather than spending cost on training employees in ethnography, corporates think that hiring an external research company expert in qualitative research on a case-by-case will ensure reliable results within the deadline. In other words, conducting ethnographic research at a laboratory assumes that an external research company is part of its research team.

As one researcher stated, "We are acting under the leadership of the research company" (December 22, 2017), meaning that the qualitative research at the laboratory is getting a big hand from an external research company. In fact, external research companies are normally involved in nearly all qualitative research planning and implementation processes.

3.2 Division of Labor with an External Research Company

"Research" involves a variety of comprehensive and detailed preparations and adjustments, such as securing budgets, selecting informants, arranging research dates, carrying out interviews and analysis, preparing interview transcripts and writing reports. In conventional academic ethnography, it is normal for a single ethnographer to carry out all these tasks. However, each team in the laboratory conducts multiple projects and carries out research concurrently, so it is difficult to consistently perform all the work involved in qualitative research projects. For this reason, tasks are often assigned in conjunction with external research companies in order to save time and costs. As can be seen in Table 1, tasks that require a significant amount of time, such as document preparation and transcriptions are outsourced to an external research company so that laboratory members can focus on the core part of the decision-making process and come up with new ideas.

Table 1. Outsourced work/not outsourced work

Outsourced work	Not outsourced work
• Collecting and listing informant candidates	• Interview, observation,
• Adjustment of the schedule	• Comment on debriefing
• Preliminary task creation	• Team meeting related to workshop
• Interview, observation, recording	• Working out ideas
• Transcription	• Confirmation of reports
• Secretarial and debriefing proceedings	
• Workshop secretary	
• Creating a report	

The division of labor is observed in research. In academic ethnographic research, a single researcher interviews, observes, takes records and photos. However, it is difficult for a non-expert member to do all this work alone in the firm's laboratory. Therefore, when ethnography research is carried out in the laboratory, job roles of interviewers, observer, note-takers, and photographers are often assigned among the members.

For example, in the research on relationships of mothers, an external research company search for informant candidates and submit the list of candidates for the laboratory members to select from the list. The external research company then contact selected mothers and schedule in their interviews. On the day of the interview, external research company mainly conduct the interviews and observations and take recordings, while laboratory members conduct supplementary interviews and observations. In the subsequent analysis, external research company transcribe the interview recordings, perform secretarial work and facilitate debriefing. The members of the laboratory focus on the findings and voice their ideas. In response to these comments and ideas from the members, the external research company prepare a report that will be reviewed by the laboratory members. The research is completed after it is checked and approved by the laboratory members. By implementing role sharing, laboratory members can concentrate on just one aspect of the work and continue to collect data, even if they are unfamiliar with qualitative research themselves.

3.3 Eliminate Information Gaps Between Members

As mentioned above, the laboratory's research is conducted in an extremely limited time period. Each research (interview and visit) takes about two hours. If many members visit the homes of research subjects, it can be stressful on the research subjects. Therefore, about three to four members perform the visit. The members who participated in the on-site research will share the data obtained during the visit with the members who did not participate in the research at a later information-sharing meeting. Members who are not actually conducting research are unlikely to share the same perspective, viewpoints and access to information as those at the research site. Therefore, information is shared to ensure that no information gap develops between the members. In the information-sharing stage, all transcribed interview data are read out and photographs and video or film footage are displayed. Sometimes members play back the video recorded by the 360-degree camera to recreate a more realistic research site. Members comment on the interview and video contents using Post-it notes, and collate opinions from other members. This is all carried out using the affinity diagram (or K-J method). Here, as in the other situations described above, the roles are shared between team members. Members of the research company read out all the text derived from the interviews, then laboratory members listen and give out their opinions. Opinions outputted are then summarized by the research company.

4 Discussion

Next, we summarize some of the differences between conventional academic ethnography and the type of ethnography carried out in laboratories.

4.1 A Single-Researcher, Long-Term Ethnography Versus Team-Based, Short-Term Ethnography

In conventional academic ethnographic research, a researcher who is familiar with the target research field collects diverse, rich, and detailed data by conducting independent, long-term participant observation and by forming rapport with the subjects in the field. The collected data is then analyzed holistically from multiple perspectives based on the researcher's own experiences. On the other hand, in a laboratory, due to various constraints from the company (especially time constraints), it is not possible to invest a long time on a single research project. Therefore, it is necessary to focus on obtaining as much data as possible within the time limit. However, as mentioned above, the research members are not experts trained in ethnography. Therefore, in many cases it is difficult for them (unless they are trained) to collect information and conduct multifaceted analysis on par with the ethnography experts. Therefore, they work with members of external research companies who are experts in qualitative research and share the workload to fill in their deficiencies. In addition, they use recording media such as 360-degree cameras, to fill in the information gaps among members and make sure that all members are informed as equally as possible. This enables to acquire a large amount of data efficiently and allows all members to analyze events with the same amount of information. Thereby they attempt to replicate the research process that academic ethnographer conducts alone, but with "multiple eyes" (multiple people).

In conventional academic ethnography, the ethnographer is involved with all aspects of the research from research design through to output, and so they respond to all interfaces, and take on risks by themselves. However, since business ethnography research is carried out by a team, the burden placed on one person is reduced. This means reduction of time in making adjustments and a more efficient collection of necessary data. For example, as mentioned above, by using 360-degree camera and a head-mounted display, it is now possible to carry out observations that are no longer limited to the photographer's fixed point of view and thus acquire more information on the environment (field) than was previously possible. There is an attempt to use VR technology while still at an experimental stage, for allowing multiple people to view videos of the same research field and communicate with each other while sharing information.

4.2 Holistic Scope Ethnography Versus Partial Scope Ethnography

Academic ethnographers strongly believe in the importance of cultivating "holistic human understanding", which is a characteristic of conventional ethnography. They seek to build relationships with the subjects living in the field to gain a comprehensive understanding of them and their entire culture. However, this is not necessarily required in business ethnography. In order to collect in-depth information and obtain a holistic understanding of the way other people live, it is at times necessary to enter a field different from one's own culture. This requires mastering their language and building rapport within the target group. However, it takes a lot of time to do this successfully. In laboratory ethnography, it is not normally required to expend costs to achieve this degree. A research with partial and limited scope was conducted on the target group. There are several reasons for this.

First, if there is no prior in-house knowledge of the target field, this can be solved by outsourcing to external experts who are familiar with that area. In other words, money can be the solution to the time-cost issue. Then, if the result can be achieved by joint analysis between multiple members and experts, it is not necessary to cultivate ethnography skills; hence, costs can be reduced. For example, the academic anthropological ethnographers begin their research by learning to speak the local language and gather information. In contrast, in the example of an overseas research, the laboratory hired academic experts who are familiar with the local language and information and thereby reduced the cost of learning this way.

Second, laboratories urge ethnographers to obtain just the information they need in their research, no more and no less, rather than acquire holistic knowledge. In the laboratory, researchers are not engaged in a single field continuously for a long period. Sometimes, a research project is initiated without waiting for detailed information to be collected in the field. For this reason, there is no time to spare in acquiring the holistic knowledge.

Third, the laboratory's product development and service creation targets are often in Japan or within the laboratory's own organization. In the case of product development and service creation directed towards the domestic market, a corporate is targeting users who share the same culture; even if certain attributes are somewhat different, at least the background culture is similar and there is no need to understand its entirety, only certain aspects of it. For instance, a research on the friendship network among child-rearing mothers targeted only Japanese mothers. Since all the research members were Japanese, they already knew what Japanese mothers are like and what role they are expected to play in the Japanese society.

5 Conclusion

This paper indicated that even though ethnography used in the firm's laboratory was primarily based on conventional/academic ethnographic methods, often it was restructured to accommodate corporate culture for the ease of handling. The absence of expert ethnographers induced ethnographic studies to be conducted in teams to obtain useful results. Faced with the challenge to conduct the researches in a very limited amount of time, and without the know-how of ethnography, they relied on external research companies to manage the researches. In this way, the events were analyzed through "multiple eyes", in an attempt to reproduce something close to the expert skills which could not have been obtained otherwise through long-term training. It also revealed cases where ethnographic methods were being adapted and customized through active experiments with technological gadgets such as 360-degree cameras and VR goggles which are rarely used in academic ethnography.

This paper has shown that some of the restrictions that exist in Japanese companies have created the characteristic features of business ethnography and influenced the way in which it is currently used. However, in this paper, other organizations' cases have not been explored and this will need to be researched in the future.

Acknowledgments. This work is based on the collaborative research between KDDI Research Inc. and Professor Yasunobu Ito of JAIST for 4 years from 2015. This work was supported in part by JSPS KAKENHI Grant Numbers JP16H01968/18H00782 and a grant-in-aid from KDDI Research Inc.

References

1. Ito, Y.: 'Ethnography' in Japanese corporate activities: a meta-anthropological observation on the relationship between anthropology and the outside. In: Nakamaki, H., Hioki, K., Mitsui, I., Takeuchi, Y. (eds.) Enterprise as an Instrument of Civilization. TSS, vol. 4, pp. 55–72. Springer, Tokyo (2016). https://doi.org/10.1007/978-4-431-54916-1_5
2. Ito, Y.: How ethnography infiltrated the Japanese business scene: a case study. Paper presented at the PICMET 2018: technology management for interconnected world, Waikiki Beach Marriott Resort & Spa Honolulu, Hawaii, USA, August 2018
3. Ito, Y.: Contact zone of anthropology of and in business: inspiring synergy between anthropology and industry in Japan. Jpn. Rev. Cult. Anthropol. (JRCA). **20**(2), 7–25 (2020)
4. Ohara, S.: Application of observation and ethnography to new product development. Doshisha Bus. Rev. **63**(5), 879–891(2012). (in Japanese)
5. Jordan, B.: Advancing Ethnography in Corporate Environment: Challenges and Emerging Opportunities Challenges and Emerging Opportunities. Left Coast Press Inc., Walnut Creek (2013)
6. Hamaguchi K.: Addressing the problems with Japan's peculiar employment system. Nippon.com. https://www.nippon.com/en/currents/d00088/
7. Kishimoto, K., Terasawa, M., Hirata, S.: Innovation of working style through business ethnography and organizational monitor. Fujitsu Sci. Tech. J. **46**(2), 199–208 (2010)
8. Ikeya, N., Vinkhuyzen, E., Whalen, J., Yamauch, Y.: Teaching organizational ethnography. In: Ethnographic Praxis in Industry Conference Proceedings, pp. 270–282. American Anthropological Association (2007)

Brand Image Building of Intangible Cultural Heritage Wulo Based on Service Design Thinking

Huai Cao, Jihao Xu(✉), Zining Liu, and Juanjuan Feng

Industrial Design Department, School of Mechanical Science and Engineering, Huazhong University of Science and Technology, Wuhan 430074, Hubei, China

Abstract. In this article, based on service design thinking, I will sort out the common methods in service design theory with the design process of the double diamond model proposed by the British Design Institute, such as stakeholder relationship diagrams, user journey diagrams, service blueprints, etc. At the same time, combine the summarized method with the brand image building process which is suitable for Chinese national conditions to explore the brand image building method suitable for Chinese intangible cultural heritage brands. This will also provide a new direction for the research on intangible cultural heritage branding.

Keywords: Service design · Intangible cultural heritage · Intangible cultural heritage Wulo · Brand image building

1 Introduction

At present, the intangible cultural heritage inheritance system has gradually lost its supposed effectiveness under the modern development situation. There are many problems in the existing intangible cultural heritage system and inheritance system. For example, the policy does not match the actual situation of intangible cultural heritage, and traditional policies cannot be adapted to local conditions, leading to the stagnation of many non-hereditary inheritance and development; traditional intangible cultural transmission methods can no longer meet the current market development needs; the pressure and responsibility of the inheritors are too heavy. Skills and publicity cannot be balanced. All these reasons lead to the intangible cultural heritage that cannot really survive and develop in modern society [1].

2 Service Design and Intangible Cultural Heritage Brand Image

2.1 Introduction and Methods of Service Design

Service design aims to ensure service interfaces are useful, usable and desirable from the client's point of view, and effective, efficient and distinctive from the supplier's point of view [2].

C. Leitner et al. (Eds.): AHFE 2021, LNNS 266, pp. 444–455, 2021.
https://doi.org/10.1007/978-3-030-80840-2_52

There are many methods and tools for service design. And they have various uses. This article sorts out the 9 tools commonly used in service design and some methods commonly used in research according to the four stages of the double diamond model (Fig. 1) as shown in Table 1.

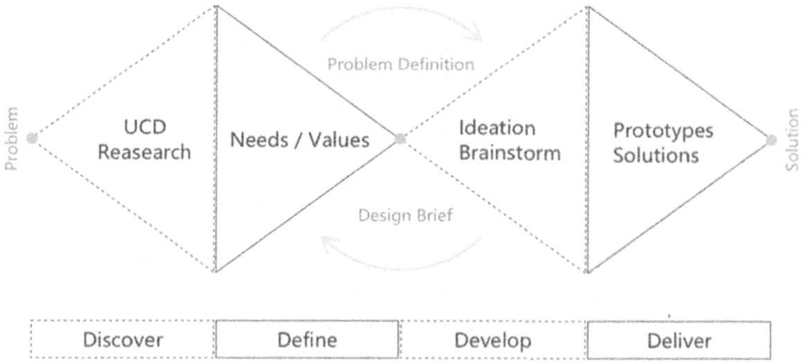

Fig. 1. Double diamond model

Table 1. Service design method tool category diagram

Stage	Tool	Method
Discover	Desktop research	Research through existing second-hand materials such as books and the Internet to understand the service background
	Interview method	Interview employees and users to understand the stakeholders in existing services and find problems
	Stakeholder map	Map the relationship between stakeholders
	Constructive interaction	Observe the actual user experience the existing service process and find problems
	Service walkthrough	Personally, experience each process of existing services and discover problems
	Focus group	Interview stakeholders to gain in-depth understanding of some issues through group discussions
Define	Issue card	Use pictures, drawings, and text to describe the problem for easy communication

(continued)

Table 1. (*continued*)

Stage	Tool	Method
	Customer journey map I	Map the user experience in the existing service system, visually express touch points, user behavior, emotions, etc.
	Service blueprint I	Present the process and various links visually of the existing service system, as well as the behavioral interactions of various stakeholders in the front and back ends of the service system
Develop	Co-design workshop	Involve stakeholders in the process of conceptual design through workshops and brainstorming
	Customer journey map II	Map the user experience in the redesigned service system
	Service blueprint II	Present the process and various links visually of the redesigned service system
Deliver	Storyboard	Turn the user's usage into a narrative story, and express it visually through photos or hand-drawing
	Service prototyping	Establish a quick model of the service design concept, and use simulation scenarios for user testing before the formal implementation

2.2 Intangible Cultural Heritage Brand Image Building Process

There are many theories related to brand image. But most of them are the digestion and absorption of western brand theories. This article quotes some theories mentioned in "Theoretical Research on Brand Image Shaping" by Shang Hongwei of Sichuan University. And combined with some of my own opinions and viewpoints, these theories are more in line with Chinese specific national conditions, and more suitable for shaping Chinese intangible cultural heritage brand image.

Brand image is a comprehensive concept. The establishment of brand image is from the perspective of consumers and reflects customers' perception and evaluation of the brand.

A complete brand image shaping process is as follows Show (Fig. 2) [3]. In the initial stage of brand image shaping, the establishment of brand positioning and personality, the establishment of the expression elements of the brand image and the integrated communication of the brand are the three most basic links.

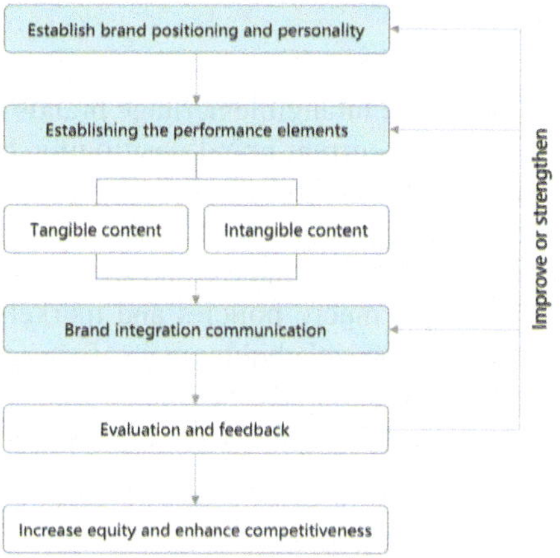

Fig. 2. Process framework of brand image building

Intangible cultural heritage brands need to focus on their own particularities when building their brand image. Here I combine the characteristics of the intangible cultural heritage brand with the brand image performance elements described above to summarize the image performance elements of the intangible cultural heritage brand, as shown in the figure below (Fig. 3).

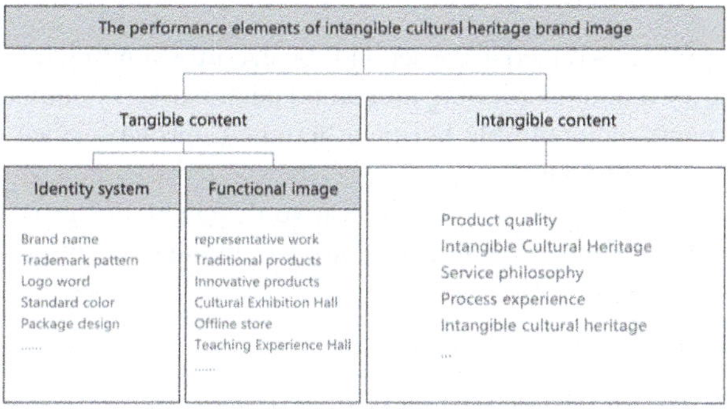

Fig. 3. The performance elements of intangible cultural heritage brand image

2.3 Building Method of Intangible Cultural Heritage Wulo Brand Image Based on Service Design Thinking

Based on the service design process and methods mentioned above, the brand image shaping process, combined with the characteristics of intangible cultural heritage brands, the author proposes the following intangible cultural heritage brand image shaping methods.

I **Discover**

Designers must fully understand the intangible heritage brands they serve. Specifically, you need to understand the brand's current brand positioning, brand personality, brand performance elements and publicity methods.

At this stage, designers can use desktop research and other methods to gain in-depth understanding of the intangible cultural heritage brand and successful brands of the same type, and understand the country's macro policies and market environment related to this direction; use interview methods, stakeholder maps, focus groups, and construction Sexual interaction, service walkthrough and other methods, analyze and summarize the brand's current brand positioning, brand personality, brand performance elements and publicity methods.

II **Define**

Based on the results of the first stage, designers need to map the existing service process, define and summarize the current problems of the intangible cultural heritage brand.

At this stage, designers can use question cards, user experience maps and service blueprints to map the existing service system processes of intangible heritage brands and the service processes of successful brands, and define the current brand positioning, brand personality, brand performance elements and publicity the problem.

III **Develop**

In response to the problems raised in the second stage, corresponding innovative solutions are proposed. Clarify brand positioning and brand personality, systematically design the performance elements of intangible heritage brands, improve service procedures, and formulate communication plans.

At this stage, designers can use collaborative design workshops, user experience maps, and service blueprints to involve stakeholders in the conceptual design process.

IV **Deliver**

Based on the design results of the third stage, the service prototype is made, the workload and effect are estimated, and the specific implementation plan is designed.

At this stage, designers can simulate and practice the entire service process and communication effects through methods such as storyboards or service prototypes, so as to find problems early and improve them, saving trial and error costs.

3 Design Practice of Wulo Brand Image Building

3.1 Discover

At this stage, I have a comprehensive understanding of the current Wu Luo brand, and specifically need to understand its current brand positioning, brand personality, brand performance elements and publicity methods.

1) Introduction of Wulo

The Wulo brand of intangible heritage (hereinafter referred to as Wulo) was born in Suzhou and was founded by Mr. Li Hailong, the inheritor of traditional intangible cultural heritage "Wu Luo". The predecessor of the enterprise was the "Jiangnan Network Silk Factory", the first private enterprise established in Guangfu in 1992. In order to integrate Wu Luo culture with modern life and restore its rich use value and cultural charm, Wulo will now transform into a modern brand with Wu Luo cultural creativity as its main business.

Although the Wulo brand has gradually become more active in front of the public than before. But there is still a long way to go before a complete brand image is established.

2) Introduction of Wulo Brand Image

Brand positioning is not to determine what the product really is, but to expect consumers to have what perception of the brand [3].

Based on the investigation of Wulo, it is found that it has no clear brand positioning. But based on the commercial and cultural activities it is currently engaged in, the brand positioning can be briefly summarized as: precious silk fabrics in ancient memory.

Like brand positioning, brand personality also answers the question of what the brand is. Brand personality can be specifically divided into three types: self-expression, appreciation of target consumer groups, and emotional symbolism [3].

At present, based on positioning of Wulo, its brand personality belongs to the first type of self-expression. Clothing of Wulo often symbolizes rich and authoritative personality characteristics.

I divided stakeholders of Wulo based on the stakeholder map of the service design tool. The stakeholder relationships of Wulo are shown in Fig. 4. The main value of Wulo is divided into two parts, commercial value and cultural value, so the stakeholders of the brand can also sort out from these two directions.

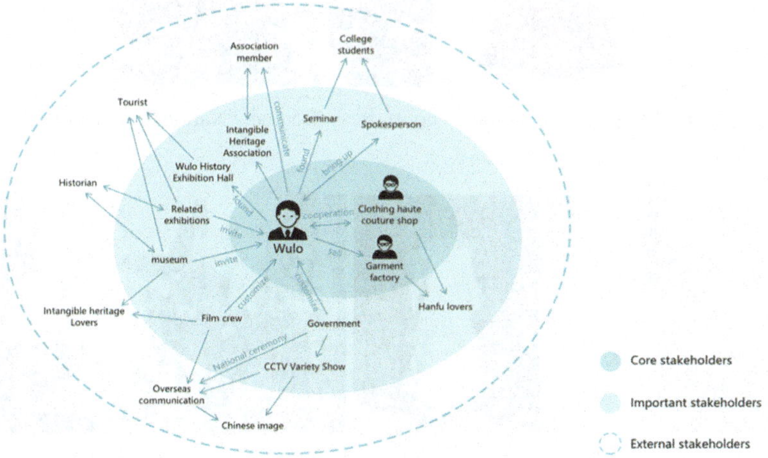

Fig. 4. Stakeholder relations of Wulo

In the process of analyzing stakeholders of Wulo, I sorted out the current Wulo performance elements. Intangible content of Wulo is mainly based on the craftsman spirit based on the intangible cultural heritage Wu Luo weaving skills. The tangible content is introduced below.

Previous brand recognition system of Wulo is not perfect. The standard color is red. The standard characters have no specific definition. The brand's trademark pattern and packaging are shown in the figure below (Fig. 5).

Fig. 5. Brand trademark and packaging of Wulo

Functional images of Wulo mainly include: historical representative works-Mingyan Shenggong's court clothes mentioned in "National Treasures" (Fig. 6), intangible heritage products-two-color peony bright flower, Luo (Fig. 7), flower and bird flower, Baoxiang flower, etc. (Fig. 8).

Fig. 6. Mingyan Shenggong's court dress (Source: "National Treasure")

Fig. 7. Mingyan Shenggong's court dress (Source: "National Treasure")

Fig. 8. Jinluo Art Museum

3.2 Define

The main task of this stage is to sort out and define the problems of Wulo brand image based on the current understanding of Wulo in the first stage. Based on current service process of Wulo, the service journey diagram drawn is as shown in the figure below (Fig. 9).

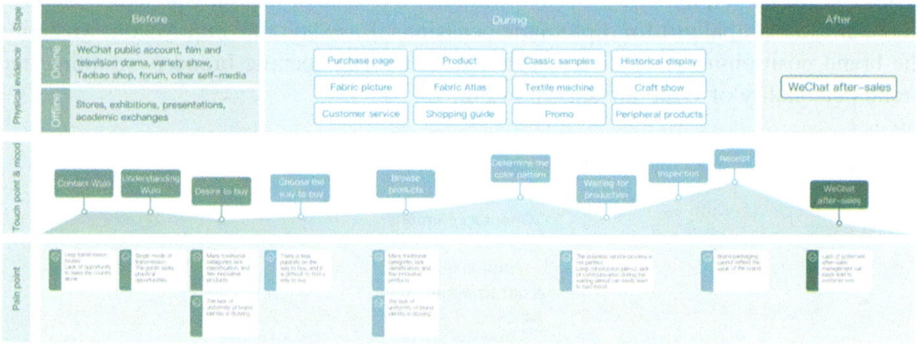

Fig. 9. User journey diagram of Wulo

1) The Problem of Wulo Brand Positioning and Brand Personality
Wulo currently lacks a clear brand positioning and brand personality. The current position of "precious silk fabric in ancient memory" is difficult to resonate with the public.

2) The Problem of Wulo Brand Performance Elements
Current brand identity system of Wulo is not complete. Traditional products of Wulo have many categories but lack classification. Wulo lacks innovative products. Current offline stores and Taobao stores of Wulo have relatively outdated visual design, and the service process needs to be improved urgently. Lack of systematic after-sales management can easily lead to customer loss.

3) The Problem of Wulo Brand Communication Method
There are fewer ways to spread Wulo. As one of the representatives of my country's intangible cultural heritage, Wulo has been invited to participate in various exhibitions and

CCTV variety shows, but most of them are currently passively participating. Moreover, most of the time Wulo appears at the same time as other intangible cultural heritage, does not highlight its own brand and culture, and does not actually build its own reputation.

The way to spread Wulo is relatively simple. Most of Wulo's current communication channels are video and graphic introductions, which lack interactivity. It is difficult for people to recognize the value of Wulo with abstract introductions.

3.3 Develop

In this stage, we should propose corresponding innovative solutions to the problems raised in the second stage. Designers should clarify the brand positioning and brand personality, systematically design performance elements of Wulo, improve service processes, and develop communication plans.

1) Brand Positioning and Brand Personality of Wulo

At this stage, the deepest impression of Wulo is that at the APEC meeting, leaders wore clothing made mainly of Wu Luo fabrics and Wu Luo silk scarves, which were given to foreign leaders as diplomatic gifts. Wulo created a classic product image in the hearts of the public, a national gift for diplomatic occasions. Based on this impression, we define the brand positioning of Wulo as "Oriental Gift". The specific brand positioning and brand personality of Wulo are shown in Fig. 10.

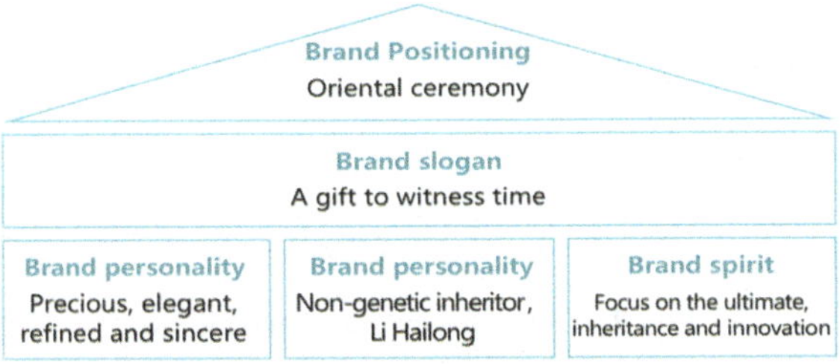

Fig. 10. Brand positioning and brand personality of Wulo

2) Brand Performance Elements of Wulo

After discussing with all stakeholders, the brand name is stipulated as "Wulo", and only "Wulo" is displayed on the logo of the trademark, and other words are not added to increase the burden of memory on the public. After numerous screenings, we determined the trademark pattern shown in the picture below (Fig. 11).

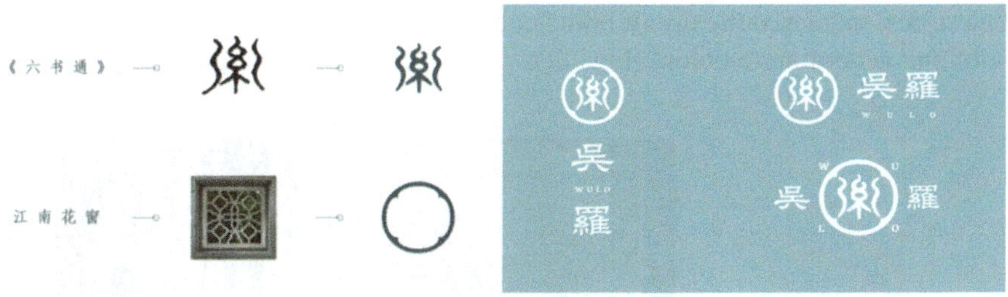

Fig. 11. The new logo of Wulo

The traditional Chinese color "Sky Green" is selected as the Wulo standard color, and the fresh oriental color is used to match it to highlight precious, elegant, refined, and sincere brand personality of Wulo (Fig. 12).

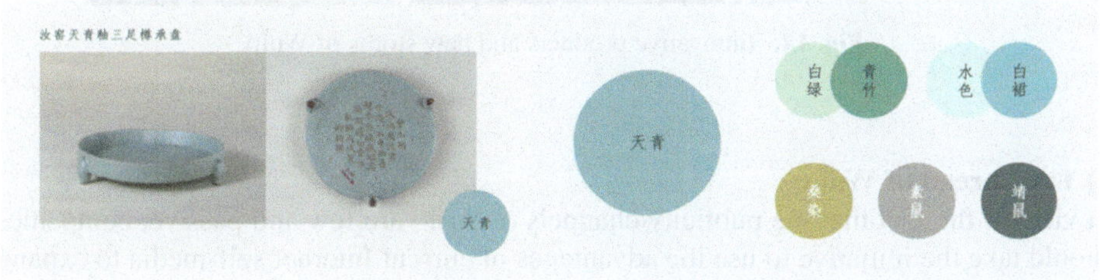

Fig. 12. The new logo of Wulo

We also redesigned the packaging according to the new trademark pattern and brand standard color (Fig. 13).

Fig. 13. New packaging of Wulo

At present, in addition to improving the brand identity system of Wulo, we also sort out existing products and organize them into a book. At the same time, we are trying to develop some new products. We also optimized the service process of offline stores and

Taobao stores, increased the use of brand colors, and reordered the placement of fabrics and clothing according to colors (Fig. 14).

Fig. 14. Innovative products and new stores of Wulo

3) The Spread of Wulo

In view of the fact that the publicity channels of Wulo are few and passive, companies should take the initiative to use the advantages of current Internet self-media to expand publicity, such as WeChat public account, Bilibili, Douyin, Weibo, Xiaohongshu and other high-quality self-media platforms.

In response to the single publicity of Wulo, companies can use pop-up stores (Fig. 15), brand museums, teaching experience halls and other channels to increase interaction and make brand image and intangible cultural heritage more popular.

Fig. 15. Concept design of Wulo pop-up store

3.4 Deliver

The designer needs to make a service prototype based on the design results of the third stage, estimate the workload and effects, and design a specific implementation plan.

Firstly, in the implementation phase of Wulo's brand image building should be to improve the existing brand service projects. Such as: brand identification system, sort out existing products and organize them into books, optimize store experience, optimize Taobao stores, and optimize WeChat public accounts.

Secondly, innovative products. On the one hand, innovate Wulo's own products, such as Wulo silk scarf, Wulo fan, Wulo cultural and creative surroundings, etc. On the other hand, it can cooperate with other brands to innovate, such as beauty brands, fragrance brands, sports brands, etc.

Thirdly, broaden sales channels. Wulo can add offline stores, pop-up stores, etc. Wulo can also list products on more high-quality e-commerce platforms online.

Fourthly, Wulo needs to increase transmission channels and enrich the transmission methods. The fourth step can be performed simultaneously with the third step.

4 Summary

For intangible cultural heritage brands, we should position the brand image as a "user-centered" design, not just an external visual design. With systematic design concepts, the brand image is deeply integrated into the entire service process, which can achieve higher commercial and cultural values. For intangible cultural heritage, shaping a complete brand image can increase the audience's awareness of intangible cultural heritage and attract more social attention. This is a good way to coordinate development with society.

References

1. Sun, L., Ren, Y.: Liaoning intangible heritage branding based on service design thinking. Packag. Eng. **41**(18) (2020)
2. Luo, S., Yi, H.: Model innovation driven by service design. Packag. Eng. **36**(12), 1–4 (2015)
3. Shang, H.: Theoretical research on brand image creation. Sichuan University (2006)
4. Deng, Z.: Research on the importance of service design thinking and tools in corporate image recognition. Ind. Des. (2) (2018)
5. Li, H.: Wulo's five literary talents, fine art has heirs. Jiangsu Silk (1), 7 (2016)
6. Li, D., Ming, X., Kong, F., Wang, X., Wang, P.: A preliminary study on service design research. Mech. Des. Res. (6) (2008)
7. Li, Y., Guo, Q.: Research on corporate brand image shaping from the perspective of service design. Des. Art Theor. (2017)
8. Cha, S.: Service Design Micro Diary. Publishing House of Electronics Industry, Beijing (2015)

Analyzing the Relationship Between Service Innovation and Customer Value Co-creation Intention: The Case of Mobile Banking

Nu Dieu Khue Ngo(✉), Youji Kohda, and Van-Nam Huynh

Japan Advanced Institute of Science and Technology, Nomi, Ishikawa, Japan
{khuengo,kohda,huynh}@jaist.ac.jp

Abstract. Customer value co-creation plays an important role in creating desired values that can better satisfy customer needs. However, there are few studies on the relationship between service innovation and customer value co-creation. This study aims to investigate the influence of mobile banking service innovation (concept, technology, and process) on customer value co-creation intention (information sharing and feedback), and how such effects are moderated by cultural values (power distance, collectivism, masculinity, and uncertainty avoidance). Data was collected via questionnaires sent to customers using the mobile banking service of banks in Vietnam. The participants were requested to response to each question using a five-point Likert scale. The data was then analyzed by exploratory factor analysis, reliability analysis, and hierarchical regression analysis. The findings could enable appropriate strategies to facilitate customer value co-creation to be identified.

Keywords: Service innovation · Mobile banking · Customer value co-creation · Cultural values · Banks in Vietnam

1 Introduction

The concept of customer value co-creation (CVCC) has recently gained considerable attention among practitioners and researchers. CVCC refers to the joint creation of values between companies and customers during the interaction process, in which customers play an active role in co-creating value with companies [1]. Zhang and Chen found that CVCC generates customization and enhances service capabilities [2]. Furthermore, CVCC was found to have a positive influence on customer satisfaction [3]. Hence, it is necessary to determine effective methods to promote CVCC.

Most prior studies have focused on central issues, processes, and outcomes of CVCC [4]. However, the literature on the antecedents of CVCC is still scarce. Such antecedents were found to be company support in travel services [5], and physical environments in the restaurant industry [6]. Interestingly, Yen et al. found that innovativeness and CVCC behaviors were positively correlated in the food and beverage industry [7]. Innovation characteristics can motivate customers to engage in value co-creation, which increases

C. Leitner et al. (Eds.): AHFE 2021, LNNS 266, pp. 456–466, 2021.
https://doi.org/10.1007/978-3-030-80840-2_53

their adoption of technology-based services [8]. Innovation is predicted as an important antecedent of CVCC behaviors. In the banking context, very little research dedicated to CVCC toward service innovation is available, especially in the case of mobile banking.

The innovation diffusion theory indicates that different people tend to react differently to innovativeness because of their different perceptions [9]. Therefore, customers tend to have different perceptions of innovation, which leads to different intentions in co-creating value with service providers. Prior studies related to mobile banking have mainly focused on factors affecting the adoption of services with the moderating effects of demographics such as age and gender [10, 11], or country culture [12]. There is very little research into cultural values as moderators at the individual level.

This study aims to empirically examine the relationship between mobile banking service innovation and CVCC intention (CVCCI). We also test the moderating effects of individual cultural values on that relationship. Empirical evidence was collected from customers using mobile banking services at banks in Vietnam. Theoretically, this study extends the literature on the linkage between service innovation and CVCC under consideration of individual cultural values. Practically, our findings could guide bank managers to build appropriate innovation strategies according to customer segments based on their cultural values, which may lead to more active CVCC behaviors.

2 Literature Review

2.1 Service Innovation and Customer Value Co-creation Intention

Innovation refers to the adoption of new concepts, processes, or technologies to provide customers superior products or services to satisfy the changing customer needs, and thereby sustain organizational competitive advantages [13]. The greater perceived service innovativeness results in higher customer perceived values and then activates customers' motivation and participation in value co-creation [14]. Yeh suggested that innovativeness promotes customers to interact with service providers and exchange information and share knowledge, subsequently affects CVCC [15]. Morosan asserted that hotel guests intended to enthusiastically engage in the value co-creation process via mobile devices when perceiving the innovativeness of personalized hotel services [16]. Moreover, the development of mobile technology or online support tools may help customers easily engage in the co-creation process, which in turn motivates customers to co-create values. Yi and Cong proposed a scale of CVCC including information seeking, information sharing, advocacy, feedback, helping, responsible behaviors, and tolerance [17].

Most previous studies have examined CVCC in the aspects of word of mouth or recommendation to others. This study focuses on customers' intentions to share information and feedback that should help companies to develop or renew their services from customers' perspectives. We predict that when customers perceive that innovative services can meet their needs; they will be willing to share information and give feedback with service providers. This study considers three aspects of service innovation: concept innovation (CI), technology innovation (TI), and process innovation (PI). CVCCI focuses on information sharing (IS) and feedback (FE) intentions of customers using mobile banking services. We propose that perceived mobile banking service innovation is positively related to CVCCI, particularly:

H1: CI is positively associated with CVCCI.

H2: TI is positively associated with CVCCI.

H3: PI is positively associated with CVCCI.

2.2 Moderating Effects of Cultural Values

White revealed that cultural values can predict human behaviors or behavioral intentions [18]. Similarly, Fu and Elliott stated that customer perceptions of service innovativeness reflect their values and beliefs, which then affect their attitudes and behaviors [19]. As mobile banking service is an innovative self-service technology, customer perceptions about mobile banking innovation may be moderated by the personal characteristics of customers. As mentioned above, the number of moderator studies related to mobile banking service has mainly focused on demographics or country culture. Four country cultural moderators (power distance, individualism, masculinity, and uncertainty avoidance) were employed to explain the influence of factors on attitude toward using and intention to use technology-based self-services [20]. The adoption intention towards new services was found to be influenced by CVCC [21]; it is assumed that those cultural variables may also moderate the relationship between mobile banking service innovation and CVCCI.

In this study, we look at the moderating effect of cultural values dimensions at the individual level. Yoo et al. developed a cultural values scale at the individual level based on the concepts of based on Hofstede's Cultural Dimensions Theory (2001) [22]. Based on the cultural theory, people in high power distance culture rely on support from powerful members; therefore, predictors related to customer supports have more importance in high-power distance culture while predictors associated with own capabilities lose importance. The term 'process innovation' in this paper refers to the support processes of banks; thus, hypothesis 4 is hypothesized. Additionally, predictors concerning personal needs or independent actions have more importance in individualistic culture while predictors concerning group needs or interdependent actions have more importance in collectivistic culture, which leads to hypothesis 5. Masculine orientation prefers tough values while feminine orientation appreciates tender values. Hence, we hypothesized hypothesis 6. In high-uncertainty avoidance culture, it was assumed that predictors capable of reducing risks such as non-provider-related recommendations and personal experiences are more important. Technology acceptance and interaction with service providers are relevant to customers in low-uncertainty avoidance culture. Hence, hypothesis 7 was formed.

H4: The influence of PI on CVCCI is stronger in high-power distance culture.

H5: The influence of PI on CVCCI is stronger in collectivistic culture.

H6: The influence of (a) CI and (b) TI on CVCCI are stronger in masculine culture; the influence of (c) PI on CVCCI is stronger in feminine culture.

H7: The influence of (a) CI, (b) TI, and (c) PI on CVCCI are stronger in low-uncertainty avoidance culture.

3 Methodology

3.1 Sample and Data Collection

To test our research questions, data was collected via questionnaire from customers using moblike banking services of banks in Vietnam. We had one academic expert review the English original version of the questionnaire. It was then translated into Vietnamese and pre-tested on five mobile banking users to ensure correct interpretation of all questions. A revised Vietnamese questionnaire was developed based on the test participants' feedback.

Using a convenience sampling technique, the questionnaires were distributed to 120 mobile banking users in Vietnam via email and social media networks. Five employees at several Banks in Vietnam asked customers using mobile banking services to voluntarily answer questionnaires. We assured the confidentiality of all collected information and provided the participants and supporters the results on requests. Seventy-nine usable responses were obtained, with a response rate of 65.83%.

3.2 Measures

All scales included in this study have been validated in previous research. Service innovation scales (CI, TI, and PI) were adopted from [23] to determine customer perceptions of mobile banking service innovation. To measure CVCCI, the scales of IS and FE were drawn from [17]. Four dimensions of individual cultural values including power distance (PD), collectivism (CO), masculinity (MA), and uncertainty avoidance (UA) were obtained from [22]. Participants were requested to response to questions using a five-point Likert scale from 1 (strongly disagree) to 5 (strongly agree).

Table 1. Constructs and measurement items.

Constructs	Measurement items
Service innovation	
CI	CI1: My mobile banking service offers innovative combinations
	CI2: My mobile banking service provides customized options
	CI3: My mobile banking service has different concepts as compared to previous services
	CI4: My mobile banking service brings different service experiences as compared to previous services
	CI5: The concept of my mobile banking service is different when compared to competitors' services
TI	TI1: My mobile banking service updates new features
	TI2: My mobile banking service supplier adopts modern infrastructures

(*continued*)

Table 1. *(continued)*

Constructs	Measurement items
	TI3: My mobile banking service supplier is always pioneering new technology in the market
	TI4: My mobile banking service is developed based on the latest technology applications
	TI5: My mobile banking service supplier always make effort to improve service quality
PI	PI1: My mobile banking service provides online support processes
	PI2: My mobile banking service applies automated service options
	PI3: My mobile banking service supplier applies new means of media for interaction
	PI4: My mobile banking service supplier has promotional campaigns at events
	PI5: My mobile banking service supplier has fast and simple support services via call center
CVCC	
IS	IS1: I will clearly explain what I want my mobile banking service supplier to offer
	IS2: I will give the proper information
	IS3: I will provide necessary information for mobile banking service to perform efficiently
	IS4: I will answer all questions regarding mobile banking service
FE	FE1: If I have a useful idea to improve service, I will tell my mobile banking service supplier
	FE2: If I receive good service from my mobile banking service supplier, I will comment on it
	FE3: If I suffer a problem, I will tell my mobile banking service supplier
Cultural values	
PD	PD1: People in higher positions should make most decisions without consulting people in lower positions
	PD2: People in higher positions should not ask the opinions of people in lower positions too frequently
	PD3: People in higher positions should avoid social interaction with people in lower positions
	PD4: People in lower positions should not disagree with decisions by people in higher positions

(continued)

Table 1. *(continued)*

Constructs	Measurement items
	PD5: People in higher positions should not delegate important tasks to people in lower positions
CO	CO1: Individuals should sacrifice self-interest for the group
	CO2: Individuals should stick with the group even through difficulties
	CO3: Group welfare is more important than individual rewards
	CO4: Group success is more important than individual success
	CO5: Individuals should only pursue their goals after considering the welfare of the group
	CO6: Group loyalty should be encouraged even if individual goals suffer
MA	MA1: It is more important for men to have a professional career than it is for women
	MA2: Men usually solve problems with logical analysis; women usually solve problems with intuition
	MA3: Solving difficult problems usually requires an active, forcible approach, which is typical of men
	MA4: There are some jobs that a man can always do better than a woman
UA	UA1: It is important to have detailed instructions so that I always know what I am expected to do
	UA2: It is important to closely follow instructions and procedures
	UA3: Rules and regulations are important because they inform me of what is expected of me
	UA4: Standardized work procedures are helpful
	UA5: Instructions for operations are important

4 Empirical Results

4.1 Exploratory Factor Analysis (EFA)

EFA was employed to determine the key factors from the 42 initial items (see Table 1). The Kaiser-Meyer-Olkin (KMO) measure of sampling adequacy is $0.715 > 0.5$, which proves the sample was adequate for factor analysis. In addition, P-value of Bartlett's test of sphericity is $0.000 < 0.01$, there is a significant interrelationship between variables. Seven key components whose items have factor loadings exceeding 0.50 were extracted from EFA (see Table 2); these explained 65.303% of the variance. Based on these results, the hypotheses are reformulated as follows:

H1: Concept and technology innovation (CT) is positively associated with CVCCI represented by customer information sharing and feedback intentions (SF).

Table 2. Results of EFA.

Components	Items	Factor loadings	Eigenvalue	Explained variance (%)	Cumulative explained variance (%)	Cronbach's alpha coefficients
CT	CI1, CI3, CI5, TI1, TI2, TI3, TI4, TI5	0.530–0.883	11.331	26.978	26.978	0.913
MA	PD2, PD4, MA1, MA2, MA3, MA4	0.597–0.804	5.311	12.646	39.624	0.681
UA	UA1, UA2, UA3, UA4, UA5	0.734–0.856	3.587	8.541	48.165	0.888
CO	CO1, CO2, CO3, CO4, CO5	0.511–0.835	2.392	5.695	53.860	0.839
SF	IS2,IS3,IS4,FE2	0.517–0.820	1.859	4.425	58.286	0.807
PD	PD1, PD3, PD5	0.631–0.739	1.608	3.828	62.114	0.655
PI	PI1, PI2, PI3, PI4	0.510–0.802	1.339	3.189	65.303	0.800

H2: PI is positively associated with SF.

H3: The influence of PI on SF is stronger in high-power distance culture.

H4: The influence of PI on SF is stronger in collectivistic culture.

H5: The influence of (a) CT on SF is stronger in masculine culture; the influence of (b) PI on SF is stronger in feminine culture.

H6: The influence of (a) CT and (b) PI on SF are stronger in low-uncertainty avoidance culture.

4.2 Reliability Analysis

Cronbach's alpha reliability analysis was employed to test the consistency of all items of each factor extracted from EFA. All Cronbach's alpha coefficients of seven scales are higher than 0.60 [24], which suggests sufficient validity of all scales (see Table 2).

4.3 Hierarchical Regression

We employed a hierarchical regression to separately examine two models: the first model estimates the influence of service innovation (CT and PI) on SF, the second model adds cultural variables (PD, CO, MA, UA), and interactive variables between service innovation variables and cultural variables (CT × PD, CT × CO, CT × MA, CT × UA, PI × PD, PI × CO, PI × MA, PI × UA) as independent variables to test the moderation by cultural variables on the relationship between CT and PI and SF. We used mean-centered predictor variables to reduce the possibility of multicollinearity. H1, H2, H5(a) were supported, but the results show a contrast with H6(a) (see Table 3). Based on the results, CT and PI have a significant positive influence on SF. MA and UA positively moderate the relationship between CT and SF at a significant level.

Table 3. Results for hierarchical regression.

	Model 1	Model 2	Hypotheses
CT	0.344*	0.135	H1 supported
PI	0.224**	0.149	H2 supported
PD		0.019	
CO		0.436*	
MA		−0.215**	
UA		0.388*	
CT × PD		−0.198	
CT × CO		−0.221	
CT × MA		0.441**	H5(a) supported
CT × UA		0.304***	Contrast with H6(a)
PI × PD		0.154	H3 not supported
PI × CO		0.125	H4 not supported
PI × MA		−0.158	H5(b) not supported
PI × UA		−0.156	H6(b) not supported
R square	0.554	0.762	
Adjusted R square	0.307	0.581	
F	16.851*	6.340*	

*, **, *** significant at the 0.01, 0.05, 0.1 levels, respectively.

5 Discussion and Conclusion

5.1 Discussion

The results demonstrate that mobile banking service innovation in concept, technology, and process positively influence CVCCI (SF). Yen et al. also found the CVCC behaviors were enhanced when customers perceived coffee shops to have innovative menus, technologies, experiences, and promotions [7]. This study also explored that masculinity culture positively moderated the influence of CT on SF. It is because males tend to prefer new designs or technologies than females do [25]. The influence of CT was found to be stronger in high-uncertainty avoidance culture. This can be explained by the fact that advanced technology can improve detection of mistakes, make financial transactions more secure, and thereby help banks to reduce risks [26].

5.2 Conclusion and Limitation

This research aims to investigate the relationship between service innovation in concept, technology and process and SF in the case of mobile banking. Particularly, CT positively affect SF more than PI does. In addition, we also test the moderating effects of cultural dimensions on the relationship between mobile banking service innovation and SF. The result revealed that MA and UA positively moderate the association between CT and SF. Based on these findings, bank managers can tailor different innovation strategies appropriate to each segment of customers to motivate their CVCCI.

This study has several limitations. First, this study was conducted in the case of a specific service, namely mobile banking. Second, the survey was conducted using a small sample of respondents. Future research should broaden the survey context and and increase the number of respondents to assure the reliability and universality of the conclusions. Lastly, there are other CVCC behaviors and moderating variables that have not been investigated. Further research is needed to comprehensively investigate other activities of CVCC under the influence of different moderators.

References

1. Grönroos, C.: Creating a relationship dialogue: communication, interaction and value. Mark. Rev. **1**(1), 5–14 (2000)
2. Zhang, X., Chen, R.: Examining the mechanism of the value co-creation with customers. Int. J. Prod. Econ. **116**(2), 242–250 (2008)
3. Vega-Vazquez, M., Revilla-Camacho, M.Á., Cossío-Silva, F.J.: The value co-creation process as a determinant of customer satisfaction. Manag. Decis. **51**(10), 1945–1953 (2013)

4. Shamim, A., Ghazali, Z.: A conceptual model for developing customer value co-creation behaviour in retailing. Glob. Bus. Manag. Res. **6**(3), 185–196 (2014)
5. Grissemann, U.S., Stokburger-Sauer, N.E.: Customer co-creation of travel services: the role of company support and customer satisfaction with the co-creation performance. Tour. Manage. **33**(6), 1483–1492 (2012)
6. Im, J., Qu, H.: Drivers and resources of customer co-creation: a scenario-based case in the restaurant industry. Int. J. Hosp. Manag. **64**, 31–40 (2017)
7. Yen, C.H., Teng, H.Y., Tzeng, J.C.: Innovativeness and customer value co-creation behaviors: mediating role of customer engagement. Int. J. Hosp. Manag. **88**, 102514 (2020)
8. Heidenreich, S., Handrich, M.: Adoption of technology-based services: the role of customers' willingness to co-create. J. Serv. Manag. **26**(1), 44–71 (2015)
9. Rogers, E.M.: Diffusion of Innovations. Simon and Schuster, New York (2010)
10. Karjaluoto, H., Riquelme, H.E., Rios, R.E.: The moderating effect of gender in the adoption of mobile banking. Int. J. Bank Market. **28**(5), 328–341 (2010)
11. Chawla, D., Joshi, H.: The moderating effect of demographic variables on mobile banking adoption: an empirical investigation. Glob. Bus. Rev. **19**(3), 90–113 (2018)
12. Goularte, A.D.C., Zilber, S.N.: The moderating role of cultural factors in the adoption of mobile banking in Brazil. Int. J. Innov. Sci. **11**(1), 63–81 (2019)
13. Tajeddini, K., Trueman, M.: Perceptions of innovativeness among Iranian hotel managers. J. Hosp. Tour. Technol. **5**(1), 62–77 (2014)
14. Hoyer, W.D., Chandy, R., Dorotic, M., Krafft, M., Singh, S.S.: Consumer cocreation in new product development. J. Serv. Res. **13**(3), 283–296 (2010)
15. Yeh, Y.P.: Corporate social responsibility and service innovation on customer loyalty. Int. J. Bank Mark. **33**(6), 823–839 (2015)
16. Morosan, C.: An empirical analysis of intentions to cocreate value in hotels using mobile devices. J. Hosp. Tour. Res. **42**(4), 528–562 (2018)
17. Yi, Y., Gong, T.: Customer value co-creation behavior: scale development and validation. J. Bus. Res. **66**(9), 1279–1284 (2013)
18. White, C.J.: Culture, emotions and behavioural intentions: implications for tourism research and practice. Curr. Issue Tour. **8**(6), 510–531 (2005)
19. Fu, F.Q., Elliott, M.T.: The moderating effect of perceived product innovativeness and product knowledge on new product adoption: an integrated model. J. Market. Theor. Pract. **21**(3), 257–272 (2013)
20. Blut, M., Wang, C., Schoefer, K.: Factors influencing the acceptance of self-service technologies: a meta-analysis. J. Serv. Res. **19**(4), 396–416 (2016)
21. Sarmah, B., Rahman, Z., Kamboj, S.: Customer co-creation and adoption intention towards newly developed services: an empirical study. Int. J. Cult. Tour. Hosp. Res. **11**(3), 372–391 (2017)
22. Yoo, B., Donthu, N., Lenartowicz, T.: Measuring Hofstede's five dimensions of cultural values at the individual level: development and validation of CVSCALE. J. Int. Consum. Mark. **23**(3–4), 193–210 (2011)
23. Mahmoud, M.A., Hinson, R.E., Anim, P.A.: Service innovation and customer satisfaction: the role of customer value creation. Eur. J. Innov. Manag. **21**(3), 402–422 (2018)

24. Hair, J.F., Black, W.C., Babin, B.J., Anderson, R.E., Tatham, R.L.: Multivariate Data Analysis, vol. 6. Pearson Prentice Hall, Upper Saddle River (2006)
25. Goswami, A., Dutta, S.: Gender differences in technology usage—a literature review. Open J. Bus. Manag. **4**(1), 51–59 (2015)
26. Singh, Y.: Three ways technology can make banks more resilient (2019). https://www.bankingexchange.com/risk-management/item/8048-three-ways-technology-can-make-banks-more-resilient

Creating Value in Communities, and Society

Co-enhancing Skills of Social Entrepreneurs: A Possible Application of Mobile Social Networks

Karishma Zaman Katha(✉), Youji Kohda, and Eunyoung Kim

School of Knowledge Science, Japan Advanced Institute of Science and Technology, 1-1 Asahidai, Nomi 923-1292, Ishikawa, Japan
{kohda,kim}@jaist.ac.jp

Abstract. The main purpose of this research is to develop a framework for co-enhancing the skills of social entrepreneurs, by deeply investigating the roles played by diverse people while participating in a social business design lab. This research adopts the case study research method and selects the Yunus Centre Social Business Design Lab in Bangladesh. The data were collected from secondary sources, including archives of social entrepreneurs from Yunus Centre Social Business Design Lab in Bangladesh, and websites of the Centre. In addition, a total of 30 interviews were conducted. We identified 25 different skills of social entrepreneurs. The results of this research show that diverse people who participated in the social business design lab played different roles to support the social entrepreneurs.

Keywords: Co-creation of value · Enhancing skills · Mobile social networks · Social entrepreneurs

1 Necessary to Co-enhance the Skills of Social Entrepreneurs in Social Business

Social business is a type of business whose main purpose is purely to achieve one or more social objectives. Diverse actors are actively involved in the process of social business, and the social business model is evolving. Previously at the Grameen Social Business Lab, the meeting was presided over by Prof. Yunus. Currently, employees from Grameen Trust have also participated in the Grameen Social Business Design Lab [1]. Social entrepreneurs required various skills, which included communication, management, teamworking, learning, fundraising, reporting and evaluation, in order to solve social problems in the society through social business [2]. *However, in social business, the roles of diverse actors involve either judging the projects or financing the projects. Few studies have focused on how the social business platform could be used for enhancing the skills of social entrepreneurs.* Therefore, this research addresses the above issue.

In addition, service-dominant (S-D) logic is a perspective of new thought on the market and marketing [3]; value co-creation is one of its core concepts. In S-D logic, value

C. Leitner et al. (Eds.): AHFE 2021, LNNS 266, pp. 469–477, 2021.
https://doi.org/10.1007/978-3-030-80840-2_54

is co-created by the multiple actors through the integration of resources from different sources. Service providers, firms, customers, and other social community members, such as beneficiaries, business organizations, government organizations, public and private organizations, as well as society as a whole, are considered as actors in S-D logic. Most scholars have defined value co-creation as the joint creation of value through the interaction between customers and service providers [3]. In this research, the term "*co-enhancing*" has been used synonymously with "*co-creation*", referring to enhancing the skills of social entrepreneurs through the interactions among diverse people who participate in the social business process.

Facebook, Line, WeChat, WhatsApp and Viber are the well-known mobile social networks (MSN) used by diverse people in society. Currently, MSN-based platforms are mainly applied for enhancing the language skills, communication and collaboration skills of students and teachers, as well as for improving teachers' professional development [4]. Specifically, Facebook Group is used as a learning management system by sharing information among students [5]. Similarly, social entrepreneurs also use MSN for different purposes. However, research efforts are needed regarding the application of MSN for co-enhancing the skills of social entrepreneurs in social business. The rest of the paper is organized in the following order: Sect. 2 presents the research objectives, and Sect. 3 describes research methodology. Section 4 provides the results and discussion, followed by Sect. 5, which describes a framework for the possible application of MSNs for co-enhancing the skills of social entrepreneurs in social business. Finally, this research concludes with the theoretical implications, in Sect. 6.

2 Research Objectives and Questions

The primary objective of this research is to develop a framework for co-enhancing the skills of social entrepreneurs, by deeply investigating the roles played by different actors while participating in the Yunus Centre Social Business Design Lab in Bangladesh. The secondary objective of this research is to propose a framework of how mobile social networks (MSNs) could be applied to enhance the skills of social entrepreneurs. To achieve the above research objectives, we investigate the following research questions (RQ): **RQ1:** What skills do social entrepreneurs have? **RQ2:** What are the roles played by different participants in the social business de sign lab? **SRQ3:** How can people's different roles co-enhance the skills of social entrepreneurs?

3 Research Method

3.1 Data Collection

To achieve the research objectives, the case study method is adopted for this research. The Yunus Centre Social Business Design Lab (YCSBDL) in Bangladesh was selected as the case for this research. The data were collected from secondary sources, which include the website of YCSBDL. In addition, a total of 33 social entrepreneurs were interviewed. In this paper, the interviewees are denoted as SE1, SE2, ..., SE30. Furthermore, we retrieved 226 social business projects from the archive of YCSBDL [6], in order to identify the diverse skills of social entrepreneurs (see Table 1).

3.2 Data Coding

We found broadly 12 broad categories of social business projects as shown in Table 1. We selected 20 social business projects in each category from a larger number, except the environment category, which contained only 6 projects. The corresponding author examined all 226 projects to identify what kinds of skills the social entrepreneurs had. Each project provided a PowerPoint/pdf file which explains the details of each social entrepreneur and his/her social business project, as well as his/her skills and experience. As a result, this research identified 25 skills of social entrepreneurs. In addition, the interviews were analyzed thematically.

Table 1. Categories of social business projects

Categories of social business projects	Selected number
Agriculture and forestry	20
Fisheries and animal husbandry	20
Factories	20
Microbusiness	20
Healthcare and nutrition	20
Fashion and fabrics	20
Power and infrastructure	20
Handicraft	20
Service and transportation	20
Shopkeeping	20
Information technology	20
Environment	6
Total	226

4 Results

First, we present the different skills of social entrepreneurs. Second, we indicate the roles of diverse people who participated in social business.

4.1 Social Business

We identified 25 different skills of social entrepreneurs, which include cultivation skill, farming skill, innovating new product skill, crafting skill, marketing skill, healthcare skill, repairing skill and environmental skill. Table 2 shows in detail the skills of social entrepreneurs.

4.1.1 Cultivation Skills

The result shows that social entrepreneurs have various agricultural product cultivation skills, which include rice cultivation, betel leaf cultivation, banana cultivation, lemon cultivation, and other food-related cultivation skills.

4.1.2 Farming Skills

The result shows that social entrepreneurs have fishery and animal husbandry skills, which include farming animal husbandry for dairy and meat-supplying purposes.

4.1.3 New Product-Making Skills

Social entrepreneurs have diverse new product development skills, which include furniture-making skill, bread-making skill, ice-cream-making skill, and other new product-making skills.

4.1.4 Marketing Skills

Social entrepreneurs have different types of microbusinesses. The results show that social entrepreneurs have shops related to garments, groceries, stationery, electronics and medicine.

4.1.5 Crafting Skills

The result shows that social entrepreneurs have diverse crafting skills. Some social entrepreneurs have artistic embroidery skills; others have home crafting skills. Some of the social entrepreneurs have jewelry skills.

4.1.6 Healthcare and Nutritional Skills

Social entrepreneurs provide healthcare and nutritional skills, in terms of providing first-aid medical services and nutritional services for the rural people. In addition, social entrepreneurs have their own medical stores for selling medicines and other health products, as well as nutrition-related products and services.

4.1.7 IT Skills

Social entrepreneurs have different IT skills, which include repairing mobile phones, smartphones, computers and other electronic devices. In addition, some social entrepreneurs have computer-related skills, whereby they train the local people to use computers for different purposes.

4.1.8 Environment-Friendly Business Innovation Skill

The results of show that social entrepreneurs have unique environment-friendly innovative business design skills, which include the design of mobile plant nurseries and aquariums, solid-waste management skill, water purification skill, and finally, skills related to making bricks from fly ash.

Table 2. Skills of social entrepreneurs

Categories of social business projects	Identified skills from social business projects presented by social entrepreneurs	No. of selected projects
Agriculture and forestry	Cultivation skill	19
	Gardening skill	1
Fisheries and animal husbandry	Dairy farming skill	19
	Poultry farming skill	1
Factories	Carpentry skill	16
	Baking skill	3
	Ice-cream making	1
Microbusiness	Microbusiness skill	20
Healthcare and nutrition	Healthcare skill	16
	Nutritional skill	4
Fashion and fabrics	Garment-making skill	15
	Knitting skill	5
Power and infrastructure	Engineering skill	20
Handicraft	Artistic embroidery skill	7
	Skill related to using bamboo for different purposes	6
	Home-crafting skill	5
	Jewellery-making skill	2
Service and transportation	Tailoring skill	17
	Telecom marketing skill	2
	Steel shop marketing skill	1
Shopkeeping	Marketing skill	20
Information technology	Mobile phones repairing skill	10
	Electronics repairing skill	5
	Computer repairing skill	5
Environment	Environment-friendly business innovation skill	6
Total		226

4.2 Roles Played by Diverse People Who Participated in Social Business

The following are the possible roles played by these diverse people.

4.2.1 Role of Professor Muhammad Yunus

The Nobel Laureate Professor Muhammad Yunus is the pioneer of social business. He is also the pioneer of the Yunus Centre Social Business Design Lab (YCSBDL) in Bangladesh. Hence, Muhammad Yunus has the leading role in YCSBDL. But now, social business design labs are organized throughout- Bangladesh. Therefore, employees from Grameen Trust and other Grameen-related organizations have participated in the social business design lab.

4.2.2 Role of Employees from Grameen Trust and Other Grameen-Related Organizations

The employees from Grameen Trust and other Grameen-related organizations play the most important role in the overall organization of YCSBDL, management of funds, and supports for social entrepreneurs. First, they organize the social business design lab. Second, they find out who are the new social entrepreneurs, help them to prepare their business presentation, and generally monitor their business. Third, they preside over the social business design lab. Fourth, they report the findings from each group in the social business design lab meeting. Least but not least, employees from Grameen Trust and other Grameen-related organizations have overall responsibility for all the activities of the social business design lab.

4.2.3 Role of Investors

The investors play one of the most important roles by investing in social business. In addition, they also evaluate the social business projects in the group discussion.

4.2.4 Role of Academics

The role of academics is very important in social business. First, academics conduct social business-related research as well as teach in their designated universities. Second, they also evaluate the social business projects while participating in the Grameen social business design lab.

4.2.5 Role of Civil Society

The civil society plays a highly significant role in social business; it is considered to consist of the respectable people in society. Civil society provides a way to engage social entrepreneurs with other organizations. More importantly, civil society empowers the social entrepreneurs by creating awareness of social issues and enabling them to develop new development programmes to meet their diverse needs.

4.2.6 Role of Donor Community

The donor community plays an important role in the process of social business, which it supports by financing the social business projects. In addition, the donor community provides policy formulations for social business, and also works as a connector for forming relationships with other organizations and people in society, for the development of the social entrepreneurs.

5 A Possible Application of Mobile Social Networks (MSN) for Co-enhancing the Skills of Social Entrepreneurs

In this section, we propose a framework for co-enhancing skills via the application of mobile social networks (MSN). The results of this research show that social entrepreneurs have different skills. The results also show that diverse people participate in the social business process, and that they play different roles in it. The results from interview data show that most of the social entrepreneurs used Facebook, IMO, and Viber as their MSN. The results also show that most of the social entrepreneurs have shown very positive attitude towards use of MSN for enhancing their skills. Some of the social entrepreneurs indicated that:

> *I use new technologies for learning purposes. I am already using Facebook for my business purposes. I use Facebook for contacting with my customers and sending them new designs. In addition, I learn new knowledge by using Facebook. So, I am very willing to learn new knowledge by applying Facebook (SE26, SE28, and SE29).*

The results of this paper show that diverse people from different backgrounds participate in the social business design lab. The results also state that they paly diverse roles in social business design lab. The participants are highly respectable people in the society. Therefore, these highly respectable people have the capability to motivate the young social entrepreneurs. So, as a special management strategy, our framework of co-enhancing skills of social entrepreneurs via MSN could motivate the young social entrepreneurs.

Currently, MSN-based platforms are mainly employed for enhancing the language skills, communication and collaboration skills of students and teachers, as well as for improving teachers' professional development [4, 5]. According to Internet World Stats [7], there are 33.713 million Facebook users in Bangladesh, as of February 2020. In addition, IMO, WhatsApp and Viber are also popular among Bangladeshis for instant messaging and calling purposes. In addition, our results show that Facebook, IMO and Viber—are the MSN used by the social entrepreneurs. Therefore, we propose a framework for how MSN could be used to co-enhance the skills of social entrepreneurs in social business.

Figure 1 shows the design of a framework for co-enhancing the skills of social entrepreneurs via MSN. First, an MSN group will be created that will be considered as a platform for co-enhancing the skills of social entrepreneurs. Second, social entrepreneurs will search that MSN platform for different types of skills they need. Thirdly, participants of the social business lab will share knowledge to provide the required skills. For example, if any social entrepreneurs seek the skill of "*how to write the letter in English*" on the MSN platform, then they will look for participants who are classified as having "*Language skill*".

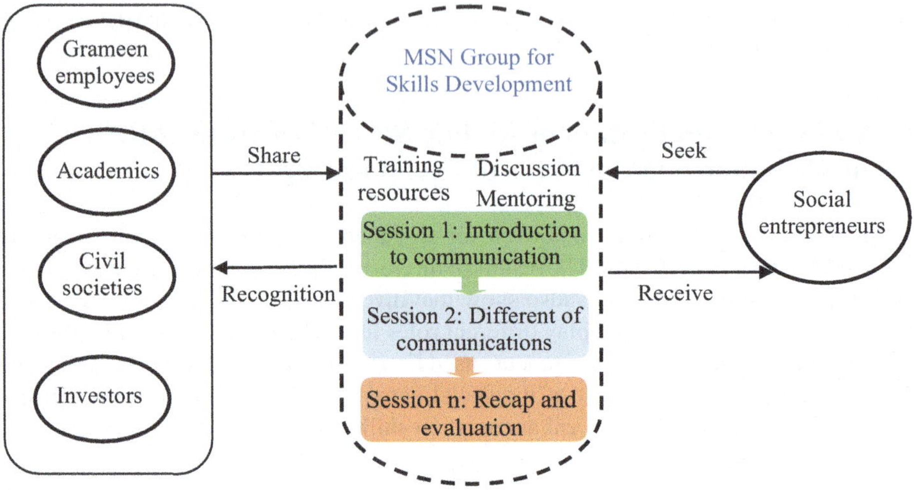

Fig. 1. How the skills of social entrepreneurs could be co-enhanced via MSN

In addition, "employees from Grameen Trust and other Grameen-related organizations" could play the most important roles for co-enhancing the skills of social entrepreneurs via an MSN group. Furthermore, "employees from Grameen Trust and other Grameen-related organizations" could organize skills co-enhancement programmes by inviting the diverse participants in the social business design lab. Through these methods, the MSN group can be used as a platform for enhancing the skills of social entrepreneurs.

6 Conclusion and Future Research Directions

In this research, we identified different skills of social entrepreneurs. We found that diverse people participated in the social business design lab; we also indicated the roles that diverse people played in the social business process. Finally, we proposed a framework for co-enhancing the skills of social entrepreneurs via MSN. By applying the framework, social entrepreneurs would be able to express their needs for the required skills on MSN. Also, the participants of the social business design lab would be able to express knowledge related to the required skills. Through applying the framework, the diverse participants of the social business lab would be able to co-enhance the skills of social entrepreneurs.

This research contributes to both S-D logic and social entrepreneurship research. First, we identified different skills of social entrepreneurs by analysing the social business projects retrieved from the archives of the Yunus Centre Social Business Design Lab in Bangladesh. Secondly, we identified diverse roles played by the participants in the social business design lab. By conceptualizing the above findings, the proposed framework for co-enhancing skills of social entrepreneurs broadens the theory of S-D logic in the context of social entrepreneurship. To the best of our knowledge, very few researchers have applied the theory of S-D logic to the theory of social entrepreneurship.

Nevertheless, this paper has limitations: (1) the framework for co-enhancing the skills of social entrepreneurs was proposed based on the results of this research. Therefore, future research should be carried out to identify the possible training programs offered by the social enterprises so that we can design the skill enhancement training programs which will be delivered via MSN (2) We have not yet identified how the enhanced skills of social entrepreneurs affect their performance; thus, this should be a topic of future research.

References

1. Yunus, M., Weber, K.: Building Social Business: The New Kind of Capitalism that Serves Humanity's Most Pressing Needs. Public Affairs, New York (2010)
2. Chell, E.: Review of skill and the entrepreneurial process. Int. J. Ent. Behav. Res. **19**(1), 6–31 (2013)
3. Vargo, S.L., Lusch, R.F.: Institutions and axioms: an extension and update of service-dominant logic. J. Acad. Mark. Sci. **44**(1), 5–23 (2015). https://doi.org/10.1007/s11747-015-0456-3
4. Cansoy, R.: Teachers' professional development: the case of WhatsApp. J. Educ. Learn. **6**(4), 285–293 (2017)
5. Siddike, M.A.K., Islam, M.S., Banna, H.: Use of social networking sites: Facebook group as a learning management system. Knowl. Manag. E-Learn. **7**(2), 232–249 (2015)
6. Social Business Pedia. Design lab projects (2020). http://socialbusinesspedia.com/designlab/projects
7. Internet World Stats. Bangladesh (2020). https://www.internetworldstats.com/asia.htm#bd

Integrating Real-World and Virtual Experiences Through eSports-Type 'Cheering'

Hisayuki Kunigita(✉) and Youji Kohda

Graduate School of Advanced Science and Technology, Japan Advanced Institute of Science and Technology, 1-1 Asahidai, Nomi, Ishikawa 923-1292, Japan
{hisayuki.kunigita,kohda}@jaist.ac.jp

Abstract. Due to COVID-19, experience-based services such as travel and dining have been scaled down significantly in 2020–2021. Such services have been shifting to online formats in an effort to integrate real-world and virtual experiences, though it is difficult for them to charge the same prices as those of real-world services. Online 'cheering' may be a potential way to enhance online experiences and bridge the real world and virtual world. Online spectating services such as Twitch have successfully incorporated cheering functions which viewers can use at low costs. This study proposes an ecosystem for online experience-based services with cheering functions such as Twitch and analyzes cheering on Twitch to examine this ecosystem. Finally, this study reveals the value of an ecosystem that provides interactive cheering functions.

Keywords: Social live streaming service · eSports · Subscription gifting · Donation · Cheering

1 Introduction

Online spectating services enable people to interact on a virtual level and watch video streams together. Viewers are able to interact with streamers while watching them. Interactive spectating may involve 'cheering,' in which viewers send streamers messages of encouragement, money, or gifts. Streamers can in turn interactively express their gratitude, for example, by saying thank you in the video, which increases the satisfaction of the viewers. Cheering can potentially create new value for online services by promoting interaction between streamers and viewers and enhancing the experiences of the viewers.

Due to COVID-19, travel, dining, and other experience-based services have become severely restricted. As a result, many service providers are turning to online services through trial and error to prevent people from gathering in person. However, it is difficult to set a high flat-rate service charge because existing real experience-based services become much more limited when converted to an online format. In addition, specialty goods and food and drink sales are a crucial part of experience-based services. While specialty goods can be displayed and sold in virtual tours, the tours are time-consuming and not often held.

C. Leitner et al. (Eds.): AHFE 2021, LNNS 266, pp. 478–484, 2021.
https://doi.org/10.1007/978-3-030-80840-2_55

Twitch and similar eSports live streaming services are examples of services that have successfully incorporated cheering into online spectating. In this service business model, in addition to watching the gameplay, it is also free for viewers to send encouraging messages to streamers in the live chat and become their followers. During or after gameplay, many viewers interact with streamers and give them money or gifts. The streamers express their gratitude, which increases the viewers' satisfaction. This service business model is upheld by cheering during or after viewing instead of charging a flat-rate from the start.

In this study, we propose an ecosystem that connects real-world and virtual experiences by applying a platform that incorporates cheering functions into spectating services, such as Twitch, to online experience-based services. The ecosystem provides interactive cheering functions that enable viewers to interact with service providers and give money or gifts when they are satisfied with the services. In addition, by linking the cheering functions with external delivery services, specialty goods and food and drink can also be provided. Interactive expressions of gratitude from the service providers increase the viewers' satisfaction, and the service is upheld by this constant cycle of exchange. Additionally, the platform makes it possible to provide services more frequently.

This study examines the ecosystem and focuses on the cheering aspect by analyzing data such as the interactions between a streamer and viewers in the chat on the streamer's channel. In particular, we looked at interactions related to donating via subscription gifting, which is one way of cheering on Twitch. Through our analysis of Twitch discourse, we reveal the value of such an ecosystem that provides interactive cheering functions.

2 Theoretical Background

2.1 Prior Research

In recent years, studies have been conducted on the interaction between streamers and viewers on social live streaming services such as viewer's engagement to streamers [1–3], "troll" gameplay [4], and gender issues in gaming [5]. However, the studies on the discourse analysis between streamers and viewers in the chat window are very limited.

Reckenwald [6] investigated the interaction between streamers and viewers on popular streamers' channels by analyzing a streamer's voice and gaze as discourse in addition to interactive commenting in the chat. The study determined that the amount of the streamer's dialogue is smaller when the amount of overall discourse is larger in the streamer's broadcast and vice versa. However, his study was limited to discourse analysis and did not address the relationship between the discourse and movement of money such as donations from viewers to streamers. There have been no studies of social live streaming services that have analyzed discourse and donation together, to the best of our knowledge.

In this paper, we investigate chat interaction data from Twitch and analyze the discourse between streamers and viewers that focus on subscription gifting as a means of donating to streamers on Twitch. Through our analysis, we reveal the value of an ecosystem that provides interactive cheering functions.

2.2 Subscription Gifting Program on Twitch

On Twitch, users can create a basic account for free, though free users are subjected to ads and cannot access various benefits exclusive to paid accounts. If viewers want to watch a particular streamer's game play, they usually join the subscription program. This program has three tiers: Tier 1 at $4.99/month, Tier 2 at $9.99/month, and Tier 3 at $24.99/month. A subscription is included with an Amazon Prime membership, but Prime members can only be subscribed to one streamer. Tier 1–3 subscribers are recognized as paid subscribers and Amazon Prime subscribers are recognized as Prime subscribers.

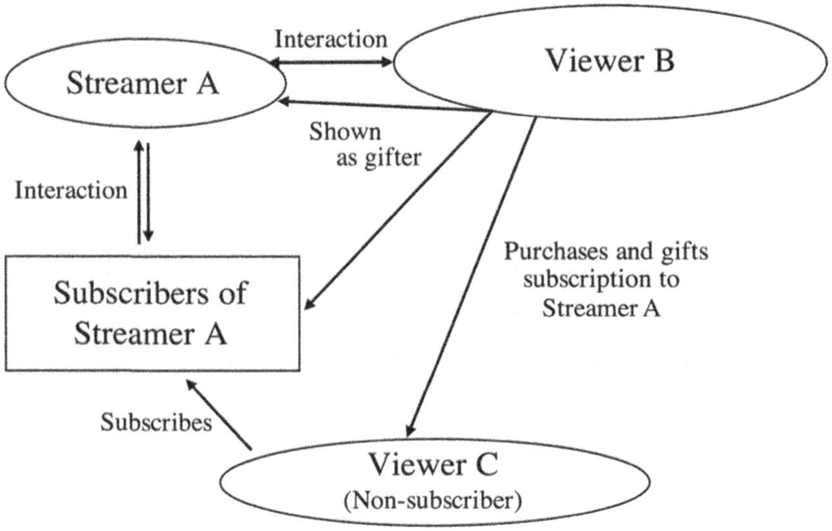

Fig. 1. Flow of subscription gifting. Viewer B purchases a subscription to Streamer A's channel from Twitch and gifts it to Viewer C. Viewer C becomes a subscriber of the channel.

Twitch has a subscription gifting program, an overview of which is shown in Fig. 1. In this figure, Viewer B first purchases a subscription to Streamer A by selecting one of the three tiers. Next, Viewer B gifts the subscription to Viewer C, who is not subscribed to Streamer A. Viewer C receives the gift and joins Streamer A's channel as a subscriber. A Gifter Badge is displayed alongside Viewer B's username in the channel's chat window when the gifting is complete, thus making Viewer B's contribution known to Streamer A and other viewers. Streamers often acknowledge users with such statuses in a message on the channel. The subscription and gifting fees are divided between Twitch and the streamer at a specific ratio.

2.3 Stakeholders in Subscription Gifting on Twitch

Stakeholders involved in subscription gifting include the following actors:

- A streamer, i.e., a game player, broadcaster, and owner of a channel.
- A moderator who acts as the chat administrator of the channel. The moderator is chosen by the streamer from Twitch users and helps the streamer with chat interactions.
- A viewer (gifter) who cheers for the streamer, purchases a subscription as donation, and gifts the subscription to a non-subscribed viewer of their choice.
- A viewer who is not subscribed to the streamer until he/she receives the gifted subscription from the aforementioned gifter.

Moderators are unique to Twitch, and a streamer usually selects multiple moderators for their channel. Moderators have a trusting relationship with the streamer and work to keep the chat safe, pleasant, and entertaining. When viewers take part in cheering, one of the moderators' main responsibilities is to respond to donations from viewers when the streamer is busy. Streamers are not always watching the chat window while playing the game, so sometimes they are too busy playing the game to respond to every donation via voice or chat. In such a case, the moderators thank the viewer for the donation on behalf of the streamer. On Twitch, conversations about subscription gifting are efficient and almost uninterrupted, in part because of the presence of these moderators.

3 Ecosystem with Interactive Cheering Functions

A cheering 'cycle' on Twitch is shown in Fig. 2. There are two main ways of contributing: one is through Twitch's donation system, and the other is by connecting to external websites such as Amazon to give physical gifts to streamers. In Fig. 2, the former is indicated by the thick solid line and the latter is indicated by the thick dotted line. Through these cheering functions, Twitch has created an ever-growing ecosystem that integrates real-world and virtual experiences.

When viewers make a donation such as subscription gifting through the donation system, Twitch's automated announcement and the conversation between the streamer and viewers in the chat share who made what donation to whom. When a donation is made, the streamer or one of moderators on behalf of the streamer usually expresses their gratitude to the donator (gifter). In the case of physical gifting, the gift is directly sent via an external website, so the communication is only between the streamer and the viewer who sent the gift.

We propose an ecosystem, based on cheering in Twitch, which incorporates cheering functions into online experience-based services, as shown in Fig. 3. By implementing a donation system and connecting to an external website to deliver food and drink and other items, cheering functions similar to those of Twitch can be provided. We believe that such an ecosystem will pave the way for an online experience-based service that integrates real-world and virtual experiences, similar to Twitch.

In this study, we analyze subscription gifting as an example of cheering by donating in Twitch and examine this ecosystem.

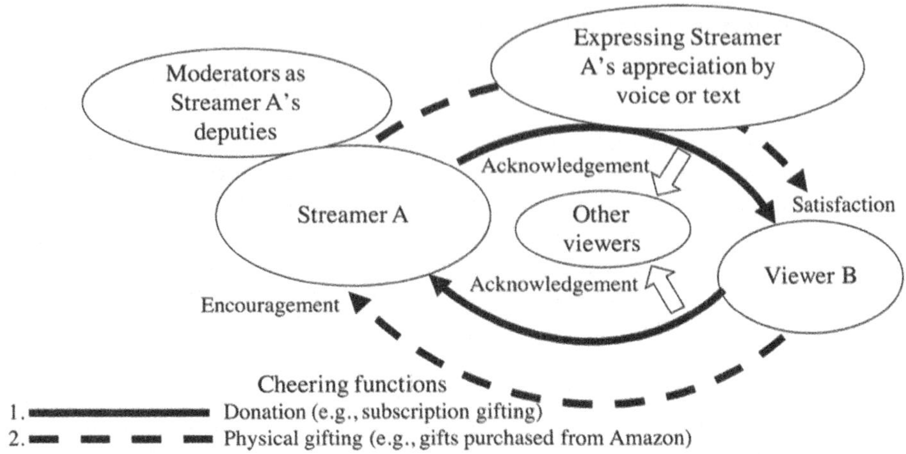

Fig. 2. Cheering cycle on Twitch. Cheering using the donation system is indicated by the thick solid line and the activity using an external website is indicated by the thick dotted line.

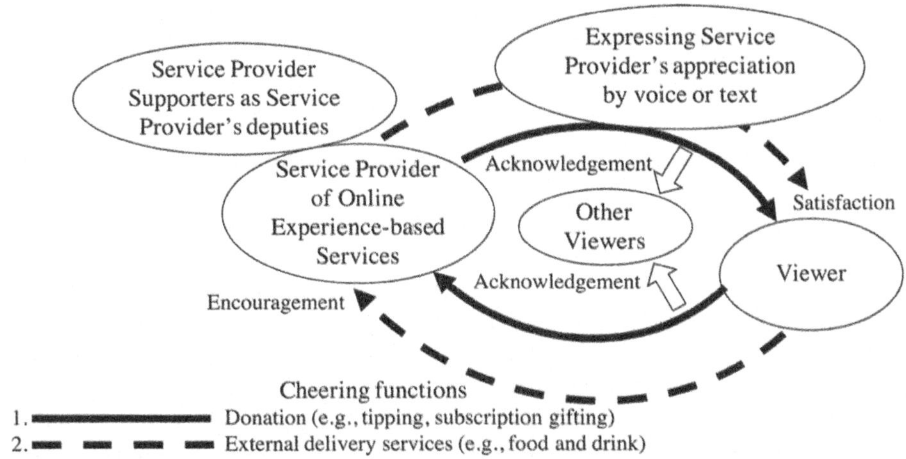

Fig. 3. Proposed ecosystem with interactive cheering functions. Cheering using the donation system is shown with the thick solid line and the activity using an external delivery service is shown with the thick dotted line.

4 Method

Livestreams are saved as videos on the Twitch website and can be rewatched for a certain period of time [7]. Chat histories from the streams are also recorded with the videos. We read and analyzed the discourse line by line from dozens of chat histories of several popular streamers. One chat history can be as long as 10 h or more. Then, we extracted the parts where a subscription was gifted as a way of cheering and analyzed the context of the conversation between the stakeholders before and after the gifting.

5 Findings

In our analysis of the discourse, we found several cases in the chat history where the stakeholders of subscription gifting conversed with each other and were satisfied with the interaction. An example of one such exchange is shown in Fig. 4.

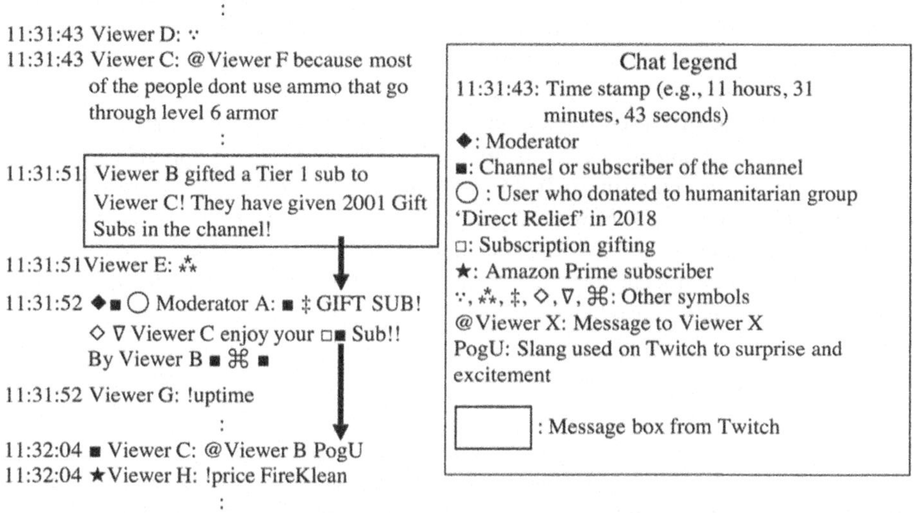

Fig. 4. Example of interaction between stakeholders in subscription gifting in the chat history on a popular streamer's channel. The left side shows the chat history before and after the subscription gifting, and the right side shows the chat legend.

The chronological sequence of the actions is as follows:

1. At 11:31:43, Viewer C is not subscribed to the streamer's channel, as there is no subscriber badge next to Viewer C's name.
2. At 11:31:51, Twitch's automated system announces that Viewer B has gifted a Tier 1 subscription to Viewer C. In addition, Viewer B has already gifted a total of 2,001 subscriptions on this channel.
3. At 11:31:52, a moderator thanks Viewer B on behalf of the streamer and tells Viewer C to enjoy the subscription.
4. At 11:32:04, Viewer C has a subscriber badge next to their name, indicating that they are now subscribed to the channel. Viewer C uses the Twitch slang term "PogU," which expresses surprise and excitement, in their message to Viewer B to express their gratitude.

The above actions from 2 to 4 are represented as arrows in Fig. 4. As a result of Twitch's automated announcements and the work of the moderator, they were able to have an uninterrupted conversation within a very short time, and all stakeholders were satisfied with the interaction.

6 Discussion

Through our analysis of the discourse related to subscription gifting as donation and viewers cheering the streamer, we found that the stakeholders were able to have uninterrupted conversations regarding subscription gifting due to the unique mechanism provided by Twitch. As a result, the stakeholders were satisfied with the exchange. The streamer is expected to thank the viewers for donations, but when the streamer is busy focusing on gameplay, the moderators can express their gratitude to the donator (gifter) on behalf of the streamer. The moderators' acknowledgement encourages viewers to continue supporting the streamer. In other words, viewers are satisfied with their act of donation to cheer for the streamer, so they will continue to do so in the future.

A cheering mechanism similar to that of Twitch could prove valuable for online experience-based services. In such an ecosystem, gratitude is clearly expressed in response to cheering. Viewers will be satisfied with their cheering, continue to cheer for a service provider such as a tour guide, and donate to the service provider. Since the donation amounts are not capped, viewers are able to donate more than the flat-rate they would be charged by the provider.

7 Conclusion

We analyzed the discourse between streamers and viewers in Twitch chats related to subscription gifting as ways of donating to and cheering for the streamers. Through this analysis, we revealed the value of an ecosystem that provides interactive cheering functions for online experience-based services. We intend to continue our research on the relationship between streamers and moderators, the activity of gifters after completing their gifting as a way of cheering on live streaming services, and ecosystems with interactive cheering functions.

References

1. Hilvert-Bruce, Z., Neil, J.T., Sjoblom, M., Hamari, J.: Social motivations of live-streaming viewer engagement on Twitch. Comput. Hum. Behav. **84**, 58–67 (2018)
2. Wohn, D., Jough, P., Eskander, P., Siri, J., Shimobayashi, M.: Understanding digital patronage: why do people subscribe to streamers on Twitch? In: Proceedings of CHI PLAY 2019, Spain (2019)
3. Lin, H., Sun, C., Liao, M.: Game streaming revisited: some observations on marginal practices and contexts. In: Proceedings of the 2019 DiGRA International Conference, Japan (2019)
4. Karhulahti, V.: Prank, troll, gross and gore: performance issues in esports live-streaming. In: Proceedings of the First International Joint Conference of DiGRA and FDG, United Kingdom (2016)
5. Siutila, M., Havaste, E.: A pure meritocracy blind to identity: exploring the online responses to all-female teams in Reddit. In: Proceedings of the 2018 DiGRA International Conference, Italy (2018)
6. Reckenwald, D.: The Discourse of Online Live Streaming on Twitch: Communication between Conversation and Commentary, Doctoral dissertation, Hong Kong Polytechnic University, Hong Kong Special Administrative Region (2018)
7. https://www.twitch.tv. Accessed 10 Jan 2021

A Possibility of Value Co-creation in the Service Ecosystem of the Paid Media in Japan: Focusing on Viewers' Life Stories

Shuji Sudo(✉) and Yasunobu Ito

Japan Advanced Institute of Science and Technology (JAIST), 1-1 Asahidai, Nomi, Ishikawa 923-1292, Japan
{sudo_shuji,itoyasu}@jaist.ac.jp

Abstract. The purpose of this study is to probe the possibility of value co-created by the media and audience in the paid media service ecosystem. This article mainly spotlights on pay television and paid Vod (Video on Demand) called OTT (Paid Vod services via the web) as paid media. The research target is on paid media of which the main area is pay television with OTT added. Inevitably, the usage or consumption of paid media is relevant to consumer value for audience. The focal points of traditional media researchers and marketing practitioners have tended to be on the current needs of the audience. Whereas this study challenged to explore untapped reasons of media consumption through the life stories of audience. The research method used was interviews with audience on their life stories. The combination of memory with current media consumption enables audience and the media to co-create new values the media firm did not provide for.

Keywords: Customer experience · Reminiscence · Human memory · Value co-creation · Media · Marketing · Service ecosystem · Life story

1 Introduction

This article mainly focuses on the paid media service which encompass pay TV (television) and paid Vod (Video on Demand) called OTT (Over the Top: Paid streaming and Vod services via the web). Pay TV is a subscription-based broadcasting service with reception devices, provided by cable television operators or direct broadcast satellite servicers. The cable TV operators include the original cable TV companies and telecommunication firms with optical communications line. OTT is a paid streaming or video-on-demand services. A user can make a contract directly with OTT service providers via the web.

Pay TV first appeared in Japan in the mid-1990s. Although the number of subscriptions increased to around 26% by 2007 (Fig. 1), it has remained flat since then. The diffusion of innovation often draws an S-curve, which means that the diffusion curve initially grows fairly linear and then flattens out [1]. Researchers have considered that the S-curve indicates "the proportion of actual adopters amongst the potential adopters

C. Leitner et al. (Eds.): AHFE 2021, LNNS 266, pp. 485–492, 2021.
https://doi.org/10.1007/978-3-030-80840-2_56

of a specific value proposition" [1]. From this point of view, the pay TV services have almost totally fulfilled the potential needs of the Japanese market.

Media usage or media consumption is tied to audience values. For the audience to expend money on certain contents in pay media, there is bound to be value in it for the audience. Conventional media researchers and marketing representatives tended to focus on the current needs of the consumers. On the other hand, this study attempts to investigate on the reasons behind media consumption through the values hidden behind audience's life stories. The method used for the study was interview towards paid media users. In essence, media, audience and memory are all equal actors in the service ecosystem.

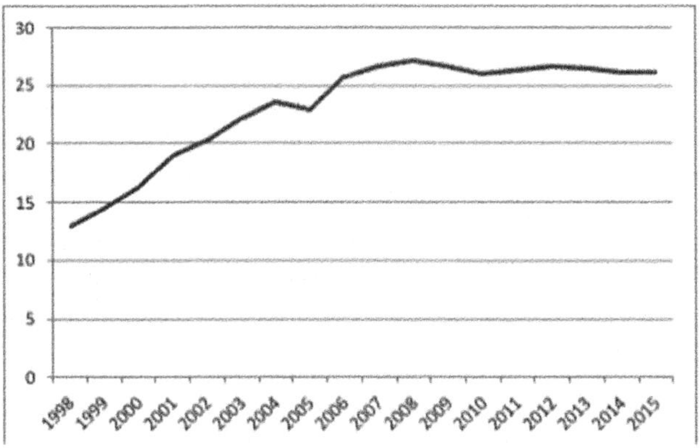

Fig. 1. Household penetration (%) of pay Television in Japan

From our viewpoint, the paid media service is considered a service ecosystem. The service ecosystem is "a relatively self-contained, self-adjusting system of resource, integrating actors that are connected by shared institutional logics and mutual value creation through value exchange" [2]. Looking through the lens of the service ecosystem, the driving power of the service ecosystem can be defined as follows: The service ecosystem which is formed through mutual interactions of multiple actors in a service is "driven by the need to exchanges resources in order to survive and thrive" [3].

The purpose of this study is to probe the possibility of value co-created between the media and audience in the paid media service ecosystem. What is the value to the audience as an actor in the paid media service ecosystem? The use or consumption of paid media is connected to the consumer value of the audience.

2 Literature Review

Finding the consumer value in pay TV contents is an orthodox approach [4, 5]. Pay TV industry in Japan attempted a survey of value analyses in pay TV programs in the past [6]. The focal points of such traditional media researchers and marketing practitioners targeted on the current needs of the audience who is a fan of specific pro baseball team or a star actor in Korean dramas. However, it is possible that these quantitatively measurable needs are only a superficial aspect of consumer value.

From a consumer-centric perspective, the problem with understanding consumers, based on consumer needs is that it focuses only on the functional aspects and does not understand why a consumer buys a service/product from the emotional and social dimensions [7]. "Jobs-to-be-done" a logic by Christensen et al. defined as "the job is the progress that an individual seeks in a given circumstance", is definitely an effective concept to understand consumer value [7,8]. In regard to the paid media consumption, it is important to note that the current need of the aforementioned contents is mainly derived from the functional aspects and often does not touch upon the emotional and social dimensional aspects.

In fact, according to Christensen et al., in a prior research, marketers and scholars had accidentally encountered hidden reasons of consumption other than the functional needs of a product from conversations by the subjects outside of the prepared flow of the focus group interview (FGI) [7]. The "job" theory approach by Christensen et al. is extremely customer-centric [9] and would be a good means to discover untapped customer value. However, finding a "job" requires patience and behavioral observations that often take a long term to acquire.

In Japan, there have been narratives from subjects related to their old memories in particular that were not in the interview flow which was designed to directly explore the subjects' needs via traditional FGI or One-on-One Interview (In-depth Interview) by practitioners regarding media use. One of the authors (Sudo) have encountered these cases in the interviews of marketing research practice of the Japanese pay TV industry. Although, ordinarily, in the practical qualitative marketing research, narratives on old memories tend to be ignored, this research focuses precisely on this point and attempts to explore the values of using paid media. Since the subject's experience of media use and activities in childhood and adolescent years from their memory appears to be linked with their current media use.

Another marketing study has focused on the time-series change detected in consumer behavior, which is called the customer journey map [9]. In the customer journey, the customer experience can be divided into three stages: Pre-Purchase, Purchase, and Post-purchase [9]. In the traditional marketing, the Pre-Purchase stage in the customer journey does not go back to the customer's childhood experience, but is the situation just before consumption, or in "prospect" which is the state of a prospective customer. In this study, in order to probe into the old memory of the audience, we adopted a method that asks for their life stories at the beginning of flow interview. In general, this approach is not used in the conventional media marketing in Japan.

In short, the uniqueness of this research is in linking customer experience with their life stories.

3 Method

The method of this research was based on interviews with audience who utilize the pay Television or OTT service. The subjects were recruited by snowball sampling. Interviews were conducted on one-on-one or in pairs with family members or friends. The number of interviewees in this study was twelve. The criteria for recruiting subjects for the interview was to prioritize those who had medium to high level involvement in pay media. The interviewees list is indicated in Table 1.

Before the formal interview, the interviewer met and corresponded via e-mail and chat with the subjects several times during the ice breaking process. In this study, using life stories meant that the subjects went as far back in their childhood as possible and talked about their life events and entertainment related experiences. Prior to the interview, subjects were asked to fill in the blanks in the chart regarding their life events and media experiences.

Interviews were conducted one to three times for each interviewee, and additional surveys were conducted separately by email or chat.

Table 1. The subjects of interviews

Subject	Age	Profile
H	36	CEO of a venture business, a fan of Sports and American TV
Q	51	Corporate legal officer, a pay TV aficionado
N	29	Businessperson, an enthusiast of musicals and pay TV
D	50	Small business owner, a sports fan of pay TV
J	84	Pensioner/K's husband, a former movie fan
K	77	Pensioner/J's wife, a fan of Go Game
Z	53	Engineer, an airplane aficionado and a professional wrestling fan
W	57	Consultant, a book aficionado
M	50	Businessperson, a lover of cats
H	57	Business Owner/Editor, a hard worker
L	44	Toy designer/Artist, a movie fan
P	32	Teacher/Artist, a movie fan

4 Results

Through the cases of interviews focusing on reminiscence of experience in media, this study presumed to have gained certain findings other than "functional needs" of media. In some cases, past experiences with the media in their life stories became important reasons for the subjects to use the media later on in their lives. In this paper, due to limited space, the remarkable cases were chosen for discussion below.

4.1 The Case of J and K

In terms of media marketing in Japan, it is common for elderly people around the age of seventy to enjoy western movies in theaters and on TV including pay TV movie channels.

Subject J, 84-year-old male, used to be an enthusiastic movie fan in his twenties through forties. Although his preference for movies maybe regarded as similar to those of his generation, that is not enough to account for his reason of consuming movie media. In the interview with J, other reasons for his love of western movies were found from his life story.

The first explanation of why he saw Hollywood films in theaters often was that he received free tickets from his elder sister who worked in an organization managing intellectual property called JASRAC (Japanese Society for Rights of Authors, Composers and Publishers), where people from show business or public entertainment were its client. The second reason why he became familiar with Hollywood movies was that he had access to read American movie magazines given to him by his relative who worked for the US military camp near his residence in Yokohama.

After marriage, J watched western movies on free terrestrial television, as reminiscence of going to theaters in his youth. TV Asahi Network of free terrestrial TV started a TV program called "Saturday Movie Theater" in October of 1966 [10], which eventually became the popular "Sunday Night Western Movie Theater". A media critic wrote, "When movie audience gathered in front of a cathode-ray tube, customers gradually turned away from going to movie theaters" [11]. At the time, J was collecting records of western film music. Without a doubt, records are one of paid media as well.

J began subscription of a paid cable television in 2007 but when he turned 74, he had already lost interest in western movies, presumably because of his age. In turn, his wife K became an enthusiastic audience of the pay TV. She loves to watch western old mystery dramas and *Go* games (traditional board game).

She is watching *Go* in '*Go* and *Shogi* (traditional board game like chess) Channel' on paid cable TV. K had started playing *Go* as a hobby when she was a high school student in western Tokyo. She had learned to play the easy *Go* game from her mother in her childhood. After that, she bought books on *Go* to learn the real *Go* game. Additionally, she joined a *Go* Game club by herself. After marriage in 1963, she took lessons from a Buddhist priest in her neighborhood for half a year.

Go is more than just a hobby for K, it is also a remnant of her youth, filled with memories of her mother. As an added note, for K, the value found in using pay TV can be linked to her childhood memory.

4.2 The Case of L and P

L is a 45-year-old toy designer and an artist. She was born in a small town near Fukushima. L, an avid fan of rock band Queen, has been listening to Queen all her life and likes watching Hollywood movies and Korean movies. She used to frequently rent videos from video rental shops. However currently, the rental video industry is fading and replaced by OTT in Japan.

L and her friend P have a somewhat odd way of watching movie on Netflix (OTT). They simultaneously watch the same movie in their own homes and talk about various scenes in the movie they just watched. They enjoy predicting which movie will become the winner of the 2000 Academy Awards, and compete with each other on making the right guesses. They also try to watch all nominated films as much as possible in order to forecast the awards. L's passion for movies showed signs at age ten. L was concerned about checking the degree of perfection of the animation pictures on TV animation programs when she was in elementary school. L continued to have a passion for painting, and during the summer vacation at age 14, she decided to become a Manga artist in the future.

Although L did not become a Manga artist, she succeeded in her career as a toy designer. On the other hand, her enthusiasm for movie contents shown in her predicting the future Academy Awards winner remains as a pleasure in her life. Watching a Hollywood movie is more than just entertainment for L. The way L enjoys movies is a customer experience, which companies 'cannot influence nor control' [9]. Lemmon et al. called such customer experiences 'customer-owned touch points' [9].

Another reason of devotion for movies was found in L's life story experience. In high school, her tutor recommended her to a British band "Queen" that had already disbanded but she fell in love with the band. According to her, her musical experience as a member of the school's brass band was linked to Queen's tune.

Table 2. General marketing interpretation in japan, reminiscence and actions regarding paid media

Case	General marketing Interpretation	Reminiscence	Actions regarding paid media
J	Many people over the age of seventies like western movies.	Used to watch Hollywood movies frequently with free tickets from his sister. Access to movie magazines from a relative working for US military.	Collect records of western movie music.
K	A senior *Go* fan loves to watch *Go* games on cable TV.	Learned simple game using *Go* stones in childhood from her mother Learned *Go* when she was in high school and after marriage.	Watch *Go* on pay TV. Enjoy *Go* reading *Go* magazines.
L	A Hollywood movie fan often goes to the theater and watches movies on Netflix.	Disliked the poor quality of pictures on a TV animation at age 10 Aspired to be a cartoonist at age 14 Became a fan after tutor recommended the band 'Queen' at age 17 Studied art in college.	Enjoy forecasting Academy Awards with a friend. Make a prediction list of the songs played in the movie 'Bohemian Rhapsody' in 2019 and sharing it with friends.

When L was 44 years old in 2019, she went to see the movie "Bohemian Rhapsody" which was modeled after Queen, with her friends. Before that, she made a predicted song list that will be played in the movie and handed it out to her friends. Here again, L conducted a customer action on her own through a paid medium. The movie company and OTT did not initiate her to take that action.

In Table 2, two limited-case analyses were depicted.

5 Discussion

Generally, in media marketing, the audience is considered to have current specific needs and reason for media consumption. These needs might as well be correct, but it is based only on external observation. However, this study challenged to find the untapped reasons in the life stories of the audience.

First, based on the case of K's *Go*, if someone is a *Go* fan, he or she will likely watch *Go* program on pay TV with a certain probability. To paraphrase in the traditional media marketing perspective, this *Go*-related activity could be defined as being related to current interests and current media consumption. Incidentally, *Go* is a traditional game that was already familiar to samurai and the common people in the 15th century Japan. For many aficionados of *Go* today, it has been continuously passed down from their grandparents and parents throughout the game's long history. When a person plays the game of *Go*, regardless of whether one is reminded of interactions with old teachers or parents, there is no doubt that the value of *Go* to that person is found in the combination of enjoying *Go* with the memory of their life stories. Therefore, by listening to life stories in interviews, it is possible to explore the untapped reasons of media consumption that goes beyond the realm of traditional marketing.

Second, in the case of K enjoying Hollywood films, there seemed to be an in-depth relationship between the events in her life and the memory of media experience. Her way of using paid media is not mere consumption, nor fanaticism, but found in the creation of new value with the media. Like "IKEA hacking" [9] in which users are voluntarily producing their original way of utilizing and designing products, audience as customer can create new value of products which cannot be influenced by media firms. In the case of L, the activities equivalent to "IKEA hacking" are forecasting Academy Awards and making her original movie pamphlet with the prediction of song order that are going to be played in the movie.

Through the interview that focused on her life story, this study gained some hints on the reasons behind L's prediction of Oscar's awards with her friend or her creating original pamphlets of 'Bohemian Rhapsody' to hand out to her friends when going to the theater. Since she was a child, she had been watching cartoons on TV with a keen interest in the works of production and drawing of animation. In traditional marketing research, L's way of enjoying movies is nothing more than a middle-aged female who likes watching American movies on OTT and in theaters. However, through her life story interview, it became apparent that she co-created a new value with the media, a value that could not be created nor managed by movie companies. Below is a summary of how value co-creation occurs in the paid media service ecosystem.

Media, audiences and memories are equal actors in the service ecosystem [2]. In the audience's life story, their encounters with media at that point in time becomes a value that remains in their memory. Combined with that memory and current media consumption, audience and media potentially co-create new value that is difficult to capture by traditional marketing research. Co-created value which was discovered by unraveling the life story not only reflects the emotional and social dimension of job theory [7], but also entails the historical dimension. Needless to say, media consumption based on current functional needs has its place in the service ecosystem, but that is easily captured in marketing research.

To some extent, it was effective to probe the value of paid media using the life stories of audience. This article may have provided informative contribution and theoretical implication on the study of consumer value in the service ecosystem. As practical implication, this study elaborated its potential as a method for exploring untapped consumer value in a short time. For the long term, this method of probing into life stories can also be useful for contents, service, and product development. Nevertheless, the life stories of customers in itself cannot adequately cover the entire spectrum of customer value and co-created value in the service ecosystem, it needs further investigation.

References

1. Vargo, S.L., Akaka, M.A., Wieland, H.: Rethinking the process of diffusion in innovation: a service-ecosystems and institutional perspective. J. Bus. Res. **116**, 526–534 (2020)
2. Lusch, R.F., Vargo, S.L.: Service-Dominant Logic: Premises, Perspectives Possibilities. Cambridge University Press, New York (2014)
3. Vargo, S.L., Akaka, M.A., Vaughan, C.M.: Conceptualizing value: a service - ecosystem view. J. Create. Value **3**(2), 1–8 (2017)
4. Ducey, R.V., Krugman, D.M., Eckrich, D.: Predicting market segments in the cable industry: the basic and pay subscribers. J. Broadcast. **27**, 155–161 (1983)
5. Hammervold, R., Solberg, H.A.: TV sports programs—who is willing to pay to watch? J. Media Econ. **19**, 147–162 (2006)
6. Sudo, S.: Thinking of customer value of Pay TV viewers: from the survey of highly involved audience of Pay TV. Bulletin Nikkei Ad Res Ins. **43**, 25–32 (2010). (in Japanese)
7. Christensen, C.M., Hall, T., Dillon, K., Duncan, D.S.: Competing against luck: the story of innovation and customer choice. Harper Business (2016a)
8. Christensen, C.M., Hall, T., Dillon, K., Duncan, D.S.: Know your customers' "jobs to be done." Harv Bus Rev. Sept **2016**, 54–62 (2016)
9. Lemon, K.N., Verhoef, P.C.: Understanding customer experience throughout the customer journey. J. Mark. **80**, 69–96 (2016)
10. Yodogawa, N.: Sunday Movie Theater, Ondori, Toyo (1997). (in Japanese)
11. Kanada, S.: Why did TV get bored? Media history made by stars. Nikkei BP, Tokyo (2006). (in Japanese)

Service Design Based on Customer Value Classification in IoT Systems

Sadaaki Shinjo(✉) and Naoshi Uchihira

Japan Advanced Institute of Science and Technology, 1-1 Asahidai, Nomi, Ishikawa 923-1292, Japan
{sadaaki.shinjo,uchihira}@jaist.ac.jp

Abstract. There are four stages for development of IoT systems identified as monitoring, control, optimization, and autonomy. The value that the IoT system can provide is constrained by the capabilities of the system, therefore it is necessary to build a service that fits the feature of the system. Based on fifty IoT business cases, the researchers classified fourteen values for four IoT systems and proposed methods to improve on services based on features of IoT system. However, the value classification had a wide range of value expressions, and therefore might not focus correctly on the provided services. This paper is reconstructed with the fourteen value classifications of IoT systems by using elements of value and suggests more appropriate service design.

Keywords: IoT system · Classification of value · Elements of value · Service design

1 Introduction

An IoT system consists of Internet-connected sensors, actuators, cloud space, and humans. The IoT system can provide different values depending on the combination of each component and the development stage of the system. In general, a business starts by setting up a customer and addressing the customer's problem, and then constructing an optimal system to solve the problem in hand, and provide value proposition. However, in the IoT business, that is created by IoT systems, the value provided can be developed by evolving and changing the components, and new value can be added to the value initially provided.

During the pandemic of Covid-19, which spread worldwide in 2020, there has been impacts and changes in the value offered to customers and required changes in business model. Service design must be able to withstand these changes in value. For this purpose, it is necessary to change the IoT system in a timely manner so that it can provide new values even when the values desired by customers change. For this purpose, it is important that the value provided match the value demanded by the customer.

This paper relies on the modified method presented by the authors in a previous research and proposes a design method that efficiently defines values and support a highly accurate service conception, by classifying the values that IoT systems could provide and supplementing the value categories with a more detailed elements of values.

C. Leitner et al. (Eds.): AHFE 2021, LNNS 266, pp. 493–502, 2021.
https://doi.org/10.1007/978-3-030-80840-2_57

2 Theoretical Overview

2.1 The IoT System

The IoT system is a system that is constructed by internet-connected components (sensors, actuators, the cloud, and humans) see Fig. 1. First, a user send commands to sensors, actuators, and the cloud through a PC or smartphone. The sensors measure and quantify the real world and send data to the cloud. The actuators influence the real world by acting based on the quantified commands from the cloud. The cloud supports user (human) thinking, making decisions, analyzing data and reporting on the status of the system and the results of the analysis to the human. The system influences the real world, and the state changed by the influence is measured by the sensor again, and the whole system is informed. The Internet connects each element and transmits data and commands.

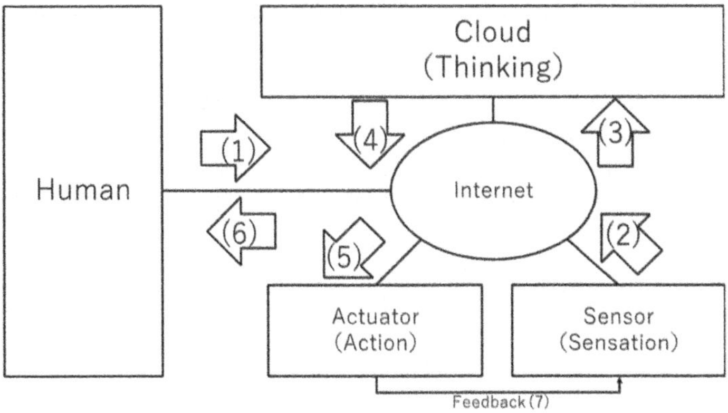

Fig. 1. The IoT system

2.2 IoT System Step Classification

Porter and Heppleman [1] explained the system development of smart products with communication functions by dividing each development stage into four steps: monitoring, control, optimization, and autonomy. Therefore, each function in the system is built on the premise of the functions of the previous step(s). Since IoT systems are smart products with communication functions, this paper applies this step classification to IoT systems and proceed with the discussion.

2.3 Provided Value Model by IoT System Step Classification

Shinjo and Uchihira [3] recapped on the value provided by 50 IoT business cases [2] based on Porter and Hepplemann's step classification, they classified the provided value and working of IoT system at each step. Furthermore, since the value that could be provided at each step of the IoT system structure was different, the researchers proposed to select a system that matches the provided value [3]. Following is the summary of the IoT system capabilities based on Porter and Hepplemann's step classification.

Monitoring. It is a system that uses sensors to take measurements of the real world, converts them into numerical values, automatically collects them, and visualize them in a way that humans can easily understand them. The system expands human perception (Fig. 2, upper left).

Control. Based on the data obtained from the monitoring, the actuators affect the real world. The operation mainly follows the human will. The system add to and expands human capabilities (Fig. 2, upper right).

Optimization. Cloud finds optimal solutions to problems and provides advice to humans. Humans work on the basis of the optimized answer and solve the problem. It is a system that assists humans (Fig. 2, lower left).

Autonomy. Cloud-based thinking level is higher than previous step and human thinking and work become un-necessary. It can replace all human information gathering, actions, and thinking (Fig. 2, bottom right).

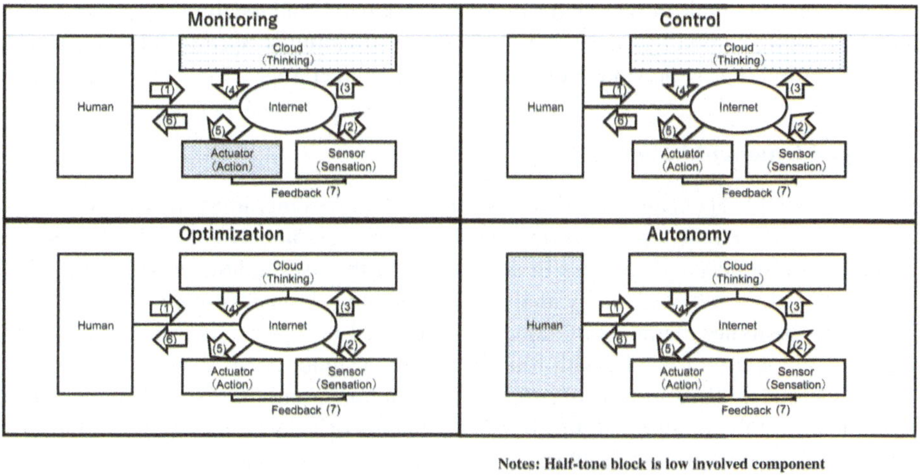

Fig. 2. IoT system work at each step

Table 1 summarizes the provided values at each IoT system step. In the classification, the functions that can be realized at each step are different, so the value provided depends on the step of the system. Since the system at each step is constructed based on the functions of the previous step, the provided value at the latter step is also added to the provided value at the previous step. In other words, the value that the system of the autonomy step can contribute to the service provider ranges from "Visualization" to "Unmanned Operation", and the value it provides for the end customer ranges from "security" to "healing".

From Table 1, we can see that when the provided value "Alarm" is provided to the customer, the IoT system must be prepared with the capability of optimization. However,

Table 1. IoT system step classification and provided value

	Monitoring	Control	Optimization	Autonomy
Value for service provider	Visualization	Augmented human	Alarm Safety Efficiency Cost saving Reduced responsibility Effective utilization of things	Unmanned operation
Value for customer	Peace of mind	Self-respect	Convenience Enjoyment	Healing

if the IoT system that can be prepared is at the monitoring step only, the gap between the optimization and monitoring systems must be recognized, and the thinking and behavior must be complemented by capability of humans. Providing service to fill this gap will give the ability to provide value beyond the system constraints.

2.4 The Element of Value

Almquist et al. [4, 5] expanded and developed on Maslow's hierarchy of needs theory, to conduct consumer analysis, and define the 30 customer elements of value by classifying the value of products and services in detail, they further defined the 40 B2B (Business to Business) elements of value by analyzing corporate customer research.

The 30 customer elements of value are organized in a pyramid structure based on Maslow's hierarchy of needs, with four levels of consumer needs from the bottom up: "(a) Functional", "(b) Emotional", "(c) Life Changing", and "(d) Social Impact" (see Fig. 3, left). On the other hand, the 40 B2B elements of value were organized into five layers from the bottom: "(e) Table Stakes", "(f) Functional Value", "(g) Ease of Doing Business Elements", "(h) Individual Elements", and "(i) Inspirational Value" (Fig. 3, right).

Almquist et al. argued that it is important to construct the services provided by considering the combination of these elements with value, and that the decision of which elements of value to include must be made strategically. For example, value of "Convenience" which is provided by banks, is combined with the elements of value "Time savings", "Decreased hassle", "Simplification", and "Reduced effort". Therefore, the value provided in other businesses needs to be realized by combining elements of values that are different from those in banks, even if the same "Convenience" is provided. The value provided in the representative services of the 50 IoT business cases in the literature [2] can be represented as a composition by elements of value (see Table 2). Expressing the value provided by each business in terms of the combination of elements of value in this way leads to a higher accuracy of the value proposition, which is the objective of the business.

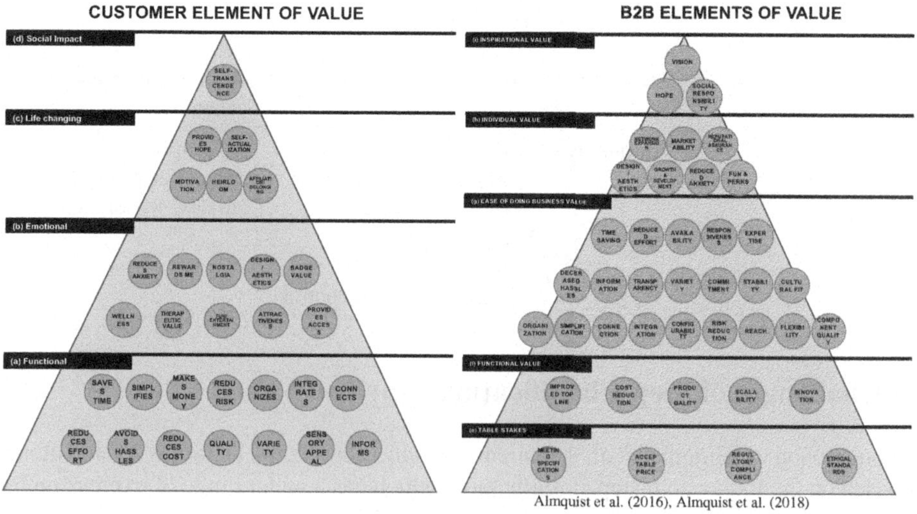

Fig. 3. The elements of value

Table 2. IoT business examples based on elements of value

IoT business	Step	B2B Element of value	Customer element of value
IoT health management system	Monitoring	Social responsibility, reduced effort, decreased hassles, information, transparency, organization, expertise, risk reduction	Reduces anxiety, wellness, reduces risk, organizes, informs
IoT care robot	Control	Social responsibility, reduced effort, decreased hassles, responsiveness, expertise, risk reduction	Provides hope, self-actualization, motivation, saves time, reduces anxiety, reduces risk, reduces effort
IoT smart hospital	Optimization	Social responsibility, reduced anxiety, time savings, reduced effort, information, transparency, integration, responsiveness, expertise, stability, risk reduction, flexibility, cost reduction	Reduces anxiety, wellness, saves time, reduces risk, quality, informs

(continued)

Table 2. (*continued*)

IoT business	Step	B2B Element of value	Customer element of value
IoT noman store	Autonomy	Reduced anxiety, time savings, reduced effort, decreased hassles, information, transparency, risk reduction, cost reduction	Saves time, reduces risk, reduces effort, avoids hassles, sensory appeal

3 Challenges of Step Classification Provided Value

In comparison to Almquist et al.'s elements of value, Shinjo and Uchihira's value model is based on the IoT system step classification, which has a wider range of expression for the provided value, and the means of realization may become ambiguous when devising a specific service. For example, for the value to be provided to the end, customer in the optimization step is "convenience", but if a combination of elements of value is considered, the design must be changed depending on whether the IoT system is to be used for"Reduced hassle" or "Saving time" in order to construct an appropriate service. In addition, because higher level step can also provide the value of lower step, these are not MECE (Mutually Exclusive, Collectively Exhaustive). Therefore, when the IoT system is chosen from provided values, these become hard to create for better system details.

4 Modification of the IoT System Step Classification Provided Value Model by Using Elements of Value

4.1 The Provided Value of Each Step by Combination of Elements of Value

As a counter-measure to the challenges pointed out earlier, this paper suggests a method to avoid ambiguity by focusing on the functional aspects of the elements of value, and to supplement the provided value by the IoT system steps with elements of value in order to develop them into a service design using the IoT system (See Table 3). Using the same IoT business as of the previous case [1], step classifications of provided value ware decomposed based on the elements of value. This paper focuses on the functional aspects of the provided value in order to discuss the service design method using the IoT system. The value for customer is constructed according to: "(a) function" hierarchy of customer elements of value, and the value for the service provider are constructed as per the "(f) Functional Value" and "(g) Ease of Doing Business Value" hierarchies of the B2B elements of value. Since each step of IoT system has various business focus, the provided value of "Convenience" or "Efficiency" contained almost all elements of value in the functional aspects, when counted by the presence or absence of elements of value. Therefore, the direction of the provided value was set largely by the function of the IoT system step, and the detailed value of the customer was set by the elements of value of the functional aspect, so that the service design of the customer value could be carried out to prevent ambiguity. When determining the VP (Value Propositions) of the business model, the customer's job, gain and pain in the customer segment profile are matched with the product and service, gain creator and pain reliever in the value map to create a value proposition [6]. By concretely representing this customer's gain and pain in the elements of value, an appropriate VP can be proposed.

4.2 Must-Be Quality

"Meeting specifications", "Acceptable price", "Regulatory compliance", and "Ethical standards", which belongs to the "(e) Table stakes" in the B2B elements of value, are functional aspects, but these elements correspond to "Must-be quality" in the Kano model. These are not listed for each provided value because they do not lead to service differentiation.

5 Proposed Service Design Procedure

From Table 2 in the previous section, when some elements of value correspond to the main object, other elements of value are set as realization methods and the provided value is constructed from these elements of value, the service is can be made better by more detailed design. In other words, the elements of value can be prioritized and utilized in design. Since it is possible to forces to the scope of the provided value using the elements of value, the problem that the range of the expression of the provided value in the step model is too wide and the means of realization becomes ambiguous can be solved. Based on these considerations, this paper proposes a procedure for constructing a service by deriving the provided value from the steps of the IoT system and setting the main provided value using elements of value. Furthermore, by applying this design method, it is possible to create a service that differs from the provided value in the past by focusing on the second- and third-priority elements, or on the elements of value that do not apply to conventional businesses (Fig. 4, left). For example, IoT smart car tires can be equipped with sensors inside the tire to measure air pressure, temperature, vibration, strain, and acceleration, and provide notification of tire and road conditions. The smart tire is sampled as an example and is showed as an example flow procedure of service design (Fig. 4, right). This procedure is summarized in Fig. 5 using Table 1 and Table 3.

Table 3. Provided value by element of value

			CUSTOMOR ELEMENT OF VALUE (FUNCTIONAL)													
	Step	Provided value	SAVES TIME	SIMPLIFIES	MAKES MONEY	REDUCES RISK	ORGANIZES	INTEGRATES	CONNECTS	REDUCES EFFORT	AVOIDS HASSLES	REDUCES COST	QUALITY	VARIETY	SENSORY APPEAL	INFORMS
Value for customor	MONITORING	Peace of mind	✓			✓			✓		✓	✓	✓			✓
	CONTROL	Self respect	✓			✓				✓	✓					✓
	OPTIMIZATION	Convenience	✓	✓	✓	✓	✓	✓	✓	✓	✓	✓	✓	✓	✓	✓
		Enjoyment				✓										✓
	AUTONOMY	Healing						✓	✓	✓			✓			✓

			B2B ELEMENT OF VALUE (FUNCTIONAL AND EASE OF DOING BUSINESSS)																									
	Step	Provided value	TIME SAVINGS	REDUCED EFFORT	DECREASED HASSLES	INFORMATION	TRANSPARENCY	ORGANIZATION	SIMPLIFICATION	CONNECTION	INTEGRATION	AVAILABILITY	VARIETY	CONFIGRABILITY	RESPONSIVENESS	EXPERTISE	COMMITMENT	STABILITY	CULTURAL FIT	RISK REDUCTION	REACH	FLEXIBILITY	COMPORNENT QUALITY	IMPOROVED TOP LINE	COST REDUCTION	PRODUCT QUALITY	SCALABILITY	INNOVATION
Value for service provider	MONITORING	Visualization	✓	✓	✓	✓	✓	✓		✓	✓				✓	✓				✓	✓			✓	✓			
	CONTROL	Augmented human	✓	✓	✓					✓					✓	✓				✓					✓			
	OPTIMIZATION	Alarm		✓		✓									✓	✓				✓								
		Safety			✓	✓	✓								✓	✓				✓	✓	✓						
		Efficiency	✓	✓	✓	✓	✓	✓	✓	✓	✓	✓	✓	✓	✓	✓	✓	✓		✓	✓	✓		✓	✓	✓	✓	
		Cost saving	✓	✓	✓	✓					✓				✓	✓	✓	✓		✓					✓	✓		
		Reduced responsibility	✓	✓	✓	✓					✓				✓	✓	✓	✓		✓					✓	✓		
		Effective utilization of things		✓		✓	✓						✓		✓					✓		✓				✓	✓	
	AUTONOMY	Unmanned operation	✓	✓	✓	✓	✓								✓					✓					✓			

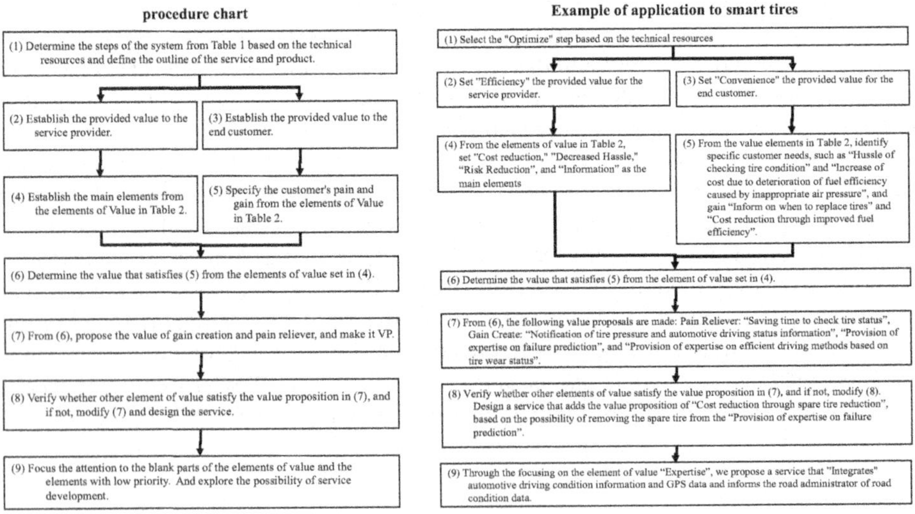

Fig. 4. Proposition chart of service design procedure by using the elements of value

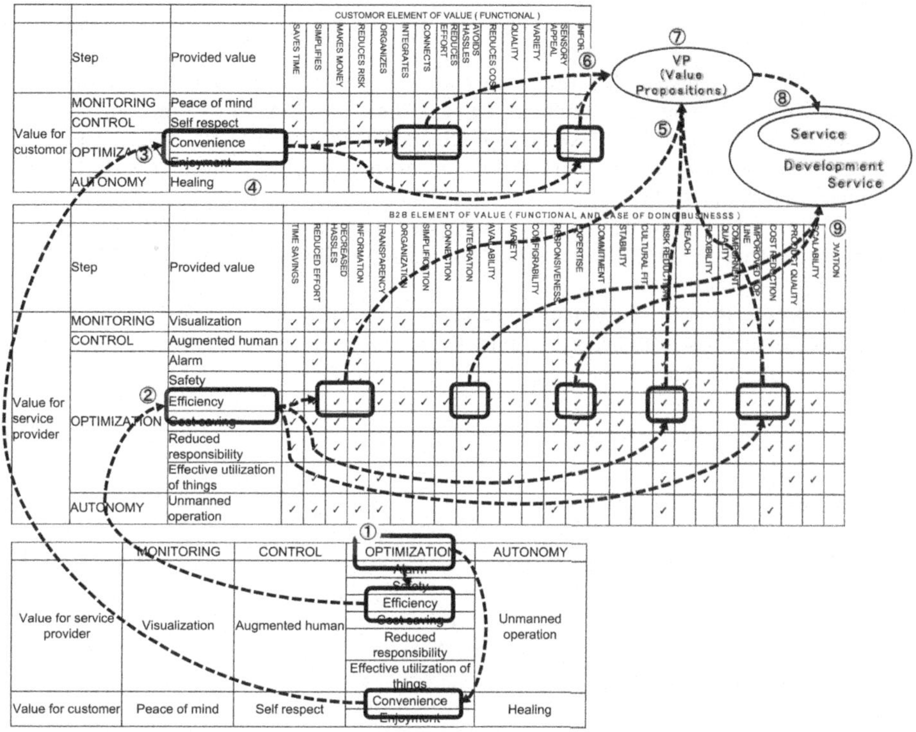

Fig. 5. Procedure of service design by using element of value

6 Conclusion

This paper proposed a method for detailed service design by focusing the value by using elements of value to address the issue of the value provided by the step classification of IoT systems. Same as the elements of physical materials, the elements of value were formed higher-level provided values by combining each. The elements of value focused on the main value of the service and verify the service design. The element of value can also be used to devise a service that provides the second level of elements of value, to develop the new service, and to differentiate it from other services. In the future, the IoT system step classification provided value model will be refined by adding the emotional aspect of the elements of value and will increase the value of service building.

Acknowledgments. The authors would like to thank Associate Professor Kunio Shirahada of the Japan Advanced Institute of Science and Technology (JAIST) for his important comments on the elements of value. The authors are deeply grateful to the working students of the Uchihira Laboratory of the Japan Advanced Institute of Science and Technology (JAIST) for their advice on the direction of the organization of the case studies. This work was supported by Grants-in-Aid for Scientific Research (C) No. 19K12234.

References

1. Porter, M.E., Heppelmann, J.E.: How smart, connected products are transforming competition. (Spotlight On Managing The Internet Of Things). Harv. Bus. Rev. **92**, 64–88 (2014)
2. Shinjo, S., CAMI & Co.: IoT Business Mirai Senryaku 2020. Nikkei BP, Tokyo (2019)
3. Shinjo, S., Uchihira, N.: Proposal of classification method of IoT Systems. Japan MOT Society Conference (2020)
4. Almquist, E., John, S., Nicolas, B.: The elements of value. Harv. Bus. Rev. **94**(9), 47–53 (2016)
5. Almquist, E., Cleghorn, J., Sherer, L.: The B2B elements of value. Harv. Bus. Rev. 2018(March-April) **96**, 72–81 (2018)
6. Osterwalder, A., Pigneur, Y., Bernarda, G., Smith, A.: Value proposition design: how to create products and services customers want. John Wiley & Sons, Hoboken (2014)

"As Normal" Co-created by Visiting Nurses and Patients: Ethnographic Study on Visiting Nurse Station in Japan

Kagari Otani(✉) and Yasunobu Ito

Japan Advanced Institute of Science and Technology, Asahidai 1-1, Nomi-City, Ishikawa, Japan
{s2040401,itoyasu}@jaist.ac.jp

Abstract. In Japan, where one out of four elderly are over 65 years old, with some 4 million visiting nurses, there is a structured system for patients to spend their final days at home instead of in the hospital. This paper aims to illustrate how collaboration between visiting nurses and patients can achieve the living style of patient's choice through co-creation. It displays methods for sustaining the normal life of the patients through detailed case studies of visiting nurses who not only provide nursing care but are proactively engaged in maintaining the "normal" life activities of the patients.

Up until now in the research of Japanese visiting nurses, there were virtually no long-term holistic investigation/research spanning several years where ethnography was being used, nor analysis conducted on co-creation between visiting nurses and patients. In that sense, the analyses in this paper are substantial.

Keywords: Visiting nurses · Co-creation · Ethnography

1 Introduction

The objective is to provide analyses of how value is co-created between visiting nurses and patients, through on-site investigation (ethnography) at the nursing station. The study is intended to contribute to the further development of visiting nurses for supporting patients in home care.

In Japan where one out of four elderly are over 65 years old, there is a structured system being developed for the elderly to spend their last years in their homes instead of the hospital. Based on the national health insurance policy, doctor, nurse, pharmacist, physical therapist, speech-language-hearing therapist, dentist, and dental hygienist can visit patient's homes to offer medical services. There are some 40,000 visiting nurses in Japan. As of date in 2019, there are 11,000 nursing stations in Japan.

Home care services are started when doctors make decision that a patient recuperating at home needs the service whereupon he makes a service request to the nursing station through prescriptions. There are two types of insurances for using this service. One is medical insurance and the other is long-term care insurance. When a patient uses the long-term care insurance, a care manager creates a care plan based on a doctor's

C. Leitner et al. (Eds.): AHFE 2021, LNNS 266, pp. 503–509, 2021.
https://doi.org/10.1007/978-3-030-80840-2_58

plan. A care manager is a long-term care support specialist and is one of the health professionals. Doctors, dentists, pharmacists, nurses or physical therapists who hold a health professional's degree and who has had more than 5 years of clinical practice is eligible to take the exam for a care manager competence test. In the case where patients use a medical insurance, nurses of visiting nurse stations create a care nursing plan for the patient based on the doctor's prescription. Based on this care nursing plan, through consultations with the patient and their family on what the services entails, visiting nurses sometimes adjusts the plan or add services for the plan to be agreeable for the patient and family. Upon agreement, home care service begins when nurses from the station visit the patient [1].

In Japan, health care service is offered to the patient according to the Act on Public Health Nurses, Midwives and Nurses. According to this law, nurses are termed as those who are licensed by Minister of Health, Labor and Welfare to provide in-home Activity of Daily Living (ADL) support and medical care assistance. In-home ADL support refers to ADL activities which involves bathing, toileting, eating, sleeping, performing daily activities, grooming, and oral care. Medical care assistance includes measuring vital signs such as blood pressure, temperature, pulse, and respiration or using stethoscopes on abdomen and chest area to perform auscultation in order to grasp patient's conditions and illness for analysis. This is also called physical assessment. It is an important nurses' task to detect any complications or aggravation of illnesses at an early stage and inform the doctors. Aside from that, nurses assist in treatment by administering intravenous drip, managing respiratory apparatus, and checking oral administration (drug compliance). To summarize, nurses' task is to perform physical assessment, assist in treatment, checkup on the patient's condition or illnesses to be stable and to support ADL. In Japanese law, performing ADL support based on medical care assistance is permitted to nurses only.

Some examples of nurses' tasks are as follows: a patient who became immobilized from a Parkinson's disease cannot go to the bathroom by herself and would never want her husband to see the excretion would ask for the support of the visiting nurse to defecate when they visit her home. Another case is a patient who has a severe heart failure condition but wishes to continue living alone and eat and drink whatever he likes. Patients' wishes are different by each individual.

Nursing is one of the health care services. According to Vargo and Lusch, who advocated Service-dominant (S-D) logic, co-created value differs by each patient and also differs by the environment, conditions, and background; in other words, it differs by its context [2, 3]. In health care terms "co-created value is unique according to the wishes of patients and their family".

There are not many who conducts research of health care services based on the perspective of S-D logic. In the few precedent studies, Batalden analyzes the co-created value between the patient and the health professionals in health care and presented a co-creation concept model [4]. In regards to nursing there is research by Ding [5]. Ding conducted questionnaire survey to nurses working in the public hospitals in northern China. It became clear that both patient and nurse satisfaction level went up when the patient and nurse cooperated in creating health care services at the hospital [5]. Zhang indicates on the importance of health care evaluation by the patient and theorized that through the co-created value system whereby health care professionals grasp the patient satisfaction level to improve health care services, will in turn contribute to the recovery of patient [6].

The aforementioned study showed that through value co-created in health care services between health care professionals and patients, it improved the quality of health care services which led to an increase in patient satisfaction. However, there is not much case study analyses that elaborate on how value is co-created between nurses and patients under various circumstances. In addition, in the study of visiting nurses in Japan there is virtually no holistic long-term research spanning several years which applies ethnography. In this context, the analyses in this study are innovative.

2 Method

The investigation took place at the X visiting nurse station in Nagoya city. X visiting nurse station offers health care services where it aims to make arrangement to patients' daily lives by enabling patients to live the life style of their choice by offering life style support. Visiting nurses use the car owned by the nurse station to visit the patient. Monthly visits amount to around 120 cases. Studies of this nursing station was conducted through participant observation and interviews. The interview subjects are six nurses from this station. The subjects are all female and include, Nurse A and Nurse B in their 60s and nurse C, D, and E in their 50s. Nurse A and B have over 40 years of working experience and nurses C, D, and E have over 30 years of experience. Aside from these nurses, there are two office administrators, two physical therapists, one occupational therapist, and two speech-language-hearing therapists.

Table 1. Patient attributes

Age	Distinction by sex	Disease	Contents of nursing
60s	Female	Parkinson's disease	Health check, Administration of medication, Massage
70s	Female	Parkinson's disease	Health check, Bed bath, Massage for bowel movement, Disimpaction, Management of Gastric fistula
60s	Female	Spinal cord injury	Health check, Massage for bowel movement
80s	Female	Chronic kidney disease, Dementia	Health check

The nurses are in charge of patients with various diseases including muscular dystrophy, Parkinson's disease, cerebral infarction sequelae, heart failure, cancer, COPD, ALS, kidney failure, and diabetes. They perform various tasks including managing respiratory apparatus, checking drug compliance, treating injuries, bathing support, cleaning body and washing hair in bed, changing diapers, disimpaction, administration of nutrients in gastric fistula, and eating support. Visiting nurses listen to the problems the patient and the family are having and offer consultation. The study conducted at the patients' home included nine patients among which four will be presented in this paper (Table 1).

3 Result and Discussion

According to the research result, the four nursing characteristics of visiting nurses are as follows:

3.1 Observe and Underst and the Daily Lives of Patient and Family

Visiting nurses makes observation from the information acquired in patients' homes. When they visit patients' homes, they observe closely the neighborhood surroundings, exterior of the house, details in the house such as smell of home, products used in daily living, how patients and family dress and how they apply make-up. They also have casual dialogue with patient and family. Through them they try to understand the daily lives of patients.

The information gathered from observation is inputted in the patients' clinical record and shared among other nurses. According to nurses, it not only provides them with the actual situation of patient and family but also life styles and values cherished by the patient. Visiting nurses respect the values and life styles and creates an image of the daily living of patient and family in order to support and sustain their preferred living style as their goal in nursing. One of the nurses mentioned, "I try to work around to fit in to their life style" The interview follows:

"The atmosphere of the homes differs by the patient. I think the exterior of the home and inside both reflect on the values and life style of the patient. I think it is important to understand the normal atmosphere of the patients' homes. The reason is, patients would like to continue on with their normal daily activities, but because of illnesses, they are no longer able to do that by themselves; that is why they are seeking nursing services and wishing to sustain their daily living. When I visit patients' home, from the moment I enter the entrance door, I compare the atmosphere from the last time I visited and assess how the patient's condition is on that day. Depending on whether the patient is in a good condition or in a bad condition, the atmosphere is different. Then based on this assessment, I apply this in nursing in order for the patient to sustain their daily lives." (Nurse A Interview)

"This patient (a woman in 60s with a Parkinson's disease) was a housewife who supported her husband and raised two children. Even when her body became immobilized due to the disease, I think she still wishes to be "a good wife who supports her husband" and "a good mother to her children". Her illness has aggravated and treatment no longer seems to be effective. Her body hardens several times a day and she cannot move. With pain, she can no longer perform housekeeping. But even under these conditions, I have not heard her complain once. I think she does not want to complain and show her weakness. I can feel her thoughts when I am massaging her body with my hands. I assess her physical condition through massaging. I would like to nurse and support her through assessing her conditions so that she can exist as she wishes. (Nurse D Interview)

3.2 Exploring Care Service Methods Preferable to the Patient and Family, and Creating It

Patients each have their own ideas on preferred care service and on how it is carried out; they have preferences in style. When drying the body, patients have their own styles and rules on where to dry first and in which order it is done, how many times to wipe, and with predilection on what products to use and the timing of care. Similarly, the family also has rules. Nurses listen carefully to what patients and family have to say. The service menu is already fixed in the nursing plan, but nursing is performed according to the conditions of the patients. Depending on the patients' physical and mental conditions or sometimes upon requests from the family, it can be changed and arranged accordingly. Below shows the field note by one of the authors (Otani).

"When the visiting nurse arrived at the patients' home she was greeted by the patients' husband. She entered the living room and asked how the husband and the patient is doing. The patient (a female in 70s with a Parkinson's disease) was lying on the bed in pajamas next to the living room. The nurse measured the patient's vital signs (blood pressure, heart beat, temperature, and respiration) and SPO2 (percutaneous oxygen saturation) and checked the chest area using stethoscope. Then she asked her husband whether there was defecation. Her husband replied anxiously that there was no defecation for several days. Because this patient was in bed most of the time, bowel movement was weak and therefore difficult to perform defecation. In the kitchen, the nurse started to warm up four towels in the microwave to wipe the body. The towel was prepared by the husband. The nurse wiped the body, massaged the abdominal area and expulsed a huge amount of feces. The patient seemed relieved. The nurse performed this while talking with her husband. The patient smiled while listening to the conversation. In the drive back to the nursing station, the nurse told me that because the husband was worried about the wife's constipation, she massaged the abdomen area more carefully than usual. In the case of this patient, four towels were used to wipe the body but the method of wiping and abdomen massage, and what to wipe the body with are different per patient. Nurses listen to the patient and the family to understand their preference on what to use, how to wipe, and perform accordingly. Patient and family makes an order each time on which method to use. Nurses take in their choice and perform as requested. (From the field note by the author)

As mentioned above, the nurse observes the body movement of the patient, listen to their words, read their facial expressions while providing care, in order to explore and create optimal care methods preferred by the patient and family. They consider patient and family to be partners in creating care together with the patient,

3.3 Using the Senses Obtained from Experience

It goes without saying that patients differ in their illnesses, its staging and the state of disease. Nurses through care also understand and memorize the patient's physical condition through their own body senses. They perform care by applying their physical senses from memory.

"When I touch the patients' body, it passes on the physical condition of the patient through my hands. In this case (a female in 60s with a Parkinson's disease), is in a good condition when her palm is cold, and in bad condition when her palm is hot. By touching the patient, I grasp the physical condition and compare this with the senses I stored in my memory to change the massage parts, pressure and timing. (Nurse C Interview)

3.4 Patient and Nurses Respond to Each Other

Nurses listen to what the patient has to say, and patient responds to the care. Through this reciprocal process, nurses dig into the needs of the patient and improves the method. The patient uses this process to propose a method to suit their needs. The fact that nurses listen and respond reduces anxiety of the patient who is suffering from illness and gives them a peace of mind. There are times when the family consult with nurses about their troubles. Nurses listen in to what the family has to say in order to seek how they can adapt their support for sustaining the life style of the patient in the home going forward.

"This patient (a female in 60s with a spinal cord injury) is constipated and the abdominal area feels heavy all the time. The patient is orally administered with a laxative. When a nurse visits this patient they always talk about constipation. The nurse listens every time and at times brings tea which is effective for constipation or introduces cooking recipes. I asked the nurse why the patient does not recover from constipation and she answered "The patient defecates. The patient wants to talk about how she is suffering from not being able to move around because of spinal cord injury. But instead of saying it straight forward, she says she suffers from constipation. So that is why I listen to her every time on how she feels heavy from constipation. I think she can sustain her daily living through this talk." (Nurse E Interview)

"This patient (a female in her 80s with chronic kidney disease who is getting dialysis) is receiving blood dialysis three times per week. The blood pressure is high and she feels lethargic. She is in bed most of the time. The patient and her husband both have dementia and their daughter lives with them to provide care. The nurse visits when the daughter is at home, listens to her story and her problems and takes care of her condition as well. Through the discussions with the daughter, the nurse continues to explore how the family can sustain their daily lives. The family cannot live together without the support of their daughter, so I place importance on how she would like to live and nurse accordingly." (Nurse C Interview)

4 Concluding Remarks

The paper clarified the above mentioned four characteristics of nursing through visiting nurses. Visiting nurses acted like an anthropologist or ethnographer by using the various data acquired, surrounding the patient and family to decipher the patients' life style, social background, and values. It became apparent that visiting nurses did not just rely on dialogue with the patient and family to perform care service but acquired various information related to the patient's home through other means. In this way, visiting nurses investigate on the life style wished by the patient and proposes methods to realize this life style. Patients respond to such care by the visiting nurses and nurses in turn make decisions based on their responses. This cycle modifies care. In this way, the value co-created by the nurses and patient formulates the "normal living style" wished to be sustained by the patient. Through the study, it became clear that visiting nurses consider patients and family as partners instead of targets, and they explore the preferred care together with the patient to create tailored care.

The interviewed nurses were all veterans with over 30 years of experience. The next agenda going forward is to investigate on how each of these nurses acquired the nursing skills.

Acknowledgments. We have had a lot of support and encouragement from patients, their families, and nurses of N Visiting Nurse Station. This work was supported in part by JSPS KAKENHI Grant Numbers JP 18H00782 /19H05469.

References

1. Japan Visiting Nursing Foundation. "Visiting Nursing System in Japan". https://www.jvnf.or.jp/homon/english/vnj2015.pdf
2. Vargo, S.L., Lusch, R.F.: Evolving to a new dominant logic for marketing. J. Mark. **68**, 1–17 (2004)
3. Vargo, S.L., Lusch, R.F.: Service-dominant logic: continuing the evolution. J. Acad. Mark. Sci. **36**, 1–10 (2008)
4. Batalden, M., et al.: Coproduction of healthcare service. BMJ Qual. Saf. **25**, 509–517 (2016)
5. Ding, B., et al.: Effect of patient participation on nurse and patient outcomes in inpatient healthcare. Int. J. Environ. Res. Public Health **16**, 1344 (2019)
6. Zang, L., et al.: A practical model of value co-creation in healthcare service. Procedia Manuf. **3**, 200–207 (2015)

Discovering Topics of Interest on Steam Community Using an LDA Approach

Yang Yu[1(✉)], Ba-Hung Nguyen[2], Fangyu Yu[1], and Van-Nam Huynh[1]

[1] Japan Advanced Institute of Science and Technology, Nomi, Japan
{s2020045,s2010185,huynh}@jaist.ac.jp
[2] Thai Binh Duong University, Nha Trang, Vietnam
nbhung@tbd.edu.vn

Abstract. Reviews from players regarding different characteristics of an esports game are one of the worthiest sources for the developers to enhance their services or adjust operating strategy. However, little research has been conducted on detecting esports players' favorite topics dealing with topic modeling. Thus, this paper aims to use a data mining approach to analyze community data in the games domain available on Steam. We collected more than 1.2 million English reviews from four esports games up to August 2020 on Steam. Our contributions in this paper are: (i) we manually build a dataset by filtering out high-quality esports reviews, (ii) we then infer and group reviews into 3 groups with 19 topics, and (iii) we add more contributions to finding the emerging opinions of esports players towards the different topics of esports reviews, which might benefit further research on understanding esports reviews.

Keywords: Esports reviews · Steam · Topic modeling · Collaborative filtering

1 Introduction

Esports (also known as electronic sports) is a rapidly growing industry, with more than 2 billion players and spectators and supports a billion-dollar revenue market. More and more players are attracted due to the competitive characteristics of esports. Compared with the traditional game mode that emphasizes the graphics and storylines, esports games are more advocating Player versus Player (PvP), which makes the competition of esports games enjoyable to watch. The competitive element also helped esports games be successfully recognized as an official medal sport at the 2022 Asian Games[1]. As a result, research related to esports has significantly progressed for the past few years [1, 2].

In general, esports games can be mainly classified into: first-person shooter (FPS)/battle royales (BO); Multiplayer online battle arena (MOBA)/real-time strategy (RTS); Fighting; Card games, and so forth based on their gameplays. According to the

[1] https://www.aesf.com/en/News-And-Media-Coverage/Esports-Returns-As-Proposed-By-The-Hagoc-For-The-2022-Asian-Games.html.

C. Leitner et al. (Eds.): AHFE 2021, LNNS 266, pp. 510–517, 2021.
https://doi.org/10.1007/978-3-030-80840-2_59

investigation of the global esports market by Newzoo[2], for the Top 25 live esports games that were watched on YouTube, Twitch, and Mixer, 72% of the games consist of FPS/BO, MOBA, and Fighting.

As new games are first released to desktop gaming clients, some gaming clients have developed rapidly in recent years. Many players share their reviews on the game on these platforms. For instance, all reviews at Steam[3] will be classified into two sentiments: *Recommended* or *Not Recommended* (refer to *Positive* or *Negative*). Players can choose whether to recommend this game when commenting, and the game judges whether the game is recommended by counting the tendency of the majority of player groups. Because of its comprehensive review system, several researchers choose Steam as their dataset to analyze player groups' behaviors [3, 4]. Game reviews on game clients are the worthiest resources to reflect esports enthusiasts' preferences and existing problems so that developers can improve the quality of an esports game. However, much existing research for game reviews consider esports games as a genre of games rather than the esports games individually, which ultimately leads to a focus on analyzing the design of the game itself, but the lack of analysis related to esports elements. In addition, little work has been paid to analyze esports games as a whole. In this study, we will use a data mining approach to analyze community data in the esports games domain available on Steam.

In the next section, we briefly present related work on topic modeling in the application and current research in the game reviews. Next, the methodological details of data collection-processing and the analytical approach are described in Sect. 3. Section 4 shows the empirical results based on the online esports reviews dataset collected from Steam. Finally, we conclude our work and provide future research directions in Sect. 5.

2 Related Work

To discover potential topics embedded in text data, the pioneering work of Blei et al. [5] proposed the original Latent Dirichlet Assignment (LDA), which uses variational Bayesian inference to infer possible topics from a large corpus. In applications of topic modeling for online reviews, for example, Heng et al. [6] found Amazon Service, Physical Feature, Flavor Feature, and Subjective Expression are four significant factors that impact the helpfulness of customer reviews at Amazon's online food shopping. Huang et al. [7] showed that Yelp reviews of restaurants could be divided into multiple potential themes, such as service, value, decoration, and health. Tran et al. [8] analyzed nearly 76 thousand reviews on TripAdvisor, a website offering travel services, and summarized them into 11 topics: food, hotel facilities, price, staff, and so forth. While the products and services mentioned above have well-explained customers' preferences, the game is a particular category of experience service. Compared with the restrictions on the number of words and symbols on platforms such as Twitter, as a particular social platform, the game platform has little to no restrictions on game reviews (only simple, sensitive word shielding), resulting in the uneven quality of review data and invalid expressions.

[2] https://newzoo.com/insights/trend-reports/newzoo-global-esports-market-report-2020-light-version/.

[3] https://store.steampowered.com/.

Game reviews could contain invalid information, such as particular characters or spam messages. As a result, the review text information obtained from game platforms lacks standardization, and high-quality reviews are scarce. This reasons the research on game reviews is different from the reviews on other platforms. In the past two decades, several studies have been made on game reviews for game user research. Gifford [9] analyze the differences of reviews between video game and film. Lin et al. [3] point out that game reviews are different from mobile app reviews along with several aspects, and both positive and negative reviews could be useful to game developers. Zagal et al. [10] demonstrate the relationship between game rating and sentiment words in which players were chosen. Bond et al. [11] identify the features of a good game by analyzing game reviews. Livingston et al. [12] discovered game reviews and ratings can be transferred to commercial success.

In general, most of the research on game reviews focus on the game rating and game selection and consider esports games as common as others while lacking the independent analysis of the esports review contents. Hence, in this study, we aim to explore esports players' potential opinions by employing a topic modeling approach for reviews from the Steam community. The discovered topics could provide esports game developers and platforms to gain deeper, useful information and feedback to enhance their services.

3 Data and Methods

3.1 Dataset

In this study, we select Steam platform as our data source due to the variety and amount of games and the comprehensive review system. Based on the survey reports of Newzoo and existing esports events, we select four representative esports games, namely CS:GO (FPS), PUBG (BO), Dota2 (MOBA), and TEKKEN 7 (Fighting). We used the API provided by Steam to obtain a new dataset from Steam, which contains all English reviews for four chosen esports games up to Aug 2020. Totally 1,297,975 reviews were collected from Steam.

Before applying NLP tools to preprocess our dataset, we found that the quality of each game reviews is scattered. For example, approximately 12% of collected esports reviews are non-English ones even though their language tags on Steam are English, with a few profanities and special characters such as Emoticons and Kaomoji were contained in our collected reviews. Therefore, we filter out non-English words and profanity besides typical preprocessing. Table 1 presents numbers of positive reviews, negative reviews, and total reviews after our prepossessing.

Table 1. Number of esports reviews per game after prepossessing.

Game	TEKKEN7	PUBG	Dota2	CS:GO
Positive	14,294	166,348	302,003	440,101
Negative	1,650	107,517	41,445	61,716
Total	15,944	273,865	343,448	501,817

To provide a high-quality esports review dataset for future research, we manually build a new dataset[4] based on our extracted esports reviews. Many studies have examined the factors that influence review helpfulness. For example, review rating [13, 14], the length of words [15, 16], positive and negative sentiment [17] have been used to explain and predict the quality of reviews. As for Steam reviews, num_voted_helpful (the number of users who found this review helpful) is the most significant factor affecting helpfulness [4]. Thus, we filtered the reviews based on the length and the num_voted_helpful, and then we manually cleaned and checked reviews. On this basis, we discard reviews that were too short, and we also refer to the playtime when the review was written. Figure 1 presents distributions of length and helpful of reviews:

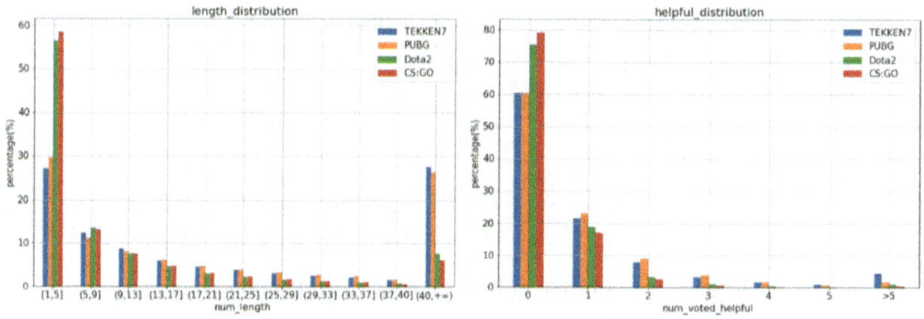

Fig. 1. Length and helpful distribution of four esports games.

3.2 Latent Dirichlet Allocation (LDA) for Topic Modelling

In this study, we applied the Latent Dirichlet Allocation (LDA) [5], a widely used topic modeling method to discover the underlining topics from the review text. LDA is a generative probabilistic model. It assumes that each document is represented as a random mixture of potential topics, and the feature of each topic is the distribution of words in a vocabulary. The algorithm as below shows the LDA generative process:

```
program Generative Process of LDA
  for each topic k∈[1, K] do
    sample mixture proportion φ_k ~ Dirichlet(β)
  end
  for each document d∈[1, D] do
    sample mixture proportion θ_d ~ Dirichlet(α)
    sample document length N_d ~ Poiss(.)
    for each word n∈[1, N_d] do
      sample topic index z_{d,n} ~ Multinomial(θ_d)
      sample a term for word w_{d,n} ~ Multinomial(φ_{z_{d,n}})
    end
  end
```

[4] Data available on request from the authors.

4 Experimental Result

For finding the optimal number of topics, we analyze the coherence value of four esports games, which is a measure used to evaluate a topic model [18]. From Fig. 2 presents the number of topics ranging from 3 to 99 with its coherence values, we could choose the optimal number of topics as 19 as the coherence values of the larger number of topics are flattened out after this.

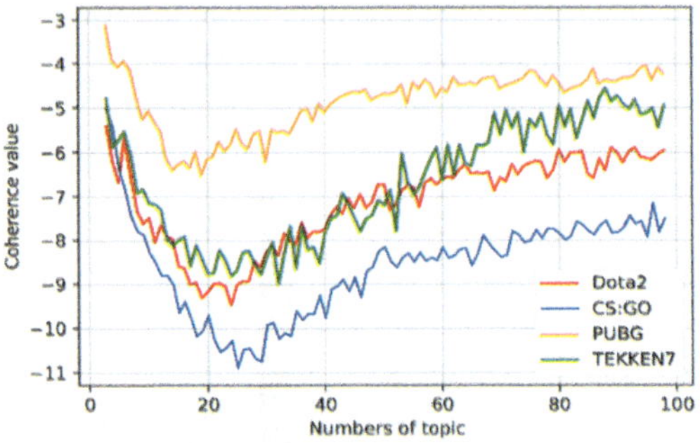

Fig. 2. Coherence values of four esports reviews.

In addition, we present a representative topic for each esports game inferred through LDA using WordCloud in Fig. 3. The size of each word corresponds to its probability in a topic:

Fig. 3. Topics discovered by LDA.

We validate the inferred topics of the LDA model by combining reading topics words from WordCloud and original review text to avoid implicit expressions or forum spam. We apply this collaborative filtering with three agents. Table 2 lists the topics we summarized from esports reviews. We discover three types of topics as follows:

Table 2. The result for inferred topics and summarization.

Category	TEKKEN7	Dota2	PUBG	CS:GO
I	Community			
	Server			
	Region			
	Ranking system			
	In-game purchases			
	Matchmaking			
	Esports competition			
	Graphics			
	Audio			
II	Character design			
	Game update			
	In-game event			
	Player skills			
	Game strategy			
	Game balance			
	Map/Scenes			
	Learning curve			
III	Combo command	Teamwork	Cheater	Teamwork
	Storyline	Hero builds	Optimization	Cheater

1. **Common Topics with Common Meanings (I):** These are topics commonly found among four esports games with the same or highly similar meanings. E.g., "Game server" includes internet stability, network delay, and so forth, and "Region" corresponds to geographical divisions such as North America (NA), Europe (EU) or Southeast Asia (SEA).
2. **Common Topics with Specific Meanings (II):** These topics can be found among four esports games but have specific meanings for the different esports game players. E.g., "Player skill" in Dota2 refers to the familiarity of champions' abilities, while it means instant reaction or spray patterns in CS:GO. "In-game event" could be daily check-in incentives or new game mode depending on different esports games.
3. **Game Specific Topics (III):** These are unique topics that do not appear in all esports games. E.g., The mechanism of fighting games determines that TEKKEN7 does not easily suffer from "Cheater" compared with other esports games. Besides, players of TEKKEN7 rely on "Combo command" to improve their offensive power, which does not exist in PUBG or CS:GO.

Based on our understanding of Table 2, Category I correspond to the fundamental elements of the esports game ecosystem, such as "Server", "Community", "Esports competitions"; Topics in Category II more reflects the unique design of each esports game, including "Character design", "Player skills", "Game balance", "Learning curve", and so forth; Category III reflects the most discussed and concerned problems of the player community. For example, PUBG is suffering from cheating behaviors, and Dota2 and CSGO have many reviews discussing teamwork in the games.

In sum, by using the LDA topic model coop with collaborative filtering, we could identify 19 topics that belong to three categories that not only represent the essential topics in esports games but also reflect each esports game's distinct characteristics.

5 Conclusion

In this study, we aimed to identify and summarize the topics of esports reviews by players on the esports communities with topic modeling. Our contributions in this paper are first by utilizing the topic modeling approaches on more than 1 million esports reviews from the Steam community, we uncover 19 topics embedded in reviews which reflect important esports aspects such as "Server", "Game update", "Cheater", "Teamwork", and so forth. Second, via simple collaborative filtering with three agents, we identify three types of topics that correspond to not only the common aspects of esports but also represent distinct characteristics of esports. Finally, we also provide an esports dataset containing manually filtered reviews from four representative esports games (TEKKEN7, Dota2, PUBG, CS: GO), in which opinion words are explicitly expressed. We hope this dataset could be helpful for further research on discovering insights from esports reviews that utilize machine learning approaches.

There are apparent limitations to our current work. Firstly, this study only focuses on the English reviews, ignoring the enormous number of reviews from non-English reviewers, mostly Chinese reviews. Moreover, the source of data was still limited. Other representative esports games such as League of Legends or Overwatch were not analyzed, which might affect our recognition of emerging opinions where esports players concern. Additionally, tracking topics overtime should also consider the game development life cycle (GDLC) to extend our study monthly and summarize the topics in different periods.

References

1. Hallmann, K., Giel, T.: eSports–competitive sports or recreational activity? Sport Manag. Rev. **21**(1), 14–20 (2018)
2. Hamari, J., Sjöblom, M.: What is esports and why do people watch it? Internet Res. **27**(2), 211–232 (2017)
3. Lin, D., Bezemer, C.P., Zou, Y., Hassan, A.E.: An empirical study of game reviews on the steam platform. Empir. Softw. Eng. **24**(1), 170–207 (2019)
4. Kang, H.N., Yong, H.R., Hwang, H.S.: A study of analyzing on online game reviews using a data mining approach: Steam community data. Int. J. Innov. Manag. Technol. **8**(2), 90 (2017)
5. Blei, D.M., Ng, A.Y., Jordan, M.I.: Latent dirichlet allocation. J. Mach. Learn. Res. **3**(Jan), 993–1022 (2003)
6. Heng, Y., Gao, Z., Jiang, Y., Chen, X.: Exploring hidden factors behind online food shop-ping from amazon reviews: a topic mining approach. J. Retail. Consum. Serv. **42**, 161–168 (2018)

7. Huang, J., Rogers, S., Joo, E.: Improving restaurants by extracting subtopics from yelp reviews. In: iConference 2014 (Social Media Expo) (2014)
8. Tran, T., Ba, H., Huynh, V.-N.: Measuring hotel review sentiment: an aspect-based sentiment analysis approach. In: Seki, H., Nguyen, C.H., Huynh, V.-N., Inuiguchi, M. (eds.) IUKM 2019. LNCS (LNAI), vol. 11471, pp. 393–405. Springer, Cham (2019). https://doi.org/10.1007/978-3-030-14815-7_33
9. Gifford, B.: Reviewing the critics: examining popular video game reviews through a comparative content analysis. Ph.D. thesis, Cleveland State University (2013)
10. Zagal, J.P., Tomuro, N., Shepitsen, A.: Natural language processing in game studies research: an overview. Simul. Gaming **43**(3), 356–373 (2012)
11. Bond, M., Beale, R.: What makes a good game? Using reviews to inform design. In: People and Computers XXIII Celebrating People and Technology, pp. 418–422 (2009)
12. Livingston, I.J., Nacke, L.E., Mandryk, R.L.: The impact of negative game reviews and user comments on player experience. In: Proceedings of the 2011 ACM SIGGRAPH Symposium on Video Games, pp. 25–29 (2011)
13. Susan, M.M., David, S.: What makes a helpful online review? A study of customer reviews on amazon.com. MIS Q. **34**(1), 185–200 (2010)
14. Cao, Q., Duan, W., Gan, Q.: Exploring determinants of voting for the "helpfulness" of online user reviews: a text mining approach. Decis. Support Syst. **50**(2), 511–521 (2011)
15. Kim, S.M., Pantel, P., Chklovski, T., Pennacchiotti, M.: Automatically assessing review helpfulness. In: Proceedings of the 2006 Conference on Empirical Methods in Natural Language Processing, pp. 423–430 (2006)
16. Hao, Y.Y., Li, Y.J., Zou, P.: Why some online product reviews have no usefulness rating? In: PACIS 2009 Proceedings, p. 100 (2009)
17. Berger, J., Sorensen, A.T., Rasmussen, S.J.: Positive effects of negative publicity: when negative reviews increase sales. Mark. Sci. **29**(5), 815–827 (2010)
18. Newman, D., Lau, J.H., Grieser, K., Baldwin, T.: Automatic evaluation of topic coherence. In: Human Language Technologies: The 2010 Annual Conference of the North American chapter of the Association for Computational Linguistics, pp. 100–108 (2010)

Dialogue Tool for Value Creation in Digital Transformation: Roadmapping for Machine Learning Applications

Naoshi Uchihira(✉)

Japan Advanced Institute of Science and Technology, 1-1 Asahidai, Nomi 923-1292, Ishikawa, Japan
uchihira@jaist.ac.jp

Abstract. With the fast spread and reliance on digital technologies in industry and society, collaboration between humans and machines (artificial intelligence and machine learning) becomes an important subject; however, it is not clear what kind of value can be specifically created by the collaboration between humans and machines. Roadmapping is effective as a dialogue tool for clarifying the value among stakeholders. However, the traditional roadmapping methods are insufficient, since collaboration between humans and machines can be considered as a socio-technical system, and hence evolves, while influencing each other side. This paper proposes the new co-evolutionary technology roadmapping method, and reports the results carried out for the roadmapping workshop for machine learning applications.

Keywords: Technology roadmapping · Value creation · Digital transformation · Machine learning application · Socio-technical system · Co-evolution

1 Introduction

It is expected that digital transformation will create new values for many industries and societies. This transformation can be realized as a "*cyber-physical system*" within which data in the real space is accumulated in cyberspace (cloud) through the Internet of Things (IoT), and value is created with the information processing in cyberspace and fed back to the real space. Here, artificial intelligence technology which includes machine learning is indispensable for information processing in cyberspace. This transformation is currently underway, but the goal and path of the transformation are not clear, and the existence of various perception gaps among the stakeholders is one of the major reasons why the transformation has not progressed as expected. There is a critical need for a "roadmap" to share with various stakeholders on how business and society should be transformed in the future and how artificial intelligence and machine learning will evolve for the digital transformation, and to align the vectors of stakeholders toward those goals.

Various roadmaps for artificial intelligence and machine learning have been proposed [1, 2]. The Japanese government also compiled the "Roadmap for R&D Goals and

C. Leitner et al. (Eds.): AHFE 2021, LNNS 266, pp. 518–526, 2021.
https://doi.org/10.1007/978-3-030-80840-2_60

Industrialization of Artificial Intelligence" [3] in 2017. In addition, the Japanese Society for Artificial Intelligence (JSAI) constructed a technology map consisting of an AI issue map and an AI technology map [4].

Roadmapping is the process of creating, sharing, and utilizing roadmaps, and roadmaps are the outputs of roadmapping. The roadmap itself should be refined as a situation change, and there is no absolute fixed roadmap. What is more important is to share the various possibilities of the current situation and the future among the stakeholders through the process of creating the roadmap, and to bridge the perception gaps. In particular, with the rapid development and spread of digital technologies, the collaboration between humans and machines (artificial intelligence and machine learning) became an important issue, but it is not clear what kind of value can be created by the collaboration between humans and machines. The roadmapping is an effective dialogue tool to clarify the value among stakeholders.

The collaboration between humans and machines is a socio-technical system, and its implementation technologies (systems engineering for machine learning application) cannot be addressed by existing roadmapping methods. In this paper, we propose a new roadmapping method considering the socio-technical systems, and apply this method in a roadmapping workshop for machine learning applications as part of the "Modelling and AI for Integration of Cyber and Physical World" project in the JST-Mirai program [5].

2 Technology Roadmapping for Machine Learning Applications

Technology roadmapping is one of the important tools for technology management, and has been investigated extensively [6]. There are two types of roadmapping: backcasting, which indicates what kind of technological development is necessary to realize the desired future state (new products and services), and forecasting, which describes the future as an extension of the current technologies. Phaal et al. [7] proposed a backcasting-type of technology roadmapping method (T-Plan). The forecasting-type roadmapping is technology-driven. In academic roadmapping, a forecasting approach is commonly used, in which it considers the future technological evolution deductively based on the existing technologies. In addition, Uchihira proposed "middle-up-down type roadmapping," which is a roadmapping method suitable for systems engineering [8].

Systems Engineering for Machine Learning Applications (SEMLA) is a rapidly progressing research field, which investigates and develops various methods and tools to support the development and operation of machine learning application systems (MLASs). In MLASs, the co-creation of value between humans and machines is an important perspective, and how to design systems that include humans and machines (socio-technical systems) is a new challenge for systems engineering in the socially-centered automation [9].

For systems that include humans and machines altogether, a new technology roadmapping method is required. In these systems, "co-evolution" is essential, in which the perceptions and roles of humans and society evolves along with the evolution of machines, and simple backcasting and forecasting types of roadmapping are not suitable for it. For example, e-commerce has been evolving in the last few decades and is giving its significant impacts on human consumption behavior. Value of e-commerce has been created by co-evolution of humans and machines. In addition, the middle-up-down type does not take into account the "co-evolution" of humans and machines.

Roadmapping in SEMLA needs to take into account the co-evolution of humans and machines. Specifically, it is necessary to set up phases in which the relationship between humans and machines evolves step-by-step, and to have the ability to design values that humans and machines can co-create for each phase (Fig. 1). For example, in the autonomous driving system, which is one of the typical machine learning applications, six levels have already been set in the relationship between the driver and the autonomous driving support system, which corresponds to the phases of the roadmap.

In the following sections, the authors propose a co-evolutionary roadmapping method for SEMLA based on current problem recognition. However, the proposed method is not a complete co-evolutionary technology roadmapping, but rather a roadmapping that considers the phases as the initial stages.

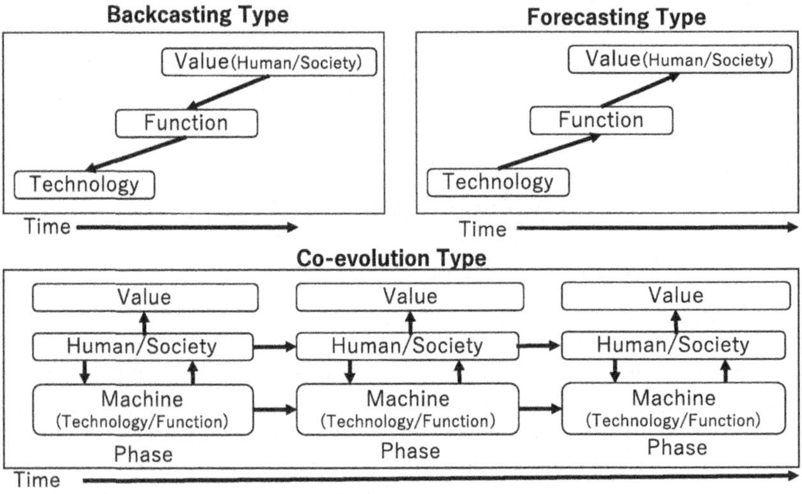

Fig. 1. Roadmapping types

3 Proposed Technology Roadmapping Method

This section proposes a new technology roadmapping method for SEMLA, which consists of seven steps (Fig. 2). In this method, a "trend map" is introduced in Step 3, which consists of a set of phases. The relationship between humans and machines evolves phase-by-phase. Each phase indicates values co-created by humans and machines as shown in the following steps.

Step 1: Issues and Needs Map. Issues and needs for the development of MLASs are extracted from various surveyed materials and interviews conducted in companies, and were classified and organized into a map. These issues and needs are for MLAS developers, not for MLAS users. Values (needs) for MLAS users are specified in the market trend map (Step 3).

Step 2: Technology Seeds and Policy Map. Related technologies and policies (non-technical solutions) that are currently considered to solve the issues and satisfy needs are surveyed from papers, articles and reports on SEMLA and interviews with experts, and are classified and organized into a map.

Step 3: Market Trend Map (by application). The market trend differs for each target application. For each application, future market and application trends are created based on published materials. Here, phases and their values of gradual evolution of the relationship between humans and machines can be introduced.

Step 4: Issues and Needs Trend Map (by application). The phases and items of the market trend map are related to the items of the issues and needs map by application.

Step 5: Needs/Seeds Matching Table (by application). A matching table of needs and seeds is created by application. In this table, linkage among values of phases (the market trend map), issues and needs (the issues and needs map) and technological seeds (the technology seeds and policy map) are specified.

Step 6: Technology Roadmap (by application). A technology roadmap is constructed by arranging the items of seeds/needs matching table on a time axis.

Step 7: Strategic Stories (by application). A strategic R&D story is prepared to explain the technology roadmap. Here, the target market and goals, important technologies and features, competitive advantages, risks, and a concrete R&D plan are proposed to the decision makers.

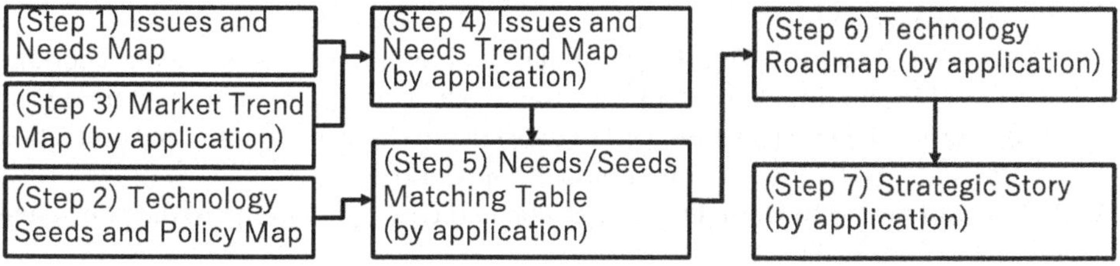

Fig. 2. Proposed Technology Roadmapping Method

4 Technology Roadmapping Workshop

Based on the proposed method in Sect. 3, the authors conducted a two-day workshop (held September 15 and 16, 2019, at Shonan Village Center, Kanagawa, Japan) with the participation of 20 experts in SEMLA and roadmapping (half of the participants were university members and the other half were company participants). This workshop was organized as part of the "Modelling and AI for Integration of Cyber and Physical World" project [5]. The main purpose of constructing this technology roadmap for SEMLA is to prepare an R&D strategy for acquiring additional funds from the Japanese funding agency.

There are several types of MLASs. The authors focused on three important target applications: autonomous driving, smart factory, and robotic process automation (RPA). Participants were divided into these three groups based on their expertise and interest.

4.1 Issues and Needs Map

The issues and needs map for SEMLA was prepared by the project core members before the two-day workshop. The map was based on the results of publicly available surveys by Japanese funding agencies NEDO (New Energy and Industrial Technology Development Organization) [10] and IPA (Information-technology Promotion Agency, Japan) [11], as well as interviews with ten companies (automobile, electronics, IT service, etc.) conducted by this project. The issues and needs can be categorized into the following 12 items, which are illustrated in a mind map [12].

1. **Reliability and Safety**: Difficulties in the quality assurance of MLAS (reliability, safety, and performance).
2. **Efficiency and Productivity**: Difficulties related to the efficiency and productivity of MLAS development (cost and delivery).
3. **Process Management**: Difficulties in process management of MLAS implementation, operation, and maintenance.
4. **Relationship between Humans and AI**: Difficulties arising from the immaturity of the human-AI relationship.
5. **Business and Monetizing**: Difficulties related to investment and return in MLAS.
6. **Standards and Guidelines**: Difficulties caused by the lack of safety standards, quality assurance guidelines, and liability recognized in society and industry.
7. **AI Awareness**: Difficulties caused by stakeholders' misperceptions of AI.
8. **AI Human Resource Development**: Difficulties related to the talent shortage capable of leveraging AI (executives, users, managers and system developers).
9. **Data and Model Distribution**: Difficulties related to the distribution and protection of data and models (ownership of models and data).
10. **Policy and Social System**: Difficulties due to the incompatibility of the policy and social system for AI.
11. **Security and Privacy**: Difficulties in ensuring security, and protecting privacy.
12. **Legal Systems and Regulations**: Difficulties due to the incompatibility of the legal and regulatory system for AI.

4.2 Technology Seeds and Policy Map

As for the technical seeds and non-technical solutions for SEMLA, the project core members extracted the following 10 items before the workshop, and this was based on interviews with experts and surveyed papers. Since technologies are constantly evolving, additional items were made accessible as necessary by participants in the workshop.

1. Quality assurance (ex. Testing, debug, and verification methods for MLAS)
2. Explainable AI (ex. Transformation from black box model to white box model)
3. Development process (ex. Development pattern and process mining for MLAS)

4. Project management (ex. Machine learning project canvas)
5. Development support environment (ex. Interactive modeling tool)
6. System-wide safety (ex. STAMP/STPA, FRAM)
7. AI-human cooperation (ex. Human-in-the-loop machine learning)
8. Security and privacy (ex. Adversarial example)
9. AI and ethics (ex. AI ethics guidelines)
10. Human resource development (ex. AI education curriculum)

4.3 Market Trend Map and Issues and Needs Trend Map

Before the workshop, the project core members prepared major trends (phases) of autonomous driving, smart factory, and RPA as a rough guide, and the participating members made additions and corrections as necessary during the workshop. The phases of autonomous driving were already defined in the industry (autonomous driving levels 0 to 5). We introduced the three phases of smart factory: "data collection and visualization," "realization of the data-driven PDCA cycle," and "automatic factory improvement," and three phases of RPA: "expansion of the automated processes," "business optimization," and "human satisfaction improvement (human happiness)." Each phase corresponds to a created value. For example, "automatic factory improvement" can reduce the huge amount of effort on engineers (data scientists) and externalize human-dependent know-how in the smart factory project.

For each application group, these phases and items of the market trend map were related to the items of the issues and needs map by using sticky notes on the white board.

4.4 Technology Roadmap from the Needs/Seeds Matching Table

Each group created a matching table of needs and seeds by application. Then, three technology roadmaps were constructed. The technology roadmap had a simple structure consisting of three layers: technological seeds and solutions, features (technological issues and needs), and market trends (phases and values). This work was completed in front of the white board as shown in Fig. 3.

4.5 Strategic Story and Presentations

The three groups (autonomous driving, smart factory, and RPA) created and presented their R&D strategic stories to explain the technology roadmap. They received feedback on the presentations from the technology roadmapping experts and other participants. The results of the workshop were used later in the proposal for the funding agency.

4.6 Evaluation of Roadmapping Workshop

A questionnaire survey for participants was conducted after the workshop, and responses were obtained from six corporate participants. Many of them said that the issues and needs map was comprehensive to some extent and could be used in roadmap creation. On the other hand, the technology seeds and policy map was not comprehensive enough because they are just work in progress or on-going, and could not be fully utilized in the roadmapping process. The six respondents, who had never experienced technology roadmapping before, commented that "roadmapping seems to be widely used in business," "I want to use it for my company's business planning," "it was helpful to understand how to grasp uncertainty," and "the three-layer structure of market trend, feature, and technology is easy to understand." On the other hand, there were opinions such as, "It was difficult due to the high degree of freedom," "There was not enough time," and "There were difficulties due to differences in the background of the collaborative work." These comments indicate the need for guidelines specific for SEMLA.

Fig. 3. Roadmapping workshop

5 Discussion

In this roadmapping workshop, the major trends in autonomous driving, smart factory, and RPA were presented with a set of phases. For autonomous driving, it was possible to derive the phases corresponding to the level of automation recognized in the industry, but for smart factory and RPA, it was difficult in the limited time of the two-day workshop to derive and deepen the phases clearly and appropriately. In fact, the RPA group members continued to discuss the issue for several months after the workshop, and were finally able to derive four phases [13] which differed from the original three phases. An additional tool may be effective to design phases more systematically by using typical patterns of value co-creation by humans and machines.

From the viewpoint of technology roadmapping as a dialogue tool, the workshop functioned to promote the mutual understanding and constructive discussions between university researchers and company engineers, as well as between engineers from different companies. On the other hand, in practice, the effectiveness of the dialogue tool would be increased by long-term continuous review of the roadmap, rather than the two-day workshop. While the roadmapping is a dialogue tool for discussing a long term strategy, other tools may become necessary to be complementary for a short term strategy like the IoT innovation design method [14] in digital transformation.

6 Conclusions

In digital transformation, the collaboration between humans and machines is inevitable. However, it is not clear what kind of value is to be created by the collaboration between humans and machines, which may bring many cases of confusion in similar digital transformation projects. This study proposed the new technology roadmapping method (co-evolution type) that considers value co-creation between humans and machines, and applied it in the roadmapping workshop on systems engineering for machine learning applications. The proposed roadmapping method can be used as a dialog tool for bridging perception gaps among stakeholders in digital transformation.

Acknowledgments. This work was supported by the JST-Mirai Program (JPMJMI20B8). The authors would like to express their gratitude to Nobukazu Yoshioka, the leader of the project "Modelling and AI for Integration of Cyber and Physical World," and other team members who participated in the preparatory stages and on the day of the workshop. We would also like to thank Dr. Robert Phaal of the University of Cambridge and Dr. Kunio Shirahada of Japan Advanced Institute of Science and Technology for their guidance and support as experts in technology roadmapping in preparation and on the day of the workshop.

References

1. UK AI Council, AI Roadmap, January 2021. https://www.gov.uk/government/publications/ai-roadmap. Accessed 16 Feb 2021
2. Yaguo, L., et al.: Applications of machine learning to machine fault diagnosis: A review and roadmap. Mech. Syst. Signal Process. **138**, 106587 (2020)
3. Artificial Intelligence Technology Strategy Council: R&D Goals and Roadmap for Industrialization of Artificial Intelligence (in Japanese), NEDO (2017). https://www.nedo.go.jp/content/100862412.pdf. Accessed 20 Feb 2021
4. The Japanese Society for Artificial Intelligence, AI Map (in Japanese). https://www.ai-gakkai.or.jp/resource/aimap/. Accessed 20 Feb 2021
5. https://www.jst.go.jp/mirai/jp/uploads/saitaku2018/JPMJMI18BA_yoshioka.pdf. Accessed 20 Feb 2021
6. Daim, T.U., Oliver, T., Phaal, R. (ed.): Technology Roadmapping. World Scientific (2018)
7. Phaal, R., Farrukh, C., Probert, D.: T-Plan: Fast Start Technology Roadmapping: Planning Your Route to Success, Institute for Manufacturing, University of Cambridge (2001)
8. Uchihira, N.: Future direction and roadmap of concurrent system technology. IEICE Trans. Fundam. Electron. Commun. Comput. Sci. **90**(11), 2443–2448 (2007)
9. Sawaragi, T.: Design of resilient socio-technical systems by human-system co-creation. Artif. Life Robot. **25**(2), 219–232 (2020). https://doi.org/10.1007/s10015-020-00598-3
10. NEDO: Survey on the Use of Artificial Intelligence and Machine Learning in Industrial Fields and the Safety of Artificial Intelligence Technologies (in Japanese) (2019). https://www.nedo.go.jp/library/seika/shosai_201907/20190000000685.html. Accessed 26 Feb 2021
11. IPA: AI White Paper 2019 (in Japanese) (2018). https://www.ipa.go.jp/ikc/info/20181030.html. Accessed 26 Feb 2021
12. Difficulty Map for Project Management of Machine Learning Application Systems. http://www.jaist.ac.jp/ks/labs/uchihira/mlas-pm-map.html. Accessed 27 Feb 2021

13. Okuda, S., et al.: Exploitation pattern for machine learning systems. In: The 36th International Technical Conference on Circuits/Systems, Computers and Communications (ITC-CSCC) (2021)
14. Uchihira, N.: Innovation design method for the internet of things: requirements and perspectives. In: 2019 Portland International Conference on Management of Engineering and Technology (PICMET), pp. 1–8 (2019)

Author Index

C. Leitner et al. (Eds.): AHFE 2021, LNNS 266, pp. 527–529, 2021.
https://doi.org/10.1007/978-3-030-80840-2